**B**ioelectrochemistry:
**P**rinciples and
**P**ractice

Volume 4

# Bioenergetics

Edited by
P. Gräber
G. Milazzo†

Coordinated by D. Walz

Birkhäuser Verlag
Basel · Boston · Berlin

Editors:

Peter Gräber, Ph.D.
Professor for Physical Chemistry
Institut für Physikalische Chemie
Universität Freiburg
Albertstr. 23a
D-79104 Freiburg
Germany

Giulio Milazzo†
formerly Professor
Istituto Superiore di Sanità
Rome
Italy

Coordinator:
PD Dr. Dieter Walz
Biozentrum
University of Basel
Klingelbergstrasse 70
CH-4056 Basel
Switzerland

Library of Congress Cataloging-in-Publication Data

**Bioenergetics** / edited by P. Gräber, G. Milazzo.
p. cm. – (Bioelectrochemistry : 4)
Includes bibliographical references and index.

1. Bioenergetics. 2. Energy metabolism. I. Gräber, Peter, 1943–
II. Milazzo, Giulio. III. Series: Bioelectrochemistry (Basel, Switzerland)
QP517.B54B56 1997
574.19'121 – dc20

Deutsche Bibliothek Cataloging-in-Publication Data

**Bioenergetics** / ed. by P. Gräber ; G. Milazzo. Coordinated by
D. Walz. - Basel ; Boston ; Berlin : Birkhäuser, 1997
(Bioelectrochemistry ; Vol. 4)

NE: Gräber, Peter [Hrsg.]; GT


Printed on acid-free paper produced from chlorine-free pulp
Softcover reprint of the hardcover 1st edition 1997

ISBN-13: 978-3-0348-9860-7 e-ISBN-13: 978-3-0348-8994-0
DOI: 10.1007/ 978-3-0348-8994-0

9 8 7 6 5 4 3 2 1

GIULIO MILAZZO (1912–1993)

Giulio Milazzo, the father of Bioelectrochemistry, died on January 6 1993 in Rome. He often pointed out that the roots of the subject go back two hundred years to Galvani and Volta, and that he had only resurrected the science. But the Bioelectrochemistry he started was certainly different from the Natural Philosophy of the past, and in many ways quite different from parallel modern developments.

At a time when science is becoming more narrowly focused and scientists more specialized, Giulio Milazzo catalyzed the formation of an interdisciplinary grouping that was broad in scope and inclusive in its organization. He envisaged Bioelectrochemistry as a discipline including all aspects of the overlap of biology and electrochemistry. He believed that science is international and that one should use all scientific means possible to foster cooperation across national barriers. His ideas catalyzed the founding of the *Bioelectrochemical Society* which to this day attempts to follow the high standards set by him.

Giulio Milazzo was convinced that communication between scientists from various fields would be greatly facilitated if a comprehensive textbook written in a common language existed. He therefore initiated the preparation of what he called a *Treatise on Bioelectrochemistry*, but his untimely death prevented him from finishing this enormous task. The Bioelectrochemical Society, under whose auspices the work had been commenced, considers it both an honor and a duty to pursue the project to completion. But with the guiding spirit no longer with us, the project of a comprehensive textbook seemed too ambitious, and the Treatise was therefore converted to the present Series of Texts. May it nevertheless not only serve the purpose envisaged by Giulio Milazzo, but also be a living memory to a great scientist and a dear friend.

*Martin Blank   Roy Caplan   Dieter Walz*

# Contents

# Contributors

Michael Blaut, Deutsches Institut für Ernährungsforschung, D-14554 Bergholz-Rehbrücke, Germany

Michael Büttner, Department of Cell Biology and Plant Physiology, University of Regensburg, D-93053 Regensburg, Germany

Wim Crielaard, E. C. Slater Institute, BioCentrum Amsterdam, NL-1018 TV Amsterdam, The Netherlands

B. M. Geier, Gustav Embden Zentrum der Biologischen Chemie, Klinikum der Uni Frankfurt, D-60595 Frankfurt am Main, Germany

Gerhard Gottschalk, Institut für Mikrobiologie der Georg-August-Universität Göttingen, D-37077 Göttingen, Germany

Elisabeth Haag, Max Volmer Institute for Physical and Biophysical Chemistry, Technical University of Berlin, D-10623 Berlin, Germany

Günter Hauska, Department of Cell Biology and Plant Physiology, University of Regensburg, D-93053 Regensburg, Germany

Klaas J. Hellingwerf, E. C. Slater Institute, BioCentrum Amsterdam, NL-1018 TV Amsterdam, The Netherlands

G. von Jagow, Gustav Embden Zentrum der Biologischen Chemie, Klinikum der Uni Frankfurt, D-60595 Frankfurt am Main, Germany

Lajos Keszthelyi, Institute of Biophysics, Biological Research Center, Hungarian Academy of Sciences, H-6701 Szeged, Hungary

Thomas A. Link, Gustav Embden Zentrum der Biologischen Chemie, Klinikum der Universität Frankfurt, D-60595 Frankfurt am Main, Germany

Paul Mathis, DBCM/SBE, CEA-Saclay, 91191 Gif-sur-Yvette Cedex, France

Bruno Andrea Melandri, Department of Biology, University of Bologna, I-40126 Bologna, Italy

A. John Moody, Glynn Research Foundation, Bodmin, Cornwall PL30 4AU, United Kingdom

Peter R. Rich, Glynn Research Foundation, Bodmin, Cornwall PL30 4AU, United Kingdom

Gernot Renger, Max Volmer Institute for Physical and Biophysical Chemistry, Technical University of Berlin, D-10623 Berlin, Germany

Dieter Walz, Biozentrum, University of Basel, CH-4056 Basel, Switzerland
Hans V. Westerhoff, Department of Microphysiology, Faculty of Biology, University of Amsterdam, NL-1081 HV Amsterdam, The Netherlands

# Introduction

Bioenergetics, the topic of volume 5 of this Series, is concerned with the energetics, the kinetics, and the mechanisms of energy conversion in biological systems. This phenomenon can be investigated on different levels of complexity. On a global level the role of biological processes for the steady state of our enviroment is considered. At the physiological level, the relation between energy input and the physiological state of an organism is of interest, while at the cellular level the biochemical pathways for degradation and synthesis of all relevant substrates is investigated. At present the majority of bioenergetic studies pertain to the molecular level. The processes in a cell are catalyzed by a large number of proteins called enzymes. The enzymes involved in energy transduction can be considered as molecular machines which transform energy from one form into another, or transfer energy from one process to another.

Living systems operate far from equilibrium and are open in the thermodynamic sense, i.e. they exchange energy and matter with the surroundings. Chapter 1 presents the principles of nonequilibrium thermodynamics applied to biological systems. About 0.05% of the energy from the sunlight which reaches the surface of the earth is used by plants and algae as well as some bacteria to synthesize organic compounds, and thus supplies all organisms with the energy necessary for life. This aspect of global bioenergetics, together with the cycles of matter, is examined in Chapter 2. Cells are divided into different compartments due to the presence of organelles. As discussed in Chapter 3, vectorial processes associated with the membranes of such structures play a crucial role in bioenergetics. Chapter 4 describes the different pathways evolved in bacteria in order to utilize the energy of various external sources. The two principal energy-converting organelles, i.e. the chloroplast and the mitochondrion, are dealt with in Chapters 5 and 6, respectively. The remaining six chapters are devoted to some important energy-transducing enzyme complexes. These include the two photosystems of chloroplasts (Chapters 7 and 8), the cytochrome b6f/bc1 complexes (chapter 9), cytochrome *c* oxidase (Chapter 10), bacteriorhodopsin (Chapter 11), and the $H^+$/ATP synthase (Chapter 12).

This volume is intended to fill the gap between textbooks of biochemistry, electrochemistry, and biophysics which usually focus on particular aspects of bioenergetics. It presents the currently accepted views in the fields, but it should be mentioned that bioenergetics is a very active field of research which rapidly produces new results and ideas.

*Peter Gräber*

Bioenergetics
ed. by P. Gräber and G. Milazzo†


# *CHAPTER 1*
# Nonequilibrium thermodynamics applied to energy conversion in biological systems

Dieter Walz

*Biozentrum, University of Basel, Basel, Switzerland*

## 1. Introduction

A fundamental property of a living system is a continuous running of numerous processes, and when the processes cease the system is dead. Therefore, life has to be supported by a regular input of energy in form of nutrients or light, which is converted into other forms of energy needed by the system in order to keep the processes going. An input of energy in form of nutrients, however, is only possible if the system is open in the thermodynamic sense, i.e. if it exchanges matter, heat, and work with its surroundings. Moreover, energy conversion between processes plays a vital role in biology. These statements apply as well to any part of a biological system. Hence, in what follows, the notation "biological system" will be used in a general sense and refers to a whole organism as well as to isolated organs, cells, and subcellular elements or organelles. Similarly a "living" system means that an organism or an isolated part of it is still functional and is investigated in that state.

In view of the features of biological systems thermodynamics dealing with equilibrium states only (i.e. "classical" thermodynamics or "thermostatics") is inappropriate for the assessment of a living system. By the same token "reversible processes", i.e. processes which proceed through a sequence of states all at or very close to equilibrium can be assumed at most for some of the processes in a living system under certain conditions. As long as the system is alive there are processes which proceed *irreversibly*[1]. Hence, thermodynamics of irreversible processes (or nonequilibrium thermodynamics) is the appropriate tool. Another important aspect concerns the fact that biological systems cannot be investigated in their natural environment. Since such systems are open changes in their surroundings will influence the system under investigation and thus introduce uncontrollable effects. Therefore, an experimental set-up appropriate for the investigation of a living system is designed in such a way that it is not subject to uncontrolled disturbances from the outside world. In this way the whole experimental set-up becomes an isolated system in the thermodynamic sense, but the biological system which forms a part of it is still open.

[1]The notations "reversible" and "irreversible" are used here in the thermodynamic sense. The common habit, particularly in biochemistry, to use these terms in order to qualify chemical reactions with respect to their equilibrium state is not followed here and should be omitted.

This chapter is intended as an introduction to the use of nonequilibrium thermodynamics for the assessment of biological systems. Special emphasis is put on the coupling between processes which is the basis of energy conversion. The term conversion is synonymous with transduction also found in the literature and includes a transfer between energies of the same type. The fundamentals of nonequilibrium thermodynamics are summarized in section 2; for a more comprehensive discussion of this subject see, for example, refs. 1 or 2.

## 2. Elements of nonequilibrium thermodynamics

### *2.1. Translation of an experimental system into a thermodynamic system*

A thermodynamic system is composed of a number of conceptual elements, each of which is well defined in terms of exchange of matter, heat, and work with its surroundings. The exchange occurs via the boundaries or walls of the elements, which may be classified according to the scheme in Table 1. In biological systems the elements consist of phases, compartments, and reservoirs. Compartments contain a single phase, separated from other compartments by boundaries which are formed by other phases such as membranes. A reservoir is an element whose capacitance for heat or matter or electric charge (or whatever else applies) is very much larger than that of any other element.

In constructing a thermodynamic system, the elements of the experimental system are substituted by equivalent elements of the thermodynamic system. The actual topology of the experimental system need not be retained, and only those elements of the experimental system which take part in the actual experiment should be included as an element in the thermodynamic system. Elements which are not involved in the pheno-

Table 1. Classification of elements in a thermodynamic system

| | Permitted* (+) or prohibited (-) exchange [through a wall] of | | |
|---|---|---|---|
| Element | matter | heat | work† |
| open | + | + | + |
| closed | - | + | + |
| adiabatic | - | - | + |
| isolated‡ | - | - | - |

*Permitted means allowed but not necessarily required.
†Work includes the action of electric and/or magnetic fields.
‡Isolation here includes electrical isolation in the most general sense. An isolated element is also said to be surrounded by rigid adiabatic walls since the rigidity of the wall prevents the exchange of work.

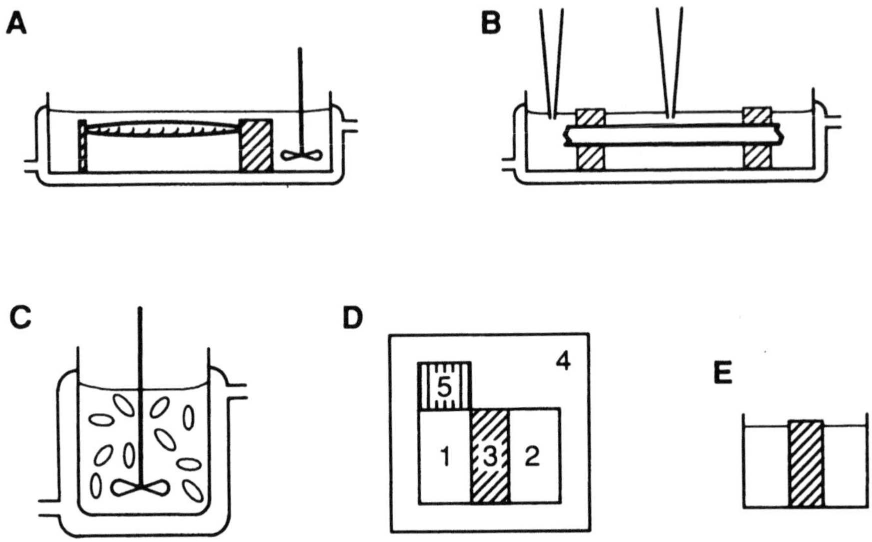


Fig. 1. Experimental systems and the corresponding thermodynamic system. In the experimental systems an isolated organ, *viz.* a muscle (A), a cell isolated from the organ, i.e. a muscle fiber (B), and organelles isolated from the cell, e.g. mitochondria (C) are investigated. The thermodynamic system (D) consists of two compartments with an aqueous phase (1 and 2) which are separated by a membrane (3) and surrounded by two reservoirs, one for heat (4) and one for mechanical work (5). The two-compartment system in E is that part of D in which the processes of interest take place. For further explanation see text.

mena investigated or whose effect is not detectable in the experimentally assessable parameters of the system should either be included in an active element of the thermodynamic system or neglected.

As examples, consider the following experimental set-ups. (i) An isolated muscle is mounted in a special holder and immersed in an appropriate bathing solution (Fig. 1A). The holder allows one to stimulate the muscle and then either keep the muscle's length constant (isometric contraction) or impose a constant force on the muscle (isotonic contraction). (ii) An isolated muscle fibre with cut-off ends is placed in a cuvette with two seals (Fig. 1B) The central chamber formed by the seals in the cuvette is filled with an aqueous solution of given composition. The solutions in the left and right chambers are identical but differ from that in the central chamber. The muscle fiber is stimulated by electrodes placed in the central and left chambers, respectively, and the thereby elicited temporal change in $Ca^{2+}$ concentration in the fiber is monitored. (iii) Mitochondria isolated from the muscle are suspended in a stirred aqueous solution (Fig. 1C). The processes performed by the organelles, say oxidative phosphorylation, are assessed by the changes in composition of the suspending medium. Each of these experimental set-ups is placed in a thermostat which should guarantee a constant temperature in the system and thus eliminate the effect of a varying ambient temperature. Since the set-ups are open the atmospheric pressure acts as a barostat, i.e. the systems are under constant pressure.

All three experimental systems are translated into the same type of thermodynamic system, i.e. two compartments separated by a membrane and supplemented with a heat reservoir and a "reservoir for work" (Fig. 1D). The heat reservoir represents the thermostat and acts as a heat source or heat sink for the rest of the system. The "work reservoir" provides the energy for the operation of the particular devices used in the three set-ups, i.e. the special muscle holder in (A), the stimulation of the muscle fiber in (B), and the stirring device in (C). The two compartments represent, respectively, the bathing solution and the interior of the muscle in (A), the cytoplasm of the muscle fiber (or myoplasm) and the interior of the sarcoplasmic reticulum in (B), and the suspending medium and the combined interior space of the mitochondria in (C). Accordingly, the membrane in the thermodynamic system is a representation of the boundary which separates the interior of the muscle from the bathing solution. It represents the membrane of the sarcoplasmic reticulum in (B) and the combined mitochondrial inner membranes in (C), respectively. The thermodynamic system thus constructed is totally isolated from its surroundings and comprises all elements of the experimental system. Neither energy nor matter need to be supplied from the outside world during the experiment. Moreover, temperature and pressure are constant throughout the system.

It should be understood that the translation of three different experimental systems into only one type of thermodynamic system is not a general rule but arises from the type of examples chosen for didactic reasons. Other examples may lead to a thermodynamic system consisting of more than only two compartments, a membrane, and two reservoirs. Moreover, the assignment of the two compartments and the membrane in the thermodynamic system to elements or combination of elements in an experimental system is essentially determined by the type of experiment carried out with the system. Thus, in the first example (Fig. 1A) the interior of the muscle as well as its boundary consists of cells which themselves are further compartmentalized. However, since none of these elements is directly assessed in the actual experiment, they need not be discriminated in the thermodynamic system. In the second example, the mitochondria in the muscle fiber as well as the solutions in the left and right chamber (see Fig. 1B) are hardly affected by the stimulation of the fiber and hence need not be included in the thermodynamic system. In the third example (Fig. 1C) it is assumed that all mitochondria are identical in composition and function. Therefore, all inner membranes are combined into one extended membrane, while the aqueous phases in the internal spaces are collected and placed into one of the compartments. The outer mitochondrial membranes are assumed not to be involved in the processes of oxidative phosphorylation and therefore neglected. The spaces between the membranes can then be combined with the suspending medium which is represented by the other compartment.

### 2.2. *Internal energy, entropy, and free enthalpy*

The state of a system is uniquely defined by its internal energy, U, which is the sum of all sorts of energies present such as mechanical energy, heat, or chemical energy. In particular, U includes the electrostatic energy arising from charges and electrical potentials. The internal energy is a unique function of the parameters entropy S, volume V, elongation $\ell$ (which is taken as a measure for performing work), and the mole numbers $n_i$ of the *i*th chemical species present in the system. The internal energy $U_k$ can be defined separately for the *k*th element of a system; in case of the system depicted in Fig. 1D the elements are the compartments (index k = 1 and 2), the membrane (index k = 3), and the reservoirs (index k = 4 and 5). The internal energy of the whole system is then the sum of $U_k$ taken over all elements

$$U = \sum_k U_k(S_k, V_k, \ell_k, n_{i,k}) \tag{1}$$

The processes occurring in a system cause a transition of the system from one state to another. In what follows this transition is divided into incremental steps which allows one to use differentials for the changes in parameters. The change in internal energy, $dU_k$, of a given element of the system due to an incremental transition amounts to[2]

$$dU_k = T\, dS_k - p\, dV_k + X_k\, d\ell_k + \sum_i \tilde{\mu}_{i,k}\, dn_{i,k} \tag{2}$$

where p and T denote, respectively, the pressure and the absolute temperature in the system which are both constant (see section 2.1.). The quantities $\tilde{\mu}_{i,k}$ and $X_k$ appearing in Eqn. 2 are, respectively, the electrochemical potential of the *i*th species in the *k*th element defined as the partial derivative of $U_k$ with respect to $n_{i,k}$ at constant S, V, $\ell$, $n_j$ (in all elements)

$$\tilde{\mu}_{i,k} = [\partial U_k / \partial n_{i,k}]_{S,V,\ell,n_{j,k}} \tag{3a}$$

and the force arising from a potential for work similarly defined as

[2]Note that the sign convention used in physics has been adopted here. According to this convention, work performed by an element and heat withdrawn from it are counted negative, while work and heat absorbed by an element are counted positive.

$$X_k = [\partial U_k / \partial \ell_k]_{S,V,\ell,n_{j,k}} \tag{3b}$$

Since the total system is isolated (see section 2.1) its internal energy and its volume are constant, hence $\sum_k dU_k = 0$ and $\sum_k p\, dV_k = 0$. It then follows from Eqns. 1 and 2 that

$$T \sum_k dS_k = -\sum_k \Big[\sum_i \tilde{\mu}_{i,k}\, dn_{i,k}\Big] - \sum_k X_k\, d\ell_k \geq 0 \tag{4}$$

The inequality $\geq 0$ is dictated by the second law of thermodynamics which states that the change in entropy of the whole system has to be positive or zero when the system changes from one state to another.

The free enthalpy (or Gibbs free energy) G can be defined as

$$G = U - T\,S + p\,V \tag{5}$$

This quantity is a measure of the system's capability to perform useful work. If temperature and pressure are constant its change in the *k*th element of the system becomes (cf. Eqn. 2)

$$dG_k = dU_k - T\, dS_k + p\, dV_k = X_k\, d\ell_k + \sum_i \tilde{\mu}_{i,k}\, dn_{i,k} \tag{6}$$

Substituting Eqn. 6 into Eqn. 4 yields

$$T \sum_k dS_k = -\sum_k dG_k \geq 0 \tag{4a}$$

It thus becomes evident that the mandatory increase in entropy of the whole system required by the second law of thermodynamics is covered by a decrease in free enthalpy of the whole system. It should be understood that only the sum of the entropy of all elements in the system has to increase in a transition, while each individual term may increase, decrease or remain constant. Similarly, only the sum of free enthalpies decreases upon a transition of the system from one state to another.

### *2.3. The dissipation function*

The rate of increase of the system's entropy is called the dissipation function, $\Phi$. It indicates how fast free enthalpy is "dissipated" into entropy. Taking the time derivative of Eqn. 4 yields

$$\Phi = T \sum_k dS_k/dt = -\sum_k \Big[\sum_i \tilde{\mu}_{i,k} dn_{i,k}/dt\Big] - \sum_k X_k\, d\ell_k/dt \geq 0 \tag{7}$$

Note that in order to determine $\Phi$ in any state of the system in the course of an experiment, it suffices to estimate the electrochemical potentials $\tilde{\mu}_{i,k}$

and the forces $X_k$ together with the pertinent time derivatives $dn_{i,k}/dt$ and $d\ell_k/dt$. No knowledge of the actual processes which take place in the system and cause the changes in $n_{i,k}$ or $\ell_k$ with time is required.

The limiting case $\Phi = 0$ is reached when $dn_{i,k}/dt = d\ell_k/dt = 0$ for all i and k. This is the equilibrium state of the system which, as discussed in section 1, is of no interest in the present context since it refers to a biological system which is dead. Classical thermodynamics (or "thermostatics") is based on so-called "reversible transitions" which occur between states at or very close to equilibrium and have to be infinitely slow. However, processes in a living biological system proceed at finite rates and therefore cannot be assessed by classical thermodynamics. They always give rise to an expenditure of free enthalpy and an increase in entropy. As a consequence, the free enthalpy available for energy conversion is always less in real processes than expected from the corresponding hypothetical reversible transitions.

No change in mole numbers occurs in the reservoirs, i.e. $dn_{i,4}/dt = dn_{i,5}/dt = 0$. Moreover, the dissipation function in Eqn. 7 includes the contribution $X_5 d\ell_5$ arising from the reservoir for work. This contribution is of no relevance to the processes of interest and can be omitted without loss of generality. Therefore, in what follows, the dissipation function

$$\Phi = -X_2\, d\ell_2/dt - \sum_{k=1}^{3} \left[\sum_i \tilde{\mu}_{i,k}\, dn_{i,k}/dt\right] \geq 0 \qquad (8)$$

will be used. This is equivalent to singling out the two-compartment system in Fig. 1D which represents the biological system in the aqueous medium.

### *2.4. Flows and forces for chemical reactions, transport and mechanical processes*

*2.4.1. Chemical reactions:* The *j*th chemical reaction, which occurs in compartment k, is given the index j,k in order to distinguish it from other reactions. It converts the initial reactants $S_{s(j,k)}$ ($s = 1, 2, \cdots$) with stoichiometric coefficients $\nu_{Ss(j,k)}$ into final reactants $P_{p(j,k)}$ ($p = 1, 2, \cdots$) with stoichiometric coefficients $\nu_{Pp(j,k)}$ and *vice versa*:

$$\sum_s \nu_{Ss(j,k)}\, S_{s(j,k)} \leftrightarrows \sum_p \nu_{Pp(j,k)}\, P_{p(j,k)} \qquad (9)$$

The notation initial and final reactants introduces the *positive direction for the reaction* when going from initial to final reactants[3]. It can be expressed

[3]It should be understood that a positive direction is not inherent in a reaction but can be chosen arbitrarily. However, once a positive direction is chosen it has to be retained consistently. Note that initial and final reactants are often called substrates and products of the reaction, respectively.

by the following sign convention: *the stoichiometric coefficients for initial reactants have a negative sign and those for final reactants have a positive sign.* The flow, $J_{j,k}$, of the *j*th reaction in compartment k is given by

$$J_{j,k} = d\xi_{j,k}/dt = (dn_{Rr(j,k)}/dt)/\nu_{Rr(j,k)} \quad (10)$$

where reactant $R_{r(j,k)}$ ($r = 1, 2, \cdots$) as indicated by the subscript Rr(j,k) now stands for initial and final reactants which need no longer be distinguished if the above sign convention is used. The quantity $\xi_{j,k}$ in Eqn. 10 is called the degree of advancement and is defined as

$$\xi_{j,k} = [n_{Rr(j,k)}(t) - n_{Rr(j,k)}(0)]/\nu_{Rr(j,k)} \quad (11)$$

The arguments t and 0 indicate the mole numbers, respectively, at time t and at the beginning ($t = 0$) of the experiment. The force of the *j*th reaction in compartment k can be expressed by its affinity, $\mathcal{A}_{j,k}$, which becomes

$$\mathcal{A}_{j,k} = -\sum_{r} \nu_{Rr(j,k)}\, \tilde{\mu}_{Rr(j,k)} = -\Delta G_{j,k} \quad (12)$$

The sum in Eqn. 12 has to be taken over all reactants with stoichiometric coefficients according to the sign convention mentioned above. The quantity $\Delta G_{j,k}$ is called the free enthalpy (or Gibbs free energy) of the reaction and emerges from classical thermodynamics. Note that free enthalpy and affinity merely have opposite signs.

*2.4.2. Transport processes:* When dealing with transport of species between compartments and/or phases, it is necessary to arbitrarily choose a *positive direction of transport which has to be the same for all transport processes* (sign convention). Here the direction from compartment 1 to compartment 2 is chosen as positive. By means of the mole numbers $n_{i,1}$ and $n_{i,2}$ of the *i*th species the flow[4] $J_i$ for the transport of this species from compartment 1 to compartment 2 can be defined as

$$J_i = -dn_{i,1}/dt = dn_{i,2}/dt \quad (13)$$

where the second part of Eqn. 13 arises from mass balance. The thermodynamic force for transport is the difference in electrochemical potential between the compartments:

$$\Delta\tilde{\mu}_i = \tilde{\mu}_{i,1} - \tilde{\mu}_{i,2} \quad (14)$$

[4]Note that the flow of a chemical reaction is a scalar, while the flow of a transport process is a vector. The direction of this vector is defined by the chosen positive direction for transport and is perpendicular to the membrane surface. Hence a scalar notation for the flows of transport processes can be used.

Similar relations for flow and force apply also to the transport of a species from a compartment to the membrane. However, for reasons given in section 3, such transports need not be considered in the present context.

*2.4.3. Redox reactions and electron transport:* Chemical reactions which consist of an exchange of electrons only are called redox reactions. The exchange occurs between two redox couples, a couple being represented by

$$R_{ox} + \nu_{e(R)}\, e^- \rightleftarrows R_{rd} \tag{15}$$

where $R_{ox}$ and $R_{rd}$ denote the oxidized and reduced species of the couple, respectively. The quantity $\nu_{e(R)}$ is the number of electrons exchanged by the couple at a time. Despite the fact that $\nu_{e(R)}$ is clearly a stoichiometric coefficient the symbol n is frequently used in the literature which is misleading because n usually denotes a mole number. Since free electrons do not exist in phases of biological relevance the reaction in Eqn. 15 is only one half of a redox reaction and therefore called the half reaction which characterizes the couple. Choosing one of the couples as the donor (denoted by $D_{ox}/D_{rd}$) and the other as the acceptor ($A_{ox}/A_{rd}$) introduces a positive direction for the electron exchange reaction. The redox reaction in the *k*th compartment *written per electron as stoichiometric unit* then reads

$$D_{rd,k}/\nu_{e(D)} + A_{ox,k}/\nu_{e(A)} \rightleftarrows D_{ox,k}/\nu_{e(D)} + A_{rd,k}/\nu_{e(A)} \tag{16}$$

The flow of this electron exchange reaction becomes

$$J_{e,k} = -\nu_{e(D)}\, dn_{Drd,k}/dt = -\nu_{e(A)}\, dn_{Aox,k}/dt = \nu_{e(D)}\, dn_{Dox,k}/dt = \nu_{e(A)}\, dn_{Ard,k}/dt \tag{17}$$

while its affinity is (cf. Eqn. 12)

$$\mathcal{A}_{e,k} = (\tilde{\mu}_{Drd,k} - \tilde{\mu}_{Dox,k})/\nu_{e(D)} - (\tilde{\mu}_{Ard,k} - \tilde{\mu}_{Aox,k})/\nu_{e(A)} = \tilde{\mu}_{e(D,k)} - \tilde{\mu}_{e(A,k)} = -\Delta G_{e,k} \tag{18}$$

The second expression in Eqn. 18 arises from the following consideration. The two species of a redox couple can be interpreted as representing "electrons", and an electrochemical potential for the electron can be defined as [3]

$$\tilde{\mu}_{e(R,k)} = (\tilde{\mu}_{Rrd,k} - \tilde{\mu}_{Rox,k})/\nu_{e(R)} \tag{19}$$

Each redox couple in compartment k thus defines its own $\tilde{\mu}_{e(R,k)}$, and the electron transfer between a donor and an acceptor is equivalent to an electron transport from $\tilde{\mu}_{e(D)}$ to $\tilde{\mu}_{e(A)}$. If both couples are present in the same compartment this transfer corresponds to the redox reaction shown in Eqn. 16 whose affinity is determined by the difference in $\tilde{\mu}_e$ as indicated in Eqn. 18. However, if donor and acceptor couple occur in different compartments the transfer is equivalent to a transport of electrons across the membrane (usually mediated by components in the membrane, for an example see section 4.3). In this case the sign convention for transport applies which means that the donor and the acceptor couple have to be in compartment 1 and 2, respectively, in order to comply with the positive direction of transport chosen in section 2.4.2. The flow of the electron transport then becomes

$$J_e = -\nu_{e(D)}\, dn_{Drd,1}/dt = \nu_{e(A)}\, dn_{Ard,2}/dt = \nu_{e(D)}\, dn_{Dox,1}/dt = -\nu_{e(A)}\, dn_{Aox,2}/dt \tag{20}$$

while its force is

$$\Delta\tilde{\mu}_e = \tilde{\mu}_{e(D,1)} - \tilde{\mu}_{e(A,2)} \tag{21}$$

in line with Eqn. 14. Note that the species electron can be assessed by means of the reduced species of a redox couple which leads to the first two terms in Eqn. 20, in line with Eqn. 13, but also in a negative sense by means of the oxidized species which yields the last two terms in Eqn. 20.

*2.4.4. Mechanical processes:* The "flow" of the *k*th mechanical process, $J_k$, is simply the velocity, $v_k$, determined by the time derivative of the pertinent length,

$$J_k = v_k = d\ell_k/dt \tag{22}$$

Since work performed by an element is counted negative (see footnote 2 and Eqn. 3b), the force for the *k*th mechanical process is

$$X_k = -F_k \tag{23}$$

where $F_k$ denotes the external mechanical force exerted on the element. In the case of a rotational movement the velocity $v_k$ and the force $F_k$ are replaced by $\omega_k$ and $\tau_k$, respectively, where $\omega_k$ is the angular velocity and $\tau_k$ denotes the torque exerted on the element [4].

### *2.5. Flows and forces in the dissipation function; energy conversion*

Chemical reactions and transport processes are connected by the effect of their respective flows on the mole number of the species involved. Thus, for the *i*th species which is transported and takes part as reactant

$R_{r(j,1)}$ in the *j*th chemical reaction in compartment 1 and as reactant $R_{r(j,2)}$ in the *j*th chemical reaction in compartment 2,

$$dn_{i,1}/dt = - J_i + \sum_j \nu_{Rr(j,1)} J_{j,1} \tag{24a}$$

and

$$dn_{i,2}/dt = J_i + \sum_j \nu_{Rr(j,2)} J_{j,2} \tag{24b}$$

Introducing Eqns. 24 into the dissipation function of Eqn. 8 yields upon rearranging and in view of Eqns. 12, 13, 14, and 23

$$\Phi = \sum_i J_i \Delta\tilde{\mu}_i + \sum_{k=1}^{2} \left[\sum_j J_{j,k} \mathcal{A}_{j,k}\right] + v_2 F_2 - \sum_i \tilde{\mu}_{i,3} dn_{i,3}/dt \geq 0 \tag{25}$$

where the sums have to be taken over all transported species and all chemical reactions occurring in both compartments. If $\Phi$ calculated by means of Eqn. 25 is identical with that according to Eqn. 8 the chosen set of definitions of chemical reactions, transport and mechanical processes is appropriate. Inspection of Eqn. 25 shows that this is only true if the last term vanishes, as is indeed the case when the membrane (k = 3) has reached a steady state (see section 3). Otherwise the transport processes and/or the chemical reactions have to be redefined in such a way that they include the membrane processes.

In a valid representation of the dissipation function in terms of flows and forces (e.g. Eqn. 25 with a vanishing last term) $\Phi$ is given by the sum of products of flow and conjugate force[5] for all processes. Thus, in terms of the general notation $J_p$ and $X_p$ for the flow and the conjugate force of the *p*th process, respectively,

$$\Phi = \sum_p J_p X_p \geq 0 \tag{26}$$

The flow of a process has the same sign as its force when it runs "downhill", i.e. when it occurs spontaneously, and the product of flow and conjugate force is positive, in line with general physical concepts. Note that this would not apply to chemical reactions if $\Delta G$ would be used as force. This is the reason why affinities instead of free enthalpies are consistently used in nonequilibrium thermodynamics. If flow and conjugate force have opposite signs the flow-force product is negative and the process is

[5]The quantities flow and force should be understood in a generalized sense. Flows comprise material transport and conversion defined as mole/time, as well as velocities in the physical sense defined as length/time. Similarly, forces comprise thermodynamic forces defined as energy/mole = mass × length$^2$/(time$^2$ × mole) as well as mechanical forces defined as mass × length/time$^2$. The flow-force product of any process yields its power defined as energy/time = mass × length$^2$/time$^3$.

driven "uphill" by another process (or processes). This is the thermodynamic expression of the energy conversion due to some form of coupling between processes. It is possible because only the sum of all flow-force products has to be positive according to Eqn. 26 and not each product by itself. Thus, in the example depicted in Fig. 1A, the muscle can contract against an external force ($v_2F_2 < 0$) if this mechanical process is coupled to a chemical reaction, e.g. the hydrolysis of ATP. The flow-force product of this reaction has then to be larger than $-v_2F_2$. In the case of the example shown in Fig. 1B the hydrolysis of ATP supplies the energy for the transport of $Ca^{2+}$ ions from the sarcoplasm into the sarcoplasmic reticulum against a force of up to 40 kJ/mol, while in the example of Fig. 1C the mitochondria can synthesize ATP ($J_{ph}\mathcal{A}_{ph} < 0$) if this reaction is coupled to another reaction or a transport process whose flow-force product has to be larger than $-J_{ph}\mathcal{A}_{ph}$. Other examples are electron transport and ATP synthesis driven by the energy of absorbed light quanta in photosynthesis, or a bacterial flagella which can rotate against a load, i.e. $\omega\tau < 0$, because its movement is coupled to the transport of $H^+$ ions across the bacterial membrane and $J_H\Delta\tilde{\mu}_H > -\omega\tau$ [4]. This list of examples is by no means exhaustive, in fact energy conversion due to coupling of processes is one of the central keys in biological systems.

### *2.6. Thermodynamic forces in terms of experimentally accessible parameters*

*2.6.1. Dependence of the electrochemical potential on concentration and electrical potential:* The force of a chemical reaction or a transport process is expressed in terms of the electrochemical potentials of the species involved. In order to relate these quantities to the experimentally accessible parameters the electrochemical potential of the *i*th species in the *k*th compartment or phase is split into a chemical and an electrical part

$$\tilde{\mu}_{i,k} = \mu_{i,k} + z_i \mathcal{F}\phi_k \tag{27}$$

Here $\mathcal{F}$ is the Faraday constant, while $z_i$ denotes the charge number of the species. The quantity $\phi_k$ is the electrical potential of the phase (Galvani potential) with respect to a suitably chosen reference point [5]. The quantity $\mu_{i,k}$ is called the chemical potential of the species and can be further split into two terms

$$\mu_{i,k} = \mu^o_{i,k} + RT\ \ln\{a_{i,k}/a^o\} \tag{28}$$

where R is the gas constant. The standard chemical potential $\mu^o_{i,k}$ is a substance-specific quantity which also depends on the type of phase (hence index k is retained) and, in general, on pressure and temperature. The quantity $a_{i,k}$ is called the activity of the species; it is normalized by the thermodynamic standard activity $a^o$ which, by definition, amounts to 1M. The activity of a species is related to its concentration $c_{i,k}$ (= $n_{i,k}/V_k$) by

$$a_{i,k} = y_{i,k}\ c_{i,k} \qquad (29)$$

The factor $y_{i,k}$ called the activity coefficient comprises the interaction of the species with all other constituents of the phase.

For the species electron associated with the redox couple $R_{rd}/R_{ox}$ in compartment k the electrochemical potential becomes (cf. Eqn. 19)

$$\tilde{\mu}_{e(R,k)} = (\mu^o_{Rrd,k} - \mu^o_{Rox,k})/\nu_{e(R)} + (RT/\nu_{e(R)}) \ln\{a_{Rrd,k}/a_{Rox,k}\} - \mathcal{F}\phi_k = -\mathcal{F}(E_{R,k} + \phi_k) \qquad (30)$$

The quantity $E_{R,k}$ is called the redox potential of the couple in the *k*th compartment and indicates the chemical potential of the electron, i.e., $\mu_{e(R,k)} = -\mathcal{F}E_{R,k}$ [2, 3]. The relation commonly used to express the dependence of $E_{R,k}$ on the activities of the redox species is called the Nernst equation and reads

$$E_{R,k} = E^{\emptyset}_{R,k} + RT/(\nu_{e(R)}\ \mathcal{F}) \ln\{a_{Rox,k}/a_{Rrd,k}\} \qquad (31)$$

When inserting Eqn. 31 into Eqn. 30 it is recognized that the quantity $E^{\emptyset}_{R,k}$ called the standard redox potential of the couple is related to the standard chemical potentials of its species by

$$E^{\emptyset}_{R,k} = (\mu^o_{Rox,k} - \mu^o_{Rrd,k})/(\nu_{e(R)}\ \mathcal{F}) = -\ \mu^o_{e(R,k)}/\mathcal{F} \qquad (32)$$

where $\mu^o_{e(R,k)}$ is the standard chemical potential of the electron associated with the couple $R_{ox}/R_{rd}$ in compartment k.

*2.6.2. Forces expressed by concentrations and electrical potentials:* Inserting Eqns. 27–29 rewritten for reactants $R_{r(j,k)}$ into Eqn. 12 yields upon rearranging

$$\mathcal{A}_{j,k} = -\sum_r \nu_{Rr(j,k)} \mu^o_{Rr(j,k)} - RT \ln\Big\{\prod_r y_{Rr(j,k)}^{\nu_{Rr(j,k)}}\Big\} - RT \ln\Big\{\prod_r [c_{Rr(j,k)}/a^o]^{\nu_{Rr(j,k)}}\Big\} \qquad (33)$$

Note that the terms of $\tilde{\mu}$ containing $\phi_k$ cancel out because a chemical reaction is *electroneutral,* i.e. $\sum_r \nu_{Rr(j,k)}\ z_{Rr(j,k)} = 0$. The sum of the standard chemical potentials of the reactants constitutes what is called the standard free enthalpy (or standard Gibbs free energy) of the reaction

$$\Delta G^{\emptyset}_{j,k} = \sum_r \nu_{Rr(j,k)}\ \mu^o_{Rr(j,k)} = \Delta H^{\emptyset}_{j,k} - T\ \Delta S^{\emptyset}_{j,k} \qquad (34)$$

It can be related to the standard enthalpy and the standard entropy of the reaction, $\Delta H^{\circ}_{j,k}$ and $\Delta S^{\circ}_{j,k}$, respectively. The standard free enthalpy determines the equilibrium constant $K^{o}_{rj,k}$ of the reaction

$$K^{o}_{rj,k} = \exp\{-\Delta G^{\circ}_{j,k}/(RT)\} = \prod_{r} [a_{Rr(j,k)}/a^{o}]_{eq}^{\nu_{Rr(j,k)}} \tag{35}$$

which in turn determines the ratio of the reactants' activities at equilibrium, as indicated by the mass action ratio[6] on the right-hand side of Eqn. 35. In many practical cases it is difficult or impossible to estimate the activity coefficients. However, conditions can often be found which cause activity coefficients to be approximately constant during an experiment. In such cases a phenomenological equilibrium constant $K_{rj,k}$ is used, which is operationally defined as

$$K_{rj,k} = \exp\{-\Delta G^{o}_{j,k}/(RT)\} = \prod_{r} [c_{rR(j,k)}/a^{o}]_{eq}^{\nu_{Rr(j,k)}} = K^{o}_{rj,k} \prod_{r} y_{Rr(j,k)}^{-\nu_{Rr(j,k)}} \tag{36}$$

where $\Delta G^{o}_{j,k}$ denotes the corresponding phenomenological standard free enthalpy. In view of Eqns. 34 and 35 the affinity of a chemical reaction becomes

$$\mathcal{A}_{j,k} = RT \ln\{K^{o}_{rj,k} \prod_{r} [a_{Rr(j,k)}/a^{o}]^{-\nu_{Rr(j,k)}}\} = RT \ln\{K_{rj,k} \prod_{r} [c_{Rr(j,k)}/a^{o}]^{-\nu_{Rr(j,k)}}\} \tag{37}$$

In most cases concentrations rather than activities are known, and the second expression in Eqn. 37 is therefore used to calculate $\mathcal{A}_{j,k}$. Due attention should be paid to what type of equilibrium constant is available and, depending on conditions, a correction in terms of activity coefficients may be necessary (see Eqn. 36).

For the transport of the *i*th species in positive direction from phase or compartment k' to phase or compartment k the thermodynamic force becomes (cf. Eqns. 27–29)

$$\Delta\tilde{\mu}_{i,k'k} = \mu^{o}_{i,k'} - \mu^{o}_{i,k} + RT \ln\{y_{i,k'}/y_{i,k}\} + RT \ln\{c_{i,k'}/c_{i,k}\} + z_i \mathcal{F}(\phi_{k'} - \phi_k) \tag{38}$$

The difference in standard chemical potentials known as the standard free enthalpy for partitioning determines the equilibrium constant $K_{pi,k'k}$, usually called the partition coefficient of the species,

$$K^{o}_{pi,k'k} = \exp\{(\mu^{o}_{i,k'} - \mu^{o}_{i,k})/(RT)\} = [a_{i,k}/a_{i,k'}]_{eq} \exp\{z_i \mathcal{F}(\phi_k - \phi_{k'})/(RT)\} \tag{39}$$

It relates the activities of the species in the two phases or compartments at equilibrium. Note that these activities are affected by a possible difference

[6]Note that equilibrium constants are dimensionless. Nevertheless, equilibrium constants with a dimension of concentration units can be found in the literature which is due to the fact that the mass action ratio is usually written without the normalization by $a^{o}$.

in electrical potential between the phases or compartments. It may be difficult to estimate activity coefficients, particularly for a membranous phase but also for an aqueous phase with many proteins as found, for example, in the internal space of mitochondria. A phenomenological equilibrium constant is therefore frequently used which is operationally defined as

$$K_{pi,k'k} = [c_{i,k}/c_{i,k'}]_{eq} \exp\{z_i \mathcal{F}(\phi_k - \phi_{k'})/(RT)\} = K^{o}_{pi,k'k}\, y_{i,k'}/y_{i,k} \quad (40)$$

In view of Eqns. 39 and 40 the force of a transport process becomes

$$\begin{aligned} \Delta\tilde{\mu}_{i,k'k} &= RT \ln\{K^{o}_{pi,k'k}\, a_{i,k'}/a_{i,k}\} + z_i \mathcal{F}(\phi_{k'} - \phi_k) \\ &= RT \ln\{K_{pi,k'k}\, c_{i,k'}/c_{i,k}\} + z_i \mathcal{F}(\phi_{k'} - \phi_k) \quad (41) \end{aligned}$$

For the transport of the *i*th species between the aqueous phases of compartment 1 and 2 $K^{o}_{pi,12} = 1$, hence

$$\Delta\tilde{\mu}_i = RT \ln\{a_{i,1}/a_{i,2}\} + z_i \mathcal{F}\Delta\phi = RT \ln\{K_{pi,12}\, c_{i,1}/c_{i,2}\} + z_i \mathcal{F}\Delta\phi_m \quad (42)$$

where $\Delta\phi_m = \phi_1 - \phi_2$ is the transmembrane electrical potential difference called the membrane potential. Due to a lack of information on the activity coefficients $K_{pi,12} = 1$ is usually assumed, but it should be understood that this assumption is more stringent than that of constant activity coefficients of species invoked in the context of a chemical reaction. Constant activity coefficients of a given species in two different aqueous phases does not necessarily imply that they are also equal. Hence, it should be kept in mind that $K_{pi,12}$ may deviate from unity.

Inserting Eqn. 30 written for donor and acceptor couple into Eqns. 18 and 21 yields, in view of Eqn. 32,

$$\begin{aligned} \mathcal{A}_{e,k} &= \mathcal{F}(E^{\emptyset}_{A,k} - E^{\emptyset}_{D,k}) + (RT/\nu_{e(A)}) \ln\{a_{Aox,k}/a_{Ard,k}\} - (RT/\nu_{e(D)}) \ln\{a_{Dox,k}/a_{Drd,k}\} \\ &= \mathcal{F}(E^{o}_{A,k} - E^{o}_{D,k}) + (RT/\nu_{e(A)}) \ln\{c_{Aox,k}/c_{Ard,k}\} - (RT/\nu_{e(D)}) \ln\{c_{Dox,k}/c_{Drd,k}\} \\ &= \mathcal{F}(E_{A,k} - E_{D,k}) \quad (43) \end{aligned}$$

for a redox reaction in compartment k, and

$$\begin{aligned} \Delta\tilde{\mu}_e &= \mathcal{F}(E^{\emptyset}_{A,2} - E^{\emptyset}_{D,1}) + (RT/\nu_{e(A)}) \ln\{a_{Aox,2}/a_{Ard,2}\} - (RT/\nu_{e(D)}) \ln\{a_{Dox,1}/a_{Drd,1}\} - \mathcal{F}\Delta\phi_m \\ &= \mathcal{F}(E^{o}_{A,2} - E^{o}_{D,1}) + (RT/\nu_{e(A)}) \ln\{c_{Aox,2}/c_{Ard,2}\} - (RT/\nu_{e(D)}) \ln\{c_{Dox,1}/c_{Drd,1}\} - \mathcal{F}\Delta\phi_m \\ &= \mathcal{F}(E_{A,2} - E_{D,1}) - \mathcal{F}\Delta\phi_m \quad (44) \end{aligned}$$

for an electron transport across the membrane. The quantity $E^{o}_{R,k}$ appearing in the second relation of Eqns. 43 and 44 is called the midpoint (standard) potential and is operationally defined as being equal to $E_{R,k}$ when $c_{Rox,k} = c_{Rrd,k}$. It is related to $E^{\emptyset}_{R,k}$ according to

$$E^{o}_{R,k} = E^{\emptyset}_{R,k} + RT/(\nu_{e(R)}\mathcal{F}) \ln\{y_{Rox,k}/y_{Rrd,k}\} \quad (45)$$

Since concentrations rather than activities are usually known, the second relation in Eqns. 43 and 44 is mostly used but, similar to the case of equilibrium constants, attention should be paid to what type of standard redox potential is available. A correction by virtue of Eqn. 45 may be necessary, whereby the activity coefficients of the redox species are assumed to be constant. According to the third relation in Eqn. 44, $\Delta\tilde{\mu}_e$ can in principle be determined by means of inert metal electrodes placed in each of the compartments [2,3]. In the case of $\mathcal{A}_{e,k}$, however, the use of such an electrode is only permissible if only one of the redox couples exchanges electrons with it. Otherwise the measured redox potential would be meaningless, but the electrode would tend to equalize $E_{A,k}$ with $E_{D,k}$ and thus to annihilate $\mathcal{A}_{e,k}$ (see the third relation in Eqn. 43). Eqns. 43 and 44 can just as well be written in terms of equilibrium constants because the following relations hold:

$$K^{o}_{e} = \exp\{\mathcal{F}(E^{\emptyset}_{A,k} - E^{\emptyset}_{D,k'})/(RT)\} \quad (46a)$$

$$K_e = \exp\{\mathcal{F}(E^{o}_{A,k} - E^{o}_{D,k'})/(RT)\} = K^{o}_{e}[y_{Aox,k}/y_{Ard,k}]^{1/\nu_{e(A)}}[y_{Drd,k'}/y_{Dox,k'}]^{1/\nu_{e(D)}} \quad (46b)$$

with k' = k for a redox reaction. Note that in general $K^{o}_{e} \neq 1$ and $K_e \neq 1$. This applies also to the transport of electrons between the aqueous phases in compartment 1 and 2, in contrast to the transport of other species for which $K^{o}_{pi,12} = 1$ and possibly $K_{pi,12} = 1$ holds true.

A comment on $\Delta\phi_m$ in Eqns. 42 and 44 is in place. Most conveniently, one of the compartments is chosen as reference point for the electrical potential [5]. According to common practice the potential of the other compartment is then called the "membrane potential" and is used instead of $\Delta\phi_m$. Moreover, a convention for the polarity of membrane potentials in energy converting systems was recently proposed [6]. It should be realized that, when following this practice or convention, the positive direction for transport processes can no longer be chosen independently. Disregarding this point can lead to an inconsistency between $\Delta\phi_m$ and the membrane potential. In fact, with the presently chosen positive direction, $\Delta\phi_m = -\phi_2$ if $\phi_1 = 0$, while $\Delta\phi_m = \phi_1$ if $\phi_2 = 0$.

*2.6.3. Ligand binding to species involved in a process; protonation:* A particular type of chemical reaction frequently encountered is the binding of a ligand L to a species R according to

$$L + R \rightleftarrows LR,\ L + LR \rightleftarrows L_2R, \cdots L + L_{m-1}R \rightleftarrows L_mR \qquad (47)$$

Usually only the total concentration, $c_{R,tot,k}$, for the species R in the *k*th compartment is known

$$c_{R,tot,k} = \sum_{i=0}^{m} c_{LiR,k} \qquad (48a)$$

If ligand binding is at (pseudo-)equilibrium (see section 3) the concentration of the individual species $L_iR$ at a given *free ligand concentration* $c_L$ can be calculated by means of Eqn. 36 written for ligand dissociation for each of the steps shown in Eqn. 47:

$$c_{LiR,k} = c_{R,tot,k} / \left[ 1 + \sum_{j'=1}^{i} \prod_{j=j'}^{i} K_{d,LjR}\ a^o / c_L + \sum_{j'=i+1}^{m} \prod_{j=i+1}^{j'} c_L / (K_{d,LjR}\ a^o) \right] \qquad (48b)$$

Here $K_{d,LjR}$ denotes the (phenomenological) equilibrium constant called the dissociation constant for the *j*th step. Eqn. 48b can serve to substitute $c_{R,tot,k}$ for $c_{LiR,k}$ in the equations expressing the thermodynamic force of a process in which $L_iR$ takes part. Note that this substitution introduces a dependence of the force on $c_L$ although the species L is not intrinsically involved in the pertinent process.

The most important ligand L is the $H^+$ ion, and the reactions in Eqn. 47 are then called the acid-base reactions or protonation of species R. Since the activity of $H^+$ ions in an aqueous phase can vary over several orders of magnitude a more convenient measure for $a_H$ is the quantity pH defined as

$$pH = -\log\{a_H/a^o\} \approx -\log\{c_H/a^o\} \qquad (49)$$

where the right-hand side of Eqn. 49 represents an often used approximation. The majority of biological systems are active only in a limited pH-range located around pH = 7. In view of this fact, the thermodynamic standard activity $a^o = 1M$ seems to be inappropriate. Hence a new standard activity for $H^+$ ions, $a_H^{\oplus} = 10^{-7}M$, was introduced which may be called the "biochemical" standard activity. This leads to a biochemical standard chemical potential for $H^+$ ions, $\mu_{H,k}^{\oplus}$, which is related to $\mu_{H,k}^{o}$ as follows (cf. Eqn. 28)

$$\mu_{H,k}^{\oplus} = \mu_{H,k}^{o} + RT \ln\{a_H^{\oplus}/a^o\} = \mu_{H,k}^{o} - 7 \ln\{10\}\ RT \qquad (50)$$

As a consequence, any chemical reaction involving $H^+$ ions is characterized by a biochemical standard free enthalpy (cf. Eqns. 34 and 35)

$$\Delta G_{j,k}^{\oplus} = -RT \ln\{K_{j,k}^{\oplus}\} = \Delta H_{j,k}^{\oplus} - T\,\Delta S_{j,k}^{\oplus} = \Delta G_{j,k}^{o} - 7\,\nu_{H(j,k)} \ln\{10\}\ RT \quad (51)$$

where $\Delta H_{j,k}^{\oplus}$, $\Delta S_{j,k}^{\oplus}$ and $K_{j,k}^{\oplus}$ are the corresponding biochemical standard enthalpy, standard entropy, and equilibrium constant, respectively. Since activities of species except $H^+$ ions[7] are rarely known in biological systems (see sections 2.6.2) the biochemical standard parameters include the activity coefficients and hence are phenomenological quantities. Thus, for $H^+$-dissociation from a carboxylic acid with $\Delta G_d^o/(RT) = 10.82$ or $K_d^o = 2 \times 10^{-5}$ the biochemical parameters are $\Delta G_d^{\oplus}/(RT) = -5.298$ or $K_d^{\oplus} = 200$, while for that from an amine with $\Delta G_d^o/(RT) = 24.23$ or $K_d^o = 3 \times 10^{-11}$ they are $\Delta G_d^{\oplus}/(RT) = 8.11$ or $K_d^{\oplus} = 3 \times 10^{-4}$. Note that the normalization of $a_H$ by $a_H^{\oplus}$ has to be used whenever calculations are done with biochemical standard parameters. Unfortunately, it is common practice to denote $\Delta X_{j,k}^{\oplus}$ (X = G, H, S) and $K_{j,k}^{\oplus}$ by $\Delta X_{j,k}^{o'}$ and $K'_{j,k}$, respectively. This should be avoided because the latter symbols are used for the phenomenological parameters at pH = 7 (see below) which emerge from the substitution of total concentrations for the concentrations of individual species mentioned above. Therefore great care should be taken when using primed quantities given in the literature since they can stand for phenomenological parameters at pH = 7 as well as for biochemical standard parameters.

*2.6.4. The phosphate potential:* Many species involved in biologically relevant chemical reactions bind $H^+$ ions. A case in point is the phosphorylation reaction where ATP is formed from ADP and phosphate (Fig. 2). Besides multiple protonation all three species can also bind $Mg^{2+}$ ions, which represent a necessary cofactor for many systems. Provided protonation and $Mg^{2+}$-binding are at (pseudo-) equilibrium the affinity defined with any combination of the species for ATP, ADP, and phosphate shown in the scheme of Fig. 2 yields the correct affinity $\mathcal{A}_{ph}$ of the phosphorylation reaction. Thus, using for example the two reactions labelled (1) and (2) in Fig. 2, and the reaction (6) HADP.Mg + $H_2P^- \rightleftarrows$ HATP.M$^-$ + $H_2O$,

$$\begin{aligned}\mathcal{A}_{ph} &= -[\Delta G^{\emptyset}_{ph,1} + RT\ln\{a_{HATP}\,a^o/(a_{HADP}\,a_{H2P})\}]\\ &= -[\Delta G^{\emptyset}_{ph,2} + RT\ln\{a_{ATP}\,a^o/(a_{ADP}\,a_{HP})\} - RT\ln\{a_H/a^o\}]\\ &= -[\Delta G^{\emptyset}_{ph,6} + RT\ln\{a_{HATP.Mg}\,a^o/(a_{HADP.Mg}\,a_{H2P})\}] \qquad (52)\end{aligned}$$

For convenience, the index k for compartment k is omitted here and in the following relations. The terms in square brackets are equal to the free enthalpy of the phosphorylation reaction which is often called the *phosphate potential.* When writing the second relation in Eqn. 52 it was assumed that the dissociation of water, $H_2O \rightleftarrows H^+ + OH^-$, is always at equilibrium, an assumption almost automatically made when dealing with an aqueous phase. In this case the following relations hold true

[7]The pH in aqueous solutions is routinely measured by means of a glass electrode which is a reversible electrode for $H^+$ ions [2] and hence determines $a_H$ according to Eqn. 49.

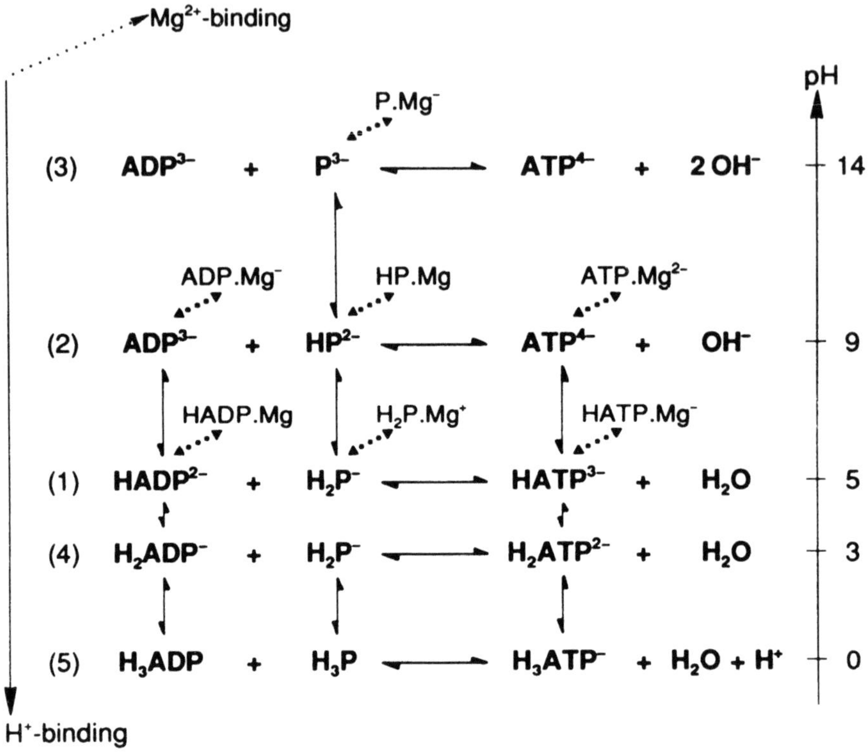


Fig. 2. Phosphorylation reactions as well as protonation and $Mg^{2+}$-binding of ATP, ADP, and phosphate. Phosphate is abbreviated by P. Horizontal and vertical double arrows represent phosphorylation reactions and binding of an $H^+$ ion, respectively, while dotted double arrows represent binding of a $Mg^{2+}$ ion. The scale on the right indicates approximately at which pH the species shown are prevalent (see also Fig. 3B). Note that $H_4ATP$ is a strong acid and hence does not occur in an aqueous phase. The labels (1)–(5) are included for reference purposes.

$$K^o_w = \exp\{-\Delta G^{\emptyset}_w/(RT)\} = \exp\{(\mu_w - \mu^o_H - \mu^o_{OH})/(RT)\}$$
$$= (a_H/a^o)\,(a_{OH}/a^o) = 10^{-14} \qquad (53)$$

where $K^o_w$ denotes the dissociation constant of water. Note that the definition of the standard free enthalpy $\Delta G^{\emptyset}_w$ includes the chemical potential of water (cf. Eqn. 28) because the activity of water is assumed to be constant. This is a reasonable approximation for not too concentrated solutions and as long as the composition of the solutions does not change too much. It should be understood that $\Delta G^o_w$ may not be constant under certain conditions but depend on composition[8].

[8]This important aspect is lost when following a common practice which assumes the aqueous phase to be in its "pure liquid standard state" for which $a_w = 1$ holds true by definition. Applying this assumption consistently would mean that no osmotic phenomena can occur.

The quantities $\Delta G^{\varnothing}_{ph,i}$ in Eqn. 52 are the standard free enthalpies of the particular reactions chosen when defining $\mathcal{A}_{ph}$. They determine the corresponding equilibrium constants $K^{o}_{ph,i}$ and are related to the standard chemical potentials of the species involved; for example

$$\Delta G^{\varnothing}_{ph,1} = -RT\ln\{K^{o}_{ph,1}\} = \mu^{o}_{HATP} + \mu_{w} - \mu^{o}_{HADP} - \mu^{o}_{H2P} \qquad (54)$$

Note that $\Delta G^{\varnothing}_{ph,1}$ includes $\mu_w$ for the reasons given above. The other standard free enthalpies can be related to $\Delta G^{\varnothing}_{ph,1}$ by means of the dissociation constants $K^{o}_{d,HiR}$ and $K^{o}_{d,HR.Mg}$ (R = ATP, ADP, P) for protonation and $Mg^{2+}$-binding, respectively:

$$\Delta G^{\varnothing}_{ph,2} = -RT\ln\{K^{o}_{ph,2}\} = \Delta G^{\varnothing}_{ph,1} + RT\ln\{K^{o}_{d,HADP}\,K^{o}_{d,H2P}/K^{o}_{d,HATP}\} \qquad (55a)$$

$$\Delta G^{\varnothing}_{ph,6} = -RT\ln\{K^{o}_{ph,6}\} = \Delta G^{\varnothing}_{ph,1} + RT\ln\{K^{o}_{d,HATP.Mg}/K^{o}_{d,HADP.Mg}\} \qquad (55b)$$

According to Eqn. 36 the phenomenological equilibrium constant $K_{ph,1}$ amounts to

$$K_{ph,1} = \exp\{-\Delta G^{o}_{ph,1}/(RT)\} = K^{o}_{ph,1}\,y_{HADP}\,y_{H2P}/y_{HATP} = [c_{HATP}\,a^{o}/(c_{HADP}\,c_{H2P})]_{eq} \qquad (56)$$

Table 2. Dissociation constants for protonation and $Mg^{2+}$-binding of ATP, ADP, and phosphate at 25°C.

| X: | ATP | ADP | P |
|---|---|---|---|
| $pK^{o}_{d,H3X}$ | —* | — | 2.12 |
| $pK_{d,H3X}$ | 1.7‡ | 1.8¶ | 1.86 |
| $pK^{o}_{d,H2X}$ | — | 4.2 | 7.18† |
| $pK_{d,H2X}$ | 4.00‡ | 4.05 | 6.94† |
| $pK^{o}_{d,HX}$ | 7.68† | 7.20† | 12.36 |
| $pK_{d,HX}$ | 7.10†‡ | 6.78† | 11.3 |
| $pK^{o}_{d,H2X.Mg}$ | | | — |
| $pK_{d,H2X.Mg}$ | | | 0.6 |
| $pK^{o}_{d,HX.Mg}$ | 3.59† | 2.45† | 2.71† |
| $pK_{d,HX.Mg}$ | 2.84†‡ | 2.00† | 1.78 |
| $pK^{o}_{d,X.Mg}$ | 5.83† | 4.27† | — |
| $pK_{d,X.Mg}$ | 4.66†‡ | 3.44† | 3.4 |

Dissociation constants are listed as $pK^{o}_{d} = -\log\{K^{o}_{d}\}$. Phenomenological dissociation constants are listed as $pK_d = -\log\{K_d\}$; they are valid for $I_c = 0.1$ M (cf. Eqn. 57) and corrected for $Na^+$- or $K^+$-binding if possible. A missing entry for both types of constants means that the pertinent binding step was not detected. Values according to ref. 10 if not indicated otherwise.
*Value not available.
†Value according to refs. 8 (same as used in ref. 7).
‡Value according to refs. 9.
¶Value was estimated from the corresponding data for ATP and phosphate.

where $\Delta G^{o}_{ph,1}$ is the corresponding phenomenological standard free enthalpy. The activity coefficients of the charged species $HADP^{2-}$, $H_2P^-$, and $HATP^{3-}$ deviate from unity mainly because of electrostatic interaction with the other ionic species in the solution. A suitable measure for the effect of this interaction is the quantity $I_c$ called the ionic strength. It is defined as

$$I_c = \tfrac{1}{2}\sum_i z_i^2\, c_i \tag{57}$$

where the sum has to be taken over all charged species present.

The concentrations of the species $HATP^{3-}$, $HADP^{2-}$, and $H_2P^-$ can be related to the pertinent total concentrations by means of Eqns. 48. This is equivalent to condensing the rections shown in Fig. 2 into the *overall phosphorylation reaction*

$$ADP + P \rightleftarrows ATP + H_2O \tag{58}$$

If $Mg^{2+}$ ions are present the $Mg^{2+}$-complexes of ATP, ADP, and phosphate have to be taken into account as well. Their concentrations are given by

$$c_{HiR.Mg} = c_{HiR}\, c_{Mg}/K_{d,HiR.Mg} \quad \text{for } R = ATP,\ ADP,\ P \tag{59}$$

where $c_{Mg}$ is the *free $Mg^{2+}$ concentration.* Including these species in Eqns. 48 when calculating $c_{HATP}$, $c_{HADP}$, and $c_{H2P}$ yields, upon substitution for these concentrations in Eqn. 56, the phenomenological equilibrium constant for the overall phosphorylation reaction (cf. Eqn. 58)

$$K_{ph}=\exp\{-\Delta G^{o}_{ph}/(RT)\} = K_{ph,1}\Sigma_{ATP}/[\Sigma_{ADP}\Sigma_P] = [c_{ATP,tot}a^o/(c_{ADP,tot}c_{P,tot})]_{eq} \tag{60a}$$

with the abbreviations

$$\Sigma_R = 1 + 1 + \left[\frac{c_{Mg}/a^o}{K_{d,R.Mg}}\,\frac{K_{d,HR}}{a_H/a^o}\right] + 1 + \frac{c_{Mg}/a^o}{K_{d,HR.Mg}} + \frac{a_H/a^o}{K_{d,H2R}} + \frac{(a_H/a^o)^2}{K_{d,H2R}K_{d,H3R}} \tag{60b}$$

for R = ATP and ADP, and

$$\Sigma_P = 1 + \left\{\left[1 + \frac{c_{Mg}/a^o}{K_{d,P.Mg}}\right]\frac{K_{d,HP}}{a_H/a^o} + 1 + \frac{c_{Mg}/a^o}{K_{d,HP.Mg}}\right\}\frac{K_{d,H2P}}{a_H/a^o} + \left[1 + \frac{c_{Mg}/a^o}{K_{d,H2P.Mg}}\right]\frac{a_H/a^o}{K_{d,H3P}} \tag{60c}$$

It is important to note that Eqns. 60 are independent of the particular reaction chosen when defining $\mathcal{A}_{ph}$, which is a consequence of the relations between standard free enthalpies indicated in Eqns. 55. As is evident from Eqns. 60, $K_{ph}$ and the corresponding phenomenological standard free enthalpy $\Delta G^{o}_{ph}$ are dependent on $a_H$ and $c_{Mg}$. By convention, their value for pH = 7, i.e. $a_H = 10^{-7}$, is denoted by $K'_{ph}$ and $\Delta G^{o'}_{ph}$, respectively. It should be understood that the condition $a_H = 10^{-7}$ does not convert $\Delta G^{o'}_{ph}$ into a biochemical standard parameter, as erroneously claimed in some textbooks. In fact, introducing total concentrations of species even eliminates $H^+$ ions from the phosphorylation reaction (cf. Eqns. 58 and 60a). Hence the standard state of these ions is irrelevant for $K'_{ph}$ and $\Delta G^{o'}_{ph}$ which, however, are valid only at pH = 7.

If only the total concentration of $Mg^{2+}$ ions is known $c_{Mg}$ has to be calculated from the mass balance

$$c_{Mg} = c_{Mg,tot} - c_{HATP.Mg} - c_{ATP.Mg} - c_{HADP.Mg} - c_{ADP.Mg} - c_{H2P.Mg} - c_{HP.Mg} - c_{P.Mg} \quad (61)$$

Introducing $c_{HiR.Mg}$ and $c_{HiR}$ (R = ATP, ADP, and P) from Eqns. 59 and 48 yields an eighth order polynomial in $c_{Mg}$ which can only be solved numerically. In this case, however, $K_{ph}$ and $\Delta G^{o}_{ph}$ become dependent on $c_{ATP,tot}$, $c_{ADP,tot}$, and $c_{P,tot}$ as well.

Rosing and Slater [7] have estimated $\Delta G^{\emptyset}_{ph,1}$ from experimental data and found it to be 31.5 kJ/mol at 25°C. By means of the dependence of activity coefficients on ionic strength used in ref. 7, one obtains $y_{HADP}\, y_{H2P}/y_{HATP} = 1.995$ and hence $\Delta G^{o}_{ph,1} = 29.8$ kJ/mol for $I_c = 0.1$M. Table 3 summarizes the standard free enthalpies which can be calculated from $\Delta G^{\emptyset}_{ph,1}$ and $\Delta G^{o}_{ph,1}$ using the relations between the $\Delta G^{\emptyset}_{ph,i}$'s (cf. Eqns. 55) and similar relations for the $\Delta G^{o}_{ph,i}$'s (not shown) together with the data for the dissociation constants listed in Table 2. Figure 3A shows the dependence of $\Delta G^{o}_{ph}$ on pH for $I_c = 0.1$ M and $c_{Mg} = 0$. $\Delta G^{o}_{ph} = \Delta G^{o'}_{ph} =$

Table 3. Standard free enthalpies for selected phosphorylation reactions at 25°C.

| | Reaction | $\Delta G^{\emptyset}_{ph,i}$ | $\Delta G^{o}_{ph,i}$ | $\Delta G^{\oplus}_{ph,i}$ |
|---|---|---|---|---|
| (1) | $HADP^{2-} + H_2P^- \rightleftarrows HATP^{3-} + H_2O$ | 31.5* | 29.8 | 29.8 |
| (2) | $ADP^{3-} + HP^{2-} \rightleftarrows ATP^{4-} + OH^-$ | -6.7 | -8.0 | 32.0 |
| (3) | $ADP^{3-} + P^{3-} \rightleftarrows ATP^{4-} + 2\,OH^-$ | -77.3 | -72.5 | 7.4 |
| (4) | $H_2ADP^- + H_2P^- \rightleftarrows H_2ATP^{2-} + H_2O$ | | 30.1 | 30.1 |
| (5) | $H_3ADP + H_3P \rightleftarrows H_3ATP^- + H_2O + H^+$ | | 41.3 | 1.3 |
| (6) | $HADP.Mg + H_2P^- \rightleftarrows HATP.Mg^- + H_2O$ | 25.0† | 25.0† | 25.0 |

Standard free enthalpies $\Delta G^{\emptyset}_{ph,i}$, phenomenological standard free enthalpies $\Delta G^{o}_{ph,i}$ valid for $I_c$ = 0.1 M, and biological standard free enthalpies $\Delta G^{\oplus}_{ph,i}$ are given in kJ/mol. They were calculated by means of relations between standard free enthalpies such as Eqns. 55, and Eqn. 51 in the case of $\Delta G^{\oplus}_{ph,i}$.

*Value estimated from experimental data [7].

†The identical value for $\Delta G^{\emptyset}_{ph,6}$ and $\Delta G^{o}_{ph,6}$ is fortuitious and occurs since accidentally $pK^{o}_{d,HATP.Mg} - pK^{o}_{d,HADP.Mg} \approx pK_{d,HATP.Mg} - pK_{d,HADP.Mg} + \log\{y_{HADP}\, y_{H2P}/y_{HATP}\}$.

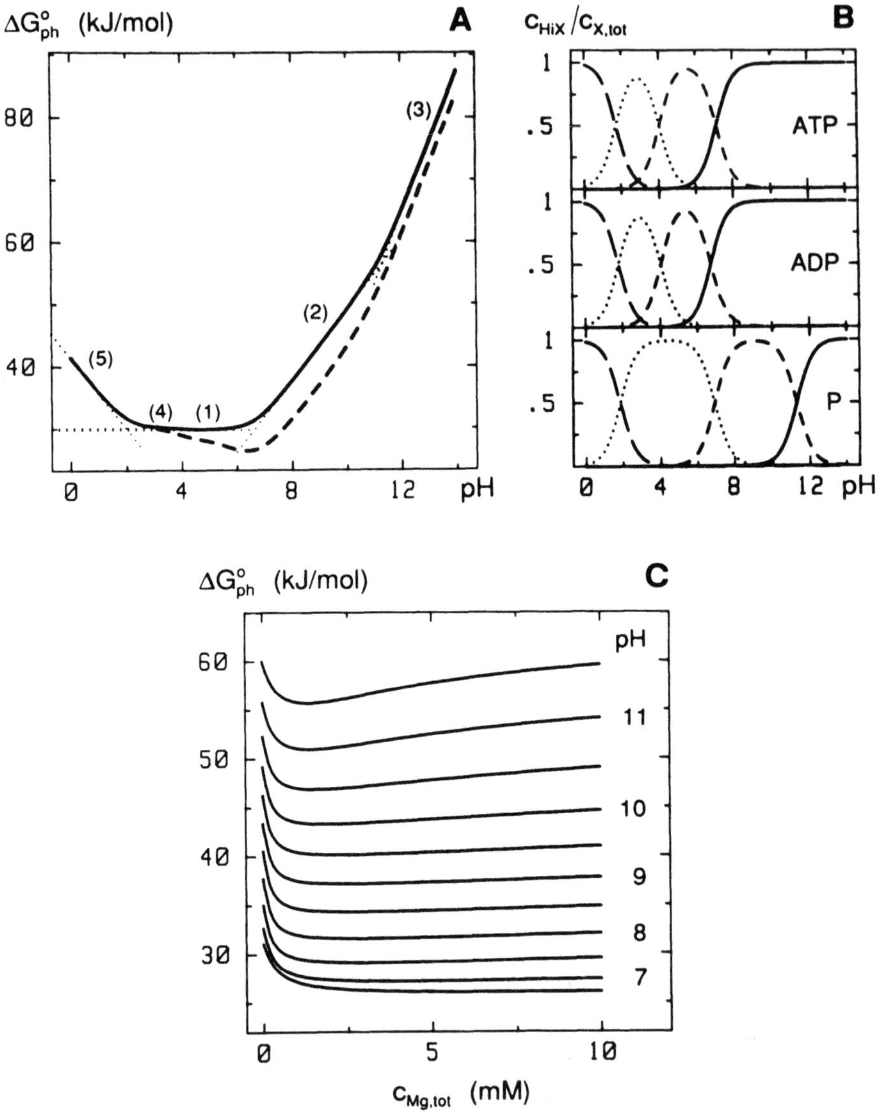


Fig. 3. Dependence of $\Delta G^o_{pH}$ on pH and on $c_{Mg,tot}$, as well as pH-dependence of the protonation of ATP, ADP, and phosphate. (A) Dependence of $\Delta G^o_{ph}$ on pH for $c_{Mg,tot} = 0$ (solid line) and $c_{Mg,tot} = c_{P,tot} = 2$ mM, $c_{ATP,tot} + c_{ADp,tot} = 0.2$ mM (broken line). The dotted lines indicate $\Delta G^o_{ph,1}$ (1), $\Delta G^o_{ph,2} + \ln\{10\}RT$ pH (2), $\Delta G^o_{ph,3} + 2\ln\{10\}RT$pH (3), $\Delta G^o_{ph,4}$ (4), and $\Delta G^o_{ph,5} - \ln\{10\}RT$pH (5); for values of $\Delta G^o_{ph,i}$ see Table 3. In (B) the pH-dependence of the normalized concentrations $c_X/c_{X,tot}$ (solid line), $c_{HX}/c_{X,tot}$ (short broken line), $c_{H2X}/c_{X,tot}$ (dotted line), and $c_{H3X}/c_{X,tot}$ (long broken line) is shown for X = ATP (top panel), ADP (middle panel), and phosphate (bottom panel). (C) Dependence of $\Delta G^o_{ph}$ on $c_{Mg,tot}$ for $c_{P,tot} = 2$ mM, $c_{ATP,tot} + c_{ADP,tot} = 0.2$ mM and different pH-values. $\Delta G^o_{ph}$ and $c_{HiX}/c_{X,tot}$ were calculated according to Eqns. 59 and 48b, respectively, $I_c = 0.1$M, temperature 25°C.

32.7 kJ/mol at pH = 7 and, as can be seen from Fig. 3B, the reactions labelled (1) and (2) in Fig. 2 contribute about equally to the overall phosphorylation reaction at this pH. $\Delta G^o_{ph}$ approaches $\Delta G^o_{ph,1}$ at pH ≈ 5 and coincides with the linear relation $\Delta G^o_{ph,2}$ + ln{10} RT pH (cf. Eqn. 52) around pH 9 because the species of reactions (1) and (2) are prevalent at these pH-values (see Fig. 3B). For the same reason $\Delta G^o_{ph}$ approaches $\Delta G^o_{ph,4}$ at pH ≈ 3 as well as the linear relations $\Delta G^o_{ph,3}$ + 2 ln{10} RT pH for pH > 12 and $\Delta G^o_{ph,5}$ - ln{10} RT pH for pH < 2.

In the presence of $Mg^{2+}$ ions at a given total concentration $c_{Mg,tot}$, $\Delta G^o_{ph}$ is dependent also on $c_{ATP,tot}$, $c_{ADP,tot}$, and $c_{P,tot}$ besides $a_H$ and $c_{Mg,tot}$. However, under the condition $c_{ATP,tot} + c_{ADP,tot}$ = constant < $c_{P,tot}$, the dependence of $\Delta G^o_{ph}$ on the nucleotide concentrations becomes negligibly small. The pH-dependence of $\Delta G^o_{ph}$ for $c_{Mg,tot} = c_{P,tot}$ = 2mM and $c_{ATP,tot} + c_{ADP,tot}$ = 0.2mM is shown in Fig.3A. $\Delta G^o_{ph}$ with and without $Mg^{2+}$ ions is equal for pH < 3 since species present in this pH-range bind little or no $Mg^{2+}$ ions (cf. Figs. 2 and 3B). $Mg^{2+}$-binding decreases $\Delta G^o_{ph}$, but this effect levels off or is even reversed if $Mg^{2+}$ ions are present in excess as compared to the species of the phosphorylation reaction. This is evident from Fig. 3C which shows $\Delta G^o_{ph}$ as a function of $c_{Mg,tot}$ for different pH-values. It may be added that Rosing and Slater [7] present additional data on $\Delta G^o_{ph}$ (which they call $\Delta G^o_{obs}$), e.g. for 37°C or for different ionic strengths. Note that these authors did not include the species $H_2ATP^-$, $H_2ADP^{2-}$, $P^{3-}$, and $H_2P.Mg^+$ in their analysis, which is legitimate because the concentrations of these species are vanishingly small in the pH-range 6.6 ≤ pH ≤ 7.6 covered by the experiments.

*2.6.5. Dependence of midpoint potentials on pH:* The protonation of redox species is of particular interest because the dissociation constants for reduced and oxidized species can be rather different. When introducing Eqn. 45 into Eqn. 31 and substituting total concentrations for the concentrations of the unprotonated species by means of Eqn. 48b the Nernst equation is transformed into

$$E_{R,k} = E^m_{R,k} + [RT/(\nu_{e(R)}\,\mathcal{F})]\ \ln\{c_{Rox,tot,k}/c_{Rrd,tot,k}\} \qquad (62a)$$

Here $E^m_{R,k}$ denotes the now $a_H$-dependent midpoint potential which is related to $E^o_{R,k}$ by

$$E^m_{R,k} = E^o_{R,k} + \frac{RT}{\nu_{e(R)}\,\mathcal{F}} \ln \left\{ \frac{1 + \sum_{j=1}^{m} \prod_{i=1}^{j} a_{H,k}/(K_{d,HiRrd,k}\ a^o)}{1 + \sum_{j=1}^{m} \prod_{i=1}^{j} a_{H,k}/(K_{d,HiRox,k}\ a^o)} \right\} \qquad (62b)$$

where $K_{d,HiRox}$ and $K_{d,HiRrd}$ are the dissociation constants for $H_iR_{ox}$ and $H_iR_{rd}$, respectively. By convention, $E^{o'}_{R,k}$ then denotes $E^m_{R,k}$ at pH = 7.

As an example consider the redox behavior of N-methyl phenazinium methyl sulfate, in short "phenazine methosulfate" or PMS [11]. The phenazinium ion (abbreviated by $P^+$) can accept one electron and is converted to the semiquinoid form Q, which in turn can also accept one electron whereby the fully reduced form $R^-$ is formed. Alternatively, $P^+$ can accept two electrons and is directly converted to $R^-$. Both Q and $R^-$ but not $P^+$ can bind one $H^+$ ion, hence

$$P^+ + e^- \rightleftarrows Q \quad \text{with } E^o_{P1} \qquad K_{d,HRox} \gg 1 \qquad K_{d,HRrd} = K_{d,HQ} \tag{63a}$$

$$Q + e^- \rightleftarrows R^- \quad \text{with } E^o_{Q} \qquad K_{d,HRox} = K_{d,HQ} \quad K_{d,HRrd} = K_{d,HR} \tag{63b}$$

$$P^+ + 2e^- \rightleftarrows R^- \quad \text{with } E^o_{P2} = (E^o_{P1} + E^o_{Q})/2 \quad K_{d,HRox} \gg 1 \qquad K_{d,HRrd} = K_{d,HR} \tag{63c}$$

The relation for $E^o_{P2}$ in Eqn. 63c arises from detailed balancing [1,2]. With the parameter values $E^o_{P1} = -4.0$ mV, $E^o_Q = 17.4$ mV, $E^o_{P2} = 6.7$ mV, $K_{d,HQ} = 2 \times 10^{-8}$, and $K_{d,HR} = 5.5 \times 10^{-10}$ at 20°C [11], one obtains (cf. Eqn. 62b) $E^{o'}_{P1} = -41.3$ mV, $E^{o'}_Q = 103.7$ mV, and $E^{o'}_{P2} = 72.5$ mV. Figure 4 shows the pH-dependence of the midpoint potential for all three redox couples. $E^m_{P1}$

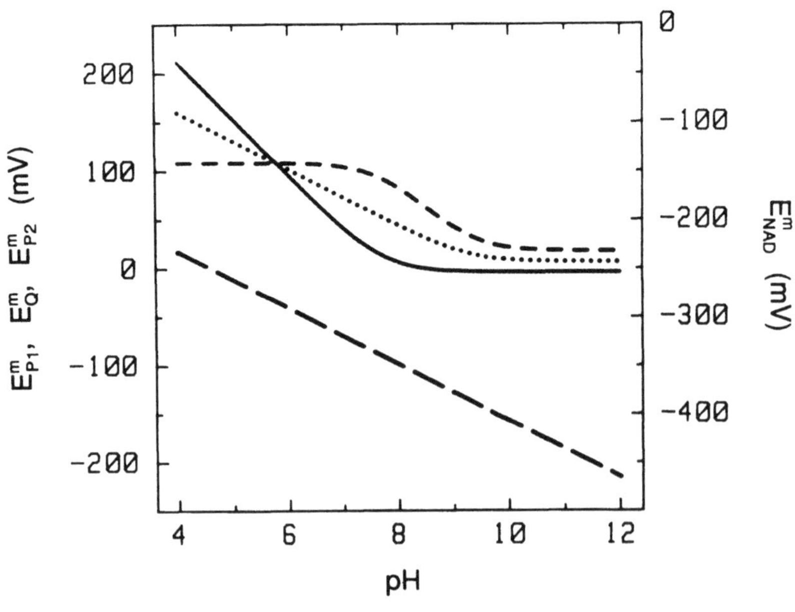


Fig. 4. Dependence of midpoint potentials on pH for the redox couples of N-methylphenazinium methyl sulfate (PMS) and for the couple $NAD^+$/NADH. Midpoint potentials of the PMS couples are (cf. Eqns. 63) $E^m_{P1}$ for P/Q (solid line), $E^m_Q$ for Q/R (short broken line), and $E^m_{P2}$ for P/R (dotted line). The long broken line indicates $E^m_{NAD}$ of the $NAD^+$/NADH-couple with $E^{o'}_{NAD} = -320$mV. For further explanation see text.

is constant and equal to $E^o_{P1}$ if $a_H \ll K_{d,HQ}$ or $pH > pK_{d,HQ}$, where pK = $-\log\{K\}$ (in analogy to Eqn. 49). It turns into a pH-dependence around $pH = pK_{d,HQ}$ and attains a linear dependence with a slope of $-\ln\{10\}RT/\mathcal{F}$ = −58.2 mV/pH-unit ($\nu_{e,P1} = m = 1$, 20°C) for $pH < pK_{d,HQ}$. Similarly, $E^m_Q$ is equal to $E^o_Q$ for $pH > pK_{d,HR}$ and turns into a pH-dependence around $pH = pK_{d,HR}$. However, it does not reach the linear dependence with slope −58.1 mV/pH-unit but turns back to a constant value around $pH = pK_{d,HQ}$. If $a_H \gg K_{d,HQ}$ and $a_H \gg K_{d,HR}$, only the protonated redox species are present, and $E^m_Q$ may then be interpreted as $E^o$ of the redox couple $HQ^+ + e^- \rightleftarrows RH$. $E^m_{P2}$ behaves similarly to $E^m_{P1}$ except that the turn into a pH-dependence occurs around $pH = pK_{d,HR}$, and that the slope attained is $-0.5 \ln\{10\}\, RT/\mathcal{F} = -29.1$ mV because $\nu_{e,P2} = 2$ while m = 1.

For several redox couples of biological relevance such as quinones $a_{H,k}/K_{d,HiRrd} \gg 1$ and $a_{H,k}/K_{d,HiRox} \ll 1$ holds true for all reasonable $a_H$-values, i.e. $0 \leq pH \leq 10$. The concentrations of the species $H_iR_{rd}$ with $i = 0,1,\cdots m-1$ and $H_iR_{ox}$ with $i = 1,2,\cdots m$ are then always negligibly small. As a consequence m $H^+$ ions are bound whenever the oxidized species accepts $\nu_{e(R)}$ electrons and *vice versa,* according to the half reaction

$$R_{ox} + \nu_{e(R)}\, e^- + m\, H^+ \rightleftarrows H_mR_{rd} \tag{64}$$

In this case Eqn. 62b simplifies to

$$\begin{aligned} E^m_{R,k} &= E^o_{R,k} - [\,RT/(\nu_{e(R)}\mathcal{F})]\ln\Big\{\prod_{i=1}^{m} K_{d,HiRrd}\Big\} + [RT/(\nu_{e(R)}\mathcal{F})]\ln\{a_H/a^o\} \\ &= E^{o'}_{R,k} - [m\, RT/(\nu_{e(R)}\mathcal{F})]\ln\{10\}\,(pH_k - 7) \end{aligned} \tag{65}$$

Since $a_{H,k} \gg K_{d,HiRrd}$ and $a_{H,k} \ll K_{d,HiRox}$ for all $a_H$-values $E^m_R$ depends linearly on pH with slope $-m\, RT \ln\{10\}/(\nu_{e(R)}\mathcal{F})$ over the entire pH-range. Moreover, none of the dissociation constants can be experimentally determined, and the only quantity characterizing such a redox couple is $E^{o'}_R$. This is illustrated by the redox couple $NAD^+ + 2e^- + H^+ \rightleftarrows NADH$ shown in Fig. 4. If $m = \nu_{e(R)}$ each electron exchanged is accompanied by an $H^+$ ion. The exchange of an electron and an $H^+$ ion together is equivalent to the exchange of a hydrogen atom, which is abbreviated by [eH], according to the half reaction $R_{ox} + \nu_{[eH](R)}\,[eH] \rightleftarrows H_mR_{rd}$ with $m = \nu_{[eH](R)}$.

### 3. Steady states

A process, a phase, or a compartment has reached a steady state when its parameters are no longer time-dependent. If under certain circumstances the time-dependence does not vanish but becomes negligibly small, a pseudo-steady state is attained. In general the processes occurring in a system may attain and thereafter remain in (pseudo-)steady states at different times in the course of an experiment, depending on the design of the experiment, the interrelations between the processes, and the topology of the system. The whole system is said to have reached a steady state when all parameters are time-independent and hence all processes occurring in it, or all its phases and compartments, have attained such a state.

Processes relax faster to steady states from given initial conditions the larger their rate constants and/or the smaller the chemical capacity of the species involved in the processes. The chemical capacity, $C_{ci,k}$, for the *i*th species in the *k*th phase or compartment relates the change in mole number, $dn_{i,k}$, to the change in concentration, $dc_{i,k}$. If the volume $V_k$ is constant it becomes [1]

$$C_{ci,k} = dn_{i,k}/dc_{i,k} = V_k\left[1 + \sum_j c_{Bj,tot,k}\, K_{d,Bj,k}/(c_{i,k} + K_{d,Bj,k})^2\right] \qquad (66)$$

The second term in square brackets in Eqn. 66 stems from buffers $B_j$ which are present at a total concentration $c_{Bj,tot,k}$ and whose binding of the *i*th species is characterized by the dissociation constant $K_{d,Bj,k}$ (cf. Eqn. 36). Acid-base reactions, i.e. binding and release of $H^+$ ions, are among the fastest processes known. Hence they usually relax to (pseudo-)steady states long before other processes reach such states. In most cases these states are (pseudo-)equilibrium states, and these reactions may then be treated as if they were at equilibrium (cf. Eqns. 36 and 39). A catalyst such as an enzyme is usually present at a much lower amount than the reactants of the process which it catalyses (see section 4.3). It binds and releases the reactants and therefore may be considered as an additional phase in the system accessible only to the reactants. Due to the small amount of the catalyst the chemical capacities for the reactants constituted by this phase are rather small. As a consequence a catalyst usually quickly relaxes to a steady state even if the rate parameters of the catalytic cycle are not very large, and independent of the catalyzed reaction itself being at such a state.

In general the parameters of a system can be constrained by "clamping", which causes at least some of the forces of the processes in the system to be clamped, too. If for technical reasons the forces cannot be totally clamped, it may nevertheless be possible to keep them approximately constant, i.e. to allow the system to reach a pseudo-steady state, by choosing appropriate chemical capacities. A clamping of flows is in general not feasible, except for the "flow" of a mechanical process. The special holder in the example of Fig. 1A may be constructed such that it imposes the velocity v on the muscle, the easiest case being v = 0 due to a rigid holder. Similarly, a bacterium may be rotated with a given angular velocity $\omega$ by means of "optical tweezers" [4].

If all forces are fixed, i.e. the maximum number of constraints is applied, the steady state is fully defined, since no more degrees of freedom are left. If some of the forces are fixed, the remainder will reach values in the steady state such that their conjugate flows become zero. For a *pair of coupled processes* this steady state is known as *static head* and is characterized by the maximal absolute value of the non-clamped force attainable under the given conditions. If the non-clamped force is zero the coupled processes are said to operate at *level flow*. This state can last only for a very short period of time unless this force is also clamped to zero. If no constraints at all are applied, the forces will all tend to decrease until the system eventually reaches equilibrium where all flows and forces vanish.

A vanishingly small change in the parameters of a compartment or a phase can also come about if the sums of the flows given on the right-hand side of Eqns. 24 are equal to zero (or at least close to zero). This happens if the chemical capacities of the species present in a given phase or compartment are considerably smaller than those of the other phases or compartments in the system. A typical example is the membrane of the system shown in Fig. 1D. Its volume is rather small as compared to the volume of the compartments, hence the condition of balancing flows is reached long before one of the compartments can attain a steady state. As a consequence, except for a short initial time period, $dn_{i,3}/dt \approx 0$ holds during the whole experiment, and the last term in Eqn. 25 vanishes. Hence, if the initial time period is not of interest, the transport of species between the membrane and the compartments as well as reactions within the membrane can be neglected. Another example where the steady state condition of a compartment proves to be very useful for the analysis of a system will be considered in section 5.3.3.

## 4. Mechanisms of coupling between processes

A coupling between two processes in general terms means that they cannot proceed independently although all necessary reactants are present. Coupling introduces a strict relation between the flows of the processes, and can occur if the processes share a common element which has attained a steady state. Note that the steady state in this statement concerns only the common element and not the whole system. The strict relation between the flows can cause the flow of one process to run against its thermodynamic force. An energy conversion then takes place from the process whose flow is in the direction determined by its force, which is called the *driving process,* to the process whose flow runs against its force called the *driven process.* Which process is driving and which one is the driven is not inherent in the coupling mechanism but depends on the conditions imposed on the system (clamping of forces, chemical capacities etc., see section 3). Inherent in the notion of coupling is the possibility of uncoupling the processes by any suitable means which causes the interrelation between the flows to be changed or totally broken. The mechanism by which such an uncoupling in general terms is brought about depends on the type of common element. This has created some confusion with respect to terminology because, for historical reasons, it has become customary to restrict the notion of uncoupling (and uncouplers as the means causing uncoupling) to a particular type of coupling (see section 4.2).

### *4.1. Electroneutrality and the electrical potential difference across membranes*

The condition of electroneutrality expresses the fact that an imbalance of charge in a phase with mobile charges cannot exist. Hence, for the bulk phase in the *k*th compartment

$$\sum_{k} z_i \, n_{i,k} = 0 \quad \text{(electroneutrality)} \tag{67}$$

where the sum includes all species in the phase. As a consequence the excess charge associated with a difference in electrical potential between two compartments separated by a membrane (usually called the membrane potential, see section 2.6.2) is confined to very thin diffuse layers on the membrane surfaces [1]. The membrane together with the surface layers thus constitutes an electrical capacitor whose capacitance for charge is given by the membrane capacity $C_m$ [1,12]. Taking the time derivative of Eqn. 67 and introducing the flows according to Eqn. 13 yields

$$\sum_{i} z_i \, J_i = 0 \tag{68}$$

The condition of electroneutrality thus introduces a strict relation between the flows of charged species across a membrane, i.e. it causes a coupling between these flows. The common element is the membrane capacitor (or the membrane potential associated with it).

Equation 68 can serve to estimate the membrane potential when the flows are replaced by relations which express the dependence of the flows on the pertinent forces. However, except for a few special cases (see, for example, ref. 2), the resulting equations are transcendental in terms of the membrane potential and hence can only be solved numerically. Note that Eqn. 68 is not valid as long as the compositions of the diffuse layers on the membrane surfaces vary substantially or, in other words, as long as a charging of the membrane capacitor takes place. As stated above, the common element has to be in a steady state before coupling becomes effective. This is evident from the results obtained with a model system to be discussed in section 5.3.2 (another example is the permeation of salt through a membrane dealt with in ref. 13).

The coupling of the flows of charged species arising from the condition of electroneutrality disappears when the membrane potential vanishes.

Hence, clamping the membrane potential to zero uncouples the flows. This can be achieved by placing a reference electrode in each of the two compartments (see, for example, refs. 2 or 3). When these electrodes are short-circuited any difference in electrical potential between the compartments is annihilated. Moreover, Eqn. 68 is replaced by

$$\mathcal{F} \sum_i z_i J_i = I \qquad (68a)$$

where I is the electric current running through the short-circuited pair of electrodes. Thus, the transport of each charged species can adopt the flow which is determined by its conjugate force, and the imbalance of charge created by these independent flows is compensated by the current I. In a system where the insertion of electrodes is impossible the short-circuiting of the compartments can be achieved by a charged species whose transport is much faster than the transport of all other charged species. Its activity in the two compartments has to be equal in order that the membrane potential is clamped to zero. Moreover, its chemical capacity in both compartments has to be sufficiently large because its flow, which has to take over the role of the electric current I, should not appreciably change its activities. It should be understood that these requirements are rather strict and often not easy to fulfil in practice.

### *4.2. Coupling due to a common thermodynamic force*

The common element required for coupling of two processes can arise from reactants which participate in both processes. As an example consider the two reactions where species A is converted into B, and species C is converted into D. Assume that these reactions involve two additional intermediate reactants X and Y according to the following scheme

$$A + X \rightleftarrows B + Y \qquad (69a)$$

$$C + Y \rightleftarrows D + X \qquad (69b)$$

$$Y \rightleftarrows X \qquad (69c)$$

The affinities of these reactions are given by (cf. Eqn. 12)

$$\mathcal{A}_1 = \tilde{\mu}_A - \tilde{\mu}_B - \mathcal{A}_3 \qquad (70a)$$

$$\mathcal{A}_2 = \tilde{\mu}_C - \tilde{\mu}_D + \mathcal{A}_3 \qquad (70b)$$

$$\mathcal{A}_3 = \tilde{\mu}_Y - \tilde{\mu}_X \qquad (70c)$$

When the concentrations of the intermediates X and Y no longer change with time, i.e. when the reaction in Eqn. 69c has reached a steady state, the two conversions of species A to B and C to D become coupled. In this state the rate of the reaction shown in Eqn. 69c as compared to the rates of the two reactions shown in Eqns. 69a and b determines how well the two processes are coupled. If the conversion of Y to X is negligible the processes are fully coupled. The more prominent the rate of the Y to X conversion becomes the less coupled are the reactions which convert A to B and C to D. Hence, any manipulation of the system which causes a change in this rate and a concomitant change in $\mathcal{A}_3$ affects the coupling.

In general, two processes which share a thermodynamic force become coupled when this force has attained a steady state. The common force can be either that of a chemical reaction (as in the case of Eqn. 69c) or that of a transport process. Thus, in a cell in which the affinity of the reaction where ATP is formed from ADP and phosphate (i.e. the phosphate potential, see section 2.6.4) has reached a constant value, any two processes which involve this reaction in either direction (i.e. synthesis or hydrolysis of ATP) are coupled. More familiar are the cases where the common element is the force of a transport process, particularly if the transported species is the $H^+$ ion. Mitchell [14] was the first who realized that $\Delta\tilde{\mu}_H$, i.e. the thermodynamic force of the transport of $H^+$ ions between two compartments[9], can serve as a common element in energy-converting organelles such as mitochondria and chloroplasts. He formulated what is known as the chemiosmotic mechanism of coupling, in short chemiosmosis. Note that chemiosmosis is not restricted to the $H^+$ ion but usually refers to it if the transported species is not specified.

Mitchell could also explain for the first time how uncouplers work in energy-converting organelles. Any agent which increases the transport of $H^+$ ions through the membrane is an uncoupler because it tends to diminish $\Delta\tilde{\mu}_H$. Among the diverse substances acting as uncouplers are agents which create a pore in the membrane (e.g. gramicidin); others operate as a carrier for $H^+$ ions and therefore are called "protonophores" (e.g. FCCP). This explanation was so elegant and so convincing that even today the notion of uncoupling is almost automatically associated with this mechanism, although it pertains only to one particular case of coupling.

### *4.3. Coupling on the molecular level by enzymes*

The reaction in Eqn. 69a (or that in Eqn. 69b) *per se* seems to represent the simplest case of coupling on the molecular level. Obviously, the conversion of A to B cannot proceed without the concomitant conversion of X to Y. However, if none of these conversions can occur independently, it

[9]Mitchell introduced the notation "proton motive force" for the quantity $\Delta\tilde{\mu}_H/\mathcal{F}$.

is meaningless to formally separate them. They have to be considered as *one* chemical reaction rather than a system of two coupled but fictitious reactions. The situation is different if both processes can occur independently. A prerequisite for coupling then is that the intrinsic rate parameters of the two processes are rather small. Otherwise the energy transferred from one of the processes to the other as the result of a coupling device would be simply dissipated by the freely running processes.

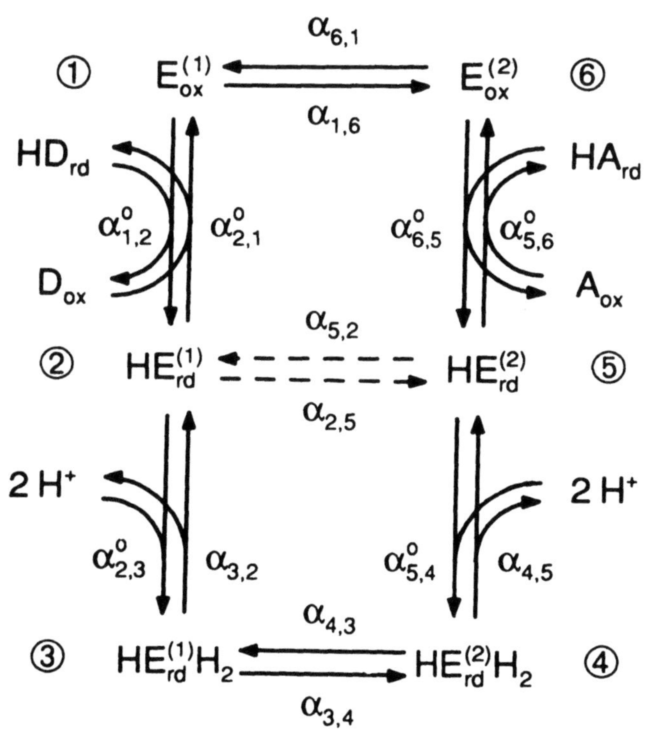


Fig. 5. Catalytic cycle for a redox-driven proton pump. The enzyme denoted by E adopts different conformations (numbered 1-6) in the course of the cycle due to binding of reactants and chemical transformations. Superscripts (1) and (2) indicate that the enzyme reacts with reactants in compartments 1 and 2, respectively. The symbol H to the left and right of E represents the $H^+$ ions involved in the redox reaction and the ion transport, respectively. The quantities $\alpha_{i,j}$ are called transition probabilities; they are comparable to the rate parameters for chemical reactions and determine the rate of the transitions between the conformations (see e.g. refs. 1 or 2). For further explanation see text.

The values for the transition probabilities used in simulations are $\alpha^{o}_{1,2} = 2 \times 10^5\ M^{-1}\ s^{-1}$, $\alpha^{o}_{2,1} = 4 \times 10^6\ M^{-1}\ s^{-1}$, $\alpha^{o}_{2,3} = 2 \times 10^{16}\ M^{-2}\ s^{-1}$, $\alpha_{3,2} = 200\ s^{-1}$, $\alpha_{3,4} = 2000\ s^{-1}$, $\alpha_{4,3} = 200\ s^{-1}$, $\alpha_{4,5} = 200\ s^{-1}$, $\alpha^{o}_{5,4} = 2 \times 10^{13}\ M^{-2}\ s^{-1}$, $\alpha^{o}_{5,6} = 2 \times 10^6\ M^{-1}\ s^{-1}$, $\alpha^{o}_{6,5} = 2000\ M^{-1}\ s^{-1}$, $\alpha_{6,1} = 200\ s^{-1}$, $\alpha_{1,6} = 200\ s^{-1}$, $\alpha_{2,5} = 0.5\ s^{-1}$, $\alpha_{5,2} = 5 \times 10^{-5}\ s^{-1}$. As is required by thermokinetic balancing [1,2] these values yield the equilibrium constant for the transport of $H^+$ ions, $K_H = 1$, and for hydrogen atoms, $K_{[eH]} = 5 \times 10^5$ (equivalent to $E^{o'}_{A,2} - E^{o'}_{D,1} = 295.79$ mV at 25°C, cf. Eqn. 46b).

The majority of processes occurring in biological systems would not proceed spontaneously even in the presence of an appropriate thermodynamic force if they were not catalyzed. An agent which accelerates (or catalyzes) a process is called a catalyst, and catalysts in biological systems are proteins called enzymes. A catalyst increases the rate parameters of a process, and it does this for the positive as well as for the negative direction (see, for example, refs. 1 or 2). It takes part in the process in a cyclic way, i.e. it is constantly regenerated while the process is running. Therefore, it has to be present only in a small (or catalytic) amount as compared to the species involved in the process.

Any two intrinsically slow processes which are catalyzed by one enzyme in a common kinetic cycle become coupled when the enzyme as the common element has attained a steady state. As an example consider an intrinsically slow redox reaction between a donor couple and an acceptor couple, both exchanging one electron (cf. Eqn. 17 with $\nu_{e(D)} = \nu_{e(A)} = 1$). The reaction is catalyzed by a membrane-bound enzyme which at the same time catalyzes the intrinsically slow trans-membranal transport of $H^+$ ions from compartment 1 to compartment 2. Figure 5 presents an example for a minimal kinetic scheme of the catalytic cycle performed by such an enzyme. According to this scheme the enzyme exchanges electrons only with the donor couple in compartment 1 and with the acceptor couple in compartment 2. Moreover, it is assumed that the reduced species of both couples as well as the reduced enzyme are present only in the protonated form due to appropriate dissociation constants (see section 2.6.4). Hence a hydrogen atom [eH] is transported by the enzyme from compartment 1 to compartment 2. However, this transport can only occur if the enzyme binds two $H^+$ ions from compartment 1 and releases them into compartment 2. This causes the coupling of the redox process to the transport of $H^+$ ions, and the enzyme is therefore said to be a "redox-driven proton pump". It may be argued that such an enzyme is hypothetical and hence has little relevance to actual systems. But it should be understood that the properties of this enzyme were deliberately chosen such that the different phenomena to be discussed in section 5.1.1, 5.3, and 6 are entirely attributable to the enzyme and not obscured by side effects.

The scheme in Fig. 5 includes also a transition between the reduced forms of the enzyme without additionally bound $H^+$ ions, as indicated by the broken arrows between states 2 and 5. The effect of this transition on the coupling of hydrogen atom and $H^+$ ion transport is analogous to the effect of the reaction in Eqn. 69c on the coupling of the reactions in Eqns. 69a and b. If the transition $2 \rightleftarrows 5$ is not operative, i.e. if its rate parameters are zero, the two transport processes are fully coupled. Non-vanishing rate parameters cause a partial uncoupling of the transport processes whose extent is determined by the rate of this transition as compared to the rate of the overall catalytic cycle (for a more detailed discussion see ref. 1). This partial uncoupling is an intrinsic property of the enzyme and hence is sometimes called "intrinsic uncoupling". In order to clearly distinguish this case from the "classical uncoupling" discussed in the preceding section, the term *slip in the enzyme* (or *slipping*) is more appropriate. It refers to the fact that, besides going around the fully coupled cycle 1–2–3–4–5–6–1, the enzyme can also "slip" through the cycles 1–2–5–6–1 and 2–3–4–5–2. Accordingly, an agent which causes a relative increase in the rate parameters of transition $2 \rightleftarrows 5$ is called a slip-uncoupler.

## 5. Phenomenological description of processes and systems

### *5.1. Kinetics of processes*

Thermodynamics cannot tell how flows are related to their conjugate forces and, in the case of energy conversion due to coupling, to the forces of other processes. Answers to this question can only be obtained on the basis of molecular and kinetic schemes. The analysis of such schemes as discussed in refs. 1 and 2 for non-coupled processes applies as well to coupled processes. But the results are considerably more complicated (see, for example, ref. 1 or 15) and therefore beyond the scope of this chapter. For the present purpose it suffices to use phenomenological and linear flow-force relations which approximate the explicit relations to a satisfactory degree.

*5.1.1. Linear flow-force relations:* The flow-force relations for processes at steady state as obtained from the analysis of kinetic schemes are in general ambiguous. However, if certain constraints apply, they can be converted to unambiguous relations [1, 2]. In the case of non-coupled and fully coupled processes these relations have the form of a hyperbolic tangent. In the case of highly coupled processes the flow-force relations are close to a hyperbolic tangent. This is demonstrated by the examples shown in Fig. 6 which pertain to the redox-driven proton pump depicted in Fig. 5. With the positive direction of transport being from compartment 1 to compartment 2 (see section 2.4.2) the conjugate force for the transport of hydrogen atoms becomes (cf. Eqns. 19, 21, 44, 64, 65, and 71b)

$$\begin{aligned}\Delta\tilde{\mu}_{[eH]} &= \tilde{\mu}_{[eH](D,1)} - \tilde{\mu}_{[eH](A,2)} = (\tilde{\mu}_{HDrd,1} - \tilde{\mu}_{Dox,1}) - (\tilde{\mu}_{HArd,2} - \tilde{\mu}_{Aox,2}) \\ &= (\tilde{\mu}_{Drd,1} - \tilde{\mu}_{Dox,1}) - (\tilde{\mu}_{Ard,2} - \tilde{\mu}_{Aox,2}) + \Delta\tilde{\mu}_H = \Delta\tilde{\mu}_e + \Delta\tilde{\mu}_H \\ &= \mathcal{F}(E_{A,2} - E_{D,1}) - \mathcal{F}\Delta\phi_m + RT\ln\{10\}(pH_2 - pH_1) + \mathcal{F}\Delta\phi_m \\ &= \mathcal{F}(E^{o'}_{A,2} - E^{o'}_{D,1}) + RT\ln\{c_{HDrd,1}\, c_{Aox,2}/(c_{Dox,1}\, c_{HArd,2})\} \qquad (71a)\end{aligned}$$

while that for the transport of $H^+$ ions is (cf. Eqns. 14, 42, and 49)

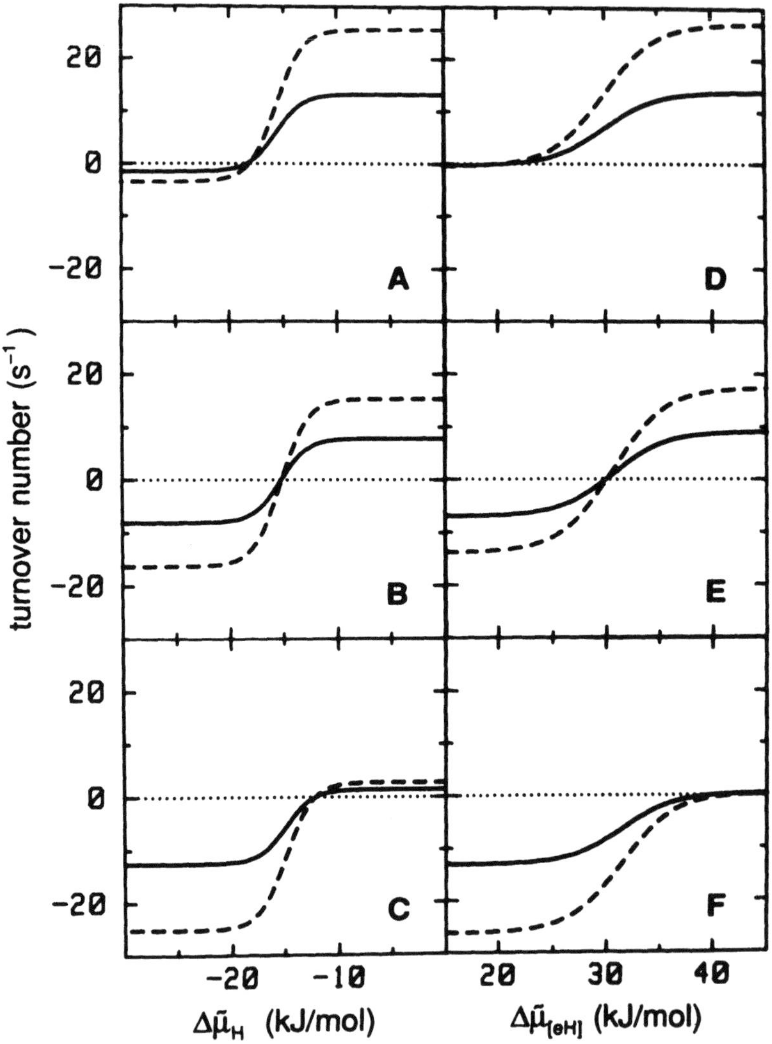


Fig 6. Flow-force relations for the redox-driven proton pump. Dependence of the flow for hydrogen atoms, $J_{[eH]}$ (solid line), and for pumped $H^+$ ions, $J_{H(p)}$ (broken line), on the force $\Delta\tilde{\mu}_H$ at a constant force $\Delta\tilde{\mu}_{[eH]}$ (A–C), and on the force $\Delta\tilde{\mu}_{[eH]}$ at a constant force $\Delta\tilde{\mu}_H$ (D–F). The flows are divided by the mole number of the enzyme which yields turnover numbers. The constraints applied to $\Delta\tilde{\mu}_{[eH]}$ are a constant value for $c_{HArd,2}$ = 100 mM and for $c_{Aox,2}$ = 1 mM, while $c_{HDrd,1}$ and $c_{Dox,1}$ were varied under the constraint $c_{HDrd,1} + c_{Dox,1}$ = 10.02 mM; $E^{o\prime}_{A,2} - E^{o\prime}_{D,1}$ = 295.79 mV (see Eqn. 71a). The values of the constant $\Delta\tilde{\mu}_{[eH]}$ are (in kJ/mol) 36.5 (A), 30.5 (B), and 24.5(C). The constraints applied to $\Delta\tilde{\mu}_H$ are a constant value for $pH_1$ = 7.5 and $\Delta\phi_m$ = 0, while $pH_2$ was varied (see Eqn. 71b). The values of the constant $\Delta\tilde{\mu}_H$ are (in kJ/mol) −10 (D), −15 (E), and −20 (F). Temperature 25°C.

$$\Delta\tilde{\mu}_H = \tilde{\mu}_{H,1} - \tilde{\mu}_{H,2} = RT \ln\{a_{H,1}/a_{H,2}\} + \mathcal{F}\Delta\phi_m$$
$$= RT \ln\{10\}\,(pH_2 - pH_1) + \mathcal{F}\Delta\phi_m \qquad (71b)$$

Note that the term $\mathcal{F}\Delta\phi_m$ cancels out in Eqn. 71a because the hydrogen atom has no charge, which is one of the reasons for choosing the hypothetical enzyme. Figure 6 then shows the dependence of the flows of the transported species on the force $\Delta\tilde{\mu}_H$ at constant $\Delta\tilde{\mu}_{[eH]}$ and on the force $\Delta\tilde{\mu}_{[eH]}$ at constant $\Delta\tilde{\mu}_H$ for different values of the constant force. The forces were varied under the constraints stated in the legend to Fig. 6 which give rise to the unambiguous flow-force relations.

In a limited range of forces the flows of the coupled processes can be approximated by a linear dependence on the forces. As is evident from Fig. 6 the most extended range of approximation of not constant flows is located around the inflection point of the hyperbolic tangent function. In general terms, for the *j*th pair of coupled processes, a linear approximation of the flows $J_{j,1}$ and $J_{j,2}$ depending on the forces $X_{j,1}$ and $X_{j,2}$ reads

$$J_{j,1} = L_{j,11}\,X_{j,1} + L_{j,12}\,X_{j,2} + J_{oj,1} \qquad (72a)$$

$$J_{j,2} = L_{j,21}\,X_{j,1} + L_{j,22}\,X_{j,2} + J_{oj,2} \qquad (72b)$$

Similarly, the dependence of the flow $J_i$ of the *i*th non-coupled process on its conjugate force $X_i$ can be approximated by

$$J_i = L_i\,X_i + J_{oi} \qquad (73)$$

The coefficients L in Eqns. 72 and 73 are called generalized permeabilities of the processes and comprise the pertinent rate parameters [1, 2]. The coefficients $L_{j,11}$ and $L_{j,22}$ relate the flows to the conjugate forces and hence are called straight coefficients, while $L_{j,12}$ and $L_{j,21}$ express the dependence of a given flow on the force of the coupled process and therefore are called coupling coefficients. The quantity $J_o$ denotes the intercept of a given linear approximation on the J-axis and causes an offset of the linearized flow from the origin. Hence, $J_o$'s have to be zero in linear relations which approximate the flows in a range of force encompassing the points $X_{j,1} = X_{j,2} = X_i = 0$ because all flows vanish at equilibrium (see section 3). This also applies to linear relations which approximate highly coupled flows around static head.

Many experimental systems were found to comply with linear relations even with $J_o = 0$ and the symmetry condition $L_{j,12} = L_{j,21}$ [16]. It then appears that the phenomenological relations given in Eqns. 72 and 73 can serve as useful approximations, as postulated by the formalism of linear nonequilibrium thermodynamics (linear NET). Westerhoff and van Dam [17] have used linear NET to describe a large number of systems and thus demonstrated the versatility of the algorithm. It has to be stressed, however, that linear flow-force relations cannot be taken for granted, and their validity has to be checked carefully for every new system or whenever the conditions for a known system are altered.

*5.1.2. Degree of coupling and phenomenological stoichiometry:* Kedem and Caplan [18] have introduced a normalization of the generalized permeabilities in the flow-force relations of coupled processes. In terms of the permeabilities in Eqns. 72, the normalized quantity called the *degree of coupling*, $q_j$, is defined as

$$q_j = [L_{j,12}\, L_{j,21}/(L_{j,11}\, L_{j,22})]^{1/2} \tag{74}$$

It is bound by the upper limit 1, which indicates full coupling of the processes, and the lower limit 0 in the case of no coupling at all. The second normalized quantity called the *phenomenological stoichiometry*, $Z_{j,21}$, for coupling of process 2 to process 1 is defined as

$$Z_{j,21} = [L_{j,22}/L_{j,11}]^{1/2} \tag{75}$$

The coupling coefficients need not necessarily be identical, and an existing asymmetry can be assessed by the quantity

$$\gamma_j = [L_{j,21}/L_{j,12}]^{1/2} \tag{76}$$

For $q_j = 1$, $\gamma_j = 1$ and $Z_{j,21}$ is equal to the molecular stoichiometry of coupling. For $q_j < 1$, the value of $Z_{j,21}$ essentially indicates the relative contributions of slipping (see section 4.3) to the coupled flows and therefore can be larger or smaller than the molecular stoichiometry. This interpretation holds for processes near equilibrium which also includes static head for highly coupled processes. A careful investigation of the meaning of q and Z in the ranges of linearity far from equilibrium showed that the interpretation of both parameters is in general no longer valid [15].

### *5.2. Efficiency and efficacy of energy conversion*

The *efficiency* $\eta$ of energy conversion between two coupled processes can be expressed by means of the flow-force product of the driven process (index out for output) and the driving process (index in for input) [16,18]. It is defined as

$$\eta = -\, J_{out}\, X_{out}/(J_{in}\, X_{in}) \tag{77}$$

The negative sign in Eqn. 77 is introduced in order to obtain a positive efficiency because $J_{out}X_{out} < 0$ (see section 2.5). The efficiency $\eta$ relates the output power to the input power, and represents the fraction of the entropy production of the driving process which is consumed by the driven process due to energy conversion.

If the force of the driven process is not clamped static head is attained (see section 3) in which case $J_{out} = 0$ and, according to Eqn. 77, $\eta = 0$. A more appropriate quantity for this state is the *force efficacy* $\varepsilon_X$ [16] defined as

$$\varepsilon_X = - X_{out}/(J_{in}\ X_{in}) \tag{78}$$

It measures how much input power is needed to support the maximal force of the driven process. In the case of a tight coupling the flow of the driving process also vanishes at static head, and the force ratio $|X_{out}/X_{in}|$ then becomes equal to the molecular stoichiometry of coupling. If the force of the driven process is clamped to zero, level flow is attained (see section 3). The appropriate quantity for this case is the *flow efficacy* $\varepsilon_J$ [16] defined as

$$\varepsilon_J = J_{out}/(J_{in}\ X_{in}) \tag{79}$$

It indicates how much input power is needed to sustain the maximal flow of the driven process at this state. Except for two coupled transports the positive directions for the coupled processes may be chosen independently. Hence, depending on the mechanisms involved, the coupling between the processes can be positive or negative. For positive coupling $J_{in}$ and $J_{out}$ have the same sign which causes opposite signs for $X_{out}$ and $X_{in}$, while for negative coupling the forces have the same sign and the flows have opposite signs. As a consequence $\varepsilon_X$ and $\varepsilon_J$ can be positive or negative depending on the sign of coupling.

To end this survey on efficiency of energy conversion the idealized case of reversible transitions will be considered. No free enthalpy is then spent on entropy production and all can be used in energy conversion (see section 2.3). However, this has to occur at an infinitely small rate which means that $J_{in} \approx 0$ and $J_{out} \approx 0$. Therefore, despite the maximal conversion in terms of forces, the efficiency of a reversible energy conversion is effectively zero or, in other words, no power can be gained from such a conversion. Needless to say no biological system could survive under such conditions.

### *5.3. Systems*

The analysis of a system requires that the concentration of all relevant species in all compartments and the electrical potential difference between compartments are known. By means of these quantities the forces can be calculated (see section 2.6). Moreover, the change in concentration of a given species in a given compartment, which arises from the flows of the processes in which the species takes part, can be estimated by means of Eqns. 24 and 66. In practice the determination of these concentrations quite frequently proves to be a rather difficult and tedious task. Besides the fact that the analytical procedures employed should not severely interfere with the processes in the system[10] the task is hampered by the fact that one compartment of the system is usually not easily accessible due to its size and/or topology.

If a suitable spectroscopic technique is available the time-course of the concentration of a species can be recorded. Thus, using the system depicted in Fig. 1B, Klein *et al.* [19] monitored the change in concentration of $Ca^{2+}$ ions which occur after excitation in an isolated muscle fiber. Similarly the electrical potential difference between the compartments may be monitored by so-called potential-indicating dyes. In many cases, however, the parameters in one of the compartments can be estimated only by indirect techniques which are based on the partitioning of suitable marker substances. Their concentration is monitored in the accessible compartment and the concentration in the inaccessible compartment is calculated from mass balance. Alternatively, the two compartments are physically separated and analyzed independently (see ref. 20 for a review of these techniques employed for the estimation of $\Delta\tilde{\mu}_H$ in energy-conversion organelles). The time required for the analysis is determined by the partitioning of the markers (which have to attain equilibrium) and/or the separation technique. It usually exceeds the characteristic time needed by the inaccessible compartment to relax to a steady state. Hence meaningful results can only be obtained if the compartment is at least in a pseudo-steady state where changes in the parameters to be determined are negligibly small.

*5.3.1. A model system:* If the properties of a system and all processes occurring in it are well defined the system's behavior can be simulated [13]. Such a system which may be called a model system has the advantage that, in contrast to an experimental system, *all* parameters are known. It is chosen here for didactic reasons because the assumptions made in an analysis and the results thereby obtained can be readily checked against the intrinsic properties of the system. The model system used is identical with the system discussed in ref. 13 and consists of vesicles suspended in an appropriate medium similar to the suspension of mitochondria in Fig. 1C. The membranes of the vesicles contain the redox-driven proton pump (cf. Fig. 5). They are permeable to the species of the acceptor redox couple, to $H^+$ ions and other univalent cations, and to univalent anions, all

[10] Note that any analytical procedure necessarily implies a certain interference with the system because otherwise no information could be obtained.

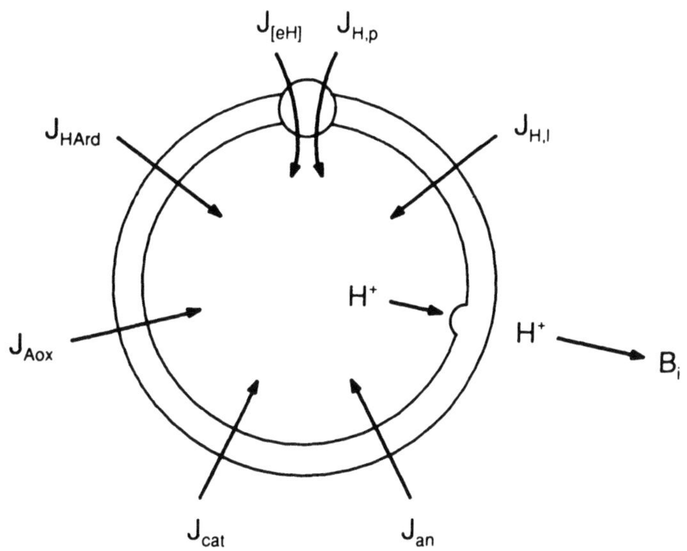


Fig. 7. Processes occurring in the model system. Sketch of a vesicle from the model system with arrows indicating the positive direction of the processes. The circle in the membrane represents the redox-driven proton pump which gives rise to the flow of hydrogen atoms, $J_{[eH]}$, and the flow of pumped $H^+$ ions, $J_{H(p)}$. Permeation of species causes the flows $J_{H(l)}$ for $H^+$ ions ("leak flow"), $J_{HArd}$ and $J_{Aox}$ for the redox species of the acceptor couple, as well as $J_{cat}$ and $J_{an}$ for the univalent cations and anions, respectively. The semicircle on the membrane represents the $H^+$-binding group, while $B_i$ stands for the impermeable buffers for $H^+$ ions in the suspending medium.

present in the suspending medium (see Fig. 7). Moreover, a group on the membrane surface facing the inside of the vesicles (or compartment 2) can bind $H^+$ ions and thus acts as a buffer for these ions in compartment 2. The suspending medium in compartment 1 includes impermeable buffers for $H^+$ ions which cause an approximately constant chemical capacity $C_{cH,1}$ (cf. Eqn. 66).

In order that the model system has some relevance to a biological system the values of the parameters are chosen according to data which are typical for isolated thylakoids. It is common practice in photosynthesis research to relate the extensive parameters of the system to the mole number of chlorophylls, $n_{chl}$, present in the membrane which represents the amount of membrane used. Vesicles according to a chlorophyll concentration of 100 μM then yield a volume of $10^4$ l/(mol chl) for the suspending medium (compartment 1), while that of the inner space (compartment 2) is 20 l/(mol chl). The mole number of enzyme per chlorophyll, $n_E/n_{chl}$, is chosen to be 0.005 mol/(mol chl) which corresponds to 1 redox-driven proton pump per 200 chlorophylls. The mole number of the membrane-bound buffer is 1 mol/(mol chl), while its dissociation constant is $10^{-5}$. The constant $C_{cH,1}$ due to the impermeable buffers yields a formal volume of $2 \times 10^7$ l/(mol chl) for $H^+$ ions. The permeability of the total membrane has units of l/(mol chl × s) and is chosen to be 20 for $A_{ox}$, 1000 for $HA_{rd}$, 200 for $H^+$ ions, 0.4 for univalent cations, and 0.2 for univalent anions, respectively. The membrane capacity $C_m$ amounts to 0.01 F/(mol chl) as calculated from 1 μF/cm² for a biological membrane [12] and 1 m²/(mol chl) for the specific surface of thylakoid membranes. The kinetic parameters for the pump are listed in the legend to Fig. 5. The units for the flows become mol/(s œ mol chl) which will be abbreviated by $s^{-1}$.

*5.3.2. Approach to steady state:* The analysis of the kinetic schemes describing the processes yields equations which express the dependence of the parameters of the system on time. In most cases these relations are so complex that the time-course of a system on its way to a steady state can only be simulated with a computer. Hence the techniques which are used when analyzing data of a time-dependent system belong to the field of nu-

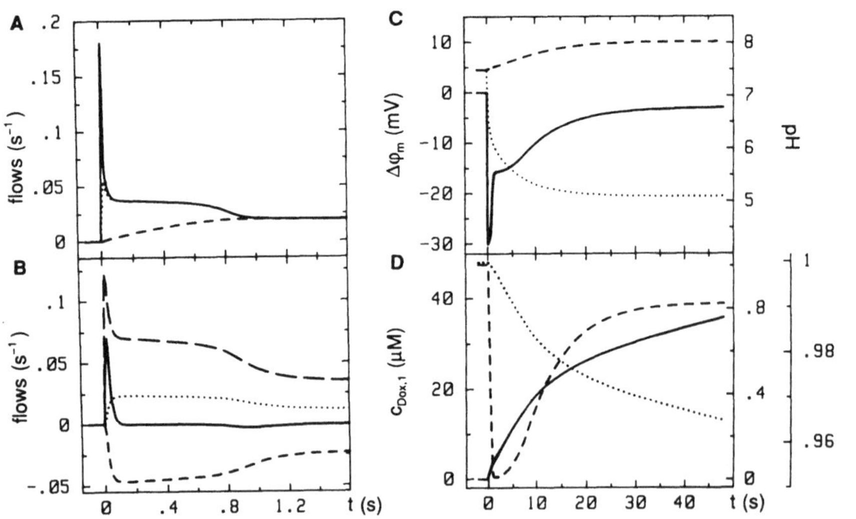


Fig. 8. Time-dependence of various parameters in the model system. In (A) the flow of hydrogen atoms from compartment 1 into the pump, $J_{[eH],1}$, and that from the pump into compartment 2, $J_{[eH],2}$, are represented by the solid line and the dotted line, respectively, while the broken line indicates $J_{Aox}$. In (B) $J_{cat}$, $J_{an}$, $J_H = J_{H(p)} + J_{H(l)}$ are represented, respectively, by the short broken line, the dotted line, and the long broken line. The solid line indicates the sum of all these flows multiplied by the charge number of the species which, when multiplied by the Faraday constant, yields the electric current (cf. Eqn. 68a). In (C) $\Delta\phi_m = \phi_1 - \phi_2$ (solid line, left scale) is shown together with $pH_1$ and $pH_2$ (broken and dotted line, right scale). In (D) the concentration of $D_{ox}$ in compartment 1 (solid line, left scale) together with that of $A_{ox}$ in compartment 2 (broken line, right scale) and in compartment 1 (dotted line, outer right scale) is shown. Temperature 25°C. Initial conditions: $c_{HArd,1} = c_{HArd,2} = 100$ mM, $c_{Aox,1} = c_{Aox,2} = 1$ mM, $c_{HDrd,1} = 0$ but raised to 10 mM at time zero, $c_{Dox,1} = 0$, $c_{cat,1} = c_{cat,2} = c_{an,1} = c_{an,2} = 100$ mM, $pH_1 = pH_2 = 7.5$, $\Delta\phi_m = 0$.

merical mathematics (for an example see ref. 19) which is beyond the scope of this chapter. The following results illustrate the path of the model system to a steady state. They were obtained as described in ref. 13, using the values of parameters listed in the preceding section.

Initially the system is at equilibrium; it contains all reactants except for the species of the donor redox couple (see legend to Fig. 8). At time zero $HD_{rd}$ is added to compartment 1 within 10 ms, and to a final concentration of 10 mM. The first event to occur is the adaptation of the pump to the changed redox conditions. It is evident from the difference between the flow of hydrogen atoms from compartment 1 into the pump, $J_{[eH],1}$, and that out of the pump into compartment 2, $J_{[eH],2}$. After about 100 ms the pump has relaxed to a steady state, and the two flows are essentially equal (Fig. 8A). A transfer of positive charge across the membrane is associated with the concomitant flow of pumped $H^+$ ions which disturbs the initially present charge balance between the compartments. Accordingly, a difference in electrical potential is built up which gives rise to a driving force for the transport of all charged species (including $H^+$) and to corresponding flows as determined by the pertinent permeabilities. All flows together constitute an electric current (cf. Eqn. 68a) which charges the capacitor associated with the membrane. This charging relaxes to a steady state within 140 ms (see Fig. 8B); from there on the electric current essentially vanishes and the flows are coupled as discussed in section 4.1. It thus becomes evident that the elements of the system with the smallest capacities are the first to reach a (pseudo-)steady state (see section 3).

The next element to reach a steady state is compartment 2. The convergence of the flow $J_{Aox}$ with the two equal flows $J_{[eH],1}$ and $J_{[eH],2}$, and the approximately constant value for all flows after about 1.2 s (see Fig. 8A) indicates that $c_{Aox,2}$ has reached an approximately constant value (see Fig. 8D). Similarly $J_{cat}$, $J_{an}$, and $J_H$ are approximately constant after about 1.4 s (Fig. 8B) thus indicating a steady state for ion transport. The transmembrane electrical potential difference $\Delta\phi_m$ initially drops very fast and, after passing through a minimum, attains a more or less constant value after about 1.4 s (Fig. 8C) again seemingly indicating a steady state. However, this state lasts only for about 2 s whereafter all parameters in compartment 2 start to change again (see Figs. 8C and D). The parameter responsible for this is $pH_2$ which has not yet reached a constant value because the chemical capacity for $H^+$ ions in compartment 2 is considerably larger than that for all other species due to buffering (cf. Eqn. 66). A longer lasting steady state is eventually reached after about 40 s where all parameters in compartment 2 become time-independent and the pump has attained static head.

*5.3.3. Analysis at steady state:* If the model system were an experimental system the only parameters readily accessible would be the concentrations of reactants in compartment 1 such as $c_{Dox,1}$, $c_{Aox,1}$ or $c_{H,1} = 10^{-pH,1}$. From these parameters the flows $J_{[eH]}$ and $J_H$ can be calculated. The constant value of $J_{[eH]}$ and the vanishing $J_H$ after 40 s (cf. Figs. 8C and D) then indicate that compartment 2 has reached a steady state. At this point an estimation of $\Delta\tilde{\mu}_H$ can be performed as outlined in section 5.3. Thus the thermodynamic quantities obtainable are $J_{[eH],s}$ and $\Delta\tilde{\mu}_{H,s}$, where the subscript s refers to compartment 2 in steady state[11]. The goal of the analysis is to obtain relations between these quantities and the parameters of the processes, whereby the steady-state condition of compartment 2 will prove helpful. An example of such an analysis applied to data from an experimental system can be found in ref. 20.

Inspection of Fig. 7 shows that the dissipation function of the model system is given by (cf. Eqn. 26)

$$\begin{aligned}\Phi &= J_{[eH]}\,\Delta\tilde{\mu}_{[eH]} + J_H\,\Delta\tilde{\mu}_H + J_{HArd}\,\Delta\tilde{\mu}_{HArd} + J_{Aox}\,\Delta\tilde{\mu}_{Aox} + J_{cat}\,\Delta\tilde{\mu}_{cat} + J_{an}\,\Delta\tilde{\mu}_{an}\\ &= J_{[eH]}\,\mathcal{A}_{[eH]} + J_H\,\Delta\tilde{\mu}_H + J'_{HArd}\,\Delta\tilde{\mu}_{HArd} + J'_{Aox}\,\Delta\tilde{\mu}_{Aox} + J_{cat}\,\Delta\tilde{\mu}_{cat} + J_{an}\,\Delta\tilde{\mu}_{an}\end{aligned} \tag{80}$$

with the abbreviations

$$J_H = J_{H(p)} + J_{H(l)} \tag{81a}$$

$$J'_{HArd} = J_{HArd} - J_{[eH]} \tag{81b}$$

$$J'_{Aox} = J_{Aox} - J_{[eH]} \tag{81c}$$

Terms pertinent to the binding of $H^+$ ions are omitted in Eqn. 80 because these reactions are always very close to equilibrium and their affinities may then be approximated by zero (see section 3). The quantity $\mathcal{A}_{[eH]}$ appearing in the second equality of Eqn. 80 is the affinity of the redox reaction between the couples in compartment 1. It can be defined irrespective of the fact that this reaction is intrinsically very slow, and amounts to (cf. Eqns. 14, 18, 42, and 71a)

$$\begin{aligned}\mathcal{A}_{[eH]} &= \tilde{\mu}_{[eH](D,1)} - \tilde{\mu}_{[eH](A,1)} = \Delta\tilde{\mu}_{[eH]} - \Delta\tilde{\mu}_{HArd} + \Delta\tilde{\mu}_{Aox}\\ &= \mathcal{F}(E^{o'}_{A,1} - E^{o'}_{D,1}) + RT\,\ln\{c_{HDrd,1}\,c_{Aox,1}/(c_{Dox,1}\,c_{HArd,1})\}\end{aligned} \tag{82}$$

When using linear approximations for the flow-force relations (cf. Eqns. 72 and 73) and when substituing for $\Delta\tilde{\mu}_{[eH]}$ by means of Eqn. 82, $J_{[eH]}$ and the flows in Eqns. 81 can be related to the forces as follows:

$$J_{[eH]} = L_e\,\mathcal{A}_{[eH]} + L_{eH}\,\Delta\tilde{\mu}_H + L_e\,\Delta\tilde{\mu}_{HArd} - L_e\,\Delta\tilde{\mu}_{Aox} + J_{o,e} \tag{83a}$$

[11]Note that compartment 1 need not be in a steady state. In fact, if it were in such a state, $J_{[eH],s}$ could not be determined because $dc_{Dox,1}/dt = dc_{HArd,1}/dt = 0$.

$$J_H = L_{He}\mathcal{A}_{[eH]} + (L_{H(p)} + L_{H(l)})\Delta\tilde{\mu}_H + L_{He}\Delta\tilde{\mu}_{HArd} - L_{He}\Delta\tilde{\mu}_{Aox} + J_{o,H(p)} + J_{o,H(l)} \quad (83b)$$

$$J'_{HArd} = L_e\mathcal{A}_{[eH]} + L_{eH}\Delta\tilde{\mu}_H + (L_{HArd} + L_e)\Delta\tilde{\mu}_{HArd} - L_e\Delta\tilde{\mu}_{Aox} + J_{o,HArd} + J_{o,e} \quad (83c)$$

$$J'_{Aox} = -L_e\mathcal{A}_{[eH]} - L_{eH}\Delta\tilde{\mu}_H - L_e\Delta\tilde{\mu}_{HArd} + (L_{Aox} + L_e)\Delta\tilde{\mu}_{Aox} + J_{o,Aox} - J_{o,e} \quad (83d)$$

where $L_e$, $L_{H(p)}$ and $L_{eH}$, $L_{He}$ denote, respectively, the straight and the coupling coefficients for the pump. Inspection of Eqn. 24b shows that $dn_{H,2}/dt = J_H$, $dn_{HArd,2}/dt = J'_{HArd}$, and $dn_{Aox,2}/dt = J'_{Aox}$. Hence, at steady state of compartment 2, $J_{H,s} = J'_{HArd,s} = J'_{Aox,s} = 0$. Solving Eqns. 83b–d for $\Delta\tilde{\mu}_{H,s}$, $\Delta\tilde{\mu}_{HArd,s}$, and $\Delta\tilde{\mu}_{Aox,s}$ under this condition, and introducing the result into Eqn. 83a yields

$$J_{[eH],s} = \left\{L_e[1 - \frac{L_{o,A}}{L_A}][1 + \lambda_H - q^2] + L_{o,e}[1 + \lambda_H] - L_{o,H}\frac{q}{\gamma Z_{He}}\right\}\frac{\mathcal{A}_{[eH],s}}{D} \quad (84)$$

$$\Delta\tilde{\mu}_{H,s} = -\left\{\frac{\gamma q}{Z_{He}}[1 - \frac{L_{o,e} + L_{o,A}}{L_A}] + \frac{L_{o,H}}{L_e Z_{He}^2}[1 + \lambda_e]\right\}\frac{\mathcal{A}_{[eH],s}}{D} \quad (85)$$

$$\Delta\tilde{\mu}_{HArd,s} - \Delta\tilde{\mu}_{Aox,s} = -\left\{L_e[1 + \lambda_H - q^2] + [L_{o,e} + L_{o,A}][1 + \lambda_H] - L_{o,H}\frac{q}{\gamma Z_{He}}\right\}\frac{\mathcal{A}_{[eH],s}}{L_A D} \quad (86)$$

with the abbreviations

$$\lambda_e = L_e/L_A \qquad \lambda_H = L_{H(l)}/[L_e Z_{He}^2] \quad \text{and} \quad D = [1 + \lambda_H][1 + \lambda_e] - \lambda_e q^2 \quad (87)$$

When writing Eqns. 84–87 the normalization of L-coefficients introduced in section 5.1.2 was used. Moreover, $L_{o,i} = J_{o,i}/\mathcal{A}_{[eH],s}$, in particular $L_{o,H} = (J_{o,H(p)} + J_{o,H(l)})/\mathcal{A}_{[eH],s}$. The quantities $L_A$ and $L_{o,A}$ are defined as

$$L_A = [1/L_{HArd} + 1/L_{Aox}]^{-1} \quad (88a)$$

$$L_{o,A} = [L_{Aox}\, J_{o,HArd} - L_{HArd}\, J_{o,Aox}]/[(L_{HArd} + L_{Aox})\,\mathcal{A}_{[eH],s}] \quad (88b)$$

and represent average parameters for the two species of the acceptor couple. Equations 84 and 85 provide the basis for the analysis. They relate the experimentally determinable quantities $J_{[eH],s}$ and $\Delta\tilde{\mu}_{H,s}$ to the parameters $L_e$, $q$, $Z_{He}$, $\gamma$, $L_{Hl}$, $L_A$, $L_{o,e}$, $L_{o,H}$, and $L_{o,A}$ which characterize the processes in the system. The quantity $\mathcal{A}_{[eH],s}$ can be estimated[12] but also eliminated when forming the ratio $J_{[eH],s}/[-\Delta\tilde{\mu}_{H,s}]$.

[12]Since $V_1 \gg V_2$, it follows from mass balance that at time t $c_{HDrd,1}(t) \approx c_{HDrd,1}(0) - \Delta c_{Dox,1}$, $c_{HArd,1}(t) \approx c_{HArd,1}(0) + \Delta c_{Dox,1}$, and $c_{Aox,1}(t) \approx c_{Aox,1}(0) - \Delta c_{Dox,1}$, where $\Delta c_{Dox,1} = c_{Dox,1}(t) - c_{Dox,1}(0)$ and the argument zero indicates initial composition at time zero. $\mathcal{A}_{[eH]}$ can then be calculated from Eqn. 70. For the feasibility of a redox potential measurement see section 2.6.2.

*5.3.4. Estimation of parameters:* Since it is impossible to determine the value of nine parameters from a single value for $J_{[eH],s}/[-\Delta\tilde{\mu}_{H,s}]$ it is necessary to increase the number of experimental data and/or to estimate the value of at least some of the parameters independently. If the permeabilities for the redox species are large enough the transport of these species operates around equilibrium. In this case $J_{o,HArd}$ and $J_{o,Aox}$ vanish (see section 5.1.1) and hence $L_{o,A} = 0$ (cf. Eqn. 88b). Moreover, $\lambda_e \ll 1$ is a reasonable approximation. Similarly, the $J_o$'s in a linear approximation for a highly coupled pump at static head vanish (cf. Fig. 6), hence $L_{o,e} = L_{o,H(p)} = 0$. However, $J_{o,H(l)}$ is certainly finite because the permeation of $H^+$ ions through the membrane occurs far from equilibrium, as indicated by the large value of its driving force $\Delta\tilde{\mu}_H$. Finally $\gamma = 1$ may be a reasonable approximation for a highly coupled pump.

Several different values for $J_{[eH],s}$ and $\Delta\tilde{\mu}_{H,s}$ are obtained when the processes in the system are manipulated by suitable means such as an inhibitor of the redox-driven proton pump or an uncoupler. An inhibitor when bound to an enzyme molecule totally blocks its catalytic activity. With increasing values of the inhibitor concentration $c_I$ more and more enzyme molecules are blocked. As a consequence the pump permeabilities $L_e$, $L_{eH}$, $L_{He}$, and $L_{H(p)}$, as well as $J_{o,e}$ and $J_{o,H(p)}$ if not zero, are proportionally decreased but the normalized quantities q, $Z_{He}$, and $\gamma$ remain unaffected. The decrease in $L_e$ can be described by the relation

$$L_e(c_I) = L_e(0)/[1 + (c_I/K_I)^h] \tag{89}$$

where $K_I$ denotes an average dissociation constant for the binding of the inhibitor to the pump. The coefficient h, which need not be an integer[13], may be larger than 1. If so it indicates that more than one bound inhibitor molecule is required in order to block the enzyme. An uncoupler usually does not interfere with the pump. It essentially increases the membrane's permeability for $H^+$ ions (cf. section 4.2) which is assessed by the parameters $L_{H(l)}$ and $L_{o,H(l)}$.

Figure 9A shows the effect of increasing inhibitor concentrations on the ratio $J_{[eH],s}/[-\Delta\tilde{\mu}_{H,s}]$ in the absence and in the presence of an uncoupler. The data are taken from simulations of the model system preformed with different values for the mole number of the pump, $n_E$, which then were converted to $c_I$ by means of the relation $c_I = K_I[n_E(0)/n_E(c_I) - 1]^{1/h}$ (cf. Eqn. 89). In a second set of simulations the membrane permeability was increased 7.5 fold which mimics the effect of an uncoupler. Since the chemical capacities for the redox species which determine $\mathcal{A}_{[eH]}$ belong to the largest in the system it can be assumed that $\mathcal{A}_{[eH],s}$ is approximately

[13]The name "Hill coefficient" is frequently used for h despite the fact that h is usually not identical with a Hill coefficient. The latter is an integer and indicates the number of sites involved in a cooperative binding.

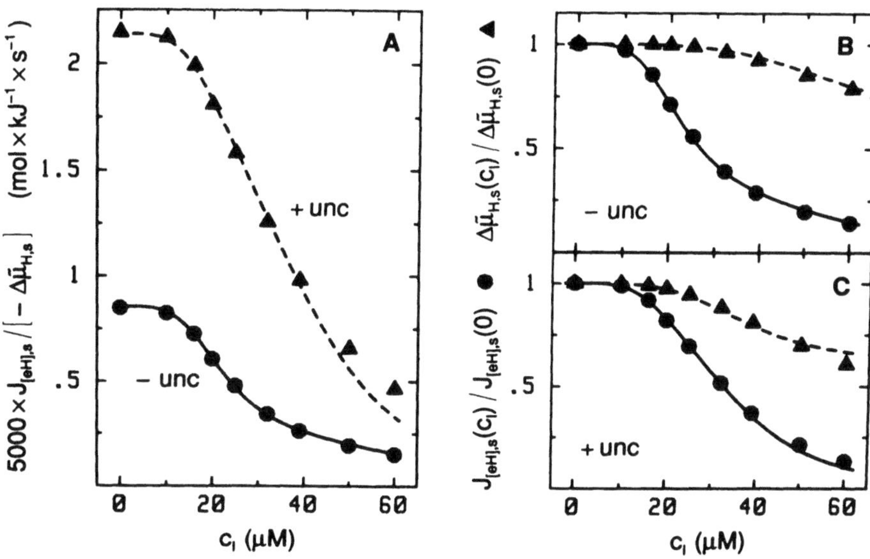


Fig. 9. Dependence of $J_{[eH],s}$ and $\Delta\tilde{\mu}_{H,s}$ on the inhibitor concentration $c_I$ in the absence and in the presence of an uncoupler. The ratio $J_{[eH],s}/[-\Delta\tilde{\mu}_{H,s}]$ in (A) was calculated from the simulated data in the absence (● and solid line) and in the presence (▲ and broken line) of an uncoupler: The ratios $J_{[eH],s}(c_I)/J_{[eH],s}(0)$ (● and solid line) and $\Delta\tilde{\mu}_{H,s}(c_I)/\Delta\tilde{\mu}_{H,s}(0)$ (▲ and broken line) were similarly calculated; they are shown in (B) for the case without uncoupler [$J_{[eH],s}(0) = 0.00290\ s^{-1}$, $\Delta\tilde{\mu}_{H,s}(0) = -17.06$ kJ/mol, $\mathcal{A}_{[eH],s}(0) = 35.20$ kJ/mol] and in (C) for the case with uncoupler [$J_{[eH],s}(0) = 0.00675\ s^{-1}$, $\Delta\tilde{\mu}_{H,s}(0) = -15.71$ kJ/mol, $\mathcal{A}_{[eH],s}(0) = 34.92$ kJ/mol]. The curves were calculated by means of Eqns. 84–89 and the parameter values listed in the text; temperature 25°C.

independent of $c_I$. Hence, when forming the ratios $J_{[eH],s}(c_I)/J_{[eH],s}(0)$ and $\Delta\tilde{\mu}_{H,s}(c_I)/\Delta\tilde{\mu}_{H,s}(0)$, $\mathcal{A}_{[eH],s}$ is eliminated (cf. Eqns. 84 and 85). The dependence of these ratios on $c_I$ is shown in Figs. 6B and C. It is found that $J_{[eH],s}$ is not reduced in proportion to the inhibition of the pump, and that $\Delta\tilde{\mu}_{H,s}$ is much less sensitive to inhibition than $J_{[eH],s}$. Thus, at $c_I = 50$ μM where the pump is inhibited by 96% (see Eqn. 89 with values for $K_I$ and h given below) $J_{[eH],s}$ is decreased by about 80% but $\Delta\tilde{\mu}_{H,s}$ is reduced only by 15% without uncoupler and by 30% with uncoupler. The time required to reach the steady state increases from 40 s at $c_I = 0$ to 540 s at $c_I = 60$ μM and from 30 s to 180 s without and with uncoupler, respectively.

A non-linear fitting procedure, which is based on Eqns. 84, 85 and 87–89 rewritten for three ratios formed with $J_{[eH],s}$ and $\Delta\tilde{\mu}_{H,s}$, allows one to estimate values for the parameters. When using all data points except for those with $c_I > 40$ μM in the presence of the uncoupler, the values thus determined are $L_e(0) = 0.00249$ [s × kJ/mol]$^{-1}$, $q = 0.988$, $Z_{He} = 2.05$, $K_I = 22.55$ μM, $h = 3.93$, $L_{H(I)} = 2.17 \times 10^{-4}$ [s × kJ/mol]$^{-1}$ and $15.7 \times 10^{-4}$ [s × kJ/mol]$^{-1}$, $L_{o,H(I)} = 6.05 \times 10^{-5}$ [s × kJ/mol]$^{-1}$ and $43.7 \times 10^{-5}$ [s × kJ/mol]$^{-1}$

without and with uncoupler, respectively. The dependence of the three ratios on $c_I$ can be calculated and is represented by the curves in Fig. 9. The reasonable fit between data points and calculated curves indicates that the phenomenological description worked out in section 5.3.3 is satisfactory approximation of the system. The value for q is smaller than 1 as expected for a pump with a slip (see section 5.1.2). Because of the high degree of coupling the value of $Z_{He}$ is close to 2, i.e. the molecular stoichiometry of coupling. The values found for $K_I$ and h essentially coincide with the respective values 22.6 μM and 4 used when converting $n_E(c_I)$ into $c_I$.

The affinity of the redox reaction in compartment 1 varies less than 0.1 kJ/mol without uncoupler, but by 1.7 kJ/mol in the presence of the uncoupler; the corresponding average value $\langle \mathcal{A}_{[eH],s} \rangle$ are 35.24 ± 0.04 kJ/mol and 35.34 ± 0.58 kJ/mol. Neverthless, the assumption made in the analysis that $\mathcal{A}_{[eH],s}$ is only weakly dependent on $c_I$ is justified. Inspection of Eqns. 82, 84, and 86 shows that

$$\mathcal{A}_{[eH],s} - \Delta\tilde{\mu}_{[eH],s} = -[\Delta\tilde{\mu}_{HArd,s} - \Delta\tilde{\mu}_{Aox,s}] = J_{[eH],s}/L_A \quad \text{for} \quad L_{o,A} = 0 \qquad (90)$$

which means that $J_{[eH],s}$ should be proportional to $\mathcal{A}_{[eH],s} - \Delta\tilde{\mu}_{[eH],s}$. A plot of the corresponding data taken from the simulations supports this conclusion (see Fig. 10A), but the proportionality is not perfect due to the linear approximations used when describing the transport of the redox species. The slope of the straight line shown in Fig. 10A has a value of 0.00681 $[s \times kJ/mol]^{-1}$ and represents an average proportionality coefficient which can serve as an estimate of $L_A$ (see Eqn. 90). With this value it follows that $\lambda_e(c_I) < 0.37$ but not small compared to 1 as assumed in the analysis. However, this is not crucial since estimating parameters with $L_A$ set to 0.00681 again yields a good fit to the data. In this case $L_e(0)$ = 0.00222 $[s \times kJ/mol]^{-1}$, $q = 0.986$, $Z_{He} = 2.00$, while the values of the other parameters are at most 6% smaller than those listed above.

The values for $L_{H(i)}$ and $L_{o,H(I)}$ with uncoupler are 7.2 times larger than those without uncoupler, in fair agreement with the 7.5 fold increase in permeability for $H^+$ ions used to simulate the uncoupled action. The values for $J_{H(I),s}$ obtained from the simulations are plotted versus $\Delta\tilde{\mu}_{H,s}$ in Fig. 10B. The straight lines with slope $L_{H(I)}$ and intercept $L_{o,H(I)}\langle \mathcal{A}_{[eH],s} \rangle$ represent the linear relations employed in the analysis. They are a satisfactory approximation except for $|\Delta\tilde{\mu}_{H,s}| < 12$ kJ/mol. For this reason the data points with $c_I > 40$ μM in the presence of the uncoupler were excluded from the estimation of parameter values. In an experimental system data for $J_{H(I),s}$ are not available but also not needed in order to detect a failure of a linear approximation in a certain range of force. In fact, if the parameter estimation is performed with all data points the fit is poorer. More important, however, is that the values found for $L_{H(I)}$ and $L_{o,H(I)}$ with uncoupler are, respectively, 14.3 times and 23.9 times larger than the corresponding values without uncoupler. This inconsistency is a strong indication of the failure of the linear approximation demonstrated in Fig. 10B. It has to be expected that, in contrast to what was assumed in the analysis, $L_{o,e}$ and $L_{o,H(p)}$ are not exactly zero. However, their contribution to $J_{[eH],s}$ and $\Delta\tilde{\mu}_{H,s}$ is too weak to be estimated in the fitting procedure and hence is included in the values for $L_e(0)$, $L_{H(I)}$ and $L_{o,H(I)}$ (cf. Eqns. 84 and 85).

## 6. Control and regulation of processes

According to the unambiguous flow-force relations obtained under suitable constraints the flows of processes are uniquely determined by the forces. Therefore, the flows are said to be under *termodynamic control.* As a consequence, if the force of a non-coupled process is clamped its flow is constant. For a pair of coupled processes, however, the flows are constant only if both forces are clamped simultaneously. This is due to the fact that the flow of each process is under termodynamic control of both forces. If only one force is clamped the corresponding process becomes the driving process, and its flow runs in the direction determined by the clamped force. But this flow also drags the flow of the coupled (or driven) process

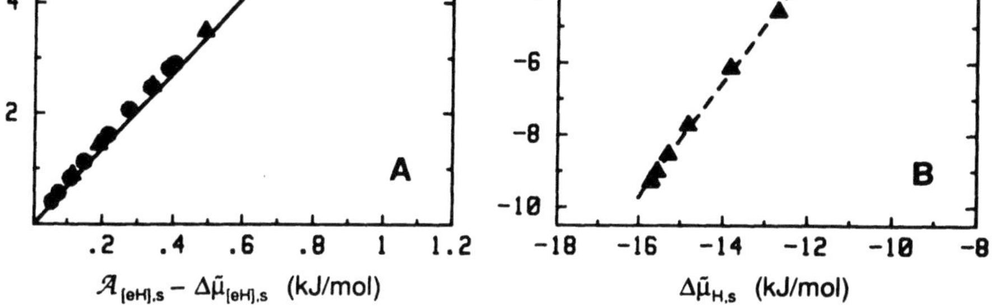


Fig. 10. Dependence of $J_{[eH],s}$ on $A_{[eH],s} - \Delta\tilde{\mu}_{[eH],s}$ (A) and of $J_{H(I),s}$ on $\Delta\tilde{\mu}_{H,s}$ (B). Data were taken from the simulations without (●) and with uncoupler (▲). The straight line in (A) represents a linear approximation (see Eqn. 90), while those in (B) were calculated according to $J_{H(I),s} = L_{H(I)}\Delta\tilde{\mu}_{H,s} + L_{o.H(I)} \langle \mathcal{A}_{[eH],s} \rangle$ with values of parameters as listed in the text; temperature 25°C.

in the same direction if coupling is positive or in the opposite direction if coupling is negative (see section 5.2), which in turn causes the force of the driven process to build up against the direction of its flow. As is evident from the flow-force relations (Fig. 6) or their linear approximations (Eqns. 72) this increasing force slows down both processes until static head is reached where the flow of the driven process vanishes[14].

For very large and very small forces the flows are independent of the forces (see, for example, Fig. 6) and no thermodynamic control seems to exist. This is due to the fact that the kinetics of enzyme-catalyzed processes, but also of non-catalyzed processes under certain constraints, displays a saturation phenomenon [1]. Accordingly, the flows in these ranges of force have reached the respective maximal values in the forward ana backward direction attainable under the given conditions, and the flows are said to be under *kinetic control* which overrides thermodynamic control. Since thermodynamic forces depend on ratios of the concentrations of reactants (see Eqns. 37, 41–44) the same force arises from a multitude of values for the concentrations. In contrast, the kinetics of processes is determined by the concentrations separately (see, for example, ref. 1 or 2), hence an effect of kinetic control has to be expected whenever the concentration of a reactant in a given system is changed. This is demonstrated by the model system if $pH_1$ is kept constant but given different values. With increasing $c_{H,1}$, i.e. decreasing $pH_1$, $\Delta\tilde{\mu}_{H,s}$ attained at steady state is diminished although $\Delta\tilde{\mu}_{[eH],s}$ is approximately equal in all cases (see Fig. 11A). Owing to kinetic control $H^+$ ions permeate faster through the membrane the larger $c_{H,1}$ at a given $\Delta\tilde{\mu}_H$ (Fig. 11B). Similarly, for small absolute values of $\Delta\tilde{\mu}_H$ the redox-driven proton pump operates faster the large $c_{H,1}$ but just the opposite is true around static head (Fig. 11C). Hence $\Delta\tilde{\mu}_{H,s}$ becomes smaller with increasing $c_{H,1}$.

It may be argued that a $pH_1$ of 6.5 or 5.5 and the considerably smaller values for $pH_2$ reached in steady state are unphysiologically small. Although in principle correct, this argument should not be mistaken for a proof that kinetic control is detectable only in model systems and has no relevance to biological systems. Kinetic control can also be demonstrated with the model system if $pH_1$ is kept at 7.5 but the membrane is made impermeable for the univalent cations and anions. In this case $\Delta\tilde{\mu}_{H,s}$ = −15.35 kJ/mol with $pH_2$ = 6.82 and $\Delta\phi_m$ = −118.9 mV, as compared to $\Delta\tilde{\mu}_{H,s}$ = −17.36 kJ/mol with $pH_2$ = 4.55 and $\Delta\phi_m$ = −5.4 mV for permeable ions. The reduced value of $\Delta\tilde{\mu}_{H,s}$ is not caused by a faster permeation of $H^+$ ions (which is even slower since $J_{H(l),s}$ amounts to $-1.4 \times 10^{-4}$ $s^{-1}$ as compared to $-6 \times 10^{-3}$ $s^{-1}$) or a smaller $\Delta\tilde{\mu}_{[eH],s}$ (36.2 kJ/mol as compared to

[14]The flow of the driven process does not vanish but becomes minimal if this process can also proceed independently. In this case the sum of the flows of coupled and independent process vanishes (for an example see $J_H = J_{H(p)} + J_{H(l)}$).

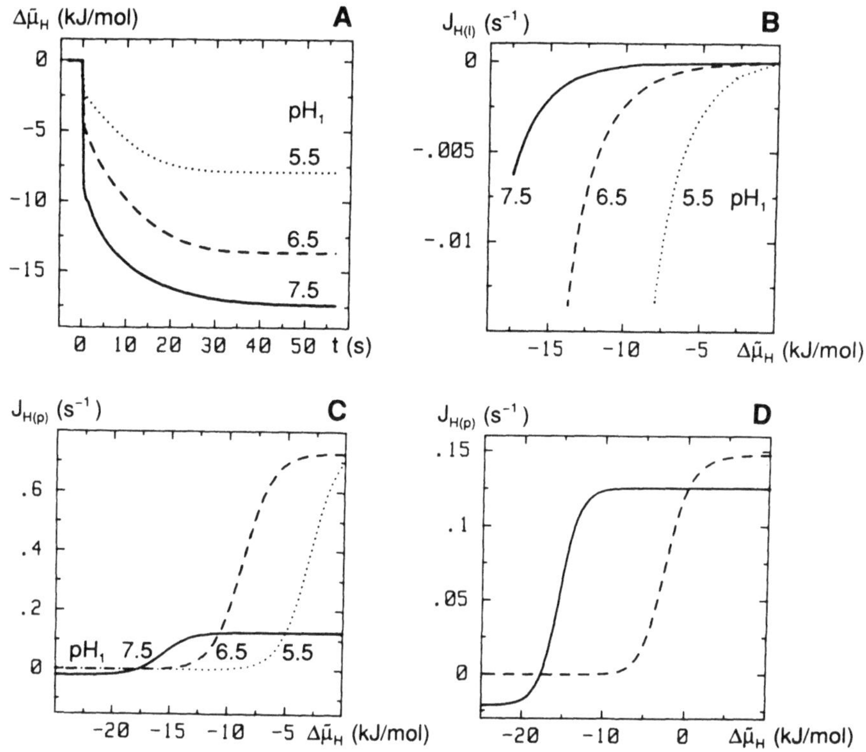


Fig. 11. Effect of kinetic control on the redox-driven proton pump and the permeation of $H^+$ ions. Time-dependence of $\Delta\tilde{\mu}_H$ (A) and dependence of $J_{H(i)}$ on $\Delta\tilde{\mu}_H$ (B) in the model system with a constant $pH_l = 7.5$ (solid line), 6.5 (broken line), and 5.5 (dotted line). The value of $\mathcal{A}_{[eH]}$ was kept constant at 36.5 kJ/mol ($c_{HDrd,l} = 10$ mM, $c_{Dox,l} = 0.02$ mM, $c_{HArd,l} = 100$ mM, $c_{Aox,l} = 1$ mM), and $\Delta\tilde{\mu}_{[eH],s}$ was found to be 35.9 kJ/mol for $pH_l = 7.5$, and 35.5 kJ/mol for $pH_l = 6.5$ and 5.5. (C), (D) Flow-force relations, i.e. the dependence of $J_{H(p)}$ on $\Delta\tilde{\mu}_H$, at a constant $\Delta\tilde{\mu}_{[eH]} = 35.9$ kJ/mol ($c_{HDrd,l} = 10$ mM, $c_{Dox,l} = 0.025$ mM, $c_{HArd,2} = 100.025$ mM, $c_{Aox,2} = 0.975$ mM). Constraints for $\Delta\tilde{\mu}_H$ in (C) are $\Delta\phi_m = 0$ and a constant $pH_l = 7.5$ (solid line), 6.5 (broken line), 5.5 (dotted line). Constraints for $\Delta\tilde{\mu}_H$ in (D) are a constant $pH_l = 7.5$ and either a variable $pH_2$ with $\Delta\phi_m = 0$ (solid line) or a constant $pH_2 = 7.5$ with a variable $\Delta\phi_m$ (broken line). Temperature 25°C.

35.9 kJ/mol). It results from the fact that the parameters which determine $\Delta\mu_H$ and $\Delta\phi_m$ (see Eqns. 27 and 71b) give rise to rather different kinetic control of the pump (see Fig. 11D). This phenomenon is referred to as the *kinetic inequivalence of the (thermodynamically equivalent) chemical and electrical potential* in the force of a transport process.

Besides the mandatory thermodynamic and kinetic control, processes may also be subject to a *regulation.* An enzyme which catalyzes a process may be inhibited or activated by one or more parameters of the system, including those involved in the process itself. Such a regulation can prove vital for the performance of a system in terms of energy conversion, a typical example being photophosphorylation which takes place in chloroplasts (see chapter 5). Light quanta absorbed by the system provide the energy to drive an electron transport and a transport of $H^+$ ions across the thylakoid membrane, both against the respective thermodynamic force. The $\Delta\tilde{\mu}_H$ thus formed constitutes the common element for the coupling between the electron transport and the synthesis of ATP from ADP and phosphate. The phosphorylation reaction is catalyzed by the $H^+$/ATP-synthase which also couples this reaction to the transport of $H^+$ ions across the thylakoid membrane. For good reasons this enzyme is regulated in that it is activated by $\Delta\tilde{\mu}_H$ and a redox process (see chapter 12). In fact, if such a regulation did not exist, the processes occurring in the light would be reversed in the dark because, in principle, all processes can run in both directions (see section 4). Thus the phosphorylation reaction, which is a driven process in the light, would become the driving process in the dark and would hydrolyse ATP according to the affinity $\mathcal{A}_{ph}$ formed in the light. Although the reverse direction in the reaction centers of the electron transport chain is extremaly slow (see chapters 7 and 8) the hydrolysis of ATP would still proceed because the $\Delta\tilde{\mu}_H$ thereby built up would be continously degraded by the permeation of $H^+$ ions through the membrane. Hence all of $\mathcal{A}_{ph}$, and at least part of the affinity of consecutive processes which depend on it in the light, would eventually be dissipated in the dark. As a consequence, the overall efficiency of this energy-providing machinery of green plants would be severely compromised.

## Acknowledgement

Financial support by the Swiss National Science Foundation is gratefully acknowledged.

## References

1. D Walz, Biochim. Biophys. Acta 1019 (1990) 171-224.
2. D Walz and SR Caplan in *Bioelectrochemistry: Principles and Practice,* Vol. 1, *Bioelectrochemistry: General Introduction,* SR Caplan, IR Miller and G Milazzo (eds.), Birkhäuser Verlag, Basel, 1995, pp. 1-48.
3. D Walz, Biochim. Biophys. Acta 505 (1979) 279-353.
4. SR Caplan and M. Kara-Ivanov, Intern. Rev. Cytology 147 (1993) 97-164.
5. D Walz and SR Caplan, Bioelectrochem. Bioenergetics 28 (1992) 5-30.
6. TP Silverstein, Biochim. Biophys. Acta 1183 (1993) 1-3.
7. J Rosing and EC Slater, Biochim. Biophys. Acta 267 (1972) 275-290.
8. RC Phillips, P George and RJ Rutman, Biochemistry 2 (1963) 501-508; J. Am. Chem. Soc. 88 (1966) 2631-2640.
9. H Sigel, R Tribolet, R Malini-Balakrishnan and RB Martin, Inorg. Chem. 26 (1987) 2149-2157; R Tribolet and H Sigel, Eur. J. Biochem 170 (1988) 617-626.
10. E Högfeldt, *Stability Constants of Metal-Ion Complexes, Part A: Inorganic Ligands,* IUPAC Chemical Data Series, No. 21, Pergamon Press, Oxford, 1982; DD Perrin,

*Stability Constants of Metal-Ion Complexes, Part B: Organic Ligands,* IUPAC Chemical Data Series, No. 22, Pergamon Press, Oxford, 1979.
11. J Rüsch, Diploma Thesis, Biozentrum, University of Basel, 1991.
12. P Müller and DO Rudin in *Current Topics in Bioenergetics,* Vol. 3, DR Sanadi (ed.), Academic Press, New York, 1969, pp. 157-276.
13. D Walz, SR Caplan, DRL Scriven and DC Mikulecky in *Bioelectrochemistry: General Introduction, Principles and Practice,* Vol. 1, SR Caplan, IR Miller and G Milazzo (eds.), Birkhäuser Verlag, Basel, pp. 49-131.
14. P Mitchell, Biol. Rev. 41 (1966) 445-502.
15. D Pietrobon and SR Caplan, Biochemistry 24 (1985) 5764-5776.
16. SR Caplan and A Essig, *Bioenergetics and linear nonequilibrium thermodynamics: the steady state,* Harvard University Press, Cambridge, MA (1983).
17. HV Westerhoff and K van Dam, *Thermodynamics and Control of Biological Energy Transduction,* Elsevier, Amsterdam, 1987.
18. O Kedem and SR Caplan, Trans. Farad. Soc. 61 (1965) 1897-1911.
19. MG Klein, L Kovacs, BJ Simon and MF Schneider, J. Physiol. 441 (1991) 639-671.
20. GF Azzone, D Pietrobon and M Zoratti in *Current Topics in Bioenergetics,* Vol. 13, CP Lee (ed.), Academic Press, Orlando, 1984, pp. 2-66.
21. D Walz in *Biological Structures and Coupled Flows,* A Oplatka and M Balaban (eds.), Academic Press, New York, 1983, pp. 45-60.

## Appendix A. Glossary

| | |
|---|---|
| $a_{i,k}$ | activity of *i*th species in *k*th compartment or phase |
| $a_o$ | thermodynamic standard activity (1 M) |
| $a_H^{\oplus}$ | biochemical standard activity for $H^+$ ions ($10^{-7}$ M) |
| $A_{ox}$, $A_{rd}$ | oxidized, reduced species of acceptor redox couple A |
| $\mathcal{A}_{j,k}$ | affinity of *j*th chemical reaction in *k*th compartment |
| $\mathcal{A}_{e,k}$ | affinity of redox reaction in *k*th compartment |
| $c_{i,k}$ | concentration of *i*th species in *k*th compartment or phase |
| $C_{ci,k}$ | chemical capacity for *i*th species in *k*th compartment or phase |
| $D_{ox}$, $D_{rd}$ | oxidized, reduced species of donor redox couple D |
| $E_{R,k}$ | redox potential of couple $R_{ox}/R_{rd}$ in *k*th compartment |
| $E^{\varnothing}_{R,k}$ | standard redox potential of couple $R_{ox}/R_{rd}$ in *k*th compartment |
| $E^{o}_{R,k}$ | midpoint potential of couple $R_{ox}/R_{rd}$ in *k*th compartment |
| $E^{m}_{R,k}$ | $a_H$-dependent midpoint potential of couple $R_{ox}/R_{rd}$ in *k*th compartment |
| $E^{o'}_{R,k}$ | $E^{m}_{R,k}$ at pH = 7 |
| F | external mechanical force |
| $\mathcal{F}$ | Faraday constant |
| G | free enthalpy (Gibbs free energy) |
| $\Delta G^{\varnothing}_{j,k}$ | standard free enthalpy (standard Gibbs free energy) of *j*th chemical reaction in *k*th compartment |
| $\Delta G^{o}_{j,k}$ | phenomenological standard free enthalpy of *j*th chemical reaction in *k*th compartment |

$\Delta G^{\oplus}_{j,k}$ biochemical standard free enthalpy of *j*th chemical reaction in *k*th compartment

$\Delta G^{o}_{ph}$ phenomenological standard free enthalpy of the overall phosphorylation reaction

$\Delta G^{o'}_{ph}$ $\Delta G^{o}_{ph}$ at pH = 7

h phenomenological stoichiometry for multiple binding

$\Delta H^{\varphi}_{j,k}$ standard enthalpy of *j*th chemical reaction in *k*th compartment

$\Delta H^{o}_{j,k}$ phenomenological standard enthalpy of *j*th chemical reaction in *k*th compartment

$\Delta H^{\oplus}_{j,k}$ biochemical standard enthalpy of *j*th chemical reaction in *k*th compartment

I electric current

$I_c$ ionic strength

$J_{j,k}$ flow for *j*th chemical reaction in *k*th compartment

$J_{e,k}$ flow for redox reaction in *k*th compartment

$J_i$ flow for transport of *i*th species

$J_p$ flow of *p*th process

$J_o$ intercept of linear flow-force relation

$K^{o}_{d,LiR}$ equilibrium constant for dissociation of complex $L_iR$

$K_{d,LiR}$ phenomenological equilibrium constant for dissociation of complex $L_iR$

$K^{o}_{e}$ equilibrium constant for a redox process

$K_e$ phenomenological equilibrium constant for a redox process

$K^{\oplus}_{j,k}$ biochemical equilibrium constant

$K^{o}_{pi,k'k}$ equilibrium constant for partitioning of *i*th species between phase or compartment *k'* and *k* (partition coefficient)

$K_{pi,k'k}$ phenomenological equilibrium constant for partitioning of *i*th species between phase or compartment *k'* and *k*

$K_{ph}$ phenomenological equilibrium constant for the overall phosphorylation reaction

$K'_{ph}$ $K_{ph}$ at pH = 7

$K^{o}_{rj,k}$ equilibrium constant for *j*th chemical reaction in *k*th compartment

$K_{rj,k}$ phenomenological equilibrium constant for *j*th chemical reaction in *k*th compartment

$K^{o}_{w}$ dissociation constant of water

$\ell$ elongation

L generalized permeability in linear flow-force relation

$n_{i,k}$ mole number of *i*th species in *k*th compartment or phase

p pressure

$P_{p(j,k)}$ *p*th final reactant (or product) of *j*th chemical reaction in *k*th compartment

q degree of coupling

| | |
|---|---|
| R | gas constant |
| $R_{r(j,k)}$ | *r*th reactant (initial or final) of *j*th chemical reaction in *k*th compartment |
| $R_{ox}$, $R_{rd}$ | oxidized, reduced species of redox couple R |
| S | entropy |
| $\Delta S^{\ominus}_{j,k}$ | standard entropy of *j*th chemical reaction in *k*th compartment |
| $\Delta S^{o}_{j,k}$ | phenomenological standard entropy of *j*th chemical reaction in *k*th compartment |
| $\Delta S^{\oplus}_{j,k}$ | biochemical standard entropy of chemical reaction |
| $S_{s(j,k)}$ | *s*th initial reactant (or substrate) of *j*th chemical reaction in *k*th compartment |
| t | time |
| T | absolute temperature |
| U | internal energy |
| v | velocity (mechanical process) |
| V | volume |
| X | force (thermodynamic or mechanical) |
| $y_{i,k}$ | activity coefficient of *i*th species in *k*th compartment or phase |
| $z_i$ | charge number of *i*th species |
| Z | phenomenological stoichiometry of coupling |
| $\alpha_{i,j}$ | first-order probability for transition from state *i* to state *j* of an enzyme |
| $\alpha^{o}_{i,j}$ | second-order transition probability for transition from state *i* to state *j* of an enzyme |
| $\gamma$ | asymmetry of coupling coefficients in linear flow-force relation |
| $\varepsilon_J$ | flow efficacy of energy conversion |
| $\varepsilon_X$ | force efficacy of energy conversion |
| $\eta$ | efficiency of energy conversion |
| $\tilde{\mu}_{i,k}$ | electrochemical potential of *i*th species in *k*th compartment or phase |
| $\mu_{i,k}$ | chemical potential of *i*th species in *k*th compartment or phase |
| $\mu^{o}_{i,k}$ | standard chemical potential of *i*th species in *k*th compartment or phase |
| $\tilde{\mu}_{e(R,k)}$ | electrochemical potential of electron associated with $R_{ox}/R_{rd}$ in *k*th compartment |
| $\mu_{e(R,k)}$ | chemical potential of electron associated with $R_{ox}/R_{rd}$ in *k*th compartment |
| $\mu^{o}_{e(R,k)}$ | standard chemical potential of electron associated with $R_{ox}/R_{rd}$ in *k*th compartment |

| | |
|---|---|
| $\mu^{\oplus}_{H,k}$ | biochemical standard chemical potential for $H^+$ ions in *k*th compartment |
| $\nu_{Rr(j,k)}$ | stoichiometric coefficient for reactant $R_r$ of *j*th chemical reaction in *k*th compartment |
| $\nu_{e(R)}$ | number of electrons exchanged by redox couple $R_{ox}/R_{rd}$ |
| $\tau$ | torque |
| $\xi_{j,k}$ | degree of advancement of *j*th chemical reaction in *k*th compartment |
| $\phi_k$ | electrical potential (Galvani potential) of *k*th compartment or phase |
| $\Delta\phi_m$ | transmembrane electrical potential difference (membrane potential) |
| $\Phi$ | dissipation function |
| $\omega$ | angular velocity |
| $[\ ]_{eq}$ | equilibrium value of quantities given between brackets |

Bioenergetics
ed. by P. Gräber and G. Milazzo†


# *CHAPTER 2*
# Global bioenergetics

Hans V. Westerhoff*, Wim Crielaard and Klaas J. Hellingwerf

*E. C. Slater Institute, BioCentrum Amsterdam, University of Amsterdam, Amsterdam, The Netherlands*

**2nd Department of Microbial Physiology, Faculty of Biology, Free University, Amsterdam, The Netherlands*

## 1. Introduction

The surface of the earth is subject to a continuous input of free enthalpy originating in the sun. A small, but important part of this free enthalpy is diverted, transiently, to physical chemical structures that perform work. We conceive of these structures as life. The emergent creatures reproduce, transform input free enthalpy to convertible forms and adjust themselves to changing environments. Through rare changes in their DNA and selection pressure these organisms evolve into increasingly complex forms.

The maintenance and reproduction of these organisms is fully dependent on the conversion and dissipation of free enthalpy in elementary reactions. Moreover, these organisms constitute a partly cyclic free-enthalpy chain. In a simplification, this chain may be viewed as consisting of two parts (Fig. 1). In the first, phototrophic organisms convert radiant energy into the chemical potential of (macro) molecules. In this photosynthesis electrons are brought to a more negative redox potential, i.e.: $H^+$ in $H_2O$ is reduced to bound H "[H]". This transfer of electrons from a midpoint potential at pH 7 of +0.82 V to −1.14 V is energetically uphill and requires input of photon free enthalpy. The ["H"] is transiently bound to the redox carrier $NAD^+$, leading to $NADH_2^+$. The bound H ($NADH_2$) then reduces $CO_2$ to carbon, in the form of macromolecules such as carbohydrates (cf. the left-hand part of Fig. 1). Next, chemotrophic organisms use the chemical free enthalpy of these compounds for growth. This boils down to the reoxidation of the fixed carbon to $CO_2$ and $H_2O$ (Fig. 1). The chemical constituents of the organisms involved in this cycle contribute a

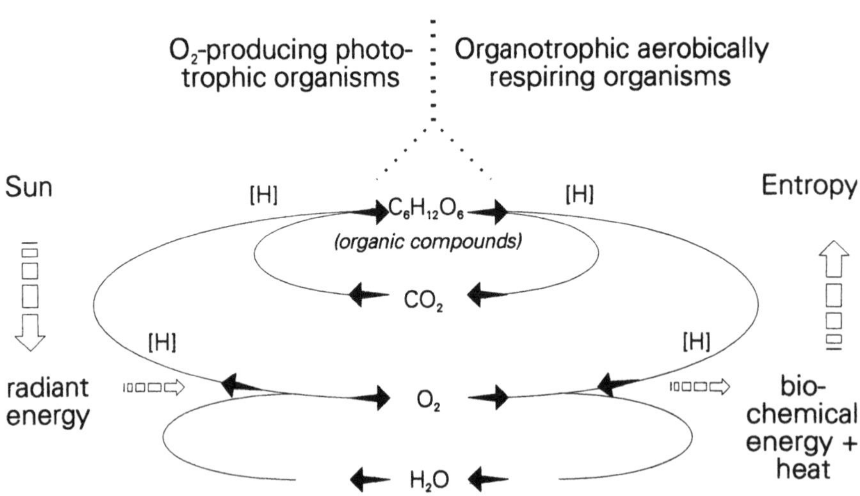


Fig. 1. A simplified scheme of global bioenergetics.

major part to the cycling of the elements on the surface of the earth, like carbon, nitrogen and sulfur.

In this chapter we first discuss the thermodynamic and energetic aspects of these reactions. Chapter 5 deals with the equally fascinating mechanisms of these processes. Subsequently, we will describe the global mass flows that accompany these global bioenergetic processes. Finally, we will discuss the evolution of living organisms to their contemporary form.

## 2. Thermodynamic principles

### *2.1. From solar heat to earthly work*

Ultimately, all biological activity is driven by nuclear energy: The radiation of light by the sun is due to the heating of the latter by nuclear fusion reactions. The nuclear reactions include

$$2\,(2 \times {}^{1}_{1}H \rightarrow {}^{2}_{1}H + e^{+} + \nu)$$

$$2\,({}^{2}_{1}H + {}^{1}_{1}H \rightarrow {}^{3}_{2}He + \gamma)$$

$$2\,{}^{3}_{2}He \rightarrow {}^{4}_{2}He + 2\,{}^{1}_{1}H$$

with as the net reaction:

$$4\,{}^{1}_{1}H \rightarrow {}^{4}_{2}He + 2e^{+} + 2\,\nu + 2\,\gamma$$

${}^{1}_{1}H$, ${}^{2}_{1}H$, He and $e^{+}$ represent the proton, the deuteron, the Helium nucleus ($\alpha$ particle) and the positron, respectively. $\nu$ and $\gamma$ represent photons of light and gamma radiation, respectively. The energy liberated in this reaction is $2.4 \times 10^{12}$ J per mol of protons. This process is astonishingly slow, it takes 3000 million years to be completed. Even at its core, the sun produces less heat per gram (i.e., $10 \times 10^{-6}$ W/g) than the free-enthalpy power a human being consumes as food ($1.5 \times 10^{-3}$ W/g). The sun is as effective as it is mainly because it is so big ($2 \times 10^{33}$ g).

In essence, all the energy liberated in these nuclear reactions heat the sun. The sun's interior has a temperature of some $20 \times 10^{6}$ K. The temperature of its surface is approximately 6000 K. $45 \times 10^{28}$ W of heat is radiated into space from the solar surface.

Some of this reaches the surface of the earth and drives processes such as the water cycle (evaporation, rain), the weather, windmills and photosynthesis (Fig. 2). Windmills and photosynthesis perform mechanical- and chemical work, respectively. Ostensibly, work derives from heat. How does this fit in with thermodynamics?

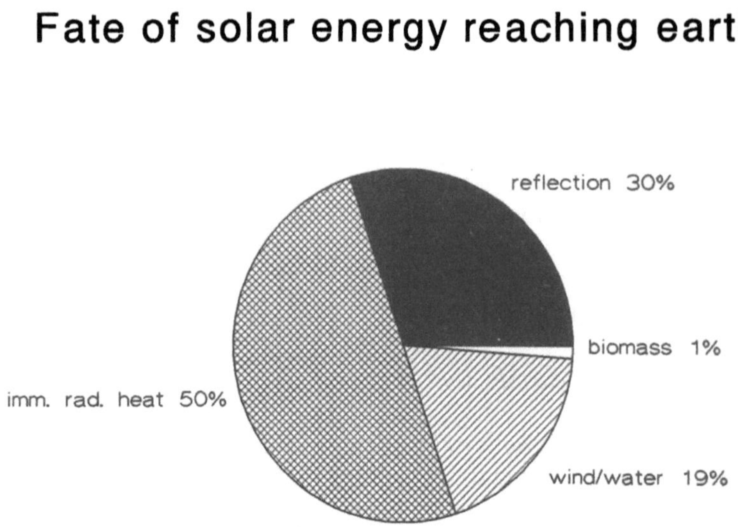


Fig. 2. Fate of solar energy reaching planet earth.

In the sense that it is transducing heat energy into work, a phototrophic organism is a heat engine. Accordingly, we consider the photosynthetic machinery as a machine that carries out a cyclic process during which it absorbs solar heat $dQ_s$ and carries out chemical or electric work $-dW$ at ambient temperature. Equation 2 of Chapter 1, when formulated for an open system away from equilibrium, reads:

$$dU = T\, d_eS - dW \tag{1}$$

$-dW$ is shorthand for the three types of work done by the system (volume, $p \times dV$; mechanical, $-X_k \times dl_k$; and electrochemical, $-\Sigma\tilde{\mu}_i \times dn_i$) considered in Chapter 1. $d_eS$ represents the entropy import into the system, which is directly related to the heat import ($dQ$):

$$d_eS = \frac{dQ}{T} \tag{2}$$

The photosynthetic machinery may be conceived to import solar heat ($dQ_h$) of a temperature equal to that of the solar surface ($T_h$), perform work $dW$ and release some heat ($dQ_l$) to its surroundings at the lower temperature of the latter ($T_l \approx 300$ K). After the cyclic process, the machine returns to its original state. Since the energy (U) is a function of state and cannot be produced or destroyed (first law of thermodynamics):

$$0 = dU = dQ_h + dQ_l + dW \tag{3}$$

i.e., the amount of work done equals the difference between the absorbed heat and the heat released to the environment. The entropy of the machine, S, can increase due to entropy import ($d_eS$) or entropy production ($d_iS$):

$$dS = d_eS + d_iS \tag{4}$$

Because also entropy is a state function, it does not change in a cyclic process:

$$0 = d_{eh}S + d_{el}S + d_iS = \frac{dQ_h}{T_h} + \frac{dQ_l}{T_l} + d_iS \tag{5}$$

Combining Eqn. 3 and 5, one finds the amount of work done by the heat engine as a fraction of the amount of heat imported at the high temperature:

$$\frac{-dW}{dQ_h} = 1 - \frac{T_l}{T_h} - \frac{T_l \times d_iS}{dQ_h} \tag{6}$$

The second law of thermodynamics dictates that to allow processes to proceed, entropy must be produced:

$$d_iS \geq 0 \tag{7}$$

In an adiabatic system, there is no exchange of heat or entropy with the environment, such that then Eqn. 7 reads as Eqn. 4 in Chapter 1 (see, however, Westerhoff and Van Dam, 1987, for accurate interpretation of entropy and heat).

At equilibrium, the equality sign of Eqn. 7 is valid and the well-known Carnot efficiency is retrieved, which expresses how much heat a heat engine can convert to work:

$$\eta_{Carnot} = \left(\frac{-dW}{dQ_h}\right)_{reversible} = 1 - \frac{T_l}{T_h} \tag{8}$$

Considering photosynthesis as a machine that uses heat from the solar surface (T = 6000 K) to perform work at an ambient temperature of some 300 K (27°C), this equation indicates that at most 100% × (1 – (300/6000)) = 95% of the solar heat could be transformed to work. This is the upper limit for the case when the machine would work reversibly. Because photosynthesis is an irreversible process (in the sense of Chapter 1), $d_iS$

must exceed zero, and the efficiency of this energy conversion may become much smaller.

## *2.2. From solar heat to earthly photons*

The above has shown that the paradoxical statement that solar heat ultimately drives photosynthesis is in keeping with the second law of thermodynamics: as in a heat engine the heat is taken from a reservoir at high temperature and then used and partly released at a lower temperature. However, although formally in order, this argument seems a bit unrealistic. The photosynthetic machinery resides in green leaves or microorganisms, remote from the sun's surface. And, if the solar heat reaching the phototrophic organisms were to have a temperature of 6000 K, would it not heat up the proteins around the absorbing chromophore to that temperature much higher than compatible with protein function? Indeed, we know that in actual fact it is not the heat that comes from the sun, but rather light that drives photosynthesis. Where then does this light appear in the thermodynamics?

When iron gets very hot it obtains a reddish glow. This is because of the heat it radiates. Indeed, any object at a temperature exceeding that of its surroundings radiates heat and this heat consists of a population of photons of various wavelengths. For so-called black bodies (these are objects without essentially any color preference) both the spectrum and the intensity of this light are completely defined by their temperature ($T_{bb}$). As a function of the wavelength the energy density ($u_\lambda$) of the black body radiation is:

$$u_\lambda = \frac{8\pi\ h\ c}{\lambda^5 \times (e^{hc/\lambda k_B T_{bb}} - 1)} \tag{9}$$

$u_\lambda \times d_\lambda$ is the amount of energy per unit volume in black body radiation between wavelength $\lambda$ and $\lambda + d\lambda$. $\int u_\lambda d\lambda$ is the total energy per unit volume in black body radiation. $T_{bb}$ is the absolute temperature of the black body, h, $k_B$ and c are Planck's constant, Boltzmanns constant and the velocity of light, respectively. Figure 3 shows the spectra of black body radiation of objects at temperatures of 6000 K, 1100 K and 300 K. The heat radiated by the sun corresponds to black body radiation of a characteristic temperature of 6000 K. It peaks in the visible region (Fig. 3). Black body radiation of 1100 K and 300 K peak in the far infra red range. For these characteristic temperatures the intensity in the visible region is much smaller than for 6000 K. The light intensity at the place where photosynthesis occurs does not generally have the spectrum of black body radiation. Its spectrum depends on its history of scattering and absorbance.

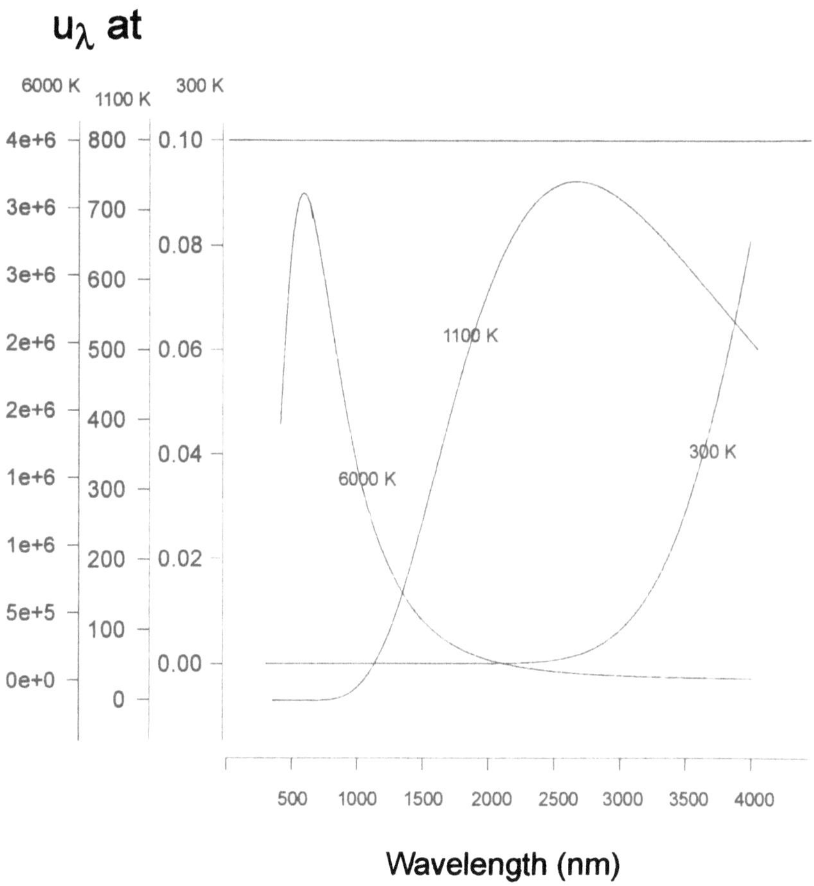


Fig. 3. Spectra of black body radiation of various temperatures.

In the corpuscular view of radiation, the latter consists of photons each with a frequency $\nu$, an energy $h \times \nu$ and moving with the velocity c. The wavelength of the corresponding radiation is

$$\lambda = \frac{c}{\nu} \tag{10}$$

Consequently, black body radiation can also be considered as a reservoir of photons at equilibrium with a black body of a certain temperature. Equations 9 and 10 make it possible to calculate the concentration of photons ($c_\nu \times d\nu$; in molar) with frequencies between $\nu$ and $\nu + d\nu$:

$$c_{\nu_{bb}} d\nu = \frac{8\pi \nu^2}{N_A c^3 \, 1000} \quad \frac{e^{-h\nu/k_B T_{bb}}}{(1 - e^{-h\nu/k_B T_{bb}})} \, d\nu \tag{11}$$

$N_A$ represents Avogadro's number and the factor 1000 derives from expressing concentrations in mol/l (Molar) rather than mol/m$^3$. The ability of a chemical substance to do work is given by its chemical potential, as defined in Eqn. 3a of Chapter 1. This ability increases with the concentration of the substance according to Eqn. 28–29 of Chapter 1:

$$\mu_i = \mu_i^o + RT\ \ln(y_i\ c_i/a^o) \tag{12}$$

In this equation, ln represents to the natural logarithm, $a^o$ is the thermodynamic standard activity, defined at 1M, $y_i$ is the activity coefficient, $\mu_i^o$ is the standard chemical potential and $R = N_A \times k_B$ is the gas constant. For photons, we write:

$$\mu_v = \mu_v^o + RT\ \ln(y_v\ c_v/a_v^o) \tag{13}$$

The chemical potential of a photon of a given frequency also increases with its concentration. $a_v^o$ is taken as $1M \times s$ and for the standard chemical potential, $\mu_v^o$ and the activity coefficient of the proton, $y_v$, we write (for a different apportionment of nonideality between $y_v$ and $\mu_v^o$, see Westerhoff and Van Dam, 1987):

$$\mu_v^o = N_a\ h\ v - RT\ \ln(8\pi^2\ v^2/(1000\ N_A\ c^3)) \tag{14}$$

$$\gamma_v = \frac{8\pi^2\ v^2}{8\pi^2\ v^2 + c_v\ c^3\ N_A 1000} = 1 - e^{-hv/k_BT_v} \tag{15}$$

where $T_v$ is the characteristic temperature of the radiation defined by the inverse of the formula for black body radiation (Eqn. 11):

$$T_v \stackrel{\text{def}}{=} \frac{h\ v}{k_B \ln\left(1 + \dfrac{8\pi^2\ v^2}{N_A\ 1000\ c^3\ c_v}\right)} \tag{16}$$

$c_v$ is the concentration of photons of frequency, v, in mol per liter times seconds (i.e., in mol per liter per unit frequency).

Whenever the photon energy exceeds random fluctuation energy at the temperature characteristic for the radiation for the cases of interest here (and this is virtually always so), $hv \gg k_BT$ and the activity coefficient $y_v$ of the photons equals 1. Then:

$$\mu_v = N_a\ h\ v - RT\ \ln(8\pi^2\ v^2/(N_A\ 1000\ c^3)) + RT\ \ln(c_v) \tag{17}$$

Inserting Eqn. 11 into Eqn. 13 or Eqn. 17 shows that the chemical potential of black body radiation at the temperature of that black body is zero: at that temperature the photon cannot do any work. This is in line with the definition that black body radiation is at equilibrium with the black body of that temperature. Equation 17 shows that at temperatures above the absolute zero not all energy contained in the photon can be used to do work. Even at a photon concentration of 1 M × s there is a loss term of RT $\ln(8\pi^2\nu^2/1000N_Ac^3)$.

Both Eqn. 8 and Eqn. 17 give the amount of work that may be done by an absorbed photon. What is the correspondence between these two expressions? The answer is given by Eqn. 16, which gives the temperature of the black body that should be at equilibrium with photons of frequency $\nu$ and concentration $c_\nu$. Indeed, combining Eqn. 16 and 13, one finds for the chemical potential of the photons:

$$\mu_\nu = N_A\, h\, \nu \left(1 - \frac{T}{T_\nu}\right) \tag{18}$$

This equation gives the correspondence between considering solar radiation as photons and looking at it as if it were heat. When one wishes to calculate the work that may be done by light of a given frequency or wavelength, the simplest procedure is to calculate the apparent temperature of that light by using Eqn. 16, and then insert that temperature into Eqn. 18.

The alternative of inserting the photon concentration into Eqn. 13 or 17 should lead to the same answer. Equation 17 simplifies the calculation of the free enthalpy loss due to dilution (i.e., reduction in light intensity):

$$\mu_\nu(c_\nu) - \mu_\nu(c'_\nu) = \frac{k_B\, T \ln(c_\nu/c'_\nu)}{h\, \nu} \tag{19}$$

If one chooses to conceive of photons, absorbed by photosynthetic organisms, as diluted photons stemming from black body radiation of the temperature of the solar surface, then one may write for the chemical potential at ambient temperature of those photons:

$$\frac{\mu_\nu}{N_A\, h\, \nu} = 1 - \frac{T}{T_s} + \frac{k_B\, T}{h\, \nu} \ln\left(\frac{c_\nu}{c_{\nu\,sun}}\right) \approx 1 - 0.05 - 0.033\, {}^{10}\log\left(\frac{c_{\nu\,sun}}{c_\nu}\right) \tag{20}$$

This equation reveals the reduction of the free enthalpy content of photons first by the Carnot efficiency and second by the lower concentration of the photons at the point of absorption as compared to the solar surface. The latter part of the equation has inserted a temperature of photosynthesis (300 K), the solar surface (6000 K) and 700 nm for the wavelength of the radiation absorbed. The equation should be used for photons of that particular wavelength only. For photons of $590 \times 10^{-9}$ m 0.027 should be substituted for the 0.033.

Because the earth only occupies 5.5 ppm of the solar sky, the above expression works out as 1 – 0.05 – 0.17 for the light intensity of the stratosphere (taking $T_s$ = 6000 K, T = 300 K, and $c_v/c_{v\ sun} = 5.5 \times 10^{-6}$). For a photon of $700 \times 10^{-9}$ m, this means that only $133 \times 10^3$ J/mol of its energy of $171 \times 10^3$ J/mol is available to do work at the light intensity of the stratosphere. For photons of $570 \times 10^{-9}$ m the energy and free-enthalpy amount to $210 \times 10^3$ J/mol and $170 \times 10^3$ J/mol, respectively. Where photosynthesis actually takes place the photon concentration is even lower than in the stratosphere. For the diffuse light in an algal suspension the characteristic temperature of the light may be only some 1100 K, leading to a free enthalpy content of the photons of only some 73% of their energy content i.e., $125 \times 10^3$ J/mol for $700 \times 10^{-9}$ m photons and $153 \times 10^3$ J/mol for $570 \times 10^{-9}$ m photons. Figure 4 shows the enthalpy and free enthalpy of photons in the stratosphere and at an effective temperature of 1100 K, as calculated from the above considerations. Clearly, not all the enthalpy of the photon can be used to do work, yet photons do have a very significant work potential (cf. Fig. 5).

It should be noted that the fact that only part of solar radiation reaches the green plant has two, rather than one, consequences. First, not all solar

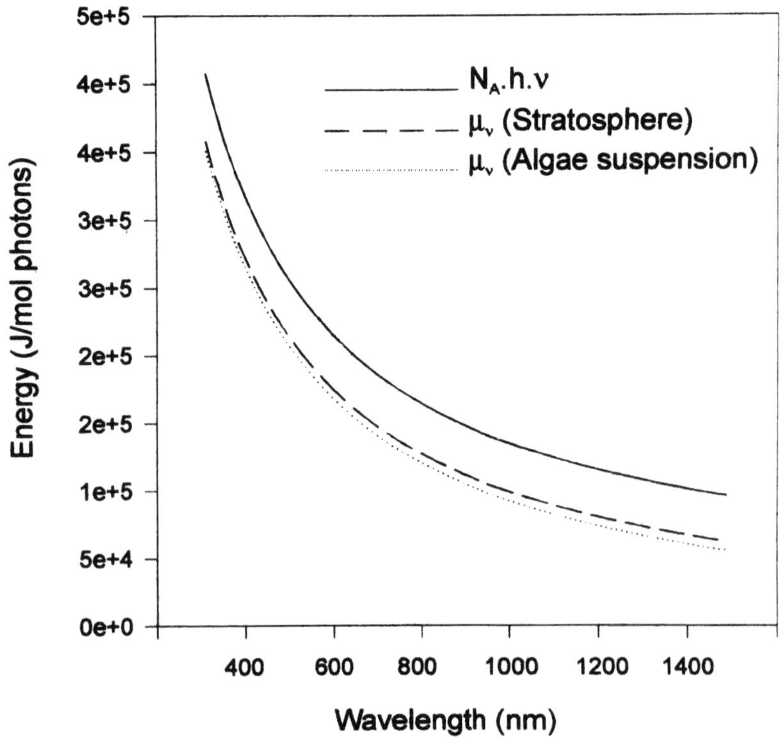


Fig. 4. Energy and free energy of black body photons as a function of the wavelength.

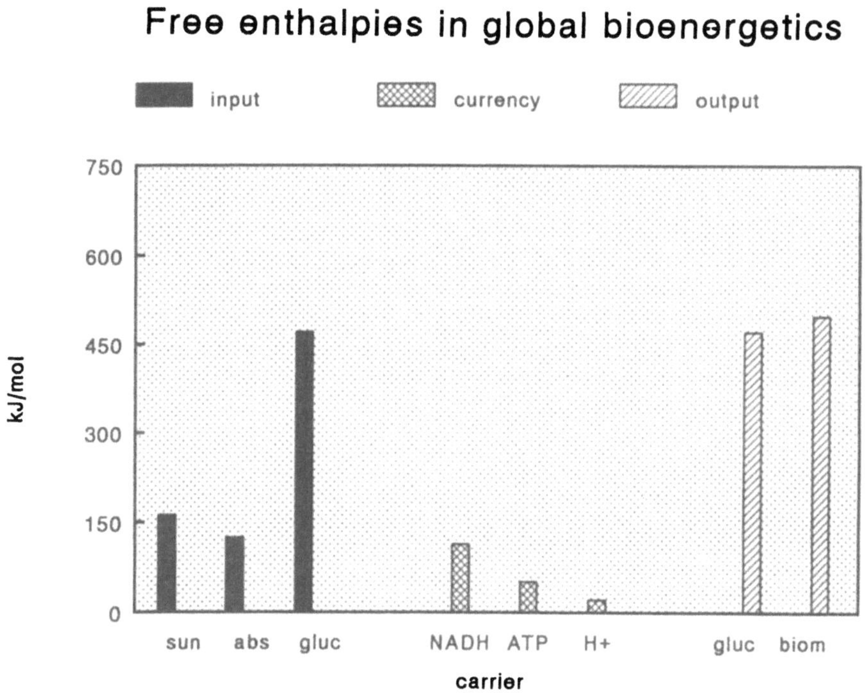


Fig. 5. Free enthalpies in global bioenergetics. Shown are the chemical potential of a $700 \times 10^{-9}$ m photon at the sun, in an algal suspension as well as the combustion enthalpies per C-mole of glucose and biomass, the redox potential difference between $NADH_2$/NADH and the $O_2/H_2O$ redox couples per mol electron, the free enthalpy of hydrolysis of ATP and the electrochemical potential difference for protons across energy coupling membranes.

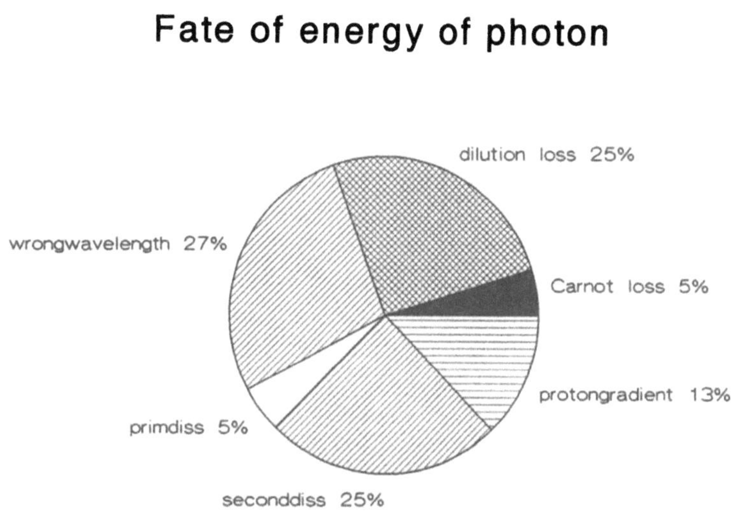


Fig.6. Fate of photon energy.

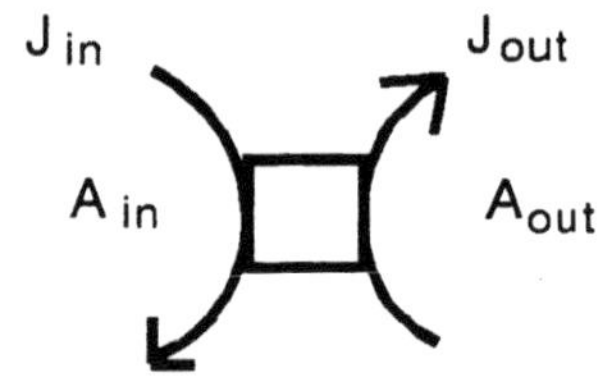


Fig. 7. Free enthalpy transducer.

radiation can be used in photosynthesis, and second, the free enthalpy of the photons that can be used is smaller than it is at the surface of the sun, because of the lower concentration of these photons (cf. Fig. 6).

### *2.3. From earthly photons to life*

From the absorption of the photon onwards, bioenergetics occurs at virtually constant temperature and pressure. Consequently, the thermodynamics as formulated in Chapter 1 from section 2.2 onwards applies.

In principle, the photon that is absorbed by a chromophore can do the amount of work calculated in the preceding section as the chemical potential of the photon. In practice much less work is done.

As discussed in section 4.2 of Chapter 1, processes may be coupled when they share conversions. This may be indicated by a diagram such as Fig. 7. An input process at flow $J_{in}$ and with free-enthalpy difference $A_{in}$ is coupled to an output process at flow $J_{out}$ and free-enthalpy difference $-A_{out}$. The input process runs downhill in the thermodynamic sense; $A_{in}$ is positive. The output process is often uphill thermodynamically, with a negative $A_{out}$. The system shown in Fig. 7 transduces part of the free enthalpy that flows into it as $J_{in} \times A_{in}$ to $J_{out} \times A_{out}$ at a thermodynamic efficiency given by (cf. Eqn. 77 of Chapter 1)

$$\eta = \frac{(J_{out}/J_{in})}{(J_{out}/J_{in})_{\eta=1}} \times \frac{(-A_{out}/A_{in})}{(-A_{out}/A_{in})_{\eta=1}} \tag{21}$$

where $\eta = 1$ refers to the condition of 100% efficiency. It should be noted that this efficiency, $\eta$, is a ratio of two free-enthalpy flows. Contrary to the Carnot efficiency of the previous sections, which was a ratio between free enthalpy and energy and had the maximum indicated by Eqn. 8, this efficiency reaches a maximum of 100% close to equilibrium.

This equation shows that there are two reasons why the thermodynamic efficiency $\eta$ may fall short of 100% and free enthalpy is dissipated: (i) the flow ratio or yield $J_{out}/J_{in}$ is low because of incomplete coupling and (ii) the ratio of the output force to the input force is lower than maximally possible. The latter may serve the purpose of making the process run faster. The former phenomenon may improve regulation or stem from mechanistic restrictions. One such restriction is caused by the coupling mechanisms. In linear electron flow in oxygenic photosynthesis at least 8 photons are required to produce the 4 NADPH that are necessary to reduce $CO_2$ to the reduction level of carbon in biomass. Fixed $CO_2$ has a free enthalpy (of combustion) of some $0.5 \times 10^6$ J/mol C, whereas the absorbed photons have a free enthalpy of some $0.125 \times 10^6$ J/mol. This leads to a maximum efficiency (in the absence of incomplete coupling) of only 50% ($0.5/8 \times 0.125$).

Global bioenergetics corresponds to a sequence of free–enthalpy transducers, hooked up in series and in parallel. Figure 8 gives a scheme for a phototrophic organism (A) and for a chemotrophic organism (B). Free-enthalpy transducers in phototrophic organisms include transducers from photon free enthalpy to redox potential ($-\Delta G_e$), from photon free enthalpy to transmembrane electrochemical potential difference for protons ($\Delta\tilde{\mu}_{H^+}$), from $\Delta\tilde{\mu}_{H^+}$ to phosphorylation potential ($-\Delta G_{ph}$), from $-\Delta G_{ph}$ to the free enthalpy of biomass (relative to the substrate, $-\Delta G_{biomass}$). Chemotropic organisms start from biomass or from chemical products made by other organisms and convert these through $-\Delta G_e$, $\Delta\tilde{\mu}_{H^+}$

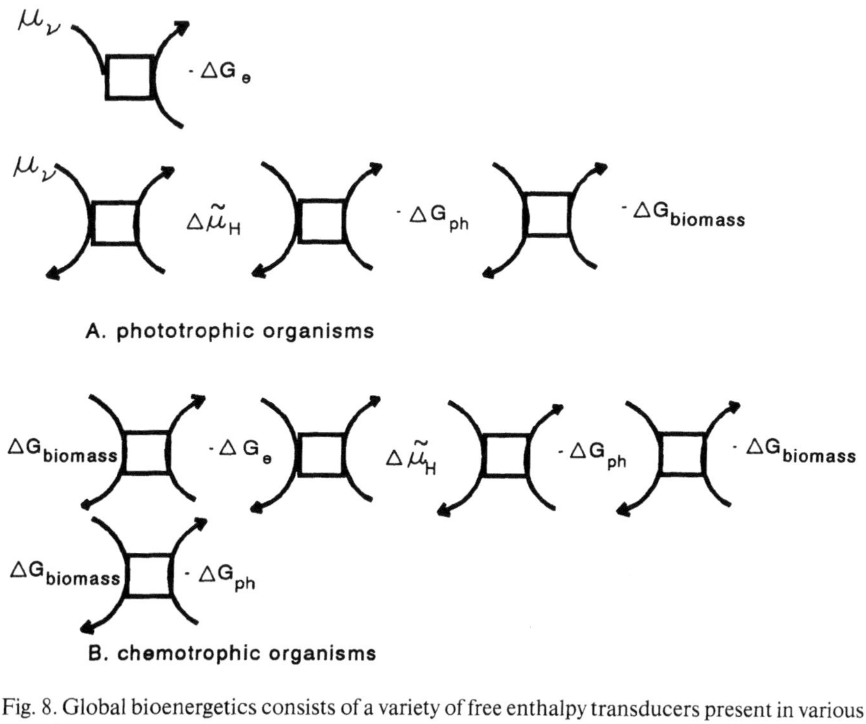


Fig. 8. Global bioenergetics consists of a variety of free enthalpy transducers present in various organisms.

and $-\Delta G_{ph}$ to the biomass of themselves and their offspring. A complete description of global bioenergetics contains the flows, forces, flow ratios, force ratios and thermodynamic efficiencies of all these processes. This is too much for the present context. In what follows we shall focus on some of the major flows and efficiencies.

Figure 5 gives some of the relevant free-enthalpy differences to consider and their usual magnitudes. As input free-enthalpy differences one has the chemical potential of the absorbed photon, the combustion free enthalpy of oxidizable substrates (more in general the free-enthalpy difference of the catabolic processes). As output free enthalpies there are the free enthalpy of the biomass synthesized (which may again serve as substrate for other organisms) minus that of the substrates used for biomass synthesis), as well as work performed (muscle work, locomotion). In between are the free-enthalpy currencies of the living cell. These include the redox potential of redox coenzymes such as NADH ($-\Delta G_e$) relative to that of oxygen, the free enthalpy of hydrolysis of ATP ($-\Delta G_{ph}$), the electrochemical potential difference for protons across energy coupling membranes ($\Delta\tilde{\mu}_{H^+}$) and the supercoiling free enthalpy of DNA. These free- enthalpy currencies can be exchanged at exchange rates given by the stoichiometry of the mechanism of interconversion, where at any exchange a loss occurs due to free-enthalpy dissipation (entropy production). They are generated by catabolic processes and consumed in anabolic processes and in other operations performed by the organism.

Of the three main free-enthalpy currencies indicated in Fig. 5, the value of two, i.e., $-\Delta G_{ph}$ and $\Delta\tilde{\mu}_{H^+}$, is well below the value of the input free enthalpy of the photon. The free-enthalpy value of the redox coenzyme NAD(H) however (which may be as high as 123 kJ/mol; at pH 6 and an NADH/NAD$^+$ ratio of 10) is fairly close to the free enthalpy of an earthly 700 nm photon (125 kJ/ml; 105 kJ/mol for a 840 nm photon). The corresponding thermodynamic restrictions imposed by the fact that uphill redox reactions are necessary for life (unless the substrates for growth are much more reduced than biomass), may have had strong implications for the evolution of the photosynthetic machinery.

The coenzyme NAD basically functions as a carrier of redox equivalents at the redox potential of hydrogen (−0.42 at pH 7); it carries "bound hydrogen". Indeed, at pH 7 the midpoint potential of the NADH/NAD$^+$ couple is only 100 mV (10 kJ/mol electrons) higher than that of hydrogen. In green bacteria, cyanobacteria and green plants, the primary electron acceptor has a midpoint potential that is approximately 100 mV more negative than that of hydrogen. Reduction of that primary acceptor by water to yield molecular oxygen would require a driving force of 133 kJ/mol.

These numbers show that the thermodynamic discussion of the chemical potential of the photon in the preceding sections is not just of academic interest. The energy content of the 700 nm photon absorbed by green plants of 171 kJ/mol, might seem sufficient to power photosynthesis, but it is not: Its free enthalpy content of only 125 kJ/mol is what counts, and falls short by some 8 kJ/mol. Less than 4% of the primary electron acceptor could be reduced under optimal conditions.

Evolution has found three different solutions (cf. subsequent chapters for their precise mechanisms) to the problem of producing redox currency at the value of bound hydrogen with light as the sole free-enthalpy source. The green sulphur bacteria use $H_2S$, which has a much lower midpoint potential (–0.24 V at pH 7) than $H_2O$ (0.82 V at pH 7) as electron donor; sulphur is less electronegative than oxygen. The $H_2S$ reduces bacteriochlorophyll a (midpoint potential of approximately 400 mV). The 840 nm photon these bacteria absorb gives sufficient free enthalpy to power the reduction of primary acceptor by the $P_{840}$ (a reaction requiring some 90 kJ/mol electrons).

Also the purple bacteria use electron sources other than water. They however forego the immediate production of bound hydrogen. Their primary electron acceptor has a midpoint potential of only -200 mV, such that the 870 nm photon only has to drive an endergonic reaction of some 65 kJ/mol electrons. Cyclic electron transport is then coupled to the formation of $\Delta\tilde{\mu}_{H^+}$, which subsequently drives electrons uphill towards reducing $NAD^+$, in a process called reverse electron transport.

The third solution is that of cyanobacteria and green plants. They use two photons instead of one. First, a 680 nm photon is used to transfer electrons from water to a primary electron acceptor with a midpoint potential around 0 mV (at pH 7; a reaction requiring 90 kJ/mol electrons, as it runs through $P_{680}$). These electrons are then allowed to flow downhill to a midpoint potential around +500 mV, meanwhile conserving part of their free enthalpy by driving proton translocation and generating $\Delta\tilde{\mu}_{H^+}$. Subsequently, a 700 nm photon brings the electron to a midpoint potential around -500 mV, sufficient to reduce $NAD^+$. This reaction requires 100 kJ/mol, again less that the free enthalpy of the photon.

A fourth solution has not been chosen, as far as we now know, that is, to absorb photons of shorter wavelength and use these to drive the reduction of NAD by water. Certain archaebacteria absorb 570 nm photons, but they use these directly to drive ion pumps. Low wavelength light is also absorbed by the carotenoids of photosynthetic organisms and then channeled to the $P_{680}$, $P_{700}$, $P_{840}$ and $P_{870}$ reaction centers. The development of antenna pigments widened the spectrum of photons that could be absorbed by the organisms. It may have proven to be a better strategy to increase the wavelength of the reaction center photon, thereby allowing a wider range of photons to drive photosynthesis in a two-step process, than to decrease the wavelength of the reaction center photon to, say, 570 nm allowing a single-step photochemical process, but foregoing much of the available light (photons of longer wavelength than absorbed by the reaction center are useless).

## *2.4. Bioenergetics and regulation: Back pressure and its relevance for control and regulation*

The second law of thermodynamics imposes a restriction to processes. This is one reason why it is of interest to keep track of the energetics of biological processes. Because many of the processes that are important for living cells are thermodynamically uphill, the rates of many such processes depend on the status ("exchange rate") of the free enthalpy currencies of the cell. This then provides control by which the cell may regulate its physiology and it is a second important reason why one may have an interest in bioenergetics. It is as yet unclear however to what extent the free-enthalpy currencies of the living cell (cf., Fig. 5) are involved in the regulation of its physiology.

In one point of view the cell has evolved so as to keep the exchange rates of these currencies homeostatically controlled, so that its functions are not readily compromised when there is a free-enthalpy shortage. Indeed, the $H^+$-ATPase (cf. Chapter 5) in *E. coli* appears to be virtually without control on the growth rate and growth yield of this organism, even under conditions where virtually all ATP synthesis runs through this enzyme.

Primary photosynthesis is a virtually irreversible process in the sense that the reverse photosynthesis reaction hardly occurs. For instance, in anoxygenic photosynthesis only two protons are pumped per photon, due to mechanistic limitations. At a photon chemical potential of $125 \times 10^3$ J/mol and a proton electrochemical potential difference ($\Delta\tilde{\mu}_{H^+}$) of $21 \times 10^3$ J/mol, the free-enthalpy difference of this reaction becomes $125 - 2 \times 21 = 84 \times 10^3$ J/mol. The ratio of the reverse to the forward rate is then $10^{-84/5.7} = 2 \times 10^{-15}$. Consequently, photosynthesis seems to correspond to what used to be considered as the prototype of a metabolic pathway with a first, irreversible, step controlling the overall flux. If indeed the photon absorption step were the single rate-limiting step, then the rate of photosynthesis could not be regulated at all by the biology that comes after this.

It is of interest to notice that this situation, perhaps expected on the basis of the simplest principles, does not actually occur. The reason is that extent of control residing in a step in a pathway is not determined by the extent to which that step is displaced from equilibrium, but rather by the sensitivity (in exact terms "elasticity") of that step with respect to its substrate and product concentrations, in this case $\Delta\tilde{\mu}_{H^+}$. Primary reactions in photosynthesis appear to have a considerable such elasticity. For the retinal protein bacteriorhodopsin the situation is perhaps clearest: the rate at which this protein pumps protons is strongly dependent on the transmembrane electrochemical potential difference for protons. This so-called back pressure implies a high elasticity coefficient of this pump with respect to $\Delta\tilde{\mu}_{H^+}$. For the photosynthetic reaction center the situation has not yet been quantified. In chemotrophic growth one finds similar phenomena; the turnover rate of the essentially irreversible enzyme cytochrome-oxidase is strongly sensitive to changes in $\Delta\tilde{\mu}_{H^+}$ and so is mitochondrial respiration. The effect of all this appears to be that control is shifted from the free-enthalpy input reaction to the reactions that consume the free enthalpy. Clearly, control has implications for thermodynamics and *vice versa.*

## 3. Thermodynamic efficiencies and global bioenergetics

### *3.1. What comes before photosynthesis?*

According to the United Nations (UNIDO), the minimum free-enthalpy requirement for a human being amounts to approximately $1 \times 10^3$ W. This includes free-enthalpy expenditure for food, transportation and the building of necessary facilities. At present, the actual per capita free-enthalpy expenditure amounts $2.1 \times 10^3$ W on average; 200 W in South East Asia and almost $10 \times 10^3$ W in Canada, leading to a global free-enthalpy expenditure of $1.4 \times 10^{13}$ W.

This is very little compared to the total energy radiated by the sun ($45 \times 10^{28}$ W). However, the planet earth has to share this solar energy with the rest of the universe; "only" $0.18 \times 10^{18}$ W strikes the earth (at its outer reaches; at an intensity of $1.4 \times 10^3$ W/m$^2$). Only $0.9 \times 10^3$ W/m$^2$ is in the range that may give rise to photosynthesis and we have to share this with $5.3 \times 10^9$ (1990 estimate; 2050 estimate is $11 \times 10^9$) other human beings, leaving $34 \times 10^6$ W/person ($16 \times 10^6$ W/person in 2050). Yet, this is more than 3000 times the amount of power we would need if the global standard were increased to the North American level: 0.3‰ of solar irradiation should suffice and 0.03‰ should provide the bare minimum. Interestingly, it is also 17000 times the power man presently harvests from fossil free-enthalpy stores, which also derives from free enthalpy radiated by the sun, be it that most of it was formed a long time ago. Moreover, earth receives two free-enthalpy bonuses; $3 \times 10^{12}$ W of tidal energy, stemming from the gravitational force of the moon, and $30 \times 10^{12}$ W of geothermal origin.

However, of the $0.18 \times 10^{18}$ W, 30% is reflected back into space, and 50% is immediately converted to heat and reradiated as such (black body radiation of some 300 K) into space, leaving $36 \times 10^{15}$ W to first power the water cycle, winds and photosynthesis (Fig. 2). Of course, also the latter power eventually becomes heat and is reradiated into space. Only some 1% of the solar energy impacting on planet earth is actually absorbed by photosynthetic pigments. Photosynthesis then produces approximately $2 \times 10^{17}$ g of biomass per year. Valued at a free enthalpy of combustion of $22 \times 10^3$ J/g biomass, this still corresponds to a little over $100 \times 10^{12}$ W, some 7 times the total human consumption of free enthalpy.

Why then do humans use fossil fuel to cover more than 90% of their free-enthalpy needs? One factor is that the recoverable fossil fuel is thought to amount to 140 times the amount used in 1993. Such fossil fuel is presently readily available and cheap compared to biomass (natural gas: 1.7, oil: 3, solar energy: 20, wind energy: 14, $/$10^{12}$ J in 1990). After conversion to domestic electricity this becomes: coal: 4, wind: 5, biomass: 5, solar thermal: 12 and photovoltaic: 18, US$cts/kWh; for domestic electricity applications, the renewable free enthalpy sources may already be competitive). A second factor is that of the solar free-enthalpy fixed by photosynthesis only some 3% is fixed on cultivated land and 8% in forests. Most is fixed in the oceans and reaches us only partly through a long food chain. One should expect however that population growth, increasing free-enthalpy consumption per person, as well as pollution caused by the use of fossil fuel, will increase the use of renewable fuels, among which is the biomass or electricity produced by photosynthesis and photovoltaic cells, respectively.

Overall free-enthalpy fixation as biomass by photosynthesis is rather inefficient. Corn crops fix some 1% of the incandescent solar energy. For the entire earth only some 0.05 to 0.2% of impacting solar energy is fixed by photosynthesis (cf. Figs. 2 and 6). This is due to the facts that not all the incandescent radiation is actually absorbed by plants and microorganisms and that part of the free-enthalpy is dissipated within the living cells; in photosynthesis itself 95% of the incoming free enthalpy is dissipated (some 90% in the reactions themselves; see below and some 5% because of side and maintenance reactions that do not lead to biomass production; see below).

### 3.2. The efficiency of photosynthesis

The first biological steps of the conversion of electromagnetic radiation into useful chemical energy takes place in the so-called Reaction Centers and antennae of the photosynthetic machinery of phototrophic organisms (Chapter 5, section 5.2–5.4). Two types of photosynthesis can be distinguished: a minor one which uses retinal as the absorbing (and operating) pigment and a major one that uses the pigment (bacterio)chlorophill. Oxygenic and anoxygenic photosynthesis employ the latter. In oxygenic photosynthesis two photochemical reaction centers cooperate according to the so-called Z-scheme (see Chapter 5). In this way they convert the free enthalpy from sunlight into an electrochemical proton gradient (which subsequently can be stored in the phosphorylation potential of ATP) and into the redox potential of NADPH. Both ATP and NADPH are subsequently used in the so-called Calvin-cycle for $CO_2$ fixation ($H_2O + CO_2$ —(hv)—> $CH_2O + O_2$; storage of chemical free enthalpy into sugars).

It is evident that in all conversion steps free enthalpy will be lost. First of all, only part of the solar radiant energy is absorbed: only photons with a wavelength between 400 and 700 nm are used in photosynthesis. Depending on the absorbing organism, this is some 65–35% (we shall use the optimistic 65% below) of the solar spectrum (cf. Fig. 6). The second major free enthalpy loss takes place during light harvesting. Although light harvesting itself is often a very effective process mechanistically (almost all photons absorbed can be used for photosynthesis), much free enthalpy from the photons captured by the antenna pigments ($400–500 \times 10^{-9}$ m), is lost for photosynthesis since the actual free-enthalpy conversion occurs at a $700 \times 10^{-9}$ m chromophore and only the free-enthalpy equivalent of a $700 \times 10^{-9}$ m photon is stored. Obviously, the amount of free enthalpy lost in this step is dependent on the absorption characteristics of the organism and on the solar spectrum, but on the average only 60% of all free *enthalpy* absorbed is used in the initial step of photosynthesis.

The efficiency of energy transduction in photosynthesis is first discussed by taking anoxygenic photosynthesis with its cyclic electron-transfer chain as the example. The initial step of this type of photosynthesis proceeds in two phases in which free-enthalpy of the photon is stored in the charge separated state of the Reaction Center ($[Chl]_2^+$-$Q_A^-$). At any single excitation event, the energy of the $700 \times 10^{-9}$ m photon ($h \times \nu = 1.77$ eV $= 2.82 \times 10^{-19}$ J $= 171 \times 10^3$ J/mol; actual free enthalpy may be as low as $125 \times 10^3$ J/mol) equals the energy difference between the excited state and the ground state of the "special pair" of chlorophylls. Subseqently, the excited electron effects charge separation across the (thylakoid) membrane. Almost all absorbed $700 \times 10^{-9}$ m photons give rise to charge separation (quantum yield $\approx 1$). The standard free enthalpy of the charge separated state (some $120 \times 10^3$ J/mol) is, however, only about 70% of that of the primary $[Chl]_2$-excited state. The loss (dissipation) of free enthalpy in this step helps to prevent back reactions.

In terms of standard free enthalpies over photon energy, the efficiency of this "primary reaction of photosynthesis", starting from (sun) light is $65\% \times 60\% \times 70\% = 27\%$. This, however, is not the proper thermodynamic efficiency (e.g., not bounded by the second law of thermodynamics); one should implement actual rather than standard free enthalpies. Because the concentration of photons around the reaction center is finite and because it should be compared to the photon concentration of black body radiation of ambient temperature, the free enthalpy of a $700 \times 10^{-9}$ m photon is smaller than $171 \times 10^3$ J/mol. For an algal suspension, we estimated this (above) at $125 \times 10^3$ J/mol. The free enthalpy of charge separation of the $[Chl]_2^+ Q_A^-$ state actually attained by the illumination may be some 11 × $10^3$ J/mol smaller than the *standard* free enthalpy for this separation (taking the concentration ratio $[Chl_2^+ Q_A^-]/[Chl_2Q_A]$ equal to 0.01). Consequently, the actual thermodynamic efficiency of the primary reaction of photosynthesis may amount to 109/125 = 87% for 700 × $10^{-9}$ m photons (in terms of charge separation free enthalpy/photon free enthalpy; in terms of charge separation free enthalpy/700 × $10^{-9}$ m photon energy it is 64%). Taking a transmembrane electrochemical potential difference for protons of some 21 × $10^3$ J/mol as the result of photosynthetic electron transport and assuming a flow ratio of 2 protons per photon, the efficiency becomes 100% × 21 × 2/125 = 34% (25% free enthalpy per 700 × $10^{-9}$ m photon energy). Starting from sun light of arbitrary wavelength, these efficiencies should be premultiplied by 0.65 × 0.60 = 0.4 leading to 25 and 13% respectively. Because of incomplete coupling further reductions in the efficiency occur. Figure 6 gives an approximate overview of the efficiency of conversion of photon energy to proton free enthalpy in cyclic electron transfer in anoxygenic photosynthesis.

Free enthalpy is ultimately stored in biomass and not in the charge separated state or proton gradient. On the other hand, in oxygenic photosynthesis part of the photon free energy is captured as an increased chemical potential of an electron (i.e. the low midpoint potential of NAOH) and ultimately biomass. Experimentally it has been shown, for instance, that for oxygenic photosynthesis at least 8 photons of 700 × $10^{-9}$ m are required to convert 1 molecule of $CO_2$ to one C-mole of biomass ($-\Delta G = 460 \times 10^3$ J/mol): which leads to an overall efficiency of at most 46% (460/8 × 125) (33% in terms of output free enthalpy/input photon energy). In view of free-enthalpy dissipation in the processes and incomplete coupling, efficiencies of only 5% seem realistic: About 95% of the incoming free enthalpy is dissipated in photosynthesis. In the real world efficiencies are also lower, due to reflection of light, fluorescence, dark-respiration and/or photoinhibition. Losses due to reflecton of light can increase up to 20%. In algal cultures, however, reflection of light can be reduced to almost zero. These cultures also have the advantage that they can reach a crop yield of 5% (in higher plants crop yields do not exceed 3.5%).

### *3.3. What comes after photosynthesis?*

For photosynthetic microorganisms and plants, photosynthesis is where the story of thermodynamic efficiency ends. For some chemotropic bacteria and animals, this is only the beginning (Fig 1). Others are only reached by part of the solar energy, after an additional passage of the free enthalpy through a food chain. The transmission of free enthalpy through every organism in such a food chain is again accompanied by dissipation of part of the free enthalpy. Often the efficiencies of these organisms are not high, probably because organisms have been optimized for properties other than thermodynamic efficiency, such as growth rate, or growth yield. In non-equilibrium thermodynamic models of linear free-enthalpy converters, optimizaton for both output flow and efficiency leads to an efficiency of only 24%; optimization for output power and efficiency gives an efficiency of 41%. During aerobic microbial chemotropic growth, 24% and lower free-enthalpy transducing efficiencies were found. In rat liver free-enthalpy transduction efficiencies of 48% have been observed, but mitochondria are only part of a eukaryotic cell and overall efficiency is a product of partial efficiencies and can hence only be lower than the efficiencies of the partial processes.

In judging the above efficiency numbers, one should heed that there are various ways of defining the thermodynamic efficiency of microbial growth on a single substrate. For determining whether an efficiency corresponds to an optimum state, the definition in which anabolism is taken to start from the growth substrate, is appropriate. For insight in to how much free enthalpy is retained when the carbon source is consumed by such an microorganism, the definition that uses $CO_2$ as the substrate for anabolism is more appropriate. In terms of the latter type of definition, the efficiencies observed for chemotropic microbial growth ranged from 60 to 30%.

Ultimately, once the free enthalpy reaches the end of the food chain, what is the actual thermodynamic efficiency with which it is converted to the appropriate form of cellular free enthalpy? The latter may be the free enthalpy of ATP hydrolysis ($-\Delta G_{ph}$, see Chapter 1), which may be used to perform muscle work, or the electrochemical potential difference for protons across the mitochondrial inner membrane (cf. Fig. 5). Let us consider anaerobic glycolysis as an example. Assuming that both glucose and lactate are present at a concentration of $10 \times 10^{-3}$ M, the free enthalpy of glycolysis amounts to $2.1 \times 10^5$ J/mol glucose. Glycolysis generates 2 molecules of ATP per molecule of glucose (at a phosphorylaton potential of some $50 \times 10^3$ J/mol); the thermodynamic efficiency of this process is approximately 47%. Glucose oxidation to $CO_2$ liberates $2.8 \times 10^6$ J/mol glucose and may generate up to 36 molecules of ATP, assuming high P/O ratios for oxidative phosphorylation; this leads to an efficiency of 65%.

All these inefficiencies have the effect that man requires a free enthalpy intake in terms of food of some 120 W for his own maintenance. Spending an additional 100 W for transportation and other non-nutritional activities, this brings his minimum requirement to some $0.2 \times 10^3$ W.

### *3.4. Photosynthesis as free-enthalpy source beyond food*

Even considering all loss processes, described above, it is clear (keeping in mind the number $2 \times 10^{17}$ g of carbohydrates fixed annually, section 3.1) that using photosynthetic biomass as a free-enthalpy source is something to be exploited (fixed carbon which is today converted to $CO_2$ by natural fires every year, would already cover the annual consumption of fossil fuels!). It is therefore not strange that, after countries such as Brazil, now even the governments of the Northern and Western world are investigating the possibility of agrification (the use of agricultural products for industrial/non-food applications). At present some 15% of the total territory of the European Union is used for food production in agriculture. At this point a major disadvantage of free enthalpy conservation through biomass becomes apparent. It clashes with food production.

Since most *products* from agriculture have a petrochemical alternative, which up till now is relatively cheap and has a constant quality, European policies focus on the agricultural "production" of "energy". This can be achieved by production of solid free-enthalpy carriers (biomass) or as liquid/gas free-enthalpy carriers (refined oils, ethanol, methanol hydrogen etc.). At present, biomass production cannot win the competition with the relatively cheap fossil fuels. An increase of oil prices to $55 per barrel (from $16 per barrel in January 1995) would be needed to make bioethanol/methanol profitable. The main costs for biomass production are the use of vast quantities of fertilizers and water.

Two other important issues are the application of known techniques into "operational management" and the $CO_2$-emission into the atmosphere and global warming. All three issues could be dealt with by using (*new*) photosynthetic biomass, or by implementing more direct methods of converting solar energy into electricity. Designs for the latter could benefit from analogies with biology.

The efficiency of photovoltaic cells is usually determined under standard conditions, which are: a radiation intensity of $1 \times 10^3$ W/m$^2$ (cf. section 3.1) and a (cell) temperature of 25°–28°C. (Note that this is different from the above: while in photosynthesis only one wavelength is *optimally* used for energy conservation, in photovoltaic cells, by using several different materials, more of the solar spectrum can be harvested). Not counting free enthalpy lost before the light reaches the photovoltaic cell, the best performing cell in 1991 reached a conversion efficiency (free-enthalpy output/photon-energy input) of 28% under standard conditions. Using concentrated solar light, efficiencies of up to 35% could be achieved. These efficiencies should be compared with the numbers 64, 25 and 33% estimated for biological photosynthesis (section 3.2).

Research on photovoltaic cells focuses on increasing efficiency and lowering production cost. Whereas the highest efficiencies are being reached with (mono/poly) crystalline silicon, the most promising development is the construction of amorphous silicon cells. These cells, which are constructed in the form of thin films, have a far lower efficiency than the cells made from crystalline silicon (<10%), but are much cheaper in construction (no crystallization). The major problem with these cells is, however, their stability.

The application of photovoltaic cells is restricted by storage and transport (of electricity). Therefore most applications of photovoltaic cells can be found in so-called "autonomous systems"; besides pocket calculators, there are other applications such as railway signalling- and other communication systems, e.g., emergency phones near remote highways. Larger "autonomous systems" can be found in China, which is one of the developing countries, most exploiting photovoltaics. In China, hundreds of millions of people are still without electricity. In areas where the cost of connection to the national network is higher than the price of photovoltaic cells (and sunlight is available) "autonomous systems" are implemented. At the moment China produces $4.5 \times 10^6$ W photovoltaic electricity, using 10 separate power plants.

In recent years many attempts have been made to mimic the process of photosynthesis as it occurs in living organisms. Artificial photosynthetic systems could not only advance the current knowledge of actual photosynthesis but could also participate in exploiting the large amount of available solar energy. Although scientists have not yet been able to synthesize light driven DA (donor-acceptor)-molecules that reach the light-conversion and storage efficiency of natural photosynthesis, impressive progress has been made. The simplest DA-molecules are molecular dyads, of which the best studied are the P-Q, porphyrinquinone, dyads. In these systems, the tetraarylporphyrin (P) plays the role of chlorophyll as the excited state electron donor and a quinone (Q) acts as the acceptor. In some systems light-induced generation of $P^+$-$Q^-$ proceeds with a yield of 100%. Unfortunately, the recombination to the ground (P-Q) state in these systems is very fast, so that the light-induced $P^+$-$Q^-$ state never exist longer than several $100 \times 10^{-12}$ s. This is far too short to direct the free enthalpy stored in the charge-separated state to a useful (storage) process.

Longer charge separation times might be accomplished by enlargement of the geometrical distance between the negative and positive point-charge in a DA molecule, perhaps by linking a carotenoid molecule (C) to a PQ dyad, forming a CPQ triad. Charge separation proved stable (depending on the solvent used) for up to $2 \times 10^{-6}$ s in this $C^+$-P-$Q^-$ state. However, a gain in stabilization was always accompanied by loss in the overall charge separation yield, sometimes even down to 4%.

Information obtained from these studies with relatively simple complexes has proven useful in design and synthesis of molecules with higher efficiencies and longer charge separation times. In 1989, one of the best performing artificial photosynthetic systems consisted of 1 carotenoid, 2 porphyrins and 1 quinone. Light-induced charge separation in this molecule, to form $C^{+}$-P-P-$Q^{-}$, occured with an overall yield of 25% and was stable for some $2.9 \times 10^{-6}$ s. A very recent development is the use of derivatized $C_{70}$ molecules as DA-molecules. It has been claimed that after reconstitution in bilayers these molecules function 40 times better than the "conventional" dy/triads.

## 4. The current steady state: Global cycles of matter associated with global bioenergetics

### *4.1. Photosynthesis*

Together, the living organisms of the earth consitute what is called the biosphere. A primary activity for these organisms is free-enthalpy transduction. Except for the primary event of photon absorption, this free-enthalpy transduction involves physical and chemical transformations of compounds composed of various chemical elements such as carbon, nitrogen, sulfur and phosphorus. Because most of the chemical compounds involved are volatile or water soluble, this leads to global fluxes of these elements that are quite substantial. We shall first discuss carbon and the most intriguing step of the global carbon cycle, i.e., $CO_2$ fixation in photosynthesis.

In photosynthesis, an electron donor is separated into reducing equivalents and the oxidized form of the electron donor (i.e., in oxygenic photosynthesis electrons in the form of NADPH and oxygen, respectively, from the electron donor water; see also Fig. 1), with subsequent utilization of the reducing equivalents for reduction of $CO_2$ (or other carbon sources) to biomass. The overall reaction equation is:

$$CO_2 + H_2O \rightarrow \text{organic material} + O_2$$

The free enthalpy that is required, in order to allow this reaction to proceed, is obtained from sunlight. Thus, in photosynthesis, radiation energy is ultimately converted into chemical-bond energy that is present in biomass.

Photosynthesis comes in three different types, to be discussed in detail in subsequent chapters. Quantitatively, by far the most important is oxygenic photosynthesis, which occurs in cyanobacteria, algae and plants. This process (section 1 and Fig. 1) produces the oxygen that nearly all higher forms of chemotrophic life on earth ultimately depend on. In plants and algae this process takes place in a specific organelle of the eucaryotic cell, the chloroplast.

The $CO_2$ fixation coupled to oxygenic photosynthesis, particularly in plants, is one of the most important partial reactions of the global carbon

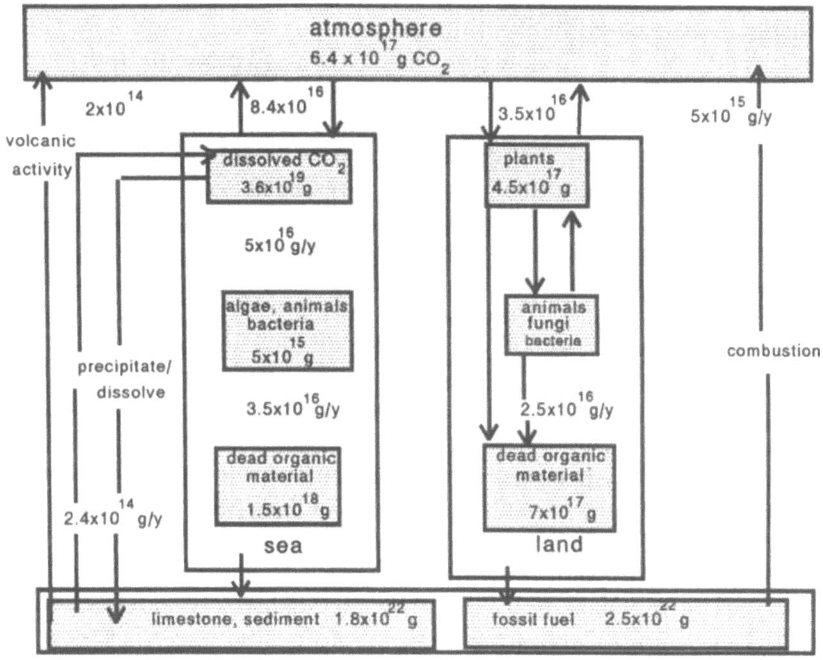


Fig. 9. Global carbon cycle.

cycle. More than $10^{17}$ g of $CO_2$ are fixed annually via this route corresponding to some $5 \times 10^{16}$ g carbon per year (Fig. 9). Oxygenic photosynthesis in algae is at the center of the carbon cycle that proceeds in the oceans (cf. Fig. 9). The end-result of oxygenic photosynthesis is the formation of large amounts of biomass and/or storage polymers (like polysaccharides) and a complementary amount of oxygen. The reversal of this storage of free enthalpy in chemical oxidoreduction potential allows for most of the chemotrophic way of life.

A second type of photosynthesis is present in a large number of representatives of one of the two kingdoms of procaryotic microorganisms (see below): the Eubacteria. This type of photosynthesis is called anoxygenic photosynthesis. It is encountered in the so-called purple and green bacteria (with representatives like *Rhodobacter sphaeroides* and *Chlorobium limicola*) and in a limited number of recently phylogenetically unrelated bacteria (e.g., in *Heliobacteria*). The transformations of free enthalpy in anoxygenic photosynthesis are different from those in oxygenic photosynthesis. In the former process the electron, which is excited by the radiation energy, travels through a cyclic chain of redox intermediates, with coupled conversion of redox free enthalpy into an electrochemical potential gradient for protons. This proton gradient subsequently provides all free enthalpy that is required to allow formation of biomass of these organisms. Two free-enthalpy transfer reactions are important in this respect: ATP synthesis and the process of "reversed electron transfer". Reversed electron transfer is required to bring the redox level of electrons, subtracted from the electron donor, at the redox level of NAD(P)H, so that –if necessary– biomass can be formed from $CO_2$. Many bacteria that perform anoxygenic photosynthesis use organic acids (acetate, succinate, etc.) or sulfide as an electron donor. Particularly with sulfide, it is easy to see that the overall reactions of oxygenic- and anoxygenic photosynthesis are analoguous: Either water ($H_2O$), an organic "acid" ($CH_2O$), or sulfide ($H_2S$) is used as an electron donor, to reduce $CO_2$ to biomass, under simultaneous consumption of radiant free enthalpy and production of oxygen or sulfur (or sulphate). This is one of the reasons why anoxygenic photosynthesis has been instrumental in providing insight into the mechanism of oxygenic photosynthesis.

The absorbance spectrum of bacteria carrying out anoxygenic photosynthesis enables these organisms to harvest light throughout the visible spectrum. However, these organisms can also absorb significantly in the near infrared (many can do this up to $900 \times 10^{-9}$ m and some, containing bacteriochlorophyll b, up to $1100 \times 10^{-9}$ m). These wavelength characteristics allow many of the anoxygenic phototrophic bacteria to thrive in stratified water columns in places, in which oxygenic photosynthesis has already filtered out most of the visible and near-infrared light.

A third type of photosynthesis is encountered in the second kingdom of the procaryotes: the Archaebacteria. This type of photosynthesis is quantitatively of minor importance and mechanistically very different from the two types of (bacterio)chlorophyll-based photosynthesis described above. In halophilic archaebacteria like *Halobacterium salinarium,* two different proteins, intrinsic to the cytoplasmic membrane of the organism, are present. A retinal chromophore is complexed to these proteins via a protonated Schiff-base linkage. This complex of protein plus chromophore strongly absorbs visible light ($570 \times 10^{-9}$ m) and can use the radiation energy to translocate protons (for bacteriorhodopsin) or chloride ions (for halorhodopsin) across the cytoplasmic membrane. This type of photosynthesis is generally considered to be auxiliary, although halobacteria are able to grow (slowly) with light as the only free enthalpy source.

### *4.2. The chemotrophic mode of life*

The chemotrophic mode of life implies that one or more chemical reactions provide the free enthalpy that is required for growth of such organisms. The rate of dissipation of free enthalpy in chemotrophic organisms, due to their endogenous metabolism, is surprisingly high. This is particularly true for microorganisms. Due to their small size, these organisms can maintain metabolic rates that are at least 100-fold higher per unit mass than the maximal metabolic rates observed in higher organisms (mammals).

During chemotrophic growth, products of photosynthesis of other organisms, such as sugars, fatty acids and proteins are oxidized to yield $CO_2$ and $H_2O$, under conservation of some of the free enthalpy in the form of phosphorylated nucleotides, ion gradients (like the proton motive force) and redox intermediates (like NADH, etc.) (Fig. 1). Many different organisms, from higher mammals to prokaryotic microorganisms, contribute to this mineralization of biomass and consequently to the closing of the global carbon cycle. Two-thirds of the flux of $CO_2$ into biomass is returned to the atmosphere in this way; the other third is released through the so-called process of photorespiration (Fig. 9).

However, this mineralization may proceed through several consecutive steps, in which the free enthalpy is stored in an ever decreasing amount of biomass. Some of the microorganisms involved in this mineralization process have fully specialized on partial reactions of this complex process, like on the hydrolysis of polymers or on the fermentation of complex organic carbon sources (like sugars, etc.). Fermentation can be understood as a process of mineralization in which an intracellular metabolite fulfills the role of electron donor and another metabolite the role of electron acceptor. In this type of metabolism, free enthalpy is primarily conserved via the synthesis of ATP.

The dominant type of free-enthalpy conversion in this group of organisms, however, proceeds via oxidation of the monomers of biopolymers via an electron-transport chain, with oxygen as terminal electron acceptor. This electron transfer gives rise to coupled proton translocation, thus forming an electrochemical proton gradient as the central free-enthalpy intermediate (Fig. 5). A large variety of electron donors from intermediary metabolism can donate electrons to these electron-transfer chains, only limited by the catalytic specificity of the dehydrogenases that funnel the electrons into the chain and by redox potential in the range of −0.4 to +0.4 V. Various metabolic pathways interconvert the metabolites in the cell to successive forms in which electrons are separated from the organic electron donors, prior to their feeding into the electron transfer chain. The most important of these pathways are glycolysis and the citric acid cycle.

Quantitatively minor flows of carbon are due to competing forms of mineralization, in microorganism ecosystems in which oxygen is not present in sufficient amounts. Depending on the electron acceptor present, electron-transfer pathways are utilized with, for instance, $NO_3^-$ (in soil) or $SO_4^=$ (in marine environments) as the terminal electron acceptor. Additional electron acceptors may be utilized, when available in specific ecosystems; $Fe^{3+}$, fumarate and sulfur. In fact, it may be speculated that microorganisms have evolved that can specifically utilize the free enthalpy contained in every major redox transition that takes place in the various ecosystems that are relevant. Regulatory mechanisms, in particular in bacteria, have undergone a thorough evolution. These mechanisms dictate that when multiple electron acceptors are present simultaneously, the acceptor has preference, which provides the largest free-enthalpy span with respect to the electrons available in the carbon/energy source of the organism. When all external electron sinks are exhausted, methanogenic bacteria may reshuffle the reducing equivalents in organic carbon sources (and hydrogen), to form $CO_2$ and methane ($CH_4$), of which particularly a large part of the latter may escape into the atmosphere. The products of anaerobic respiration, like $NO_2^-$ in soil, sulfide in sediments and methane, can function again as substrates for aerobic chemotrophic growth. When the first two of these are used, one is dealing with the so-called lithotrophic mode of life. These reactions form an essential step in the global cycles of the elements nitrogen (Fig. 10) and sulfur. Figure 11 illustrates the quantitive importance of the biological contributions to the global nitrogen cycle.

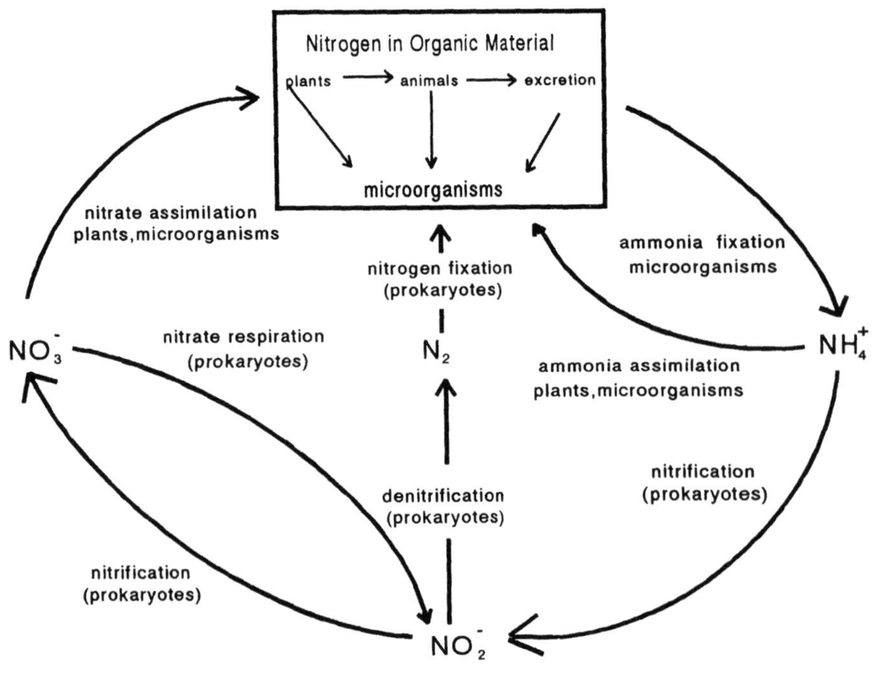


Fig. 10. The biological nitrogen cycle.

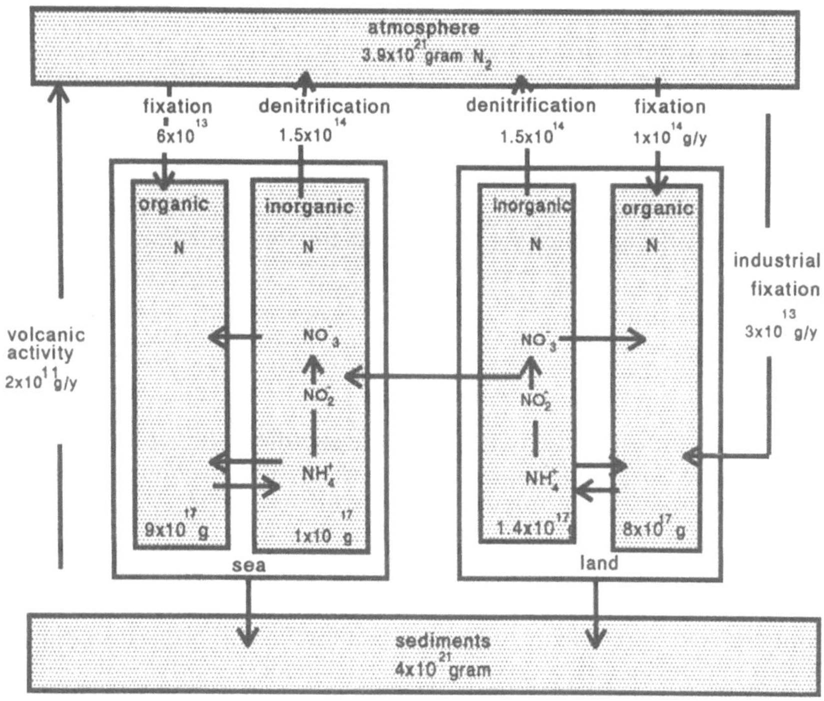


Fig. 11. Global nitrogen cycle.

## *4.3. Hydrothermal vents*

At the bottom of the deep sea, near the edges of so-called hot vents (or: black smokers) a dense community of microorganisms and invertebrates can be found, for which the free enthalpy required for growth is not provided by radiation energy. This ecosystem is maintained by a cyclic process of oxidation/reduction reactions, driven by the strongly negative redox potential and the high temperature of the contents of the magma of the earth.

The hot vents originate in places where the magma is near the sea floor, so that cold sea water, containing oxidized compounds, predominantly sulfate, can enter the hot magma. This leads to reduction of the sulfate to sulfide and its subsequent emission from the hot springs. At the edges of these hot springs lithotrophic bacteria oxidize the sulfide to sulfate, using oxygen as electron acceptor and $CO_2$ as carbon source for growth. Besides sulfide, other lithotrophic organisms can use reduced compounds like $Mn^{2+}$, hydrogen or ammonia, that are formed in parallel.

The invertebrate animals that are present in this ecosystem are long (up to 2 meters) tube worms. They do not eat the lithotrophic microorganisms, but rather live in a very specialized form of symbiosis with them. They have specialized organs to provide the lithotrophic bacteria, which live intracellularly in these tube worms, with sulfide and oxygen (and a special type of hemoglobin, to transport both substrates). Via their excretory products and their own biomass, the microorganisms provide the tube worms with substrate for a heterotrophic mode of life, in which organic substrates are oxidized with oxygen as the terminal electron acceptor. As a consequence, this type of ecosystem depends on radiant energy from the sun exclusively via its dependence on the presence of oxygen as a terminal electron acceptor for both the microorganisms and the invertebrates.

### *4.4. Man*

Man and his industrial activity contribute significantly, although not yet dominantly to the global cycles. Figure 9 shows that combustion of fossil fuel brings some $5 \times 10^{15}$ g/y of $CO_2$ into the atmosphere. This might be compared with the atmospheric $CO_2$ generation of $2 \times 10^{15}$ g/y and the $120 \times 10^{15}$ g/y due to exchange with the biosphere (see Fig. 9). Such a comparison is inappropriate, however, because the quoted flux from the biosphere is counterbalanced by an equal flux in the opposite direction, whereas the former two fluxes are not balanced by other fluxes. They should therefore be compared with the $CO_2$ reservoir present in the atmosphere ($6 \times 10^{17}$ g), suggesting that the latter could well double in 90 years. This suggestion is not quite realistic as the $CO_2$ dissolved in the ocean is likely to function as a buffer, extending the doubling time to more than 1000 years. This positive note is turned around however when one considers the possibility of combusting as little of 1% of the total fossil fuel. This would outweigh the $CO_2$ buffer in the oceans by a factor of 6 (Fig. 9). Especially in view of the free-enthalpy shortage, which may induce man to utilize much of that fossil fuel, $CO_2$ overproduction and associated global warming is a potential danger. Cycling free-enthalpy sources are to be favored.

## 5. Approach to the current steady state: Evolution of free-enthalpy transducing systems

### *5.1. The tree of life*

In the free-enthalpy conversions in the biosphere a large array of organisms play a role (cf. section 4). In the following we will discuss the phylogenetic relations between these groups of organisms, i.e. picture their evolutionary descent. Until recently, systematic classification, particularly of microorganisms, was based on morphological and physiological criteria.

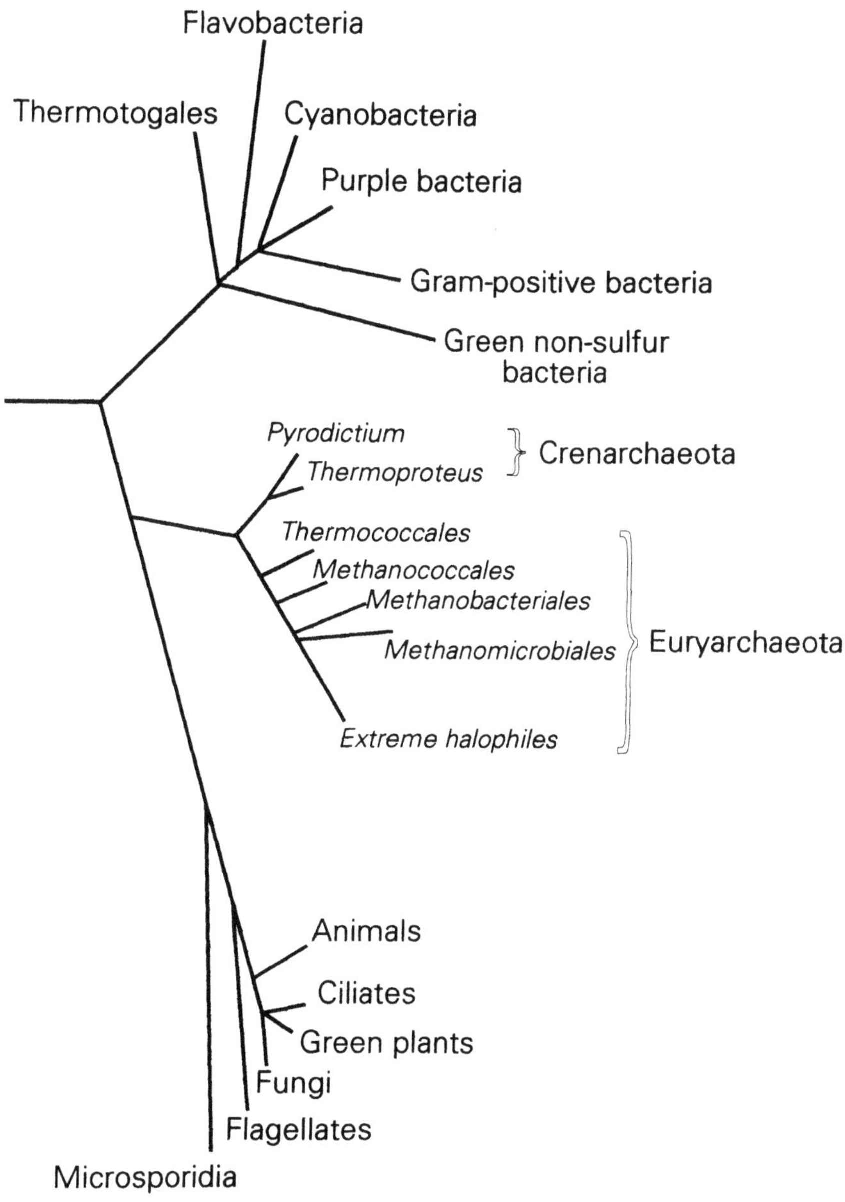


Fig. 12. Evolution of organisms and classification of bacteria.

This had the inherent disadvantage that an arbitrary weight had to be assigned to the criteria used. With the strong advance of molecular genetics, means have become available for the classification of living organisms on a phylogenetic basis, i.e. reflecting the evolutionary development of all living organisms. Their application has led to dramatic changes in the generally accepted view of bacterial classification (see Fig 12).

Procaryotic organisms are now thought to belong to either one of two bacterial kingdoms, the Bacteria and the Archaea (or: the Eubacteria and Archaebacteria in a slightly older terminology), with very characteristic differences between organisms belonging to these separate groups. Furthermore, all eucaryotic organisms are supposed to have evolved from the archaeal lineage. However, the organelles within the eucaryotic cell specialized in free-enthalpy conversion (i.e. mitochondria and chloroplasts) are supposed to have originated via a symbiosis between a primitive eucaryotic cell and specific eubacteria (a chemoheterotrophic bacterium like *Paracoccus* and a cyanobacterium, respectively). Within each of the three kingdoms, the evolutionary relations have been analyzed too (see for instance Fig. 13 for the Bacteria). This has dramatically changed the classification of these organisms, particularly in the eubacteria. It has revealed for instance that there are many close parallels between photosynthetic and non-photosynthetic organisms; organisms that were completely separately classified in the older schemes.

A major problem with setting up a phylogenetic tree is the determination of its root. This can be tackled mathematically, but the resulting conclusions have long been controversial. Determination of the root is relevant in order to be able to discuss the properties of the so-called "progenote". The term progenote was introduced to describe a hypothetical organism which occured so early in evolution that fixed relationships between genotype and phenotype did not yet exist for these organisms. However, in the more recent literature the term is usually applied to describe the properties of the last common organism, from which the three

Fig. 13. Evolution of eubacteria.

kingdoms evolved. Recently, the root of the tree of life was determined more or less independently from the data on which the tree shown in Fig. 12 is based. This was possible through the availability of a large amount of sequence data of proton pumping ATPases. These enzymes have evolved in a form that makes it very likely that very early in evolution a gene duplication occured, which gave rise to multiple subunits in the enzyme, of which two (the $\alpha$ and the $\beta$ subunits) exhibit extensive homology. This allowed a sequence comparison between pairs of $\alpha$ subunits of the ATPase from a large number of organisms, but now the resulting tree could be rooted via a comparison with the $\beta$ subunit. This procedure resulted in a tree, very similar to the one shown in Fig. 12. Further analysis of the ratio at which the various subunits are present in the $H^+$-ATPases of different organisms, led to the conclusion that already in the progenote, the $H^+$-ATPase complex had obtained a function in ATP synthesis, rather than merely a function in ATP hydrolysis for the generation of a proton gradient. This conclusion will be central in our further discussion of the evolution of biological energy transduction mechanisms. However, this analysis also led to the conclusion that these ATPases originally (i.e. before the gene duplication) had evolved as ATP hydrolyzing proton pumps, with most likely a function in the maintenance of the intracellular pH.

### *5.2. Shifts in paradigm*

The start of the scientific work on the origin of life is marked by the pioneering work of Oparin and Miller in the early 1950s. They demonstrated that electrical discharges in an atmosphere containing hydrogen, ammonia, methane and water led readily to the formation of a complex mixture of organic compounds, many of which play an important role in living organisms. This observation formed the basis of the hypothesis that on the surface of the primitive earth, lightning in a strongly reducing atmosphere would generate a primordial soup, rich in the building blocks for living organisms and that life arose in a "warm little pond", filled with this soup. This is the theory of the heterotrophic or fermentative origin of life. This led to the idea that further evolution of metabolic pathways was driven by the exhaustion of intermediates from this primordial soup, which drove the organisms, via "anaerobic respiration", eventually to a photoautotrophic mode of life as proposed by Horowitz. Subsequently, the development of oxygenic photosynthesis would have led to massive development of organisms that use an aerobic respiratory type of metabolism.

An opposite scenario has been proposed by Granick. He postulated that "biological pathways recapitulate their evolution", meaning that more and more complex steps have been added to primitive pathways that evolved from pre-biotic processes. This hypothesis is gaining more and more support, for instance, because it has become clear that the primitive atmosphere was much less reducing than initially supposed in the experiments of Miller. In Granick's hypothesis photosynthesis plays a central role. It supposes that photosynthesis arose from processes that occur in common minerals as they are present in clay. His theory is therefore called the autotrophic origin of life. The initial organization of the pre-protoplasm would be a primitive free-enthalpy conversion unit, made up of iron-containing minerals and able to convert radiant free enthalpy (from the short wavelength range of the spectrum), into chemical-bond free enthalpy of complex molecules. These complex organic molecules would organize themselves into a replicating unit. Progressively, the composition of the resulting "biomass" would become more and more complex, starting with only the elements C, H and O and the successive involvement of S, N and P.

Inspired by the Granick hypothesis, Hartman described this process in a hypothetical scenario in which photosynthesis evolved in a four-stage process. Phase one would be replicating clay, using UV light to transform oxalic acid into glyoxylate and glycerate. In phase two, sulfur entered the system and ferredoxins became involved in electron-transfer reactions. Radiant free enthalpy in this phase was used for the formation of thioesters and was predominantly absorbed in the blue part of the visible spectrum by iron-quinone complexes. In phase three, nitrogen became involved, with the possibility to synthesize amino acids, tetrapyrroles, etc., which led to the involvement of more extended electron-transfer chains. In phase four, phosphate became involved. This created the possibility to synthesize phospholipids, nucleic acids, etc. In this phase photosynthesis may have been of a type that is similar to the current-day photosynthesis as it is present in *Chloroflexus.* According to this scenario, life originated in a photo(auto)trophic mode, with only subsequent development of chemotrophic-respiratory and -fermentative metabolism.

A completely different type of scenario furthered by A. Muller stresses that proteins and possibly related simpler molecules may harvest free enthalpy from oscillating electric fields or oscillating temperatures. Oscillations in temperature arise by the day/night rhythm, but also around geothermal sources. Molecules with a temperature-dependent dipole moment will transduce these temperature oscillations into locally oscillating electric fields.

### *5.3. The development of photosynthesis*

The origin of the earth dates back approximately $4 \times 10^9$ years. Chemical traces of the activity of oxygenic photosynthesis can be dated all the way to at least $2 \times 10^9$ years back, on the basis of the ferric iron deposition.

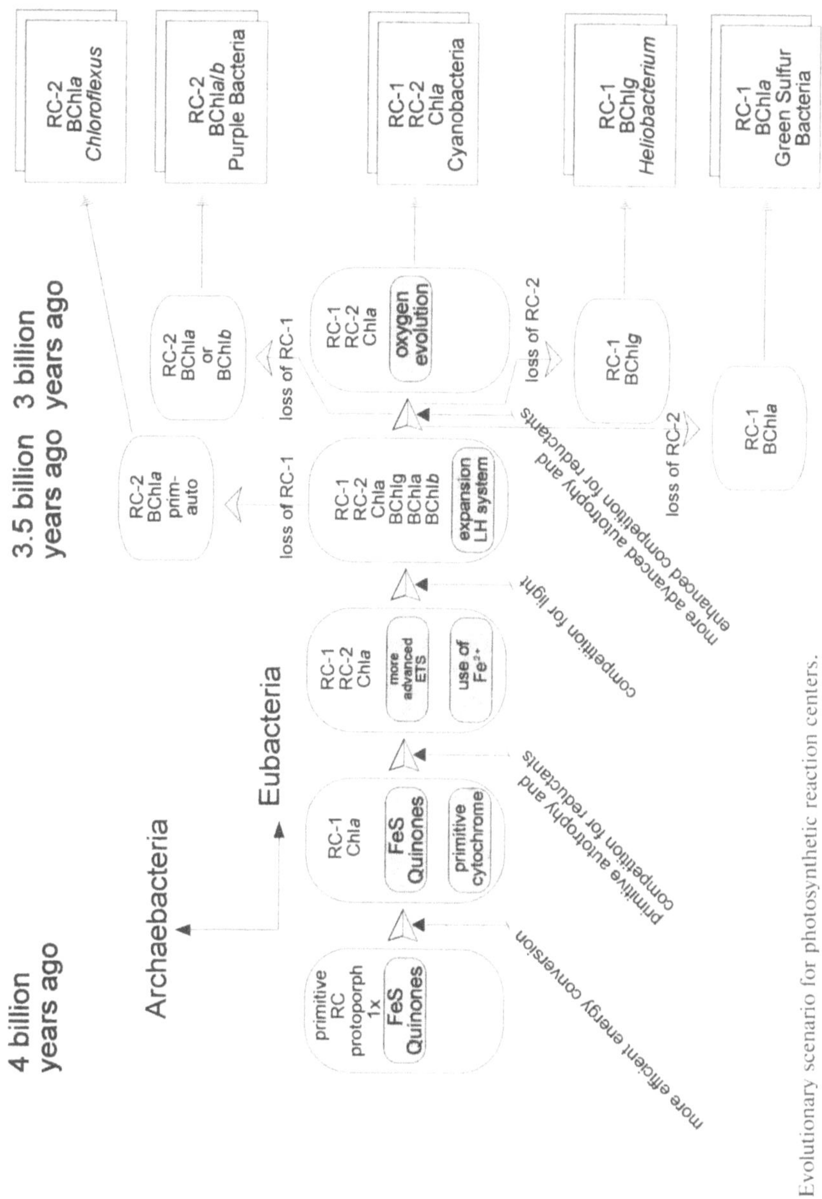


Fig. 14. Evolutionary scenario for photosynthetic reaction centers.

However, even as far back as $3.5 \times 10^9$ years, traces of the presence of oxygen can be found, although there is no guarantee of course that these are of biological origin. Analysis of fossils, from various layers of –and geographical locations on– the mantle of the earth, have led to the conclusion that the remainders of organisms, that are very reminiscent of cyanobacteria, can be found in layers that are at least $3.5 \times 10^9$ years old. If these interpertations are correct, this implies that the four-phase development depicted above, had been completed in the first $300 \times 10^6$ years of the existence of the earth. Additionally, however, also the divergence of the various types of phototrophic microorganisms that can be discriminated within the tree of the eubacteria (see Fig. 13) would have taken place in this period, because the ribosomal RNA analysis depicts the cyanobacteria among the most advanced of the eubacteria.

The presence of bacteriochlorophyll in widely different phototrophic organisms suggests that chlorophyll photosynthesis evolved from bacteriochlorophyll photosynthesis. The genesis of oxygenic photosynthesis would then be explained as being initiated by a (genetic) fusion of the photosynthetic machinery of two organisms, with FeS-type and quinone-type reaction centers, respectively. From this hypothetical organism onwards cyanobacteria and subsequently chloroplasts would have evolved.

However, following Granick's hypothesis, we come to a different scenario. That is because the biosynthetic route of chlorophyll pigments goes from δ-amino-levulinic acid, via uro- and protoporphyrin, protochlorophyllide to chlorophyll and from there finally to various forms of bacteriochlorophyll. As a consequence, we have to postulate that a primitive type of chlorophyll photosynthesis predated bacteriochlorophyll photosynthesis and that the latter may have been a development which originated due to excessive screening by the chlorophyll containing organisms.

Granick's hypothesis leads to a scheme in which several hypothetical types of organisms have to be postulated, but in which all known types of photosynthetic organisms can be accomodated (see Fig. 14). Further analysis, particularly of the genes coding for the enzymes involved in pigment synthesis, will decide whether or not this scheme will withstand the test of time. It will be particularly relevant in these analyses to try and determine whether or not horizontal transfer of genes, involved in photosynthesis, has taken place.

### *5.4. Retinal photosynthesis*

A discussion of the evolutionary origin of retinal-based photosynthesis can now be based on the same arguments that led to Fig. 14. Already in the four-phase development of photosynthesis in the "progenote", pigments evolved with a carotenoid-like structure and photosynthesis exploited the middle part of the spectrum of visible radiation. It has been argued that chlorophyll-based photosynthesis is much more efficient than retinal-based photosynthesis. This assumption, together with the complementarity in their absorbance characteristics (see above), has led to the postulate that in the course of evolution first retinal-based photosynthesis evolved and that subsequently it was outcompeted by chlorophyll-based photosynthesis, because of the much higher efficiency of the latter process.

We consider this scenario unlikely. First, it can be questioned whether or not chlorophyll photosynthesis is more efficient than retinal photosynthesis (see above). Furthermore, the primitive blue-light dependent type of photosynthesis proceeded according to a completely different mechanism than retinal photosynthesis. In the progenote, photosynthesis was dependent on redox reactions, rather than on a conformational mechanism of proton translocation. The above discussion of the rooting of the tree of life ended in the conclusion that initially the $H^+$-ATPase evolved so as to hydrolyze ATP. Consequently, we have to assume that radiation free-enthalpy was primarily converted into redox free enthalpy of preprotoplasmic metabolites, giving rise to a primitive system of substrate level phosphorylation. This type of photosynthesis is completely opposite to the light-driven proton translocation, catalyzed by bacteriorhodopsin. The conversion of radiation free enthalpy into the free enthalpy of ion (i.e. proton) gradients, can only have evolved after the development of a more advanced system of electron transfer (like cytochromes, etc.), so *after* the separation of the archaebacterial and eubacterial lineages (see also Fig 14).

On the basis of these considerations it seems more likely that archaebacteria lost the capacity to perform tetrapyrrole-based photosynthesis. They may have developed a protective mechanism against UV-induced damage of their nucleic acids, in the form of a negative phototactic response. This then may have led to the development of free-enthalpy transducing retinal proteins like bacteriorhodopsin and halorhodopsin and consequently to retinal photosynthesis.

## 6. Concluding remarks

In the above it has become clear that the position of bioenergetics in global energetics, though small in quantitative energetic terms, is important in view of life itself and its future. Also, the carbon fluxes associated with global bioenergetics are quite substantial. Man does not yet, but may begin to interfere substantially with the biological global carbon balance. The individual processes of photosynthesis are quite efficient. It is their multitude that leads to low overall efficiencies. The energetic efficiency of biological processes is far from maximal, probably because efficiency has not been the (only) property for which organisms have been optimized in evolution. Continued study of the mechanisms and energetics of biological free-enthalpy transduction is bound to further enhance our insight into coupling mechanisms and how they may be made more efficient.

## Acknowledgements

We thank Prof. Dr. K. van Dam, Prof. Dr. M. C. E. van Dam, Dr. G. Meszena and Mrs M. van der Kaaden for invaluable help in collecting substrate for and results of calculations as well as discussions, Dr. A. Vaz Gomes for reading the manuscript and Ms. J. Wijker for preparing it. This study was supported in part by the Netherlands Organization for Scientific Research (NWO).

## Further reading and key references

AYu Borisov, In: *Photosynthesis in relation to model systems.* J Barber (ed.), Elsevier, Amsterdam, 1979.

GR Davis, et al., Scientific American 263, (1990) 21–115.

LNM Duysens, Brookhaven Symp. Biol. 11, (1958) 10–25.

DO Hall, FEBS Lett. 64, (1976) 6–16.

KJ Hellingwerf, W Crielaard, and HV Westerhoff, In: *Modern Trends in Biothermokinetics,* S Schuster, M Rigoulet, R Ouabi and J-P Mazat (eds.), Plenum Press, New York and London, 1994 pp. 45–51.

DH Meadows, DL Meadows and J Randers, *Beyond the Limits. Confronting Global Collapse; Envisioning a Sustainable Future.* Earthscan, London 1991.

HJ Morowitz, *Foundations of Bioenergetics.* Academic Press, New York 1978.

G Porter, In: *Techniques and New Developments in Photosynthesis Research.* J Barber and R Malkin (eds.), Plenum Press, New York and London 1989.

HG Schlegel, *General Microbiology,* 7th edition, George Thieme Verlag, Stuttgart 1992.

VP Skulachev, *Membrane Bioenergetics.* Springer, Berlin 1988.

MCE Van Dam-Mieras and CK Leach, *In Biotechnological Innovations in Energy and Environmental Management,* 1994, Butterworth-Heinemann, Oxford.

HV Westerhoff and K Van Dam, *Thermodynamics and Control of Biological Free-enthalpy Transduction,* Elsevier, Amsterdam 1987.

HV Westerhoff and K Van Dam, *Thermodynamics and the Regulation of Cell Functions,* In: *Molecular Mechanisms in Bioenergetics,* L. Ernster, (ed.), Elsevier, Amsterdam 1992, pp. 1–35.

Bioenergetics
ed. by P. Gräber and G. Milazzo†


# *CHAPTER 3*
# Vectorial bioenergetics

Bruno Andrea Melandri

*Department of Biology, University of Bologna, Bologna, Italy*



## 1. Bioenergetics in a homogeneous compartment

### *1.1. Energetics of metabolism*

Cells are open thermodynamic systems exchanging energy and matter with their surroundings. They also operate at constant temperature and pressure. The proper thermodynamic state function suitable for evaluating the bioenergetic events in a cellular system is, therefore, the Gibbs free energy ($\Delta G$) (or free enthalpy). Any metabolic reaction taking place within a homogeneous phase, i.e. inside an intracellular compartment surrounded, but not subdivided by a biomembrane, must therefore proceed in the direction of a net decrease of its associated free energy change. The value of this change is a function of the standard free energy of the reaction (i.e. its equilibrium constant, depending on temperature and on such parameters as pH ionic strength of the compartment in which the process takes place) and the activity of the reactants and products:

$$\Delta G = -RT \ln K + RT \ln \frac{\prod a_p^{v_p}}{\prod a_s^{v_s}} \qquad (1)$$

where K is the equilibrium constant and $\Pi a_p^{v_p}$, $\Pi a_S^{v_s}$ are the stoichiometric products of the activities of products and reactants, respectively.

Intermediary metabolism consists of a network of catalytic enzyme reactions interconverting biological molecules (metabolites) in order to provide energy and building blocks for the cellular biosynthesis. The metabolic reactions generally occur spontaneously, associated to a definite decrease in free energy, and at a rate depending on the concentration and the kinetic characteristic of the enzyme involved, and on the concentrations of the matabolites at a steady state reflecting the dynamic balance of their rates of formation and utilization. The essence of that part of metabolism aimed to the production of cellular energy (catabolism) include the decomposition of complex macromolecules (proteins, polysaccarides and triglycerides) into their monomeric units and the oxidative demolition of their structure (to water and $CO_2$ in aerobic respiration). The opposite, anabolic, reactions of metabolism produce, starting from simpler molecules, the building block for the structures by which cellular life can be perpetuated.

Particularly in anabolism, but some times also in catabolism, some reactions are required that would correspond to a net increase in free energy. They would be thermodynamically impossible unless they can proceed associated to another reaction, largely exergonic, that could drive them. In this case the two reactions are *coupled,* in the sense that the reaction mechanism, often the same catalytic mechanism within the enzyme active site, renders obligatory the contemporary occurrence of the endergonic and exergonic processes. In this case, therefore, the two coupled reactions proceed spontaneously in the direction of a net overall decrease in free energy. The reaction most frequently utilized in metabolism for this purpose is the decomposition of ATP into ADP and phosphate, or into AMP and pyrophosphate. Very often the mechanism of the coupling of the two reactions is obvious, in that a portion of the ATP molecule (a phosphate group or an adenylate residue, or more rarely a pyrophosphate residue) are directly transferred to an acceptor molecule; this *activated* reaction product is subsequently used in a second reaction, *per se* thermodynamically impossible.

A typical example of such a mechanism is the formation of an amino-acyl-tRNA, the activated monomer of protein synthesis. Here a free aminoacid forms an ester bond with 3'-oxidryl residue of a tRNA, utilizing a molecule of ATP. During the actual mechanism however, a carboxy-adenylate intermediate is formed, liberating pyrophosphate; this compound in turn is the real acyl donor to the tRNA.

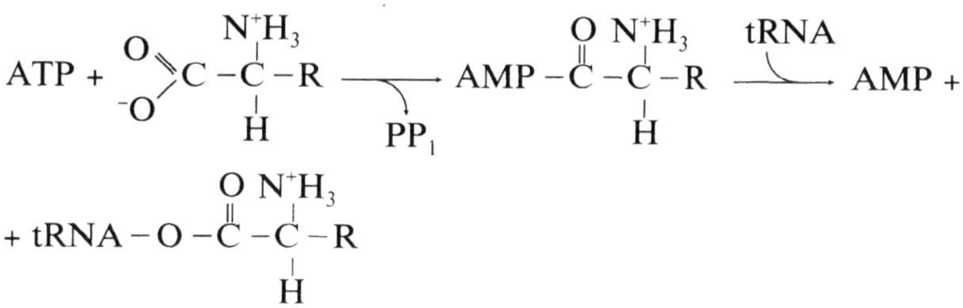


(2)

The overall free energy change of this two-step reaction is therefore formally equivalent to the summation of the free energy for ATP decomposition into AMP and pyrophosphate and the formation of aminoacyl-tRNA from a free amino acid and a free tRNA. While the latter process would be thermodynamically impossible ($\Delta G'^{0} \approx 34$kJ mole$^{-1}$), the coupled process proceeds associated to a net free energy decrease. The coupling of these two reactions in this case occurs at the active site of a single enzyme (an aminoacyl-tRNA synthetase) on which the adenylate intermediate is transiently bound.

The synthesis of an aminoacyl-tRNA exemplifies satisfactorily the role of ATP in metabolic bioenergetics. Often this role is even more evident, in that phosphorylated or adenylated intermediates are stable free metabolites, present in the cells at measurable concentration, are diffusable inside cellular compartments and are often involved in multiple metabolic reactions. Sometimes the role of ATP in driving the endoergonic processes is more indirect since its utilization takes place for the formation of a different activated metabolite. This is the case, for example, of the carboxylation and decarboxylation reactions that occur during the synthesis of fatty acids, in which malonyl-S-CoA, a carboxylated derivative of acetyl-S-CoA, is the activated intermediate for the condensation of an acyl group to yield a β-keto-acyl derivative (Fig. 1). In other cases the role of ATP is more subtle and not immediately explained by the presence of a stable covalently bound intermediate. In the reaction of succinyl-S-CoA synthetase, for example, GTP (another nucleotide triphosphate functionally analogous to ATP) is formed from GDP and phosphate during the decomposition of succinyl-S-CoA to succinate and CoA-SH; this reaction is perfectly reversible ($\Delta G = -3.4$ kJ mole$^{-1}$). In this case the reaction intermediate is a phosphorylated form of the enzyme protein; a hystidinyl residue is phosphorylated and acts as phosphate donor to GDP.

Whatever might be, the actual mechanism of the involvement of ATP

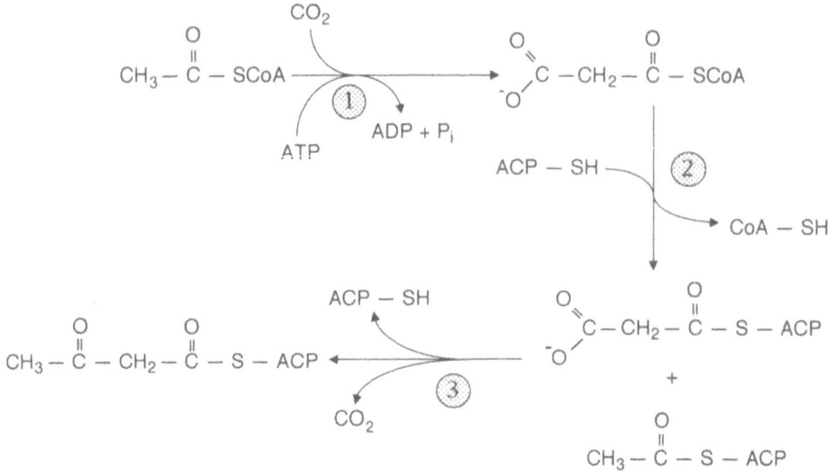


Fig. 1. Malonyl-CoA acts as an activated acyl acceptor in the synthesis of fatty acids. Acetyl-CoA is carboxilated in step ①, yielding malonyl-CoA that is transferred to the ACP (acyl carrier protein) of the fatty acid synthetase complex in step ② and again decarboxilated during the transacylation step ③. Expenditure of ATP occurs only during step ①. Although in this figure only the transacylation of acetyl-CoA is shown, step ③ can be repeated cyclically many times for the final synthesis of palmityl-CoA. In each cycle the complete reduction of the β-cheto-acyl-CoA to an acyl-CoA also occurs, utilizing NADPH as reductant.

the thermodynamic principle remains unchanged. A reaction intermediate is formed with the exoergonic scission of ATP and it is decomposed during the following endoergonic step. The overall net change of free energy becomes negative due to the contribution of the ATP decomposition step and the process can take place. The common intermediate during the two steps of the reaction embodies the molecular entity by which ATP decomposition and the driven process are coupled. Only in this case, in fact, it is allowed to compute the overall free energy change as a formal summation of those the two separated processes.

Other nucleotide-triphosphates, besides ATP, (GTP, CTP and UTP) can act as energy donating metabolites; their properties and reactivity are very similar to those of ATP and their utilization in different sections of metabolism depend on the substrate specificity of the enzymes involved. Unique to ATP however is the possibility of being regenerated from ADP and phosphate at high rate in the cell during the processes of oxidative or photosynthetic phosphorylation and of substrate level phosphorylation (the synthesis of GTP by the above-mentioned succinyl-S-CoA synthetase reaction is one of the few cases of direct synthesis of a nucleotide-triphosphate different from ATP and is relatively unimportant to the bioenergetics of the cell). The synthesis of CTP, UTP, and, to a large extent, also of GTP depends on the availability of ATP as phosphate donor to the respective nucleotide-diphosphate. This is the true meaning of the often asserted statement that ATP is the energy exchange molecule of the cell.

*1.2. Metabolic oxidoreductions*

Oxidoreduction reactions represent an important part of metabolism. Here, although the free energy changes are usually expressed as redox potential differences, the same thermodynamic considerations should be applied, taking into account that the free energy decrease corresponds to the maximal (reversible) electrical work that can be obtained from the redox reaction,

$$-\Delta G = \nu_e \mathcal{F} \Delta E \qquad (3)$$

being $\nu_e$ the number of electrons involved in the reaction and $\mathcal{F}$ the Faraday constant.

Redox reactions in cells utilize special coenzymes which act as electron donors or acceptors, the most important being the $NAD^+$/NADH couple (and the related coenzymes $NADP^+$/NADPH) and the flavin group (mainly flavin-adenine dinucleotide, FAD) associated with the flavin dehydrogenases. $NAD^+$ and $NADP^+$ (and their reduced forms NADH and NADPH) are stable metabolites that can diffuse freely in cellular compartments (flavin groups, on the contrary, are permanently associated to their apoprotein). The redox potential of the $NAD^+$/NADH couple (as well as of the $NADP^+$/NADPH couple) is sufficiently negative ($E'^o$ = -0.32 V at pH = 7.0) to allow NADH and NADPH to act as reductants in most of the metabolic reductions occurring in a cell. $NAD^+$ and $NADP^+$, on the other hand, are reduced during the catabolic oxidation of sugars, fatty acids and amino acids. The catabolic (oxidative) and anabolic (reductive) reactions are therefore coupled by functional cycles of these coenzymes that cyclically are utilized as electron acceptors or donors in the two sets of reactions. This energetic role of $NAD^+$ and $NADP^+$ takes place without any involvement of ATP.

$NAD^+$/NADH is however also an essential molecule for the respiratory metabolism. $NAD^+$ acts as a general electron acceptor for most catabolic oxidations (during glycolysis, but also in the $\beta$-oxidation of fatty acids and particularly in the tricarboxylic acid cycle), and funnels reducing equivalents into aerobic respiration. The oxidation of NADH by the respiratory chains is by far the most intense redox process in an aerobic cell, and by far the most important source of energy, being the respiratory reactions coupled to the synthesis of ATP in the process of oxidative phosphorylation.

On the other hand, $NADP^+$/NADPH play a key role in photosynthetic metabolism of higher plants. $NADP^+$ is generally the final electron acceptor of the photosynthetic chains in higher plant chloroplasts. The photosynthetic production of NADPH, therefore, represents a major mechanism by which the energy of light is stored in photosynthetic cells, and photosynthetically produced NADPH is the principal source of reductants in a photosynthetic cell. The production of ATP in photophosphorylation is the second form of metabolic energy storage for a photosynthetic organism.

Besides the process of oxidative and photosynthetic phosphorylation, oxidoreduction reactions can be directly coupled to the synthesis of ATP in a single catalytic reaction. This is the case of the substrate level phosphorylation occurring during the glycolytic degradation of sugars to pyruvate, or, less frequently, in specialized fermentative reactions involving acids, such as acetate, propionate, butyrate or lactate. In fermentative microorganisms these mechanisms of synthesis are often the only metabolic source of ATP. Prototype of these processes is the extensively studied reaction of glyceraldehyde-3-phosphate dehydrogenase, that, in association with the enzyme phosphoglycerate kinase, produces one molecule of ATP coupled to the oxidation of glyceraldehyde-3 phosphate to 3-phosphoglycerate:

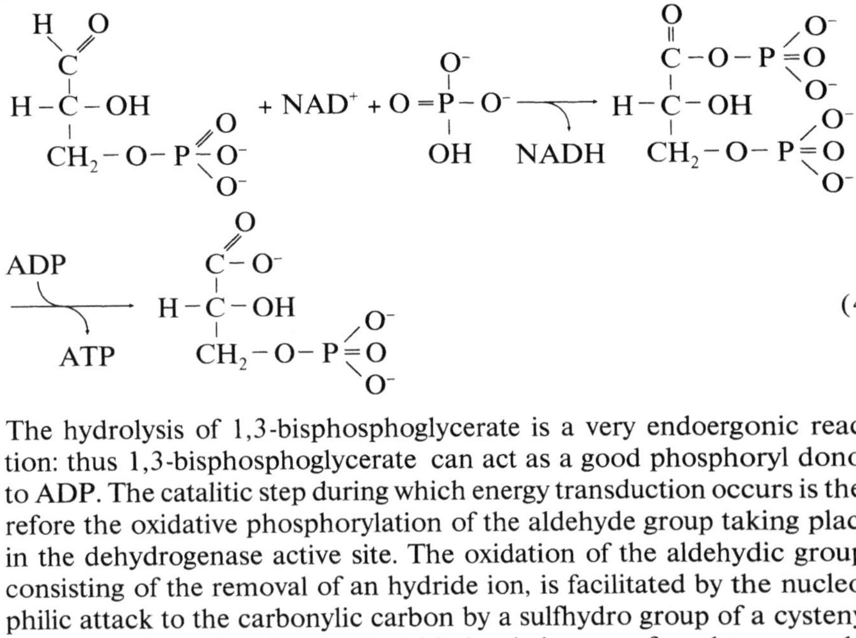


(4)

The hydrolysis of 1,3-bisphosphoglycerate is a very endoergonic reaction: thus 1,3-bisphosphoglycerate can act as a good phosphoryl donor to ADP. The catalitic step during which energy transduction occurs is therefore the oxidative phosphorylation of the aldehyde group taking place in the dehydrogenase active site. The oxidation of the aldehydic group, consisting of the removal of an hydride ion, is facilitated by the nucleophilic attack to the carbonylic carbon by a sulfhydro group of a cystenyl residue in the active site, the hydride ion being transferred to a nearby $NAD^+$ molecule (Fig. 2). The resulting acyl tioester is subsequently decomposed by a phosphate anion yielding the acylphosphate derivative (i.e. 1,3-bisphosphoglycerate). In this case, therefore, the inner mechanism of catalysis is producing directly an activated product capable of acting as direct precursor of ATP. The process of energy transduction results from the direct coupling of oxidation and condensation reactions in a single catalytic site. By this process the free energy decrease corresponding to the oxidation by $NAD^+$ of an aldehydic group to a carboxylic group is

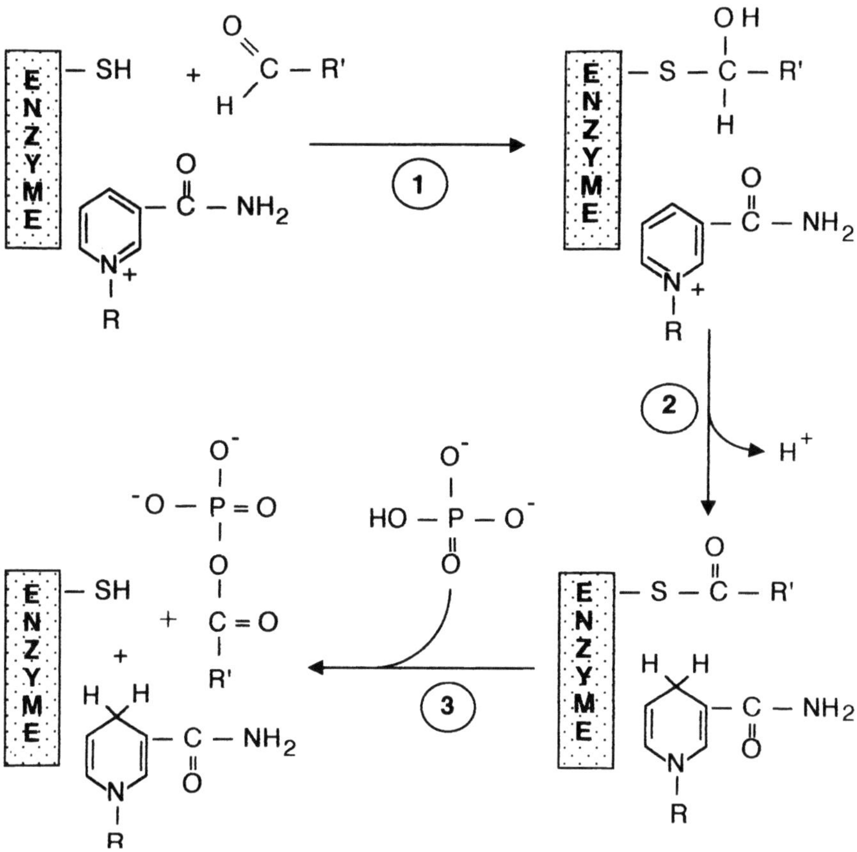


Fig. 2. Catalytic mechanism of D-glyceraldehyde-3-phosphate dehydrogenase. A cysteinyl residue in the active site plays a key role in the catalytic mechanism of this enzyme. The aldehyde residue of the substrate is attacked by this group yielding a semiacetalic derivative (step ①), thereby facilitating the oxidation of the aldehydic group by a $NAD^+$ molecule present in the active site (removal of a $H^-$, step ②). The thioester derivative thus produced is a good acyl-transferring intermediate and donate the acyl to a phosphate ion (step ③), yielding 1,3 bisphosphoglycerate. NADH in the active site is exchanged with free $NAD^+$.

almost entirely conserved in an anhydridic bond and ultimately utilized for the production of ATP.

For many years the reaction of glyceraldehyde-3-phosphate dehydrogenase was considered as a model for the mechanisms of oxidative phosphorylation. This assumption was demonstrated to be wrong, since the synthesis of ATP in respiration and photosynthesis involves more than one enzyme and occurs across the inner membrane of mitochondria or the thylakoid membrane of chloroplasts. In order to deal with the problem of photosynthetic or respiratory ATP synthesis, therefore, the discussion should be extended from a monophasic, single compartment system to multiphasic systems and to cell compartmentation.

## 2. Cellular compartmentation and bioenergetics

### *2.1. Multicompartment systems*

In the organization of eukaryotic cells (and, to a lesser extent, also in prokaryots) the intracellular volume is subdivided into several different compartments separated by biological membranes. This cellular compartmentation brings decisive advantages to the cell, in that it allows a metabolic specialization of the different compartments and the possibility of controlling at the same time processes with conflicting regulatory needs (typically catabolic versus anabolic processes).

In terms of thermodynamics the appropriate function of state for dealing with this situation is still the free enthalpy. In the evaluation of the $\Delta G$ of the process, however, the translocation of molecular species between different compartments must be taken into account. A very general relation for $\Delta G$ can accordingly be written as:

$$\Delta G = -RT \ln K + RT \ln \frac{\prod a_p^{\nu_p}}{\prod a_s^{\nu_s}} + \sum_j \left( \sum_s \nu_s \tilde{\mu}_{s,j} + \sum_p \nu_p \tilde{\mu}_{p,j} \right) \quad (5)$$

where $\tilde{\mu}_{s,j}$ and $\tilde{\mu}_{p,j}$ are the electrochemical potentials of the species in the j-phase, and the summation includes all translocation processes coupled to the process taking place for all reagents (s) and all products (p) and for all the compartments (j).

If, for simplicity, the process is restricted only to the translocation between two compartments of one single molecular species, Eqn. 5 takes the form:

$$\Delta G_{i(\text{transl})} = \nu_i (\tilde{\mu}_{i,2} - \tilde{\mu}_{i,1}) \quad (6)$$

and expresses the free energy change corresponding to the translocation of $\nu_i$ molecules of species i from compartment 1 to compartment 2.

Any translocation process occuring spontaneously must correspond to a net decrease of $\Delta G_{i(\text{transl})}$ and, therefore, to the condition $\tilde{\mu}_{i,2} \leq \tilde{\mu}_{i,1}$. If the definition of electrochemical potential is written explicitly this condition is

$$RT \ln a_{i,2} + z_i \mathcal{F}\phi_2 \leq RT \ln a_{i,1} + z_i \mathcal{F}\phi_1 \quad (7)$$

where $z_i$ is the net electric charge of the i-th species, $\mathcal{F}$ the Faraday constant and $\phi_1$ and $\phi_2$ the electrostatic potential in compartments 1 and 2 respectively. Equation 7 also defines the equilibrium relation for the distribution of i between two compartments. Imposing $\Delta G_{i(\text{transl})} = 0$ in facts yields, after rearrangement:

$$RT \ln \frac{a_{i,2}}{a_{i,1}} = z_i \mathcal{F} (\phi_1 - \phi_2) \quad (8)$$

The activity ratio of an electrically charged molecules in two phases at equilibrium reflects therefore also the electrostatic potential difference between them, being a cation spontaneously accumulated in a negatively charged compartment and an anion in a positively charged one. For a univalent cation, for example, a difference in electrostatic potential of 59 mV will induce a tenfold increase in activity in the negative compartment; likewise for divalent cation the same potential difference will equilibrate with a 100 fold activity ratio. The difference in electrostatic potential between two compartments is therefore a very effective controller of the equilibrium distribution of ions and plays a major role in cellular bioenergetics and physiology. In biological systems, in fact, different cellular compartments are often characterized by different electrical potential. The biological membrane surrounding them is the major dielectric barrier across which the electric potential difference can be measured; for this reason the electric potential difference between two compartment is frequently termed "membrane potential": it is in fact assumed that within any cellular compartment a substantial electric equilibrium exists, a reasonable assumption given the relatively high ionic concentration of biological phases (>100 mM of uni-univalent salts) and hence their high conductivity.

Any imbalance of electrochemical potential of a chemical species (e.g. an ion) across a biomembrane corresponds to a disequilibrium situation and offers a potential source of energy to produce work. If the membrane separating these compartments possesses suitable translocation apparatuses, i.e. contains a specific permease, the difference in electrochemical potential will promote an ionic flux, corresponding to an electric (ionic) current across the membrane (a form of positive electric work) and, also, to mass transport from one compartment to the other. Always a substantial amount of energy is dissipated as heat since these ionic currents must be intense in order to produce any physiological effect and the translocation process assumes the character of an irreversible process.

Often in cellular bioenergetics ion fluxes are coupled to other metabolic processes, so that part of the free energy decrease of translocation can be utilized for driving an endoergonic process. This is the case, for example, of secondary active transport system, in which a metabolite or an ion is accumulated in one compartment (i.e. is translocated in a phase where its electrochemical potential is higher than in the initial one) at expense of the negative free energy change for the translocation of a driving ion (generally $H^+$ or $Na^+$). The coupling between these two fluxes will be effected by a single transmembrane enzyme requiring both the translocated molecule and the driving ion for the catalysis of the transport process. These phenomena will be discussed further in Sections 4 and 5.

In many cases, however, ion fluxes are also coupled to non vectorial processes, mainly, but not only, the hydrolysis of ATP. In these cases the coupling mechanism involves a vectorial process (the ion flux) and a scalar one (the metabolic reaction): the coupling therefore necessarily requires a vectorial coupling mechanism (Curie-Prigogine principle) or, in more biological terms, an enzyme arranged asymmetrically in the membrane and catalyzing a coupled reaction endowed with a precise directionality of the fluxes and precise compartmentation of substrate and products. A catalytic process possessing these characteristic is defined as "vectorial catalysis".

The bioenergetic role of such vectorial membrane enzymes is that of utilizing metabolic energy for generating transmembrane differences of electrochemical potential of ions. The nature of the ion concerned is various ($Na^+$, $Ca^{2+}$, $H^+$ or $K^+$) and their physiological roles different. The driving force for the ionic fluxes is mainly ATP hydrolysis. A large number of membrane associated ATPase has been characterized (the so-called ion-motive ATPases) which promote the translocation of one ion (or sometimes two ions) coupled to the hydrolysis of ATP.

In the process of oxidative and photosynthetic phosphorylation, however, flows of $H^+$ are coupled to exergonic redox reactions catalyzed by the redox enzyme complexes present in the inner mitochondrial membrane, in the plasma membrane of prokaryots, or in the thylakoid membrane of chloroplasts and *Cyanobacteria*. In these cases the redox systems generate large electrochemical potential differences of protons across these membranes. A $H^+$-translocating ATPase present in the same membrane operates in reverse and catalyzes the synthesis of ATP from ADP and phosphate, exploiting the highly exergonic translocation of protons for driving the process. Only occasionally this type of ATPase acts as generator of $\Delta\tilde{\mu}_{H^+}$ the most prominent examples being found among fermentative microorganisms, in which a plasma membrane $H^+$-ATPase is usually the only generator of $\Delta\tilde{\mu}_{H^+}$, driving secondary active transport of other ions and of metabolites.

In oxidative and photosynthetic phosphorylation therefore the generation and utilization of protonic electrochemical potential differences embodies the basic mechanism by which the exoergonic redox processes of respiration or photosynthesis are coupled to the synthesis of ATP. In the following sections of this chapter some other aspects of these processes will be dicussed in general terms, the individual components in mitochondria, chloroplasts and bacteria being the topic of subsequent chapters of this book.

As a final example of coupling between a scalar reaction and the flux of an ion, let us consider a generic ion-motive ATPase translocating a divalent cation between two compartments: an example of such an enzyme could be the $Ca^{2+}$-motive ATPase operating across the plasma membrane of mammalian cells. Let us also assume that a membrane potential of 60 mV, negative inside, is established across the membrane and that under the cytoplasmic physiological conditions the $\Delta G$ for the hydrolysis of ATP ($\Delta G_{ATP}$) has a value of $-48.1$ kJ mole$^{-1}$.

The reaction for the plasma membrane $Ca^{2+}$-ATPase is

$$ATP^{4-} + H_2O + 2Ca^{2+}_{in} \rightarrow ADP^{3-} + HPO_4^{2-} + H^+ + 2Ca^{2+}_{out} \qquad (9)$$

and the corresponding $\Delta G$ is

$$\Delta G = \Delta G_{ATP} + 2(\tilde{\mu}_{Ca^{2+},out} - \tilde{\mu}_{Ca^{2+},in}) = \Delta G_{ATP} + RT \ln \frac{[Ca^{2+}]^2_{out}}{[Ca^{2+}]^2_{in}} + 2\mathcal{F}(\phi_{out} - \phi_{in}) \qquad (10a)$$

and substituting the numerical values

$$\Delta G = -48.1 + 2RT \ln \frac{[Ca^{2+}]_{out}}{[Ca^{2+}]_{in}} + 2 \times 96.5 \times 0.06 \ (\text{kJmole}^{-1}) \qquad (10b)$$

and at equilibrium

$$2RT \ln \frac{[Ca^{2+}]_{out}}{[Ca^{2+}]_{in}} = 48.1 - 2 \times 96.5 \times 0.06 = 36.5 \ \text{kJmole}^{-1} \qquad (10c)$$

or

$$\frac{[Ca^{2+}]_{out}}{[Ca^{2+}]_{in}} = 10^{3.2} \qquad (10d)$$

The calculated concentration ratio is of course a theoretical maximum never attained *in vivo*. The coupling of ATP hydrolysis to the excretion of two $Ca^{2+}$ can assure the maintenance of a concentration of this ion inside the cytoplasm (about $10^{-7}$ M), much lower than in the extracellular space, even against a membrane potential unfavorable to the excretion of a cation. It is in fact with such a mechanism, together with a $Na^+/Ca^{2+}$ exchange permease, that the homeostasis of $Ca^{2+}$ in cells is attained.

In this example the distribution ration of $Ca^{2+}$ is influenced by the presence of a difference in electric potential across the membrane. The quantitation of this contribution amounts to $-11.6$ kJ mole$^{-1}$ (for $\phi_{out} - \phi_{in} = 0.06$ V), corresponding to $[Ca^{2+}]_{out}/[Ca^{2+}]_{in} = 10^{-0.98}$. As stated above the value of $\Delta\phi$ negative inside is opposed to the exit of $Ca^{2+}$ from the cell and increases the energy demand for $Ca^{2+}$ excretion. It is noteworthy that the presence of a membrane potential is not due to the action of the Ca ATPase, but mainly to the Na, K ATPase, or, more generally, to an imbalance in the distribution of cations across the plasma membrane. The interaction (coupling) between the two ATPases (Ca- and Na, K-) occurs irrespectively of the nature of the transported ion: the interaction is purely electrostatic and requires a net flux of electric charges coupled to each individual process (i.e. electrogenic reactions) (see section 2.3. for further comments).

## 2.2. *Vectorial catalysis*

The coupling of exergonic reactions (redox reactions or group transfer reactions) to the translocation of ions requires that the catalytic mechanism is oriented with respect to the membrane in such a way as to impose on the coupled ion flux a precise directionality (this is generally true for every transport process coupling a scalar reaction to mass transport). This can be accomplished by means of asymmetric structures of intrinsic enzymes and their insertion into the membrane with a univocal orientation; it can also be obtained through the location of extrinsic proteins on one of the faces of the membrane, or, in the most primitive cases of bacterial systems, only with the location of soluble enzymes in two different compart-

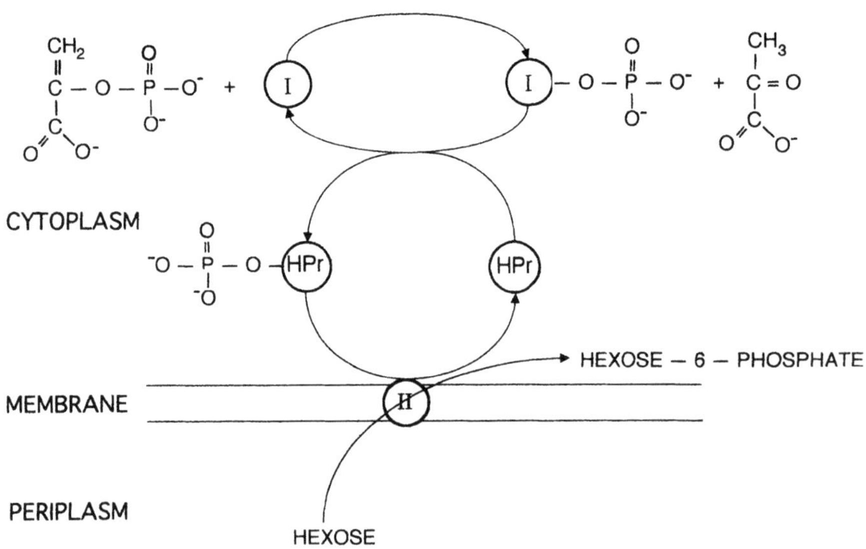


Fig. 3. Scheme of phosphoenolpyruvate-dependent sugar transport in bacteria. A small polipeptide (HPr) acts as phophosyl transfer itermediate in this transport system: this protein accepts a phosphate on a hystidinyl residue from phosphoenolpyruvate, the transphosphorylation reaction being catalyzed by enzyme I. In turn phosphorylated HPr acts as phosphoryl donor to several types of sugars, through the catalytic action of a number of transphophorylases (named Enzyme II). The transphosphorylases couple the phosphorylation step to the translocation of the sugar across the bacterial plasma membrane; in some systems for this step a third enzyme (enzyme III) is necessary. The accessibility to the active site of enzyme II is from the periplasm for sugars and from the cytoplasm for HPr-phosphate; the release of the hexose-6-phosphate is towards the cytoplasm. In this way the coupling of sugar transport and phosphorylation is achieved.

ments on both sides of the membrane (typically in the bacterium cytoplasm and periplasm) and their coupling through a common transport system. Typical examples of such systems will be schematically summarized in the following.

A first example is offered by some sugar transport systems present in *E. coli* known as PTS (Phosphoenolpyruvate-dependent Transport System) (Fig. 3); these systems couple the translocation of various hexoses across the plasma membrane to their phosphorylation on position 6, yielding the corresponding hexose-6-phosphate. The substrate sugar is assumed directly on the periplasmic face of the membrane, while the phosphorylated product is released into the cytoplasm. Both phosphorylation and translocation are catalyzed by a single transmembrane protein, called enzyme II, and the vectoriality of the reaction is conferred by the side of accessibility of substrate and product to the catalytic site. The phosphate donor to this class of translocators is a specialized cytoplasmic protein (known as HPr, Heat stable Protein), that receives the phosphate group on a hystidinyl residue from phosphoenolpyruvate. The scalar raction and the expenditure of metabolic energy (i.e. the consumption of phosphoenolpyruvate) occur entirely in the cytoplasm.

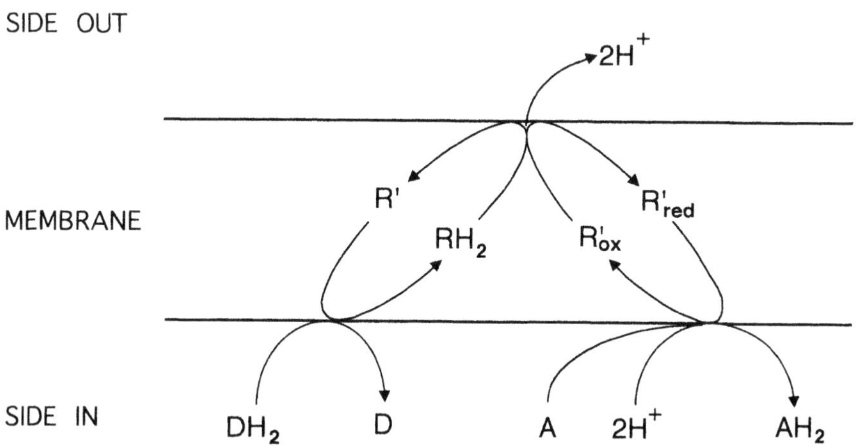


Fig. 4. Scheme of a complete protonic loop. In the first arm of the loop protons are transferred during the dehydrogenation of $DH_2$ to a protonable redox center R (e.g. a flavin group) that subsequently is oxidized on the opposite face of the membrane. In this way protons are transferred electroneutrally across the membrane (phase *out*). During the oxidation of $RH_2$ by a non-protonable redox center R' (e.g. an iron-sulphur center or a cytochrome), protons are released into the outside compartment. Negative electric charges (electrons) are transferred back to the inside by the reduced redox center R', that is reoxidized by the oxidant A in the same compartment where D and $DH_2$ is located (phase *in*). The protons for the reduction of A are taken up in this compartment. By this mechanism protons (two in this scheme) are translocated from phase *in* to phase *out* together with their electric charges; charge separation occurs in the second arm of the loop (the electrogenic arm catalyzed by center R') in the form of transfer of two electrons from phase *out* to phase *in*.

Proton translocation is probably the most basic transport event in cells. It is at the basis of the mechanism of oxidative and photosynthetic phosphorylation, it is associated with secondary active transport of many ions and metabolites and it is central to important cellular controls such as pH homeostasis and cellular turgor regulation. It is natural therefore that many different mechanisms were developped during evolution for coupling proton flows to exergonic reactions. One of the simplest is the so-called proton loop, first suggested by Peter Mitchell as a general mechanism for proton translocation coupled to respiratory redox reactions (Fig. 4). This mechanism envisages the transfer of proton from a hydrogen donor (reductant) to an electron acceptor (oxidant) through a protonable redox intermediate (e.g. a flavin group or a quinone), which is reduced and oxidized on opposite faces of the membrane. The loop is completed by the transfer of electrons in the opposite direction across the membrane. The mechanism is therefore based on the different accessibility of protonable redox centers to protons, so that these are assumed from one side of the membrane upon reduction and released on the opposite side upon oxidation. The mechanism does not require any actual movement of redox group but only the existence of proton-conductive gated structures connecting the reaction site alternatively to the two sides of the membrane.

Although no single membrane enzyme has been clearly demonstrated to operate according to this principle, the proton loop has been proved a suitable model for describing the operation of complex redox chains in which ubiquinone acts as a protonable and diffusable redox intermediate. This molecule can be reduced and protonated in sites (reductase sites) adjacent to one face of the membrane (e.g. close to the matrix face of the inner mitochondrial membrane or the cytoplasmic face in bacteria) and oxidized and deprotonized on sites (the oxidase sites) adjacent to the opposite face. The two redox steps may be or may not be catalyzed by different enzymes, provided that their orientation in the membrane is maintained, the protonable and diffusable ubiquinol providing the electrically neutral carrier for proton translocation. Probably the best known example of such a mechanism is offered by the photosynthetic chain of purple bacteria (Fig. 5); in this system quinone is reduced by the photosynthetic reaction center complex that also operates as a photoactivated electron translocator across the membrane. A quinol oxidase site is located in a different enzyme complex, the $bc_1$ complex, which also contains a second quinone reductase site and a transmembrane electron transfer chain formed by a diheme cytochrome b. The oxidase and reductase sites of both complexes are located on opposite faces of the bacterial plasma membrane so that the chain can operate cyclically as an effective two-loop system.

Simpler, more primitive and incomplete loop mechanisms have been proposed for other bacterial redox chains. In many cases only electrons are transferred across the membrane, often by multiheme intrinsic cy-

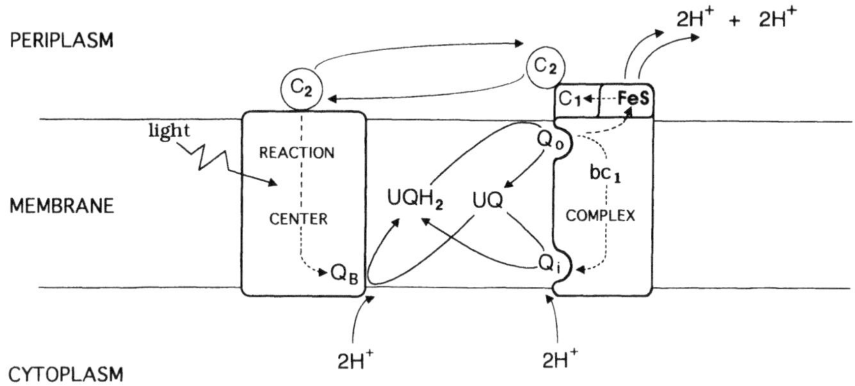


Fig. 5. Scheme of the photosynthetic electron transfer in the bacterium *Rhodobacter capsulatus* (and other *Rhodospirillaceae*). The photosynthetic reaction center oxidizes cytochrome $c_2$ and reduces ubiquinone at the site $Q_B$. This reaction occurs across the membrane so that it forms an electrogenic arm of a protonic loop. The diffuson of ubiquinol to the oxidase ($Q_o$) site of the $bc_1$ complex, located on the opposite site of the membrane with respect to $Q_B$, provides the proton translocating arm of the loop. A second electrogenic step transfers one electron per $UQH_2$ oxidized to the reductase site ($Q_i$) of the $bc_1$ complex, where ubiquinone is reduced to ubiquinol. Also this second proton loop is completed by the diffusion of ubiquinol to the $Q_o$ site. Electrons are supplied to the reaction center by cytochrome $c_2$ that is reduced by the iron sulphur protein (FeS) and cytochrome $c_1$ ($c_1$) of the $bc_1$ complex. For each electron transferred cyclically, two protons are consumed in the cytoplasm (one at the $Q_B$ site of the reaction center and one at the $Q_i$ site of the $bc_1$ complex) and two protons are released in the periplasm (at the $Q_o$ site). For simplicity the redox center within the two complexes (bacteriochlorophyll dimer, bacteriopheophitin and ubiquinones in the reaction center, an b-hemes in the $bc_1$ complex) are not indicated in the figure.

tochromes, and protons are released in oxidation reactions or consumed in reduction reactions on opposite sides of the membrane. Protons can be released from a membrane bound redox center (typically a flavin) or directly by the oxidizable substrate (e.g. $CH_3OH$ oxidized to HCHO and 2 $H^+$, $H_2$ oxidized to 2 $H^+$, etc,); protons can also arise from water when water oxygen is involved in the oxidation reaction (e.g. $NO_2^-$ + $H_2O$ oxidized to $NO_3^-$ + $2H^+$, $SO_3^{2-}$+ oxidized to $SO_4^{2-}$+ $2H^+$, etc.). Analogously proton consumption upon reduction can derive from the protonation of a reduced redox center, or from the protonation of an electron acceptor (e.g. $O_2$ + $4H^+$ reduced to $2H_2O$, $NO^{3-}$ + $2H^+$ reduced to $NO^{2-}$ + $H_2O$, fumarate + $2H^+$ reduced to succinate, etc.) (see, for example, Fig. 6). In all these cases, all that is needed to obtain a rudimental proton "semi loop" scheme is a redox connection across the membrane by an electron carrier and two extrinsic, even soluble redox enzymes located in the appropriate compartment and receiving or yielding electrons to this carrier. Many such examples have been described or proposed in heterotrophic or autotrophic bacteria.

In both the complete or semiloop mechanisms the stoichiometry between the protons translocated and the electron transferred is fixed, and strictly related to the mechanism of the reaction itself (nearly always the

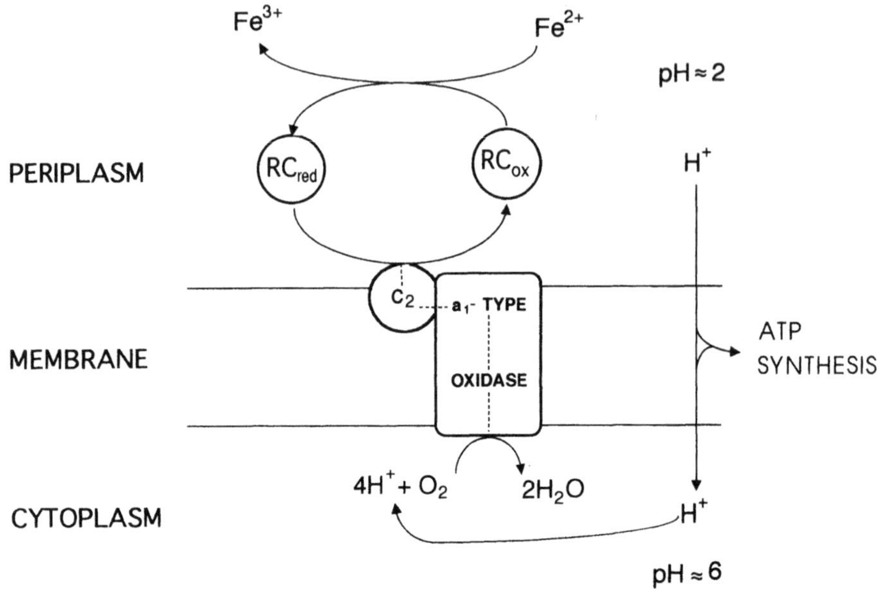


Fig. 6. Electron transfer scheme in *Thiobacillus ferrooxidans.* The electron transfer system of this bacterium forms a rudimental semiloop mechanism. This organism lives in very acidic waters (pH ≈ 2) rich in ferrous ions. The difference in pH between the cytoplasm (pH ≈ 6) and the outside medium provides most of the energy for the chemiosmotic synthesis of ATP. The consumption of the protons entering the cytoplasm for the synthesis of ATP (and also through proton symporters and proton leaks) is obtained by means of the reduction of $O_2$ to water on the cytoplasmic side of the plasma membrane. Electrons for this reduction are supplied by the external $Fe^{2+}$ ions, oxidized by the periplasmic protein rusticyanine (RC) and by a transmembrane electron transfer system formed by a membrane bound cytochrome ($c_2$) and an $a_1$-type oxidase. The transmembrane electron transfer catalyzed by the oxidase forms an electrogenic arm of a proton semiloop, and polarizes the membrane electrically, negative inside. This membrane potential contributes to the $\Delta\tilde{\mu}_{H^+}$ for ATP synthesis.

$H^+/e^-$ ratio is = 1). This rigidity of the loop model has often been criticized as unsuitable for the subtleties of physiological controls and to the variability of the energy balances in different bioenergetics systems. A fixed $H^+/e^-$ ratio equal to 1 in primary proton translocation systems appears also to conflict with the $H^+/e^-$ ratios experimentally measured (particularly in eukaryotic respiratory systems), where $H^+/e^-$ as high as 2 have been claimed for a single proton translocating complex.

The concept of proton redox pump has been accordingly introduced, in which the mechanism of coupling between redox reactions and proton translocation is less direct, being mediated by the shift in pK induced by the redox state of one redox center on nearby protonable group(s). Also in this case it is necessary to have a precise orientation of the group across the membrane and to control the proton accessibility to the protonable group(s) from one or the opposite side of the membrane, in phase with the protonation and deprotonation steps of the reaction; the model however does not pose any restriction on the number of protonable groups, whose pK's may be modulated, nor to the actual values of these pK's. The model can accordingly be applied to explain any $H^+/e^-$ ratio observed experimentally, to any pH value on both sides of the membrane and to energy balance proper to any biologycal system. The model is, in other words, very flexible, but also, unfortunately more undefined in its molecular mechanisms than the loop model.

For many complexes of the mitochondrial respiratory chain a proton pump mechanism has been proposed for explaining the experimentally determined $H^+/e^-$ ratios. For none of these systems, however, our knowledge of the protein structure is so advanced to allow the demonstration of this indirect proton pump mechanism on a molecular basis.

Notwithstanding these difficulties the proton pump model has received much support from the results of studies on bacteriorhodopsin, a small membrane protein isolated from the halophilic bacterium *Halobacterium salinarium* (formerly *H. halobium*), which pumps protons across the membrane utilizing directly the energy of light. The structure of this protein has been resolved at near atomic resolution by electron diffraction of bidimensional crystals and its function studied in great detail using site-specific mutants. In addition, the catalytic cycle is strictly associated with the spectral characteristics of a chromophore (a retinal molecule), covalently attached to the protein, that is brought to the excited state by a green photon ($\lambda$ = 570 nm) and undergoes specific changes in its absorbtion spectra in relation with the different steps of the proton translocating process. Kinetic studies with very good time resolution are therefore possible.

The structure of bacteriorhodopsin includes seven transmembrane helices, five of which form a hydrophilic pore across the membrane. The functional retinal group is positioned across this pore, approximately at the center of the membrane, thus dividing the pore itself in a cytoplasmic (CP) channel and in a extracellular (EC) channel. Retinal is attached to residue Lys216 of the protein through a Schiff base, that is normally protonated and thus positively charged. The hydrophilic wall of the pore contains protonable groups, among which the most important functionally are Asp85 and Arg82 in the EC channel and Asp96 in the CP channel (Fig. 7).

As discussed elsewhere, models for the vectorial catalysis of proton pumping by bacteriorhodopsin have been based on this structural information. In brief, the model envisages these sequential steps:

1) Isomerization of the retinal upon photoexcitation: the isomerization involves the 13,14 *trans* double bond which is converted into a 13,14 *cis* bond and compensatory rotations of the 14,15 single bond and of one or more metilene bonds of Lys216. The isomerizaton causes a repositioning of the protonated Schiff base.
2) Transfer of a proton from the Schiff base to Asp85 in the EC channel. This amino acid residue is normally deprotonated ($pK < 3.5$) and acts

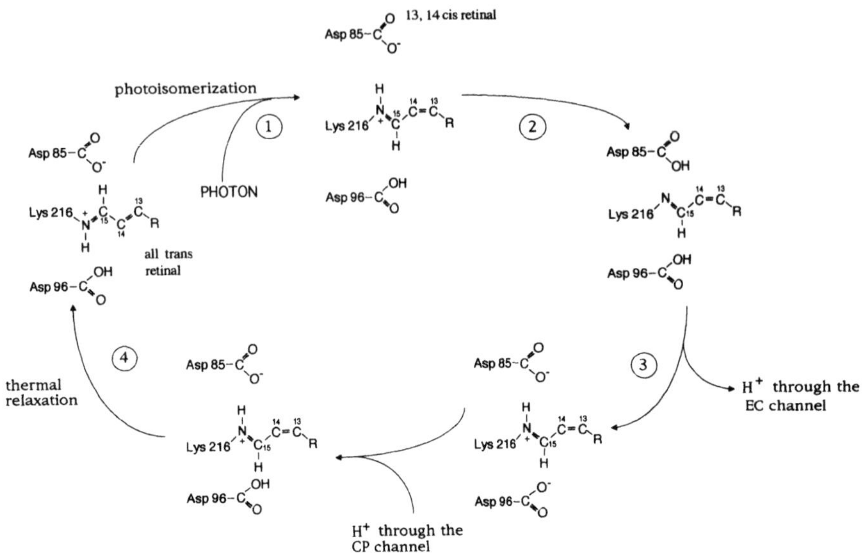


Fig. 7. Catalytic cycle of bacteriorhodopsin. This scheme outlines the main features of the catalytic cycle of bacteriorhodopsin taking into consideration only the functional role of retinal and of its proton donor and acceptor (Asp96 and Asp85 respectively). Many other amino-acid residues involved less directly in the catalytic cycle are not considered here for the sake of simplicity (see text for further details).

Retinal, attached through a Schiff base to Lys216, undergoes photoisomerization from an all *trans* to 13,14 *cis* configuration (Step ①) This causes repositioning of the protonated Schiff base and a subsequent proton transfer to Asp85 (Step ②), placed in a proton conducting channel facing the extracellular space (EC channel). Reprotonation of the Schiff base occurs from Asp96, located in a proton conductive channel facing the cytoplasm (CP channel), after a change in pK of the base, coupled to rearrangement of the protein molecule (Step ③). Thermal relaxation of the 13,14 *cis* retinal to an all *trans* conformer follows (Step ④). The whole catalytic cycle couples the translocation of one proton to the excitation of retinal by one photon.

as a proton transferring catalyst to Arg82 placed close to the mouth of the EC channel.

3) Transition of the deprotonated Schiff base from a state at low pK to one at high pK (> 10) so that it can accept a proton from Asp96, placed in the CP channel. The nature of this change of state is still obscure. It might induce further isomerizations of the retinal (with repositioning the Schiff base) or even of part of the protein structure in the retinal pocket. A well established consequence of the change in state is that the protonation route of the Schiff base is now obligatorily through the CP channel.
4) After reprotonation the retinal undergoes a thermal relaxation to the 13,14 *trans* isomer, becoming ready for a new catalytic cycle.

This mechanism of catalysis by bacteriorhodopsin conforms therefore to the general concept of a proton pump, although in essence it is not yet clear how the conformation of the apoprotein controls the proton conductivity from the cytoplasm and the extracellular space and thus confers directionality to the proton flux driven by the energy of light. Similar mechanisms may very well exist also in redox driven proton pumps, although not yet demonstrated clearly.

The concept of redox pump, consisting of the modulation of the proton binding energy due to the interaction with an adjacent group changing its redox state, can be easily generalized to other ligands and to other type of changes in state. Among bacterial rhodopsins another retinal-containing membrane protein has been isolated that is capable of utilizing the energy of light for the direct translocaton of $Cl^-$ (this protein is called halorhodopsin). The structure and the properties of this protein are surprisingly similar to those of bacteriorhodopsin. The change in ion specificity is certainly related to some specific substitution, particularly in the ion channel structure, and precisely to the lack of proton acceptor and donor groups (Asp85 and Arg96 are missing and substituted for by the cationic residues Arg200 in the CP channel and Arg108 in the EC channel). In the absence of proton acceptor and donors, the Schiff base remains permanently protonated and positively charged, while still undergoing isomerization photoreactions. It is believed that the gating function of retinal is still maintained, the protonation/deprotonation cycle being replaced by the reversible formation of an ionic pair between the positive Schiff base and the translocated Cl ion. Thus a very similar catalytic cycle can drive the translocation of an anion.

Somewhat analogously (as better explained in another section of this chapter and in Chapter 12) the proposed coupling models for some ion-motive ATPases assume that binding energy changes, as well as accessibility of the binding sites from one or the other face of the membrane, are associated with the phosphorylaton/dephosphorylation cycle of a specific amino acid residue of the apoprotein (an aspartate). Also this mechanism, therefore, conforms to the pump model being the driving event for the ion movement, rather than a change in electrical charge related to oxidation/reduction cycle of a redox center, a group transfer reaction (again associated to large changes in the local electric charge brought about by the addition of an anionic phosphate group) (Fig. 8).

The pump mechanisms are therefore thermodynamically sound and quite flexible as far as the nature of the transferred substrate and the nature of the driving reaction are concerned. It has been convincingly demonstrated in the case of bacteriorhodopsin and it offers the most plausible model for many other ion-translocating enzymes. Its molecular mechanism in every single case is however far from being elucidated.

It is probably by a similar mechanism that the $H^+$-ATPases of photosynthetic or oxidative phosphorylation operate. These enzymes are considerably more complex in structure than a cation-motive ATPase, being formed by a minimum of eight proteins, some of them present in multiple copies (overall at least 20 polipeptides). These form two sectors

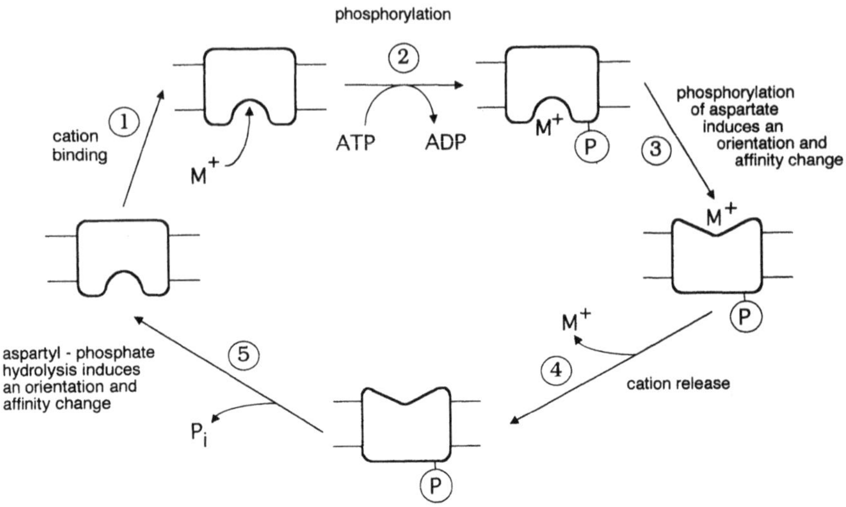


Fig. 8. Catalytic cycle of a generic ion-motive ATPase of P-type. This scheme applies to a generic P-type ion-motive ATPase without specifying the nature and stoichiometry of the ion(s) translocated. The model is based on a conformational change of the enzyme leading to a change in affinity for the ligated ion and reorientation of the ion-binding site. Phosphorylation of an aspartyl residue of the protein follows the binding of the ions on a site facing one side of the membrane (generally the cytoplasmic face) (Steps ① and ②). Reorientation of this site and a decrease in its affinity causes the release of the ion on the opposite side (Steps ③ and ④). The dephosphorylation steps (⑤) and the ensuing reorientation and change in affinity of the site is controlled by the release of the ion.

In many P-type ATPases a second ion is bound and transferred during steps ④ and ⑤. In this case the ATPase drives an active ion exchange, rather than a unidirectional transport of ions. The mechanism is intrinsically electrogenic, also in ion-exchanging ATPases provided that the stoichiometry of the ions translocated in one and in the opposite direction is not equal (this is the case, for example, of the Na,K ATPase that transfers 3 $Na^+$ and $2K^+$-per ATP hydrolyzed).

of the complex, an extrinsic one (called $F_1$), catalyzing the actual synthesis of ATP, and intrinsic one (called $F_0$), that may form a proton conducting device across the membrane. As a whole the complex contains three binding sites for ATP (or ADP), acting catalytically, and another three adenyl nucleotide binding sites believed to be non-catalytic. Although occasionally this complex can operate as a proton translocating ATPase, its main physiological function consists in ATP synthesis driven by a protonic flux. It is beyond the scope of this chapter to discuss in detail the function of such F-type ATPases (see Chapter 12). The current models for the coupling between proton translocation and ATP synthesis within this complex, conforms however again to the general mechanism of a proton pump, in that ATP binding and its decomposition into ADP and phosphate, still bound to the catalytic site, are believed to be related to the protonation of specific residues, while the release of nucleotides (either ATP or ADP) out of the site is proposed to occur as a consequence of the deprotonation of other specific groups. When considered in the synthetic direction, therefore, the protonation of groups by protons derived from the high proton-potential compartments would cause an increase in binding energy for ADP and phosphate at the catalytic site and their spontaneous condensation into tightly bound ATP. The subsequent deprotonation of the same (or of other) groups would decrease the affinity for nucleotides, inducing the release of ATP. The analogy of such model to a general pump mechanism is quite clear. What makes this class of enzymes rather unique is the presence of three catalytic sites, suggesting that cooperativity between subunits may be an important aspect of the mechanism. The trimeric structure of the catalytic portion of $F_1$ does not however fit with the presence of a single copy of three other smaller peptides in the same $F_1$ and, also, does not correspond to the structure of $F_0$, apparently lacking any tripartite symmetry. The $F_0$ portion, being the only transmembrane structure of the complex, must be the sensor of the difference in proton electrochemical potential across the membrane (i.e. must be the structure confering vectoriality to the enzyme catalysis). The entire

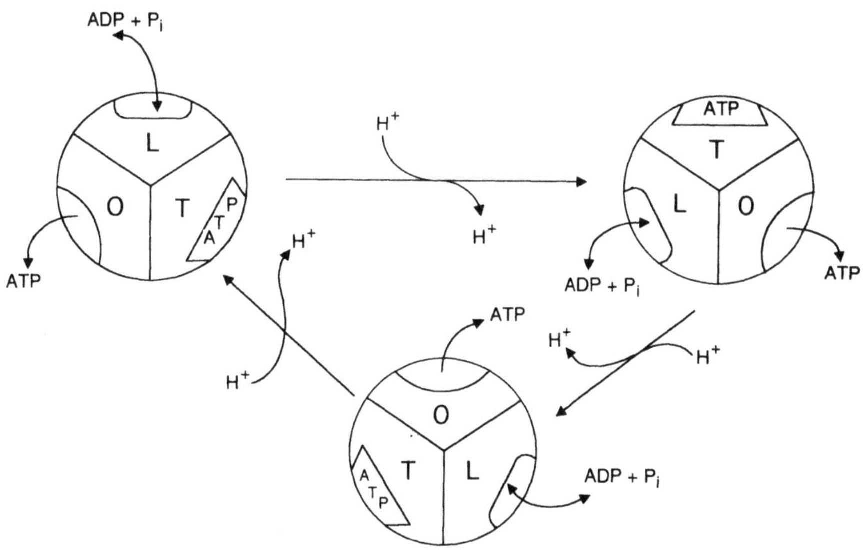


Fig. 9. Proposed catalytic mechanism of F-type ATP synthases. This mechanism is based on a change in affinity of the catalytic site that undergoes transitions of state: from a low affinity (L) state capable of rapidly exchanging ADP and phosphate with the external medium, to a closed state (T) in which ATP is formed spontaneously in a tightly bound form, and subsequently to an open (O) state that releases ATP. Since the enzyme contains three catalytic sites (one for each αβ pair), these transitions are thought to occur cyclically and separately for each site, so that the sites are never identical to each other. This asymmetry would be the consequence of the interactions with the single copy subunits of $F_1$ and $F_0$ (alternate binding mechanism). the energy supplied by the proton flow is necessary for inducing the state transition, and therefore the release of tightly bound ATP. This energy is sensed and transmitted to $F_1$ by the transmembrane subunits of $F_0$. This model has recently received decisive support by the crystal structure of bovine $F_1$ that has confirmed the asymmetric properties of the enzyme.

enzyme complex, therefore, does not possess the symmetry typical of most the "normal" cooperative proteins. Current models in fact consider the possibility of "rotational" mechanism in which a single catalytic site at a time interacts in turn with the central portion of the complex ($F_0$ and the three small subunits of $F_1$), and experiences the structural interactions that will eventually lead to the changes in the affinities for nucleotide binding (Fig. 9).

### *2.3. Coupling of different vectorial enzymes*

We have so far discussed the coupling of two different processes catalyzed by the same enzyme and have described pertinent examples of this type of phenomena. Energetic coupling can however occur also between two (or more) enzyme reactions taking place in a single phase or also across a membrane connecting two phases.

An already mentioned example is that of two metabolic cytoplasmic reactions, one reducing $NAD^+$ and the other oxidizing NADH: this is the case of many fermentative metabolisms, in which a final metabolic reduction oxidizes NADH (the reduced product being excreted) in order to allow glycolysis to proceed at high rate. In this case the mass and redox balances of $NAD^+$ and NADH impose constraints on both reactions so that the free energy change associated with either one will be affected by the substrate concentrations of the other. This effect is due to the presence of common reagents and products in the two reactions, in this case $NAD^+$ and NADH. If the system were at equilibrium the equilibrium position of the first reaction would affect that of the second and a formal equilibrium constant for the overall process could be written (see also Chapter 1, section 4, for further treatment of this situation in steady state).

Similarly, energy coupling between two reactions can take place across a biomembrane separating two cellular compartments (1 and 2) (including the external of the cell). In this case the principle of the common reaction intermediates must be applied as well; in the case of vectorial processes however, the coincidence must include also that of the membrane across which the processes occur and of the two compartments separated by it. If, for example, the coupling occurs between two ion translocators, they must translocate the same ion, be placed parallelly or antiparallelly across the same membrane and, obviously, release or take up the transported species in the same couple of compartments.

In this case the free enthalpy change for the two coupled processes will be:

$$\Delta G = \Delta G_{(1)} + \nu_{i,(1)} (\tilde{\mu}_{i,2} - \tilde{\mu}_{i,1}) + \Delta G_{(2)} + \nu_{i,(2)} (\tilde{\mu}_{i,2} - \tilde{\mu}_{i,1}) \qquad (11)$$

where $\Delta G_{(1),(2)}$ are the free enthalpy changes of the two chemical reactions and $\nu_{i,(1,2)}$ are the number of ions translocated by each translocator per mole of product of the reaction and $\tilde{\mu}_{i,1}$ and $\tilde{\mu}_{i,2}$ the electrochemical potentials of the ion *i* in compartments 1 and 2 respectively.

When the two translocators are parallelly oriented and the two active flows have the same direction, it can happen that one reaction (the driving or input reaction) will proceed spontaneously, generating an electrochemical potential difference of the ion transported, while the second one could be reversed with respect to its spontaneous direction and could proceed in the opposite, endoergonic way. This reacton therefore (the output reaction), e.g. reaction (2) will be driven by reaction (1), and an effective transduction of free enthalpy from reaction (1) to reaction (2) will take place.

In condition of steady state, in which the two opposite flows will balance $(J_{i,(1)} = -J_{i,2})_{ss}$, a difference in electrochemical potential of the transported ion will develop $(\tilde{\mu}_{i,1} - \tilde{\mu}_{i,2})_{ss}$ that will be kept constant in time. If, moreover, the concentrations of reagents and products of both reactions are allowed to fluctuate, so that $\Delta G_{(1,2)}$ will also fluctuate, eventually both reactions will equilibrate ideally with the same electrochemical potential difference and a true equilibrium will be established, for which it will be:

$$\frac{\Delta G_{(1)}}{\nu_{i,(1)}} = \frac{\Delta G_{(2)}}{\nu_{i,(2)}} = -(\tilde{\mu}_{i,2} - \tilde{\mu}_{i,1}) \tag{12}$$

Thus process (2) can maintain an equilibrium position apparently impossible thermodynamically, because it is coupled to process (1) through the common transport process.

Above equilibrium equations are, as always, ideal, in that they assume a reversible equilibration between an electrochemical potential difference of an ion and two chemical reactions coupled to the translocation of this ion. This equilibration implies therefore an ideally perfect catalytic machinery within each translocator (an assumption probably not too far from reality for the actual membrane enzymes), and , also, an ideal behavior of the interposing membrane, namely no passive flow of the transported ion across the membrane. Passive flow in fact would cause dissipation as heat of part of the free enthalpy that is stored across the membrane in the form of an electrochemical potential difference. The integrity (structural and functional) of the interposing membrane, *the coupling membrane,* becomes therefore an essential feature for the equilibration of the coupled processes and, in general, for the efficiency of energy transduction processes through the circulation of ions.

In biological systems many processes are coupled through the circulation of ions. Preeminent in this type of mechanisms are the processes of ATP synthesis coupled to the redox reactions of photosynthesis, of aerobic respiration or of many other types of redox reactions occurring in membrane of anaerobic and fermentative bacteria. The ion involved in these processes is the proton. The same ion is also involved in many mass transport processes (in bacteria, in plant plasma membranes and in organelles and endocellular membrane of all eukaryotic cells), in which the transport of one substrate, ionic or non ionic, is coupled to the translocation of one or more protons in the same or in opposite direction. This is the phenomenon of secondary active transport in which a *symporter* or an *antiporter* of protons drives the accumulation of a cotransported species across the membrane. This modality of driving an active transport by an ion implies the involvement of another, primary translocator on the same membrane, generating the necessary difference in electrochemical potential and, essentially, driving the transport process. Since many of these processes can take place across the same membrane, all of them should be taken into account simultaneously while analyzing the proton coupled systems.

A role similar to that of protons in transport is also performed by $Na^+$, particularly across the plasma membrane of mammalian cells. In this case the driving process is the hydrolysis of ATP that translocates $Na^+$ outwards from the cells (and $K^+$ inwards). The driven processes are represented by the great number of cotransport processes by which metabolites are accumulated inside the cell and also by some antitransport processes in which a molecule (e.g. $Ca^{2+}$ or $H^+$) is extruded from the cell in exchange for $Na^+$. The role of both $H^+$ and $Na^+$ in transport will be discussed further in this chapter.

As a final, but not secondary, comment to this section, it should be noted that the difference in electric potential across a membrane is nearly always an important component of the $\Delta\tilde{\mu}_i$ of all ionic species (see above). In fact, in all cases in which a net translocation of electric charges occurs, the presence of a membrane potential introduces a contribution in the energy balance of the transport process, particularly when the transported species carries multiple charges. This particular character of electrogenic reactions holds true irrespective of the chemical nature of the transported species and is only dependent on the charge stoichiometry and sign of the translocated molecule(s). It follows that when many electrogenic processes take place across the same coupling membrane, they are all coupled (at least partially) to each other, irrespective of their nature, by the energetic interaction due to charge translocation. At first sight this mechanism of coupling between transmembrane vectorial reactions appears to overcome the principle of the common intermediate. This is not the case, however, since the "common intermediates" are the net electric charges translocated in the electrogenic reactions; these electric charges are not, however, directly related to the chemical nature of the molecule involved. The electric polarization of the coupling membrane is therefore an essential component in the bioenergetics of many cellular processes and often the membrane potential is the main, if not the only, force driving endoergonic processes coupled to ion transport. This characteristic sheds a new light on the importance of intracellular compartmentation for cellular bioenergetics and emphasizes the low ionic conductivity of biomembranes as a basic feature for an efficient energy transduction.

As a final example of coupling of two processes through the circulation of protons let us consider the accumulation of lactose by cells of *E. coli:* this sugar is accumulated in cotransport with one proton by a symporter (the inducible lactose permease, product of the y gene of the *lac* operon) that translocates one $H^+$ per one molecule of lactose. Let us also assume that the $\Delta\tilde{\mu}_{H^+}$ is produced by a single redox reaction oxidizing NADH and reducing fumarate (this reaction is catalyzed by the enzyme fumarate reductase and occurs during the anaerobic respiration of fumarate) (Fig. 10).

The $E^m$ of the fumarate/succinate and of the $NAD^+/NADH$ couples at pH 7.0 are 0.03 and −0.32 V respectively. If we assume that the fumarate/succinate and the $NAD^+/NADH$ ratios are maintained constant at values of 10 and 0.1 by other metabolic reactions (i.e. the two concentration ratios are maintained in a stationary state by metabolism) and that $2H^+$ are transferred for every fumarate reduced, the equilibrium $\Delta\tilde{\mu}_{H^+}$ will be:

$$(\Delta\tilde{\mu}_{H^+})_{eq.} = \frac{\Delta G_{(fum)}}{\nu_{H,(fum)}} =$$

$$= 1/2\,[-2 \times 96.5 \times (-0.32 - 0.03) + RT \ln (10 \times 10)] = \qquad (13)$$

$$= 39.5 \text{ kJmole}^{-1}$$

being $\tilde{\mu}_{H^+,out} > \tilde{\mu}_{H^+,in}$.

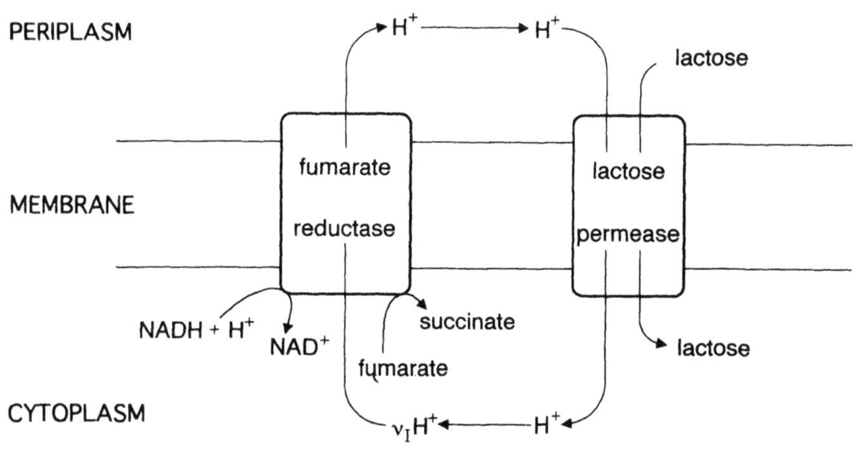


Fig. 10. Coupling of a $H^+$ symporter to a redox pump. The scheme illustrates the case of lactose transport in *E. coli* driven by the anaerobic respiration of fumarate. A redox pump (fumarate reductase) oxidizes NADH utilizing fumarate as electron acceptor (anaerobic respiration). This redox reaction is coupled to the translocation of protons toward the exterior of the cell (the $H^+/NADH$ stoichiometry, still uncertain, is not specified in the scheme). Active accumulation of lactose is coupled to the entry of one proton per lactose, and is therefore driven by $\Delta\tilde{\mu}_{H^+}$. The transport of lactose is catalyzed by the lactose permease.

The concentration ratio of lactose in equilibrium with this value of $\Delta\tilde{\mu}_{H^+}$ will be given by a second equilibrium relation

$$2.3\ RT\ \lg \frac{[\text{lact}]_{in}}{[\text{lact}]_{out}} = 39.5$$

or (14)

$$\frac{[\text{lact}]_{in}}{[\text{lact}]_{out}} = 10^{6.9}$$

At equilibrium with the fumarate reductase reaction lactose will be accumulated nearly 10 million fold inside the cell. In reality, the accumulation ratio found experimentally never exceeds 1000, since the value of $\Delta\tilde{\mu}_{H^+}$ is only about 18 kJ mole$^{-1}$ due to passive proton leaks and competing protonic fluxes.

## 3. Proton bioenergetics

### *3.1. The protonic gradient as energy store*

Following the general definition of electrochemical potential, the free energy stored across a biological membrane in the form of a difference in protonic electrochemical potential takes the form:

$$\Delta\tilde{\mu}_{H^+} = \mathcal{F}(\phi_1 - \phi_2) - 2.3\ RT\ (pH_1 - pH_2) \tag{15}$$

where 1 and 2 refer to two generic compartments placed on the opposite sides of a membrane. As such the difference in potential is expressed in energy units, referred to the translocation from compartment 2 to compartment 1 of mole of protons (kJ mole$^{-1}$). It is customary however in bioenergetics to express this force in electric units, as $\Delta\tilde{\mu}_{H^+}/\mathcal{F}$; this quantity was defined by P. Mitchell as protonmotive force, $\Delta p$, and is generally expressed in mV.

Proton accumulation and the onset of large difference in electric potential require special properties of the membrane. This has to be endowed with a low passive conductivity for protons, so that to sustain a considerable pH difference, and in general should be not permeable to ions. These are not uncommon characteristics of lipid bilayer, which forms a large part of the surface of a biological membrane. The membrane specialized in the formation of large $\Delta\tilde{\mu}_{H^+}$, such as those of chloroplasts and the inner membrane of mitochondria, are however particularly low conductors, since the presence of ions translocating proteins across them is limited and strictly compatible with their primary bioenergetic function.

The presence of ion conducting permeases in this type of membranes modulates the protonmotive force, redistributing its composition between the electric and concentration components. Thus, in chloroplast thylakoids $\Delta\tilde{\mu}_{H^+}$ consists mainly in a large difference in pH between the lumen (acid) and the stroma (close to pH 8.2). The absence of any measurable $\Delta\phi$ is due to the presence of highly conductive channels for $Cl^-$, which, by redistributing this anion, compensate for the charge movements associated to proton traslocation. On the contrary, across the inner membrane of mitochondria the difference in pH is minimal (<1 unit), since there are many active proton-cation antiporters or proton-anion symporters. This distribution of ion transport systems is related to the physiological needs (e.g. the need of very large fluxes of metabolites in and out of the mitochondrial matrix), but also to the necessity of controlling the pH and ion composition inside different compartments. Thus, since the thylakoid lumen does not have, to our knowledge, any physiological role besides photosynthetic electron transfer and ATP synthesis, its pH can be lowered to values generally incompatible with enzymatic functions; on the contrary, an intense metabolic activity is taking place on both sides of the mitochondrial inner membrane and pH must accordingly be controlled.

The amount of energy that can be delivered by a protonic potential difference across membranes is measured by the number of moles that can be transferred from a high potential compartment ($\tilde{\mu}_{H^+,2}$) to a low potential compartment ($\tilde{\mu}_{H^+,1}$). For unit time, therefore, the power extractable is given by $J_{H^+} \times A \times \Delta\tilde{\mu}_{H^+}$, where $J_{H^+}$ is the flux of protons (moles $s^{-1}$ $cm^{-2}$), and A the area of membrane involved in the process (measured in $cm^2$). It is clear that the value of $\Delta\tilde{\mu}_{H^+}$ is not the only parameter involved in proton bioenergetics, another fundamental quantity being the amount of $H^+$ that can be removed from a given compartment without depleting considerably its concentration and thus decreasing $\Delta\tilde{\mu}_{H^+}$. Of course in operating protonic circuits the value of $\Delta\tilde{\mu}_{H^+}$ is maintained at a steady state value by the input active proton translocators. A pronounced limitation of the amount of protons available in a given compartment will however expose the system to a marked instability and strong oscillation of the steady states.

Proton activity is nearly always very low inside cell compartments ($10^{-9}$ to $10^{-6}$ M). Protons appear therefore not very suitable to act as energy store, and itermediates for the coupling of membrane reactions. The low activity of protons can be however modulated quite promptly through relatively modest alteration of transmembrane flows and thus the proton coupled process can be regulated accurately. This is particularly true for the electrostatic component of $\Delta\tilde{\mu}_{H^+}$ since on average biological membranes have a rather low capacitance (1 $\mu F$ $cm^{-2}$) and large oscillation of $\Delta\phi$ can be achieved with imbalance of a small number of electric charges between the two compartments.

Several factors tend however to counterbalance excessive oscillations of $\Delta\tilde{\mu}_{H^+}$. Free protons are in continous equilibrium, within one single aqueous compartment, with many protonable groups (on proteins, phospholipds etc), that act as very effective buffers. The amount of protons available at a given $\Delta\tilde{\mu}_{H^+}$ is, accordingly, prevalently measured by the amount of bound protons in equilibrium with the free ones in those compartments. Likewise, other cations (like $K^+$ and $Na^+$ present at high concentration on either side of many plasma membranes) or anions (particularly $Cl^-$), can effectively clamp $\Delta\phi$ at a regulated value, if the membrane is equipped with transport proteins that maintain the ion distribution in equilibrium with the membrane potential. Moreover, particularly for divalent cations ($Mg^{2+}$ and $Ca^{2+}$) many binding sites are available on macromolecules that effectively buffer their concentrations. Ion-binding equilibria and ion redistribution across the membranes therefore contribute in damping the oscillations of $\Delta\phi$ during steady states.

### *3.2. Proton circuits in oxidative and photosynthetic phosphorylation*

The energetic role of the protonic potential difference across biomembranes was first suggested, and subsequently demonstrated, in relation with the synthesis of ATP in oxidative and photosynthetic phosphorylation. Now it is recognized that protons play a more general role in cellular energetics, since their flow is coupled to other active (endergonic) processes, such as metabolic and ion transport and, in prokaryots, also flagellar motility.

The coupling through proton flows of ATP synthesis to the reactions of the respiratory chain in the mitochondrial inner membrane or in the plasma membrane of heterotrophic bacteria is the consequence of the presence on the same membrane of several reversible proton transporters. The redox complexes of the respiratory chains always perform the role of proton translocators (at least when the redox potential span between their substrates and products is sufficiently large to sustain $\Delta\tilde{\mu}_{H^+}$). Their orientation is such to excrete protons from the mitochondrial matrix, or from the bacterial cytoplasm. Their structure, still largely unresolved at an atomic resolution, seems generally to conform to a Mitchellian proton loop; however in the respiratory complexes of eukaryots the stoichiometry of proton translocation (higher than one proton per one electron per complex) demonstrates that a more complex mechanism of coupling, mediated by the apoproteins of the complex, should be operative. The protonmotive force in respiratory systems reaches values higher than 220 mV and is mainly composed by the membrane potential.

ATP synthesis is performed by a reversible $H^+$-ATPase, operative in the synthetic direction, since the reaction is driven by the inverse flow of protons into the low potential compartment (mitochondrial matrix or bacterial cytoplasm). The inversion of the reaction of ATP hydrolysis obviously requires that the free energy decrease corresponding to proton translocation was higher than the free energy increase for ATP synthesis, i.e.

$$-\Delta G_{H^+} \geq \Delta G_{ATP} \tag{16}$$

Critical for this energy balance are several parameters, the most important being the stoichiometry of the ATPase reaction

$$ADP^{3-} + HPO_4^{2-} + \nu_H H^+_{out} \rightarrow ATP^{4-} + \nu_H H^+_{in} + OH^- \tag{17}$$

where *in* and *out* refer to the mitochondrial inner and outer compartments, and $\nu_{H^+}$ is the number of proton translocated for every catalytic cycle. Since $\Delta G_{H^+} = \nu_{H^+} \Delta\tilde{\mu}_{H^+}$, $\nu_{H^+}$ is higher the lower the value of $\Delta\tilde{\mu}_{H^+}$ needed to balance $\Delta G_{ATP}$.

This quantity, in turn, depends on the matrix concentrations of ATP, ADP and phosphate (i.e. by the so-called phosphate potential). When the accumulation of ATP is very high, in general when the [ATP]/[ADP] ≥ 500, $\Delta G_{ATP} \geq 62$ kJ mole$^{-1}$; under this conditions the value of $\Delta\tilde{\mu}_{H^+}$ is no longer sufficient to drive ATP synthesis and this reaction stops, having reached an equilibrium situation within the ATPase itself. Since the coupled flow of protons through the ATPase complex is also halted, the decrease of the inward proton flow is reflected in an increase in the steady state value of $\Delta\tilde{\mu}_{H^+}$, and, as a consequence, the rates of the $H^+$-translocating redox reactions are also decreased. This is essentially the mechanism by which the ATP concentration controls the rate of oxygen consumption in the respiratory systems. It must be stressed, however, that the condition of null ATP synthesis at high [ATP] is, for the whole system, a steady state in which the respiratory proton transport compensates for the passive proton flow across the coupling membrane and, also, for all flows coupled to processes other than ATP synthesis (i.e. $H^+$-linked transport processes). Moreover, even within the ATPase complex itself, the equilibrium could not be completely reached if a significant proton slip occurs during the catalytic cycle.

In photosynthetic phosphorylation the situation is quite analoguos. Photosynthetic chains act as proton translocators and generate a $\Delta\tilde{\mu}_{H^+}$ across the coupling membranes (thylakoid membranes in chloroplasts and *Cyanobacteria* and plasma membranes in photosynthetic bacteria). The direction of flow is towards the internal lumen of thylakoids or the periplasmic space of bacteria. A reversible ATPase, of the same type as that present in respiratory systems (and in the case of facultative photosynthetic bacteria coincident for oxidative and photosynthetic phosphorylation) synthetizes ATP. A protonic circuit is thus established driving ATP synthesis at the expense of energy stored by the photosynthetic reaction as $\Delta\tilde{\mu}_{H^+}$. Many of the phenomena described for oxidative phosphorylation occur also in photosynthetic membranes, including the control of the rate of the photosynthetic redox reactions by [ATP]. In addition the $H^+$-ATPases in photosynthetic systems appear to be endowed with peculiar regulatory properties that control their activity in relation to the availability of radiant energy and of the reduction state of the chloroplast stroma (compare for a detailed discussion on these complexes, see Chapter 12).

### *3.3. Efficiency of vectorial energy conversion: Leaks versus slips*

In the process of oxidative or photosynthetic phosphorylation proton fluxes mediate the transduction of energy from the redox reactions to ATP synthesis. The efficiency of this energy transducing process is measured by the ratio of the power gained in the output process (ATP synthesis) over the power expenditure in the input process (i.e. in the respiratory or photosynthetic reactions). Utilizing the terminology of non equilibrium thermodynamics, the efficiency is given by

$$\eta = \frac{-J_{ATP}\,\Delta G_{ATP}}{J_{ox}\,\Delta G_{ox}} \tag{18}$$

where $J_{ATP}$ and $J_{ox}$ are respectively the rates (fluxes) of ATP hydrolysis and of respiratory or photosynthetic electron flow and $\Delta G_{ATP}$ and $\Delta G_{ox}$ their associated free energy changes. Of course the value of $J_{ATP}$ and $J_{ox}$ should be normalized to the same membrane area, e.g. to 1 milligram of membrane protein. In the phenomenological description of the input and output processes the rate of each reaction is linearly related to the extent of its conjugated force ($J_{ATP}$ to $\Delta G_{ATP}$ and $J_{ox}$ to $\Delta G_{ox}$) and, also, linearly cross-related to the non conjugated force (being $L_{ATP,ox} < 0$):

$$J_{ATP} = L_{ATP}\,(-\Delta G_{ATP}) + L_{ATP,ox}\,(-\Delta G_{ox}) \tag{19}$$

$$J_{ox} = L_{ox}\,(-\Delta G_{ox}) + L_{ATP,ox}\,(-\Delta G_{ATP}) \tag{20}$$

This phenomenological description leads to a univocal relation between the maximal efficiency of the coupling process and a phenomenological parameter of the systems, known as degree of coupling *q* that is a function of direct and cross linear coefficients

$$q = \frac{L_{ATP,ox}}{\sqrt{L_{ox}\,L_{ATP}}} \tag{21}$$

This relation holds as

$$\eta_{max} = \frac{q^2}{(1 + \sqrt{1 - q^2})^2} \tag{22}$$

In this non formal treatment of the energy transduction processes we will not discuss further the validity of the linear phenomenological description of the processes, but rather will discuss the molecular meaning of the degree of coupling. The degree of coupling $q$ has an upper theoretical value of 1; when $q = 1$ the two processes are totally coupled, so that no reaction of the driving process can take place in the absence of a corresponding reaction of the driven process (the + or − sign of the degree of coupling refer to the direction of the flows of the two coupled processes (compare Chapter 1)). By contrast, when $q = 0$ each process can proceed at any rate, linearly dependent only on its conjugated force, without any influence by the other process which becomes totally uncoupled. For any intermediate value of q ($0 < |q| < 1$) each of the two coupled processes possesses a certain degree of freedom and can proceed at a certain rate in the absence of a corresponding proportional effect on the second process.

In chemiosmotic protonic circuits, in which protons fluxes across the coupling membrane act as intermediate in energy transduction, the thermodynamic parameter of the degree of coupling may assume a rather precise molecular meaning (see Chapter 2). If we consider, for example, the process of oxidative phosphorylation, the input process consists of the proton translocation driven by the catalytic activity of the respiratory complexes. The flux of transported protons will be therefore proportional to the respiratory rate, being the proportionality constant $\nu_{H,ox}$, an expression of the stoichiometry of the catalytic process ($\nu_{H,ox}$ is considered as constant at this stage of the discussion). It will be therefore $J_{H,ox} = \nu_{H,ox} J_{ox}$.

Likewise the rate of ATP synthesis will be proportional to the flow of protons across the ATP synthase complex and it will be $J_{H,ATP} = \nu_{H,ATP} J_{ATP}$. At steady state, when no net flow of proton across the membrane occurs, it will be

$$-J_{H,ATP} = J_{H,ox} \tag{23}$$

and therefore

$$-\nu_{H,ATP} J_{ATP} = \nu_{H,ox} J_{ox} \tag{24a}$$

or

$$-\frac{J_{ATP}}{J_{ox}} = \frac{\nu_{H,ox}}{\nu_{H,ATP}} = \text{cost} \tag{24b}$$

and the two processes will be totally coupled.

If however a passive *leak* of protons takes place across the membrane, assumed to be proportional to $\Delta\tilde{\mu}_{H^+}$ ($J_{H,\rho} = L_{H,\rho}(-\Delta\tilde{\mu}_{H^+})$), at steady state for the proton flux it will be

$$-J_{H,ox} = J_{H,ATP} + J_{H,\rho} \tag{25}$$

and therefore

$$-\nu_{H,ox}\, J_{ox} = \nu_{H,ATP}\, J_{ATP} + L_{H,\rho}\, (-\Delta\tilde{\mu}_{H^+}) \tag{26}$$

The respiratory activity will have to translocate a flux of protons compensating the fluxes utilized for ATP synthesis and for the passive proton leaks. The higher the proportion of the passive flux (i.e. the higher the conductance of the coupling membrane to protons), the higher the fraction of respiratory power dissipated as heat (in a futile proton circuit) and not converted into ATP. The degree of coupling between the respiratory reactions and the synthesis of ATP becomes therefore a function of constants expressing the stoichiometry of the catalytic mechanism and the conductivity of the coupling membrane to protons (including other proton fluxes coupled to other processes, such a transport mechanism (see below)). It is obvious, moreover, that a value of $q = 0$ corresponds to a very high conductivity of the membrane, as compared to the rates of the respiratory and the ATP synthesis/hydrolysis reactions; in this case, in fact, these latter processes cannot compete with the passive flows and the two processes proceed independently, completely uncoupled.

Many ingenious experiments have been performed in order to establish the validity of this model, the approach mostly utilized being the measure of the rate of respiration in the absence of ATP synthesis (or of ATP hydrolysis in the absence of respiraton). Following Eqns. (25) and (26), in these two cases at steady state, it should be respectively

$$-J_{H,ox} = J_{H,\rho} = L_{H,\rho}\, (-\Delta\tilde{\mu}_{H^+}) \tag{27a}$$

$$-J_{H,ATP} = J_{H,\rho} = L_{H,\rho}\, (-\Delta\tilde{\mu}_{H^+}) \tag{27b}$$

The basal rate of respiration (or ATP hydrolysis) in these cases will compensate only the passive leaks and should be unique function of the $\Delta\tilde{\mu}_{H^+}$ and of the conductivity of the membrane to protons. Such experiments have brought up an unexpected behavior of the system: the value of $L_{H,\rho}$ was found to be not constant for the entire range of $\Delta\tilde{\mu}_{H^+}$ measured ($\Delta p <$ 0.220 V), but it increased steeply with $\Delta\tilde{\mu}_{H^+}$ above a certain threshold level (around 0.180 to 0.200 V). The system was therefore behaving non ohmically, in that flux of protons (the protonic current) was not proportional to the $\Delta\tilde{\mu}_{H^+}$ (the potential difference driving the current).

The non-ohmic behaviour was explained by many authors as an intrinsic property of the membrane, whose conductivity was increased at high $\Delta\tilde{\mu}_{H^+}$. This rather trivial explanation is however not in agreement with the observation that the threshold $\Delta\tilde{\mu}_{H^+}$ for the transition to the non-ohmic behavior is not constant for a given membrane preparation, but depends upon the type of proton translocator studied, i.e. single or combination of respiratory complexes or the ATPase. The non-ohmic behavior appears therefore to be a property specific of a given proton translocator rather than of the membrane as a whole.

It has been proposed therefore that some non coupled processes can take place within the proton translocators themselves, i.e. some *slips* can occur, meaning with this definition that, during the catalytic processes, redox reactions or ATP hydrolysis could proceed without a concurrent translocation of protons, as a consequence of an intrinsic failure of the coupling mechanism. This failure is more likely to take place at a high $\Delta\tilde{\mu}_{H^+}$, when the back pressure of the protonic potential difference will exert an increasing strain on the reaction mechanism. Above a certain threshold $\Delta\tilde{\mu}_{H^+}$, therefore, characteristic for each type of proton translocator, the probability of slipping will increase progressively and consequently the system will exhibit a non ohmic behavior. Phenomenologically the presence of a slipping proton translocator will correspond to non constant values for $\nu_{H,ox}$ and $\nu_{H,ATP}$.

No agreement still exists on the interpretation of these phenomena since the concept of slip is not accepted by all researchers in the field of bioenergetics. It is this author's opinion, however, that slip mechanisms are very likely to operate in all coupled catalytic mechanisms. They might have even evolved as a control device against the onset of excessive $\Delta\tilde{\mu}_{H^+}$ across the coupling membrane, or as a means of controlling the stoichiometry of proton pumping. It is likely therefore that a combination of leaks and slips will contribute to the energy dissipation occuring at high $\Delta\tilde{\mu}_{H^+}$ and to the modulation in the utilization of $\Delta\tilde{\mu}_{H^+}$ in energy transduction.

## 4. Transport phenomena

### *4.1. Transport coupled to proton fluxes*

The transmembrane $\Delta\tilde{\mu}_{H^+}$ is a major source of energy for secondary ion and metabolite transport across the inner membranes of mitochondria and across bacteria plasma membranes. In these processes the inward flow of protons, driven by $\Delta\tilde{\mu}_{H^+}$ is coupled to the inward (in $H^+$ symport), or to the outward (in $H^+$ antiport) flow of the transported molecule. The coupling is performed within the active sites of $H^+$-permease that can perform the conformational changes implicit in every mediated transport process only when protonated. A model catalytic cycle of such permeases therefore (e.g. a symporter) includes the protonation of one (or more) functional groups on the high potential side of the membrane followed by the binding of the transported substrate; debinding and deprotonation occur after the enzyme has switched its active site on the low potential side of the membrane (Fig. 11). Although this type of scheme is logical, the actual molecular mechanism of coupling has not yet been elucidated for any symport or antiport transport system.

The role of the membrane potential in this type of transport depends on the net charge of the tertiary $H^+$/substrate/enzyme complex (or in general of the multiprotonated complex), as does the substrate distribution in equilibrium with $\Delta\tilde{\mu}_{H^+}$. The accumulation ratio of the substrate, for a general symport phenomenon

$$S^Z_{out} + \nu_H H^+_{out} \rightarrow S^Z_{in} + \nu_H H^+_{in} \quad (28)$$

will be:

$$RT \ln \frac{[S^Z]_{in}}{[S^Z]_{out}} = (\nu_H + z)\,\Delta\phi - \nu_H \times 2.3\,RT\,\Delta pH \quad (29)$$

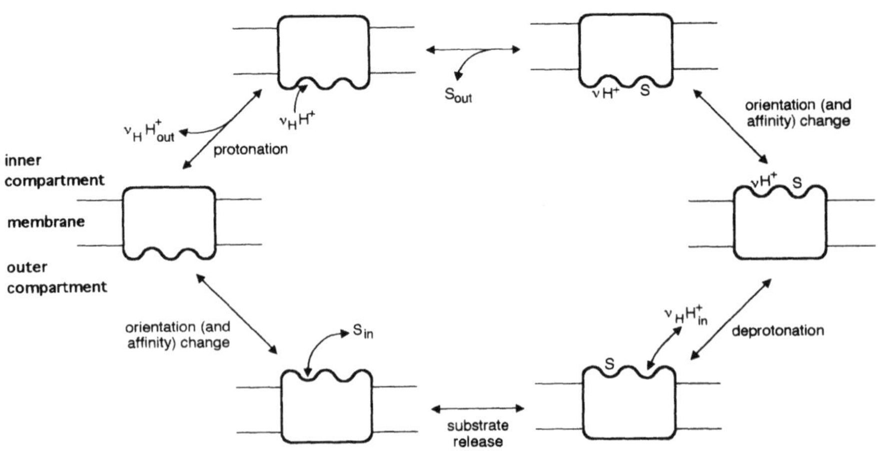


Fig. 11. Functional scheme of a proton symporter. The scheme represents the catalytic cycle of a generic permease, coupling the translocation of one molecule of substrate ($S_{in,out}$) with that of $\gamma_H$ protons. Differently from the scheme of Fig. 8, in this case the reorientation of the substrate- and proton-binding sites is due only to thermal fluctuations, but requires obligatorily the contemporary protonation of $\gamma_H$ sites and the binding of substrate. All steps of the catalytic cycle are reversible and the directions of the net flows depend only on the values of electrochemical potentials of protons and substrate on both sides of the membrane.

where $z$ is the charge of the transported substrate and $\nu_H$ is the number of proton cotransported. For an antitransporter system the direction of the flow of $S^Z$ is inverted (the sign of $\ln([S^Z]_{in}/[S^Z]_{out})$ is opposite) and the $(\nu_H + z)$ is changed into $(\nu_H - z)$.

Two opposite limit cases are of interest here:

1) the symport of a monovalent anion with a proton. In this case $(\nu_H + z) = 0$ and the transport is driven only by the difference in pH across the membrane. This is the case, for example, of the phosphate carrier of the inner mitochondrial membrane that couples the entry of one proton with that of $H_2PO_4^-$.
2) the antiport of differently charged substrates. In the case of the adenylnucleotide exchanger of the inner mitochondrial membrane, for example, the exchange of $ADP^{3-}$ and $ATP^{4-}$ corresponds to $\nu_H = 0$, but the charge difference of the two transported substrate results in the translocation of a net negative charge in parallel with the flow of $ATP^{-4}$ (opposite of that of $ADP^{-3}$). The $\Delta\phi$ across the membrane, negative in the matrix, tends therefore to favor the excretion of $ATP^{4-}$ from the matrix and the entry of $ADP^{3-}$. A limiting case of this phenomenon can also be considered an ion uniport, like the $Ca^{2+}$ transporter of mitochondria, in which this cation is actually accumulated into the matrix only by virtue of the negative electric potential with respect to the cytoplasm. These are two additional examples of the coupling between two processes exclusively through the action of the membrane potential. In this case the transmembrane $\Delta\phi$ generated by the redox electrogenic $H^+$-pumps of the respiratory chain and the same $\Delta\phi$ drives the active ATP/ADP exchange and the $Ca^{2+}$ uptake, even if $H^+$ are not explicitly involved in these two transport processes.

The most studied example of proton-coupled transport system is the lactose permease of *E. coli.* This inducible enzyme couples the entry of lactose with that of one proton in the bacterial cell. The tertiary complex is charged thus the entire $\Delta\tilde{\mu}_{H^+}$ is driving the sugar transport. In addition to the protonic circuits that promote ATP synthesis, therefore, another circuit is established that causes the lactose accumulation from the external medium (compare Fig. 10 for an example of coupling between lactose transport and anaerobic fumarate respiration). Similar proton-dependent transport systems are very frequent in prokaryots and plant cells. They include the transport of many different sugars, amino acids, anions and vitamins. Of particular interest in bacteria are the cation/proton antiporters, particularly the $Na^+/H^+$ exchanger. This permease catalyzes the excretion of $Na^+$ in exchange with more than one $H^+$ (probably two); this active transport system can therefore drive the outward flow of sodium utilizing the $\Delta\phi$, even when the outside pH is higher than the cytoplasmic one; in this case only $\Delta\phi$ plays an active role. In bacteria the $Na^+/H^+$ antiporter contributes also, in cooperation with other ion transport systems, noticeably for $K^+$, to the homeostasis of intracellular pH and to the control of turgor pressure.

In intracellular organelles of eukaryots and particularly in mitochondria, examples of $\Delta\tilde{\mu}_{H^+}$ driven permeases are quite numerous. In addition to the already mentioned phosphate carrier and adenyl nucleotide exchanger, they include the pyruvate, the glutammate and the tricarboxylic acid carriers and probably the ornitine/citrulline exchanger. Since the metabolic traffic across the inner membrane is very intense, it is certainly an oversimplification to consider the role of $\Delta\tilde{\mu}_{H^+}$ only in driving oxidative phosphorylation.

### *4.2. Protonic gradients in intracellular compartments of eukaryotic cells*

The inner mitochondrial membrane and chloroplast thylakoids are not the only intracellular membranes across which a protonic gradient is established. In many other cases $\Delta\tilde{\mu}_{H^+}$ is generated in intracellular vesicles or vacuoles; its role is generally to drive the transport, usually an accumulation of specific solutes, outward from the cytoplasms. These systems include plant vacuoles, lysosomes, secretion vesicles (including presynaptic vesicles), chromaffin granules, intracytoplasmic transport vesicles and, possibly, the cysternae of the Golgi apparatus. The membranes of these vesicles probably do not contain any redox chain able to translocate protons; the $\Delta\tilde{\mu}_{H^+}$ is generated in these cases by membrane bound ATPases, generally termed vacuolar ATPases (V-ATPases), which utilize ATP from the cytoplasms for translocating protons into the lume of the vesicles.

The low intravesicular pH may have different physiological meanings: in lysosomes it causes the release of the transported glycoproteins from their receptors; in excretion vesicle it contributes to the accumulation of the protein to be secreted. More pertinent to the specific topic of this chapter is however its role in the accumulation of solutes. The best known of these systems are chromaffin granules, in which cathecolamines are accumulated by $H^+$-antiportes, all powered indirectly by the ATP hydrolyzed by a V-ATPase and mediated by $\Delta\tilde{\mu}_{H^+}$ (which is composed in this case prevalently by a $\Delta$pH). A somewhat similar system is offered by the presynaptic vesicles in nerve cells; here neurotransmitters are accumulated, and made ready for a subsequent exocytotic release, by $H^+$-antiporter system, each specific for a neurotransmiter, but all similar in that they are driven by the $\Delta\tilde{\mu}_{H^+}$ generated by a single type of V-ATPase. Similar phenomena are also very relevant in plant vacuoles, where reserve or discard products are accumulated across the tonoplast by virtue of V-ATPases; the vacuole osmotic pressure, moreover, is also the controlling device for the turgor in plant cells.

Thus $H^+$-coupled fluxes and the formation and utilization of $\Delta\tilde{\mu}_{H^+}$ are major events in many, if not all, aspects of cell biology, the synthesis of ATP in mitochondria and chloroplasts being only the system most accessible to experimental research, due to its quantitative relevance and long standing tradition in bioenergetics.

The structure of V-ATPases, as we know today, is very much related to that of the ATPases of mitochondria and chloroplasts (termed F-type ATPases). Both types of enzyme are complexes consisting of two portions, an extrinsic one formed by five types of polipeptide subunits and an intrinsic part, possibly forming a $H^+$-conducting device across the membrane. The structure and the function of these membrane complexes will be described later in Chapter 12. The overall structure and also more subtle features (like protein sequence homology) suggest an evolutionary connection between these two classes of ATPases. Each of them appears however to have specialized in different cellular roles: the F-ATPases in operating in the synthetic direction and being localized in the ATP generating organelles; the V-ATPases in catalyzing ATP hydrolysis for the onset of a protonic driving force for transport in intracellular stores (Fig. 12).

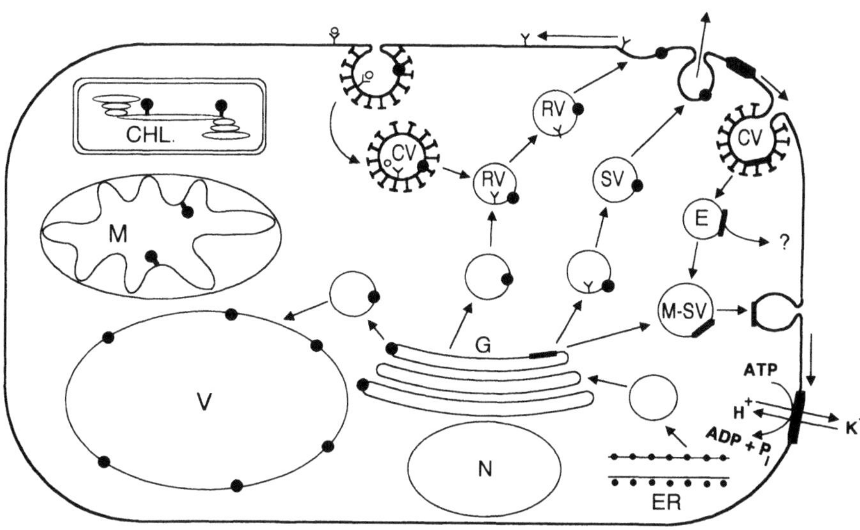


Fig. 12. Intracellular distribution of V-, P- and F-type ATPases. The figure schematically shows the internal distribution of the three forms of ion-motive ATPases in a photosynthetic eukaryotic cell. P-type ATPases (■) perform their physiological role across the plasma membrane (here a H+/K+ exchange ATPase is indicated). Occasional internalization in endosomes (E) during endocytosis can occur. V-type ATPases (●) are present in intracellular membranes, such as those of vacuoles (V), Golgi cisternae (G), secretion vesicles (SV), transport vesicles and chlatrin-coated vesicles (CV). F-type ATPases (♦) are only present in energy transducing organelles like mitochondria (M) and chloroplasts (CHL). (Reproduced with permission from N. Nelson (1992), Biochim. Biophys. Acta 1100, 109–124).

## 5. Sodium bioenergetics

### *5.1. Sodium gradients across the plasma membrane*

In the cytoplasm of most cellular systems, either prokaryotic or eukaryotic, cations are present at very different electrochemical potentials as compared to the extracellular fluids. This implies both an expenditure of metabolic energy for producing and maintaining such ionic gradients, and also, a possible bioenergetic role for the ion fluxes associated with them. In general the most significant differences in electrochemical potential are found for $K^+$, which is about 140 mM in the cytoplasm versus 5 mM in extracellular fluids, and for $Na^+$ present inside the cell at about 10 mM, but very concentrated outside (about 150 mM). Also $Ca^{2+}$, controlled at a very low concentration inside the cell (about 0.1 μM) is at much higher concentration in extracellular compartment (1 mM at least). Of course, these figures are only indicative and are mainly related to the actual situation in mammals. However, although the intracellular concentrations of these cations can vary widely from one organism to another, and can depend on the environment in which one organism lives, the figures reported above are indicative of a general situation: $Na^+$ and $Ca^{2+}$ are at a much higher potential outside the cytoplasm, while the opposite holds for $K^+$.

In order to calculate the free energy associated with the translocation of one mole of these cations, a membrane potential of 0.05 V is assumed, negative at the cytoplasmic side of the membrane (the actual value of the membrane potential across the plasmalemma ranges between 0.09 and 0.01 V depending on the cellular type in mammals). The electrochemical potential difference for the three cations will therefore be:

$$\Delta G_{K^+,(out-in)} = RT \ln \frac{0.005}{0.140} + 96.5 \times 0.05 = -3.4 \text{ kJmole}^{-1} \quad (30a)$$

$$\Delta G_{Na^+,(out-in)} = RT \ln \frac{0.150}{0.010} + 96.5 \times 0.05 = 11.5 \text{ kJmole}^{-1} \quad (30b)$$

$$\Delta G_{Ca^{2+},(out-in)} = RT \ln \frac{0.001}{0.1 \times 10^{-6}} + 2 \times 96.5 \times 0.05 = 32.5 \text{ kJmole}^{-1} \quad (30c)$$

$K^+$ therefore will tend to flow spontaneously outward from the cytoplasm, while $Na^+$ and $Ca^{2+}$ will have a strong tendency to enter into this cellular compartment.

Although $Ca^{2+}$ translocation is associated with the highest free energy change per mole, its very small absolute concentration does not allow its utilization for driving a significant metabolic work, such as mass transport processes. The role of $Ca^{2+}$, and of its redistribution in different cellular compartments, is in fact restricted to the regulation of metabolic processes, such as those associated with the secondary response to hormone or neurotransmitter stimuli and to the muscular function. $Na^+$ fluxes, on the contrary, play a pivotal role in driving endoergonic processes across the animal plasma membranes, particularly transport of metabolites or other ions, and, in association with $K^+$ in generating the transmembrane ionic currents that are at the basis of the electric responses of excitable tissues.

The electrogenic accumulation of cations across the membranes requires the expenditure of metabolic energy, mainly the hydrolysis of ATP. A whole family of ion-motive ATPases is known that couple ATP hydrolysis to the flux of one or more than one cationic species. In mammals these ATPases are mostly placed across the plasma membrane, although there are exceptions, such as the extensively studied Ca-ATPase of the sarcoplasmic reticulum. This group of membrane proteins include well known enzymes, such as the Na,K-ATPase that couples the hydrolysis of one molecule of ATP with the translocation of 3 $Na^+$ outside the cell and 2 $K^+$ inside, a plasma membrane Ca-ATPase, which extrudes $Ca^{2+}$, the gastric H,K-ATPase that excretes $H^+$ in exchange with $K^+$, and many others. Some examples of similar ATPases are also found in bacteria, the best known being the K-ATPase of *E. coli,* which drives the accumulation of $K^+$ into the cell.

All these enzyme complexes are evolutionarily related. They have in common a transmembrane protein of about 110 kDa of molecular weight, which can sometimes form a functional dimer and that is largely homologous in all the enzymes mentioned above. To these basic subunits other subunits can be associated; a dimer of β-subunits for the Na,K-ATPase, two other polipeptides for the gastric enzyme. The role of these ancillary subunits has not been well established yet. All ion-motive ATPases with this basic structure are classified as P-type ATPases.

This type of enzyme is in fact characterized by the possibility of isolating a phosphorylated intermediate of the catalytic mechanism that couples ATP hydrolysis to the translocation of the ions. The phosphate residue is donated by ATP to an aspartate residue of the active site, which is strongly conserved in all complexes studied so far. Since the stability of the phosphorylated intermediate is affected by the presence of the translocated ions, models have been proposed in which the phosphorylation influences the stability of the ion-enzyme complex, and also its orientation towards one face of the membrane or towards the opposite one. Thus, for example for the Ca-ATPase of the sarcoplasmic reticulum, the presence of $Ca^{2+}$ on the cytoplasmic side of the membrane is mandatory for phosphorylation (this reaction occurs only on a (2 $Ca^{2+}$)-enzyme complex); in turn, the phosphorylation switches the orientation of the ion-binding sites towards the opposite face of the membrane and decreases their affinity, so that $Ca^{2+}$ can be relased into inner compartment of the sarcoplasmic cisternae, where its concentration is much higher (see Fig. 8 for a general scheme). The release of $Ca^{2+}$ on the other hand destabilizes the phosphorylated intermediate, and dephosphorylation is followed by the reorientation of the $Ca^{2+}$-binding sites. Similar models have been proposed for other ion-motive ATPases.

In the case of this type of enzyme, therefore, the phosphorylated intermediate embodies a mechanism by which ATP hydrolysis and ion translocation are coupled: the transient covalent modification of the protein catalytic site corresponds to the catalytic intermediate linking the binding and the release of ions on opposite sides of the membrane to the decomposition of ATP. These enzymes are therefore one of the best understood examples of vectorial energy transduction.

### *5.2. $Na^+$ circuits in bioenergetics*

$Na^+$ gradients present all the properties that make them suitable for a bioenergetic role within cells: the ion usually presents a large difference in electrochemical potential across the plasma membrane and its absolute concentration is high enough for assuring intense ionic fluxes that can be coupled to significant mass transport phenomena. In fact, most of the active transport taking place across the plasma membranes of animal cells is driven by the co-transport of $Na^+$. A typical example of such a mechanism is the active absorption of nutrients across the apical membrane of intestinal epithelia; here, for example, glucose is concentrated more than $10^2$ fold inside the epithelial cell by cotransport of one $Na^+$ per glucose molecule. The concentrated sugar is subsequently released across the basolateral membrane merely through a passive permease. In this way the absorption of glucose is unidirectional across the epithelium and it is driven indirectly by the ATP hydrolyzed by the Na,K-ATPase, being the mechanism mediated by a $Na^+$-motive circuit (Fig. 13). Similar cotransport mechanisms are very general; they include also the transport of amino acids across all plasma membranes. These metabolites are transported by a group of permeases, selective for homogeneous groups of amino acids (e.g. cationic amino acids, acidic amino acids etc.), most of which are cotransporting sodium ions.

$Na^+$-driven and $K^+$-driven transport is also important for the control of intracellular pH in animal cells. Proton extrusion is mediated by a $Na^+/H^+$ antiporter, while proton uptake is coupled to a $K^+/H^+$ antiporter. This flux of protons coupled to $K^+$ or $Na^+$ fluxes is also at the basis of the control of cellular volume in mammals, since the coupling of the $Na^+/H^+$ or of the $K^+/H^+$ antiporter with the anion exchange permease (transporting $Cl^-$ in exchange with $HCO_3^-$) can result in a net efflux of KCl or a net influx of NaCl, controlling cellular osmolarity.

$Na^+$ transport is also one of the means by which the concentration of $Ca^{2+}$ can be controlled inside the cells, in addition to the already mentioned Ca-ATPase of the plasma membrane. $Ca^{2+}$ extrusion is also cata-

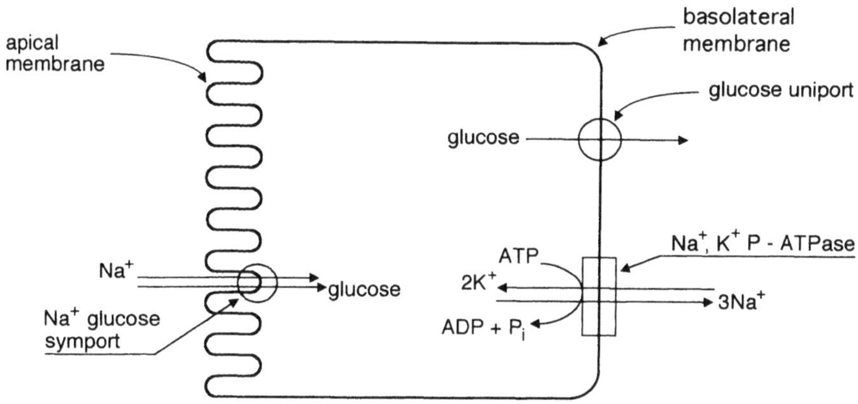


Fig. 13. Scheme of $Na^+$-driven transport of glucose across an intestinal epithelial cell. The scheme depicts an intestinal epithelial cell. On the left glucose is actively transported across the apical membrane from the intestinal lumen by a $Na^+$-glucose synporter. Excess $Na^+$ is extruded from the cell into the blood by the Na,K ATPases present across the basolateral membrane. The glucose accumulated is released to the blood by a uniport permease, since the intracellular concentration is constantly maintained higher than in the blood. A net flow of glucose and $Na^+$ across the intestinal epithelium is thus established, driven by ATP hydrolysis (i.e. by the Na,K ATPase). The coupling of the two processes is however mediated by $Na^+$ fluxes, through the action of the $H^+$/glucose synporter in the apical membrane.

lyzed by a $Ca^{2+}/3Na^+$ antiporter, and can be therefore indirectly driven through a $Na^+$-motive circuit by the Na,K-ATPase. In the case of the $Ca^{2+}/3Na^+$ antiporter the unbalance of the electric charges in the permease mechanism assures the energy contribution of the membrane potential and the net exoergonic nature of the process (compare Eqn. 29 for an evaluation of the free energy changes associated to this process).

Finally $Na^+$-currents, as well as $K^+$ and $Ca^{2+}$ currents, are fundamental for the operation of the nervous system and excitable tissues in general. The generation of action potential and its transmission along a neural axon is the result of a transient localized large ionic current mediated by voltage-gated channels present across the plasma membrane. Thus, while the voltage-gated channels are the regulatory elements of the neural function, the presence of an electrochemical potential difference across the plasma membrane for $Na^+$ and $Ca^{2+}$ (both negative inside) and for $K^+$ (positive inside) is the energetic condition for any normal activity. Again these ionic gradients are promoted and maintained by P-type ATPases (Ca- and Na,K-ATPases). Analogously $Na^+$, $K^+$ and $Ca^{2+}$ currents are at the basis of the operation of synapses (both neuronal and neuro-muscular synapses), in which the ionic channel conduction can be controlled either by voltage or by specific ligands (neurotransmitters) released into the synaptic cleff. In these cases, in addition to driving the ionic currents, $Na^+$-driven cotransport systems are also involved in the reabsorption of the neurotransmitters across the presynaptic membrane (Fig. 14) (exception to this mechanism is acetylcholine, that is hydrolysed directly within the synaptic cleff).

$Na^+$, $K^+$ and $Ca^{2+}$ fluxes are therefore involved in many, if not all, transport mechanisms taking place at the cell plasma membrane and that the very essence of fundamental processes in cellular physiology of animal cells, and to a minor degree also in all eukaryotic cells.

In prokaryots, $Na^+$-gradients play a less fundamental role. In some cases Na-cotransport systems have been described, but they are an exception rather than a widespread phenomenon. $Na^+$-gradients are in this case generated and maintained by $H^+/Na^+$-antiporters, while $K^+$-gradients can be produced by the above mentioned P-type ATPase. Their role in controlling intracellular pH and cellular turgor have already been discussed.

The exciting possibility of a much more fundamental role of $Na^+$ in the bioenergetics of some bacterial species has become evident following the discovery of direct transport systems for $Na^+$ coupled to decarboxylative enzymes and ATPases in some alkalophilic or halophilic bacteria. Prototype of these studies is the obligate fermentative anaerobic bacterium *Propi-*

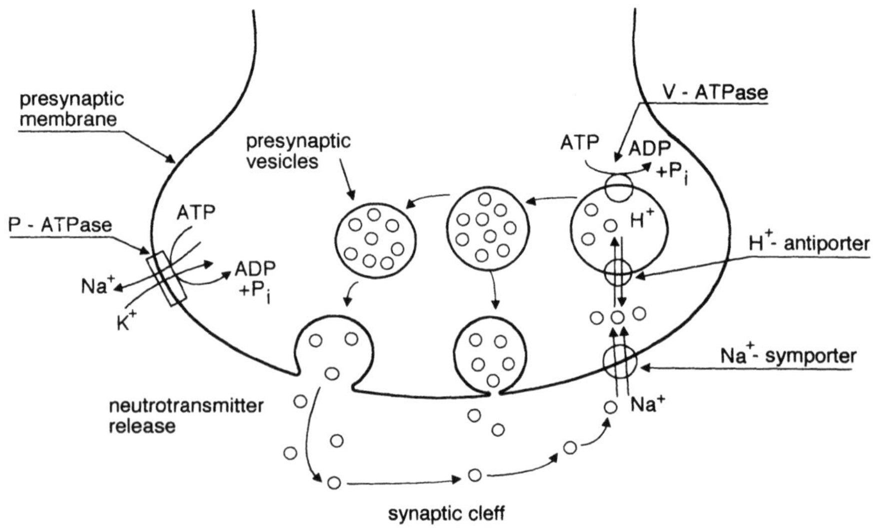


Fig. 14. Release and reabsorption of a generic neurotransmitter at the presynaptic membrane. Upon excitation of a synapse a neurotransmitter is released into the synaptic cleff through exocytosis of presynaptic vesicles (a type of secretion vesicle). The exocytotic mechanism is controlled by intracellular calcium concentration. The reabsorption of the neurotransmitter and its accumulation in new presynaptic vesicles involve the action of a $Na^+$-symporter across the presynaptic membrane and a $H^+$-antiporter across the vesicular membrane. The differences in $Na^+$ and $H^+$ activity across these membranes are maintained respectively by the Na,K ATPase (a P-type ATPase) and the vacuolar ATPase (a V-type enzyme). The present example demonstrates the different roles that $Na^+$- and $H^+$-driven fluxes perform in an eukaryotic cell, the former being involved in the extracellular and the latter in the intracellular transport. Analogously the differential role of P- and V-type ATPases is exemplified.

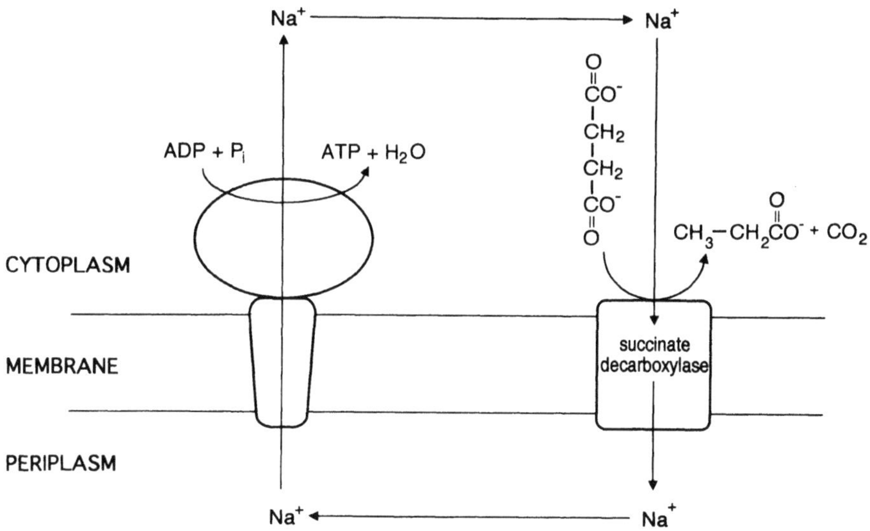


Fig. 15. Chemiosmotic $Na^+$-circuit in *Propionigenium modestum*. A chemiosmotic $Na^+$-driven ATP synthesis has been proposed to be operative in the anaerobic bacterium *P. modestum*. The only source of energy for this organism is a reaction decarboxylating succinate and producing propionate that is excreted into the external medium. This reaction catalyzed by an intrinsic membrane complex is coupled to the primary active transport of $Na^+$ out of the cell. An ATP synthase, very similar to a normal F-type ATP synthase, but specific for $Na^+$, can directly couple an inward $Na^+$ flow to ATP synthesis. The $Na^+$/succinate and $Na^+$/ATP stoichiometry, still unknown, are not specified in the scheme.

*onigenium modestum*. This organism lives anaerobically, converting succinate into propionate and $CO_2$; the decarboxylative step of this fermentation is exoergonic and is coupled to the active excretion of $Na^+$ from the cell. $\Delta\tilde{\mu}_{Na^+}$ appears therefore the only way by which the free energy of the decarboxylation is conserved across the plasma membrane of this bacterium. Recently from the same organism an F-type ATPase has been purified and ATP synthesis driven by artificially induced $Na^+$ potential difference has been demonstrated *in vitro*. It seem to be firmly established therefore that in this bacterium a $Na^+$-motive chemiosmotic circuit is operating, which produces the chemiosmotic synthesis of ATP mediated by $Na^+$-rather than $H^+$-fluxes (Fig. 15).

Although *P. modestum* is the only example so far studied at the level of purified enzyme complexes, similar $Na^+$-motive circuits have been claimed to be present also in other bacteria, noticeably in *E. coli* when grown at alkaline pH's. It is likely therefore that sodium chemiosmosis is a phenomenon relatively frequent in alkalophilic and halophilic microorganisms as an alternative solution when the induction of a protonic electrochemical potential difference of sufficient amplitude is difficult to achieve for environmental reasons.

## Further reading

*Multicompartment systems*
HV Westerhoff and K van Dam, *Thermodynamics and Control in Biological Energy Transduction,* Elsevier, Amsterdam, 1987.
D Walz, Biochim Biophys. Acta 1019 (1990) 171–224.
D Oesterheld, J Tittor and E Bamberg, J. Biomem. Bioenerg. 24 (1992) 181–192.

*Proton bioenergetics*
P Mitchell, Biol. Rev. 41 (1966) 445–502.
DG Nicholls and SJ Ferguson, *Bioenergetics 2,* Academic Press, London, 1992.
VP Skulachev, *Membrane bioenergetics,* Springer Verlag, Berlin, 1988.
D Pietrobon, M Zoratti, GF Azzone and RS Caplan, Biochem. 13 (1986) 767–775.

*Transport*
WD Stein, *Transport and diffusion across cell membranes,* Academic Press, New York, 1986.
N Nelson, Biochim. Biophys. Acta 1100 (1992), 109–124.
BP Rosen, Ann. Rev. Microbiol. 40 (1986) 263–286.
PM Cola, Curr. Top. Mem. Transp. 26 (1986) 79–100.

*$Na^+$ bioenergetics*
P Dimroth, Bioessays 13 (1991) 463–468.

Bioenergetics
ed. by P. Gräber and G. Milazzo†


# *CHAPTER 4*
# Energetics of aerobic and anaerobic bacteria

Michael Blaut and Gerhard Gottschalk

*Institut für Mikrobiologie der Georg-August-Universität Göttingen, Göttingen, Germany*

## 1. Introduction

Microorganisms play a pivotal role in the cycle of matter on our planet. Their importance lies in the decomposition of the organic material that was originally produced by the activity of photosynthetic autotrophs. Green plants and cyanobacteria capture the light and utilize its energy to synthesize organic material (biomass) from a number of simple inorganic molecules such as $CO_2$, $H_2O$, and $NH_4^+$.

Animal and most bacterial cells absolutely depend on the fixation products of the phototrophic organisms. They utilize them as starting material for the synthesis of their own cells or they degrade them in order to generate energy. The final products of their activities are $CO_2$ and $H_2O$. This is a simple description of the carbon cycle (Fig. 1).

In contrast to animal cells, bacteria may thrive under conditions that are difficult to tolerate for higher organisms. They have developed capabilities not met by higher organisms. Characteristically, bacteria occupy all possible niches in nature. They are very flexible with respect to the sub-

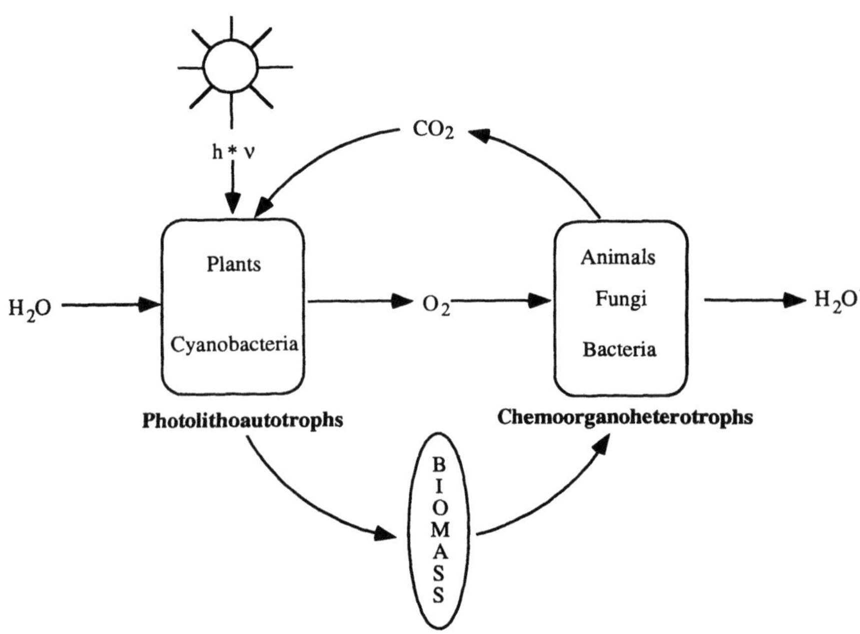


Fig. 1. The global carbon cycle.

strates utilized and to environmental changes. Due to the high surface to volume ratio, they display high rates of metabolism and reproduction. They have evolved an astonishing variety of degradative capabilities which allows the mineralization of all organic materials produced by organisms.

The principal biochemical routes are very much the same in all living cells, and the same mechanisms of energy transduction are used by prokaryotes and eukaryotes. However, the variety of biochemical pathways evolved by microorganisms is amazing. Each pathway reflects the organism's phylogenetic origin and the particular environmental conditions under which it grows. In spite of the identity or similarity of the major biochemical routes, bacteria have capabilities not shared by higher organisms.

It is impossible to give here a treatise on all metabolic variations encountered in bacteria. So it will be necessary to restrict the discussion to processes that are related to the production and utilization of energy. We will try to relate the energy-yielding reactions to the use of a particular substrate by a bacterium. Emphasis will be on pathways that are exclusively found in bacteria. Some selected examples will be treated in more details, others will only be mentioned. We will not discuss molecular mechanisms in detail.

## 2. Principles of biological energy transformations

For growth and reproduction, living cells depend on a continuous exchange of matter and energy with their environment. The cell harbors the machinery that transforms the physical or chemical energy taken up into biologically useful energy (catabolism). This energy is used for the transformation of nutrients into cellular constituents (anabolism).

The universal form of cellular energy is ATP. The generation of ATP from ADP and $P_i$ may occur by three principal mechanisms (see below). The other important molecule in the cellular metabolism is $NAD(P)^+$ which serves as a carrier of reducing equivalents. $NAD(P)^+$ cycles between its reduced and its oxidized state; its metabolic role depends on the mode of energy generation.

One useful way of distinguishing the biochemical capabilities of microorganisms is based on their nutrition (Table 1). It can be related to the sources of energy, electrons, and carbon utilized. The various modes of energy generation will be discussed along these categories.

Phototrophic bacteria exploit the light for the synthesis of ATP and in some instances for the reduction of $NAD(P)^+$. The mechanism of ATP synthesis employed by phototrophs is termed *photophosphorylation* (Fig. 2A). Light drives the energy-dependent transfer of electrons from an organic or inorganic electron donor to an acceptor molecule in the

Table 1. Modes of nutrition

| Energy source | Carbon source | Electron donor |
|---|---|---|
| *Chemical reactions:* Chemotrophic | *Organic compounds:* Heterotrophic | *Organic compounds:* Organotrophic |
| *Light:* Phototrophic | *Carbon dioxide:* Autotrophic | *Inorganic compounds:* Lithotrophic |

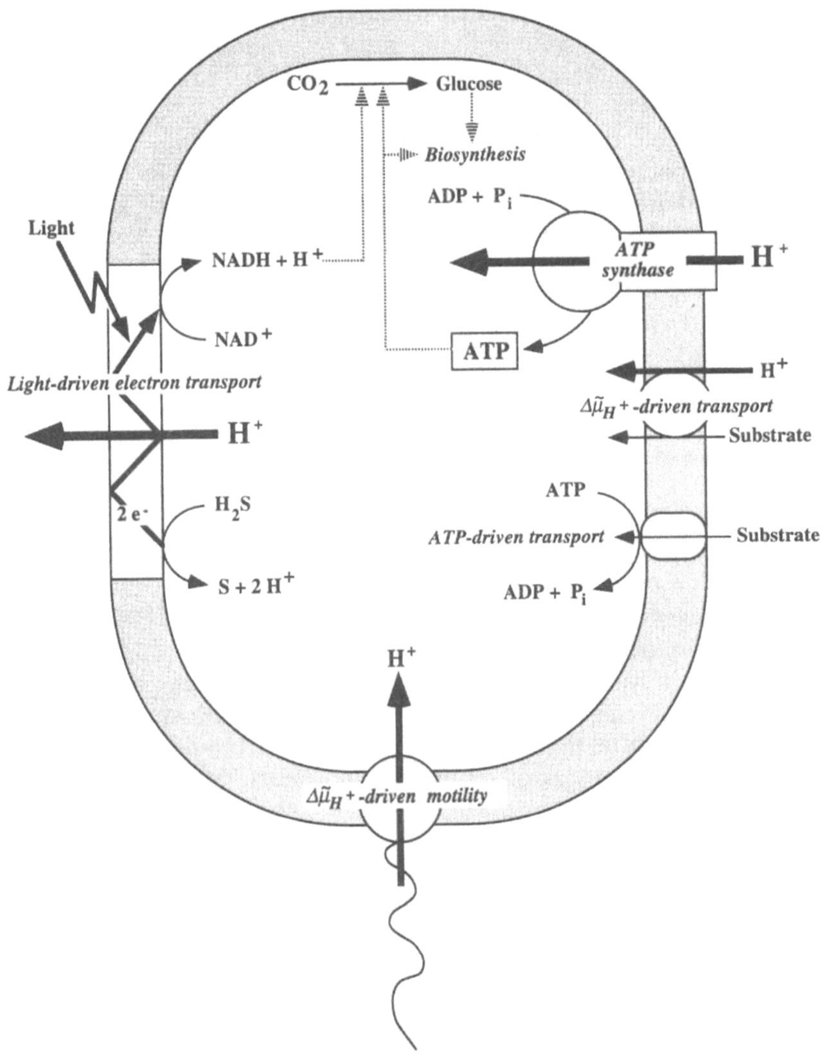


Fig. 2A. The three principal modes of energy conservation encountered in bacterial cells. A. Photophosphorylation.

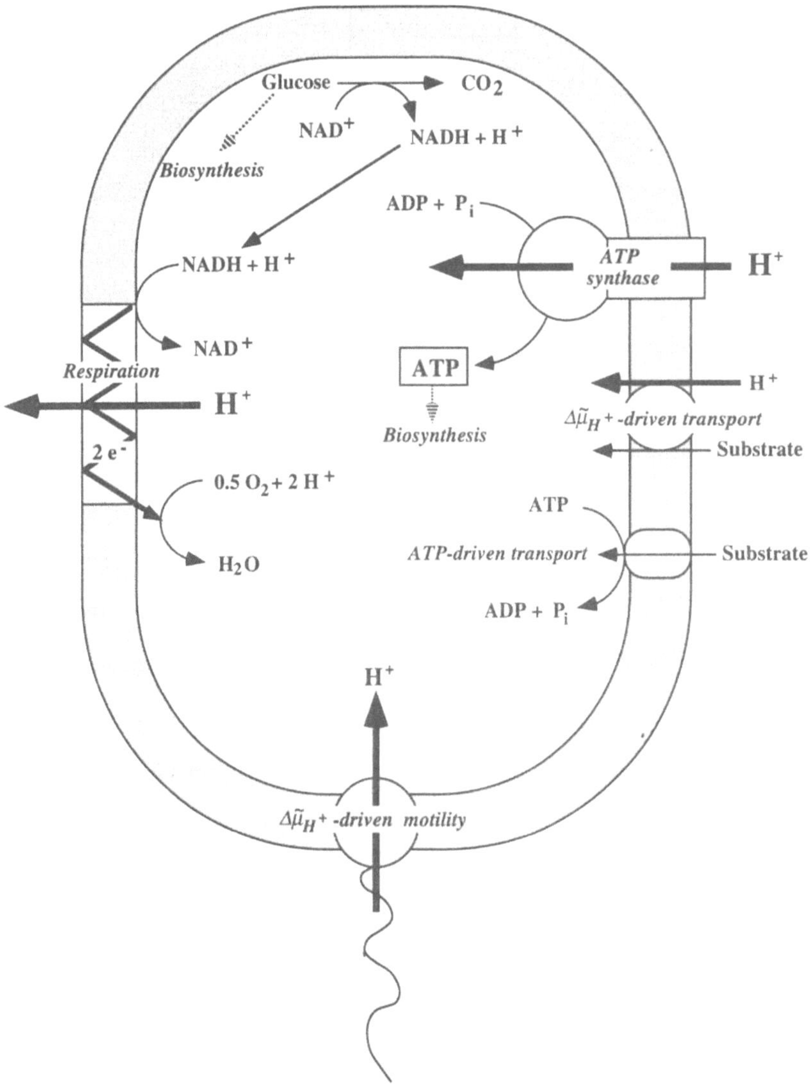


Fig. 2B. Oxidative phosphorylation or electron transport phosphorylation.

photoreaction center with a comparably negative redox potential ($E_{m,7}$). The exergonic electron transport from this acceptor molecule to a carrier with a more positive redox potential is accompanied by the extrusion of protons from the inside of the bacterial photosynthetic cell to the outside. This electrochemical potential difference ($\Delta\tilde{\mu}_{H^+}$), the protonmotive force (p), is composed of an electrical potential difference or membrane potential ($\Delta\phi$) and a chemical potential difference ($\Delta pH = pH_{in} - pH_{out}$):

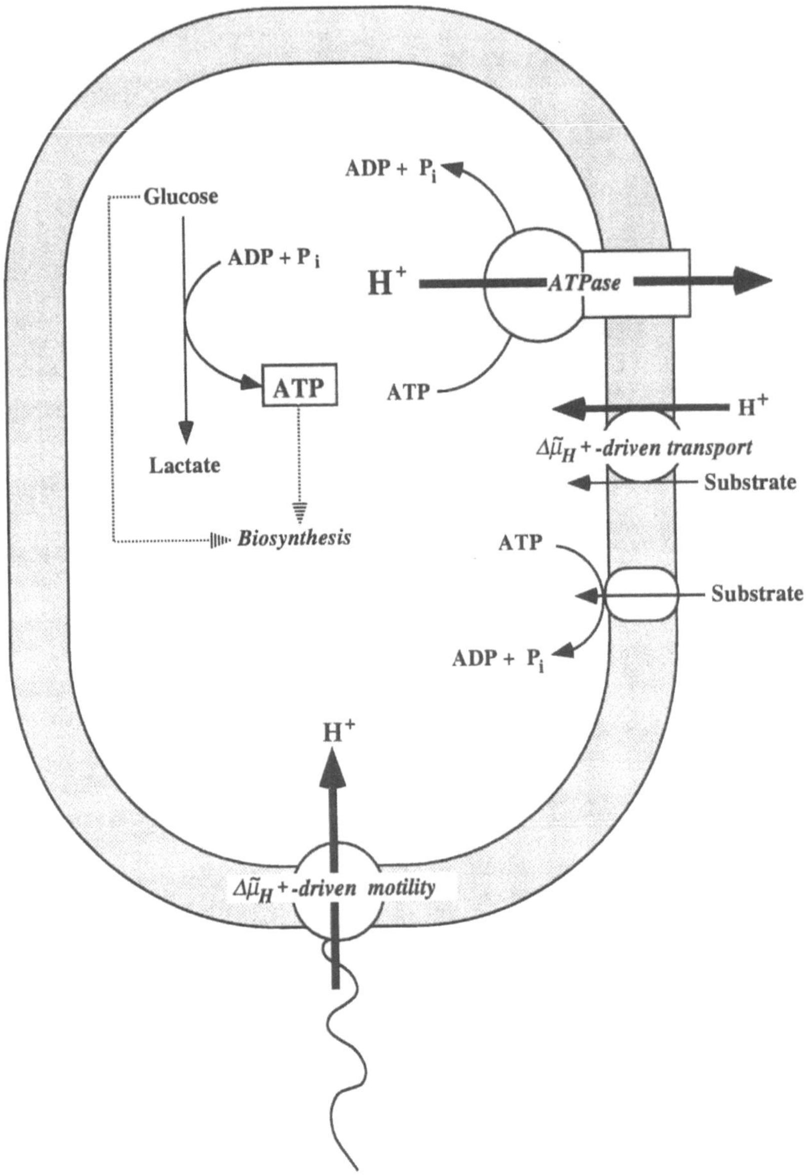


Fig. 2C. Substrate level phosphorylation. Explanations are given in the text.

$$\Delta\tilde{\mu}_{H^+} = F\,\Delta\phi - 2.303\,RT\,\Delta pH$$

$$p = \frac{\Delta\tilde{\mu}_{H^+}}{F}$$

The energy stored in this transmembrane $\Delta\tilde{\mu}_{H^+}$ drives the endergonic synthesis of ATP from ADP and $P_i$ via the membrane-bound ATP synthase and other processes such as active transport and flagellar movement. ATP synthesis is accomplished by the influx of protons along their potential gradient into the cell. At equilibrium the number of protons (n) flowing through the ATP synthase and the minimal $\Delta\tilde{\mu}_{H^+}$ needed for ATP synthesis correlate with the phosphorylation potential ($\Delta G_p'$) according to the following equation:

$$n\,\Delta\tilde{\mu}_{H^+} = \Delta G_p'$$

$\Delta G_p'$ corresponds to the Gibbs free energy change for the synthesis of ATP:

$$\Delta G_p' = \Delta G^{0'} + 2.303\ RT \log \frac{[ATP]}{[ADP]\,[P_i]}$$

The energy generated by photophosphorylation is used for anabolic reactions. Autotrophic phototrophs have to invest ATP for the *de novo* synthesis of cell constituents from $CO_2$ and NAD(P)H. Depending on the photosynthetic organism NAD(P)H may either be formed by light-driven or $\Delta\tilde{\mu}_{H^+}$-driven electron transport.

Chemotrophic bacteria derive their metabolic energy from the oxidation of organic or inorganic electron donors. The Gibbs free energy changes ($\Delta G^{0'}$) associated with these oxidation reactions is taken advantage of for energy-dependent metabolic activities. Two principal mechanisms of ATP generation can be distinguished in chemotrophic cells: substrate level phosphorylation and oxidative phosphorylation.

*Oxidative phosphorylation* is a process in which the oxidation of a substrate and the ensuing exergonic electron transport to a terminal acceptor such as $O_2$ is accompanied by the generation of $\Delta\tilde{\mu}_{H^+}$ which drives the synthesis of ATP from ADP and $P_i$ via a membrane-bound ATP synthase (Fig. 2B). Alternatively, this process may be called *electron transport phosphorylation*. NAD(P)H for biosyntheses is in most cases formed during substrate oxidation. In some instances, however, $NAD(P)^+$ reduction is achieved by $\Delta\tilde{\mu}_{H^+}$-driven reverse electron transport.

*Substrate level phosphorylation* refers to a process in which the Gibbs free energy of a reaction is conserved through the formation of a phosphorylated intermediate. The activated phosphoryl group is subsequently transferred to ADP yielding ATP. Organisms that gain ATP exclusively by substrate level phosphorylation have to generate $\Delta\tilde{\mu}_{H^+}$ for processes that depend on this form of energy (Fig. 2C). This is accomplished by the extrusion of protons from the cell at the expense of the hydrolysis of ATP. This process is the reversal of the ATP synthase reaction mentioned above, and it is catalyzed by the same type of enzyme called ATPase.

Whereas oxidative phosphorylation and photophosphorylation are intrinsically membrane-associated processes, substrate level phosphorylation is independent of membranes.

## 3. Energy-generating processes

### *3.1. Chemotrophic bacteria*

Chemotrophic bacteria exploit organic and/or inorganic chemical compounds for the generation of metabolic energy. The Gibbs free energy change ($\Delta G^{0'}$) associated with the oxidation of these compounds is taken advantage of for the synthesis of ATP.

*3.1.1. Chemoorganotrophic bacteria:* Chemoorganotrophic bacteria couple the oxidation of organic substrates with the generation of ATP. Although in principal any substance produced in nature may be degraded by bacteria, there are differences regarding the degradative versatility. Some bacteria may utilize a wealth of substrates, whereas others are restricted to a limited number or even a single substrate.

*3.1.2. Pathways of sugar degradation:* Carbohydrates play an important role as substrates for chemoorganotrophic organisms. Taking glucose as an example, the most prominent oxidative routes will be shortly summarized (Fig. 3).

Depending on the organism, one out of three pathways is used for the initial oxidation of glucose. The *Embden-Meyerhof-Parnas (EMP) pathway* (also called glycolysis) is not only found in bacteria, but occurs also in animal and plant cells. The phosphorylation of glucose with ATP as phosphoryl donor (or phosphoenolpyruvate when taken up through the phosphotransferase system) yields glucose-6-phosphate (Fig. 3A). The isomerizaton of the latter to fructose-6-phosphate is followed by another ATP-dependent phosphorylation giving rise to fructose-1,6-bisphosphate. This reaction is catalyzed by phosphofructokinase, the key enzyme of the pathway. Fructose bisphosphate aldolase splits fructose-1,6-bisphosphate into dihydroxyacetonephosphate and glyceraldehyde-3-phosphate. Triosephosphate isomerase catalyzes the interconversion of the two $C_3$ intermediates. The further degradation of the $C_3$ units to pyruvate starts with the oxidation of glyceraldehyde-3-phosphate and the concomitant reduction of $NAD^+$. The Gibbs free energy change associated with this oxidation is conserved by the formation of 1,3-bisphosphoglycerate which serves as phosphoryl donor in the ensuing phosphorylation of ADP giving rise to ATP and 3-phosphoglycerate. This is an example of a substrate level phosphorylation. Following the conversion of 3-phosphoglycerate to 2-phosphoglycerate, $H_2O$ is split off yielding phosphoenol-

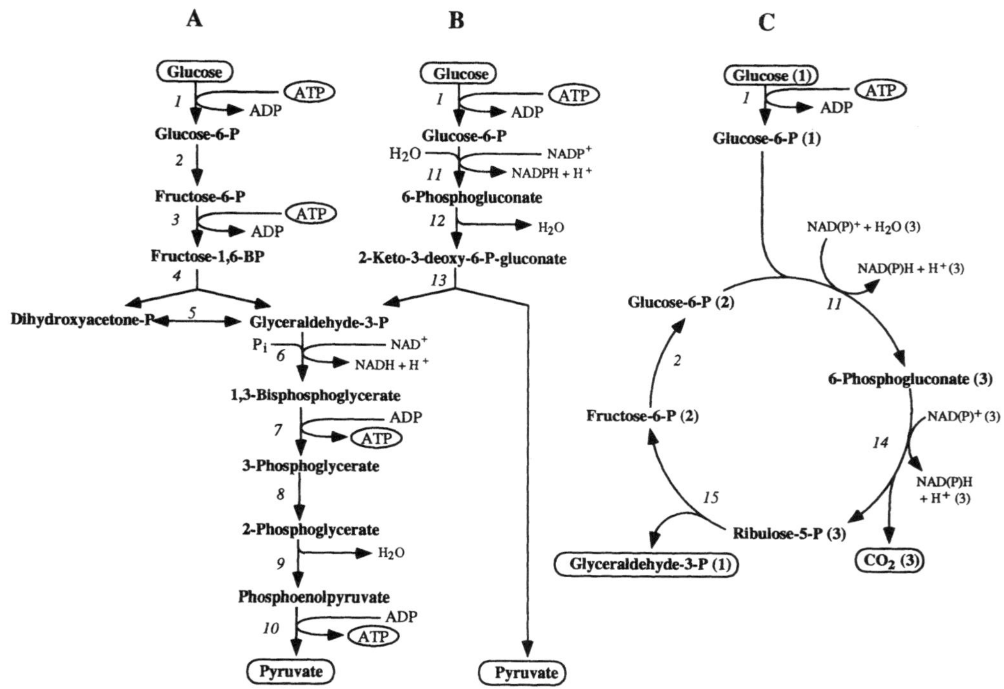


Fig. 3. The three most common pathways employed by bacteria for sugar degradation. A. Embden-Meyerhof-Parnas pathway (glycolysis). B. Entner-Doudoroff pathway. C. Cyclic pentose phosphate pathway. In C the number of molecule is given in parentheses. The italicized numbers represent the following enzymes: *1. hexokinase. 2. glucose-6-phosphate isomerase. 3. phosphofructokinase. 4. fructose-1,6-bisphosphate aldolase. 5. triosephosphate isomerase. 6. glyceraldehyde-3-phosphate dehydrogenase. 7. 3-phosphoglycerate kinase. 8. phosphoglycerate mutase. 9. enolase. 10. pyruvate kinase. 11. glucose-6-phosphate dehydrogenase. 12. 6-phosphogluconate dehydratase. 13. KDPG aldolase. 14. 6-phosphogluconate dehydrogenase. 15. Various enzymes transforming 3* $C_5$ *sugars to 2* $C_6$ *sugars and 1* $C_3$ *compound.*

pyruvate. Conversion of the latter to pyruvate as catalyzed by pyruvate kinase is associated with ATP formation by substrate level phosphorylation. In summary, the EMP pathway degrades one glucose to two molecules of pyruvate and is linked to the formation of two ATP and two NADH.

A second important route for carbohydrate breakdown is the *Entner-Doudoroff (ED) pathway* (Fig. 3B). It is found in many aerobic Gram-negative bacteria such as *Pseudomonas, Alcaligenes,* and *Paracoccus.* The degradation of one glucose along this pathway leads, just like in the EMP pathway, to the formation of two molecules of pyruvate. In an $NAD(P)^+$-dependent reaction glucose-6-phosphate is oxidized to 6-phosphogluconate. Subsequently, 6-phosphogluconate dehydratase catalyzes the transformation to 2-keto-3-deoxy-6-phosphogluconate (KDPG) which is split by KDPG aldolase into pyruvate and glyceraldehyde-3-phosphate. The latter is further degraded to pyruvate by enzymes of the EMP pathway. The key enzymes of the ED pathway are 6-phosphogluconate dehydratase and KDPG aldolase. Glucose oxidation along this pathway is associated with the formation of two NAD(P) and one ATP.

In some bacteria such as *Thiobacillus novellus* and *Brucella abortus* a third pathway for carbohydrate degradation, the *oxidative pentose phosphate cycle,* is operating (Fig. 3C). These organisms are devoid of the key enzymes of the above two pathways. Glucose-6-phosphate enters the cycle and is oxidized to 6-phosphogluconate. The latter is further oxidized to ribulose-5-phosphate and $CO_2$. Both steps are associated with the reduction of $NAD(P)^+$. Three molecules of ribulose-5-phosphate are subsequently transformed to one molecule of glyceraldehyde-3-phosphate and two molecules of fructose-6-phosphate. No ATP, but two NAD(P)H are formed in this pathway. The oxidative pentose phosphate cycle may bring about the complete oxidation of glucose to six $CO_2$ without the participation of the tricarboxylic acid (TCA) cycle.

The common end product of the EMP and the ED pathway is pyruvate. Pyruvate dehydrogenase catalyzes its oxidation to acetyl-CoA and $CO_2$ which is coupled to NADH formation (Fig. 4). The subsequent condensation of acetyl-CoA with oxaloacetate in the TCA cycle yields citrate which is converted to isocitrate. The latter undergoes an oxidation to $\alpha$-oxoglutarate and $CO_2$ under formation of NAD(P)H followed by another oxidation to succinyl-CoA, the conversion of which, to succinate, is accompanied by the phosphorylation of ADP. The ensuing oxidation of succinate to fumarate ($E_{m,7}$ = +33 mV) cannot be coupled with the reduction of $NAD^+$ to NADH ($E_{m,7}$ = −320 mV), but the electrons are fed into the respiratory chain at the level of ubiquinone. This is catalyzed by the FAD-containing membrane-bound succinate dehydrogenase (see below). The conversion of fumarate to malate and the subsequent $NAD^+$-dependent oxidation of malate to oxaloacetate closes the TCA cycle. In

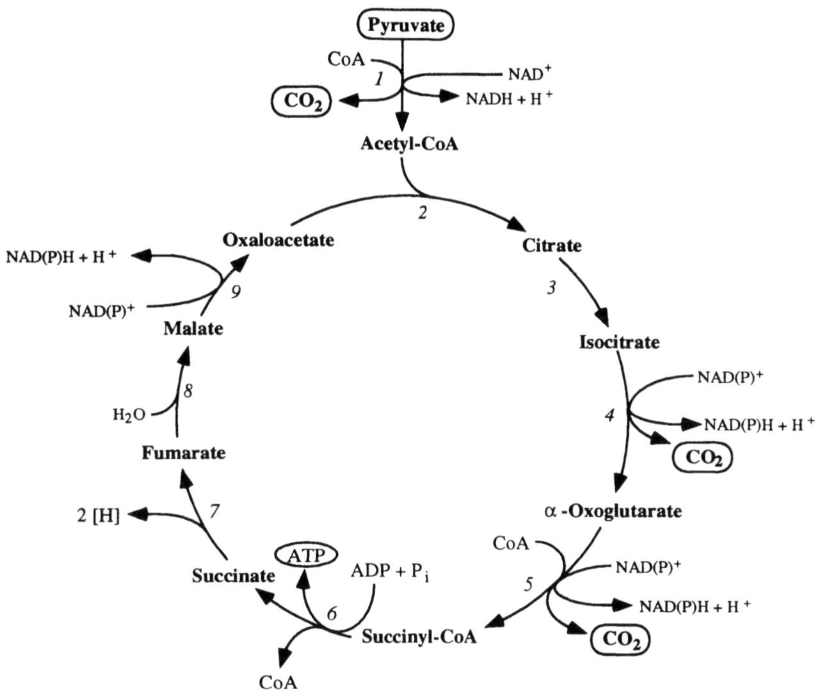


Fig. 4. The complete oxidation of pyruvate via pyruvate dehydrogenase and the tricarboxylic acid cycle. The italicized numbers represent the following enzymes: *1. pyruvate dehydrogenase. 2. citrate synthase. 3. cis-aconitate hydratase. 4. isocitrate dehydrogenase. 5. α-oxoglutarate dehydrogenase. 6. succinate thiokinase. 7. succinate dehydrogenase. 8. fumarase. 9. malate dehydrogenase.*

summary, the oxidation of pyruvate to 3 $CO_2$ as carried out by the concerted action of pyruvate dehydrogenase and TCA cycle is linked to the formation of 1 ATP, 4 NAD(P)H, and one additional pair of reducing equivalents.

*3.1.3. Energy conservation in chemoorganotrophic bacteria:* Chemoorganotrophs conserve energy by oxidative phosphorylation (respiration), by substrate-level phosphorylation, or by both. The principal differences between these two mechanisms were outlined above. Respiration allows for the complete oxidation of a given organic substrate to $CO_2$ and makes the maximal Gibbs free energy change ($\Delta G^{0\prime}$) available. If an external electron acceptor such as $O_2$ or nitrate is not present, or if an organism is devoid of an electron transport chain, the oxidation of a substrate must remain incomplete resulting in a smaller $\Delta G^{0\prime}$.

This is explained by taking the following example: let us assume glucose is degraded to pyruvate by way of the EMP pathway. During this conversion both ATP and NADH are generated. ATP formation occurs at this stage only by substrate-level phosphorylation. The ATP gain is 2 ATP/glucose. For organisms devoid of respiration two problems arise: First, the NADH produced has to be reoxidized in order to recover $NAD^+$. Since no external electron acceptor such as $O_2$ or nitrate is available, the electrons have to be transferred to an intermediate of the degradation pathway. Lactic acid bacteria for instance recover $NAD^+$ by transferring the reducing equivalents to pyruvate giving rise to their typical fermentation product lactic acid (see also 3.1.12.1.). Second, the Gibbs free energy change associated with the lactic acid fermentation is only –198 kJ/mol glucose. Instead of using pyruvate as electron acceptor, respiring organisms oxidize pyruvate further by way of pyruvate dehydrogenase and the ensuing TCA cycle. They take advantage of a membrane-bound electron transport chain which takes over the reducing equivalents from NADH and transfers them to the terminal electron acceptor $O_2$. Hence respiration not only recovers $NAD^+$ but, more importantly, leads to the generation of a transmembrane electrochemical proton potential difference which may drive the synthesis of ATP by means of an ATP synthase. The complete oxidation of glucose to $CO_2$ is accompanied by a Gibbs free energy change of –2847 kJ/mol glucose. The oxidation of the NADH produced in glycolysis and the TCA cycle is associated with the synthesis of maximally 38 ATP from ADP and $P_i$ per mol of glucose illustrating the high efficiency of respiration compared to fermentation.

*3.1.4. Aerobic chemoorganotrophs:* The number and variety of microorganisms taking advantage of $O_2$ as terminal electron acceptor is immense. The reducing equivalents generated by the oxidation of organic substrates are fed into the respiratory chain. The principal setup of the respiratory chain is similar in all aerobic bacteria. The main point of entry is the NADH dehydrogenase. The electrons are passed along membrane-bound electron carriers to $O_2$ giving rise to the formation of a transmembrane $\Delta\tilde{\mu}_{H^+}$. Simply stated, an electron transport chain is a device that serves the purpose of efficiently transforming the chemical energy of substrate oxidation into an energy form that can be used by a bacterial cell to perform work.

Although most aerobic bacteria have the capacity to oxidize sugars such as glucose, there are many more substrates that are degraded by microorganisms. Some of them are directly oxidized by membrane-bound complexes so that the entry into the electron transport chain takes place without the need for NADH formation. Typical examples found in various bacteria are succinate dehydrogenase (see also above), lactate dehydrogenase, hydrogenase, and formate dehydrogenase. Some of these examples will be treated below in more detail.

Whereas the metabolism of higher animals (which are also chemoorganotrophs) depends absolutely on oxygen as terminal electron acceptor, aerobic bacteria differ significantly in that respect. Although aerobic bacteria have a preference for $O_2$ as terminal electron acceptor, many of them may in addition utilize alternative electron acceptors such as nitrate or fumarate; many have even the capacity to gain ATP by substrate-level phosphorylation when neither $O_2$ nor alternative terminal electron acceptors are available. This flexibility enables bacteria to react properly to changes in their environment and to survive under various conditions. However, there are groups of bacteria, the metabolism of which strictly depends on oxygen. *Acinetobacter* species are a typical example of strictly aerobic bacteria. Members of this genus are well known for their capacity to degrade a large number of chemical compounds. These are initially transformed into intermediates of the major biochemical routes. Typically, these primary degradative reactions are not associated with the formation of ATP by substrate-level phosphorylation. A strict oxygen dependence may also arise from the necessity to oxygenate an inert molecule such as a hydrocarbon to allow its further degradation. This may either be realized by dioxygenase reactions which catalyze the introduction of both oxygen atoms from $O_2$ into a compound, or by a mono-oxygenase reaction. In the latter case one oxygen atom from $O_2$ is introduced into the compound being oxidized and the other is reduced to $H_2O$ (for an example see 3.1.7.). It is obvious that the reducing equivalents used in a mono-oxygenase reaction are not available for the respiratory chain and, therefore, also not for ATP generation. The conversion of catechol to *cis, cis*-muconate, both of which are key intermediates in the degradation of aromatic compounds, may serve as an example for a dioxygenase reaction:

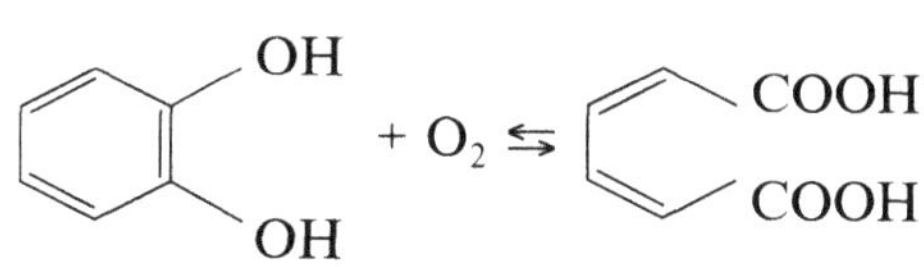


*3.1.5. The respiratory chain of* Paracoccus denitrificans: Electron transport chains are composed of flavoproteins, iron-sulfur proteins, quinones, cytochromes and metal (other than Fe)-containing proteins. The proteinaceous compounds are usually organized in large complexes which are embedded in the cytoplasmic membrane. Reducing equivalents fed into the electron transport chain are passed along a series of redox components with increasing redox potential to a terminal electron acceptor such as $O_2$. The Gibbs free energy change associated with this process is taken advantage of for the generation of a transmembrane electrochemical potential difference of protons. The spatial organization of a representative respiratory chain from the aerobic bacterium *Paracoccus denitrificans* is given in Fig. 5. The principal setup of the *Paracoccus* respiratory chain bears many similarities with the mitochondrial respiratory chain. However, the design of the bacterial counterpart is simpler. In analogy to the terminology used for the mitochondrial electron transport chain the traditional number of the complexes is given. Three complexes (I, III, IV)

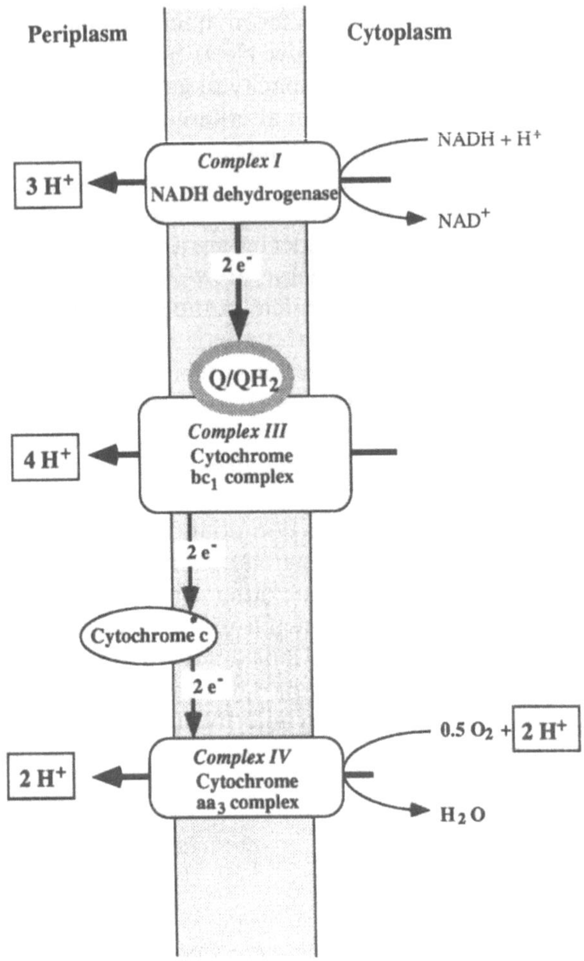


Fig. 5. Simplified scheme of the aerobic respiratory chain of *Paracoccus denitrificans*. Although present, alternative oxidases were omitted for the sake of clarity. No details concerning the mechanism of proton translocation in the $bc_1$ complex are given.

associated with proton translocation can be distinguished; complex II which corresponds to the succinate dehydrogenase of the mitochondrial respiratory chain is not included in the scheme. NADH oxidation is catalyzed by complex I, designated as NADH dehydrogenase. NADH oxidation occurs at the cytoplasmic face of the membrane and the reducing equivalents are transferred to ubiquinone which is thereby reduced to ubiquinol. The prosthetic groups involved in this downhill redox reaction are FMN and Fe-S centers. Electron transfer through this complex drives proton translocation.

Several $NAD^+$-independent dehydrogenases such as glycerol-3-phosphate dehydrogenase, malate dehydrogenase as well as succinate dehydrogenase may be found in respiratory bacterial membranes. These ubiquinone reductases may be considered as alternatives to the mitochondrial complex II (succinate dehydrogenase). The composition of the quinone reductases depends on the substrate being oxidized. The main types of quinones found in bacteria are menaquinone and ubiquinone. Whereas most bacteria contain one of the two, some few bacteria such as *Escherichia coli* contain both types. The phototrophic cyanobacteria contain plastoquinone.

The second complex of the respiratory chain involved in energy transduction is the quinol: cytochrome *c* oxidoreductase (complex III). This complex is widely distributed not only in bacteria, but finds its counterpart also in mitochondria and in chloroplasts. In the latter it is called cytochrome $b_6f$ complex. It is often referred to as $bc_1$ complex to indicate the presence of two *b*-type cytochromes and of cytochrome $c_1$; it contains in addition the Rieske iron-sulfur protein, which has a 2 Fe-2 S center bound to two cysteine and two histidine residues. The $bc_1$ complex mediates the electron transfer from ubiquinol to the periplasmic cytochrome *c* and the concomitant translocation of protons. The coupling between the electron transfer reaction and the proton extrusion (4 $H^+$/2 $e^-$) is mechanistically brought about by the so-called Q-cycle which involves the formation of a semiquinol. The details of the Q-cycle are discussed in Chapter 10. The periplasmic cytochrome *c* mediates the electron transfer between the $bc_1$ complex and the cytochrome oxidase complex (complex IV).

Cytochrome oxidase catalyzes the electron transfer from cytochrome *c* to $O_2$ which is associated with proton translocation. The $aa_3$-type cytochrome *c* oxidase contains all the major elements also found in the mitochondrial cytochrome *c* oxidase: cytochrome *a* ($Fe_A$), cytochrome $a_3$ ($Fe_B$), $Cu_A$, and $Cu_B$; $Cu_A$ and $Cu_B$ designate copper-containing redox centers. The subunit composition of the bacterial $aa_3$-type cytochrome oxidase is much simpler than its mitochondrial counterpart. It is composed of two or three subunits in contrast to the enzyme from beef heart mitochondria which contains 13 subunits (see chapter 11).

*3.1.6. Branched electron transport chains:* The scheme in Fig. 5 does not take into account that most bacteria including *P. denitrificans* possess branched electron transport chains. Branched respiratory chains are very common in bacteria and may reflect the fact that microorganisms have to be flexible with respect to changes of the substrate and the availability of $O_2$. Branched electron transport chains are characterized by alternative electron transport routes. The electron transport chain outlined in Fig. 5 represents therefore only one of three alternatives found in *P. denitrificans.*

A second route branches off in the ubiquinol region (Fig. 6). This terminal oxidase catalyzes the direct electron transfer from ubiquinol to $O_2$

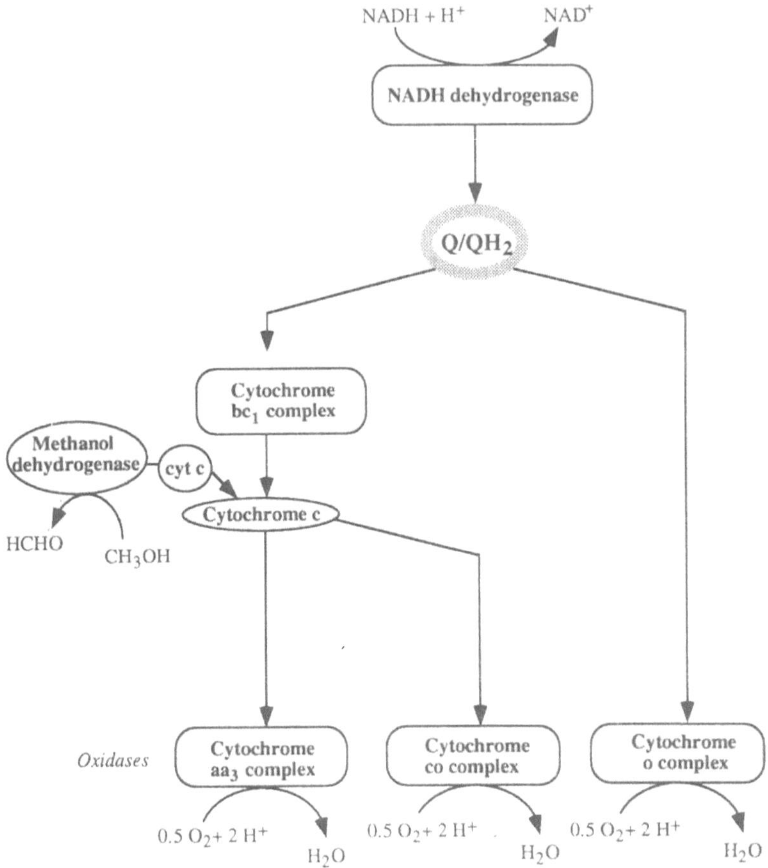


Fig. 6. Diagram of the branched respiratory chain of *Paracoccus denitrificans* during aerobic growth on methanol. The arrows represent the flow of electrons. Q/$QH_2$, ubiquinone pool. See text for further explanations.

and acts therefore as ubiquinol oxidase. It contains heme *b* and may be termed cytochrome *o* complex because of its oxidase function. Cytochrome *o* is a *b*-type cytochrome which reacts with $O_2$ as terminal electron acceptor and shows a characteristic absorption spectrum upon binding of CO. Mutants defective in the $bc_1$ complex still grow with maximal rate but with reduced growth yields (85% of the wild type). In these mutants the electrons flow from ubiquinol to the cytochrome *o*-type oxidase. However, two protons less are translocated in this pathway per two electrons than in the $aa_3$ pathway.

A third terminal oxidase branches off at cytochrome *c* (Fig. 6). This alternative cytochrome oxidase is a cytochrome *co*-type oxidase which also displays proton-translocating activity. It is produced in *P. denitrificans* during growth on substrates such as methanol or methylamine when the

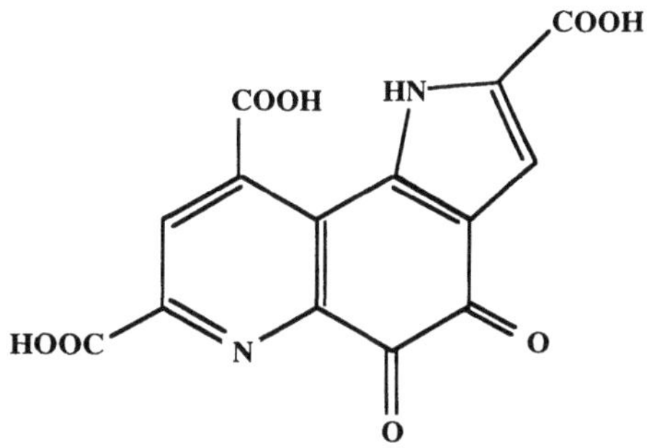


Fig. 7. The chemical structure of pyrrolo-quinoline quinone (PQQ).

$aa_3$-type oxidase is inhibited or deleted by mutation. Methanol or methylamine oxidation to formaldehyde ($E_{m,7}$ = −180 mV) is catalyzed by periplasmic dehydrogenases which contain pyrrolo-quinoline quinone (PQQ) as prosthetic groups (Fig. 7).

This coenzyme has a characteristic absorption spectrum with a maximum at 345 nm. The midpoint potential ($E_{m,7}$) of the unbound PQQ is +120 mV. PQQ occurs in a number of other bacterial dehydrogenases as well. Examples are the glucose dehydrogenase and the alcohol dehydrogenase from *Acinetobacter calcoaceticus*.

The reducing equivalents derived from methanol oxidation are then transferred to a cytochrome *c* which is especially required for methanol oxidation and subsequently to the typical cytochrome *c* of the respiratory chain. The oxidation of cytochrome *c* is brought about by the $aa_3$-type oxidase or, alternatively, by the *co*-type cytochrome oxidase. In contrast to the $aa_3$-type oxidase, electron transport via the *co*-type oxidase is not accompanied by proton extrusion. This is in accordance with the finding that mutants defective in the $aa_3$-type oxidase display 40% reduced growth rates and 14% decreased growth yields.

The production as well as the activity of all these terminal oxidases is highly regulated and depends on parameters such as oxygen and carbon supply. With *P. denitrificans*, the $aa_3$-type oxidase is expressed under conditions of optimal oxygen supply. Under these conditions, most of the electrons flow by way of this oxidase to $O_2$.

*3.1.7. Methylotrophic bacteria:* Certain chemotrophic bacteria such as *P. denitrificans* have the ability to adapt their energy-generating machinery to the conditions dictated by the available substrate or by the supply of oxygen. Not every bacterial species exhibits the same flexibility as *P. denitrificans*. Other groups of bacteria are specialists utilizing only a limited number of substrates. Examples are obligately methylotrophic organisms such as *Methylomonas methanooxidans* or *Methylosinus trichosporium*. These bacteria are only capable of growth on methane, methanol, or methylamines. Methane oxidation by these organisms absolutely depends on the presence of $O_2$. Methane oxidation is initiated by a soluble non-specific mono-oxygenase reaction (Fig. 8):

$$CH_4 + O_2 + NAD(P)H + H^+ \rightarrow CH_3OH + NAD(P)^+ + H_2O$$
$$\Delta G^{0'} = -341 \text{ kJ/mol}$$

The introduction of an oxygen atom from $O_2$ into methane gives rise to methanol and is accompanied by the reduction of a second oxygen atom to $H_2O$. Other substrates such as alkanes or $NH_3$ are oxidized as well by this mono-oxygenase. The consumption of NADH in the methane mono-oxygenase reaction decreases the ATP gain because less NADH may be oxidized in the respiratory chain.

The subsequent oxidation of methanol is catalyzed by the periplasmic methanol dehydrogenase which contains pyrrolo-quinoline quinone (PQQ) as a prosthetic group (Fig. 7). The passage of the electrons from methanol dehydrogenase to the respiratory chain occurs by way of a special periplasmic cytochrome *c* (called $c_L$) to cytochrome $c_H$ (Fig. 8). The latter corresponds to the typical soluble cytochrome *c* of the bacterial res-

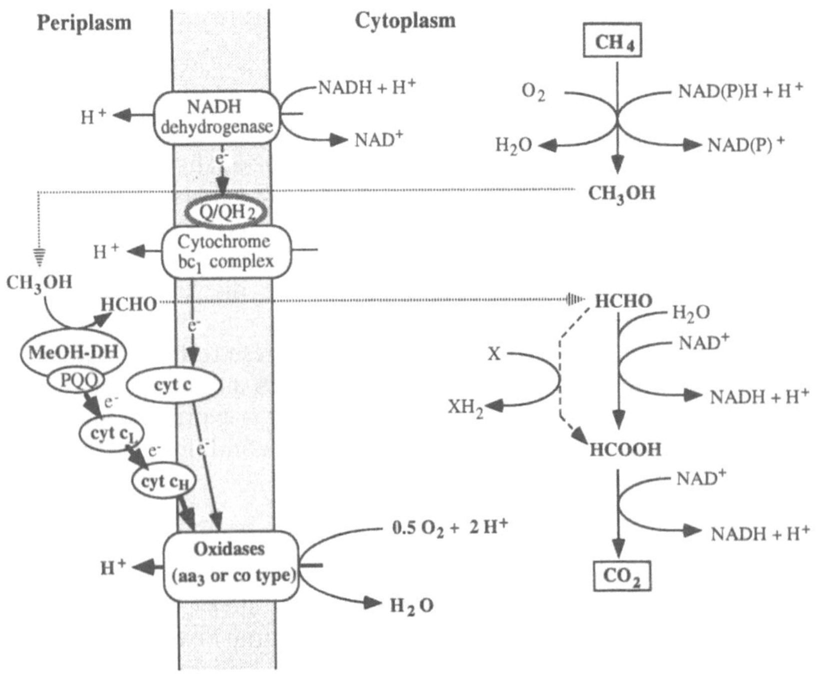


Fig. 8. Methane oxidation by methanotrophic bacteria. Stoichiometries of electron flow and proton translocation have been disregarded. X, unknown electron acceptor. Further explanations are given in the text.

piratory chain. The oxidation of cytochrome $c_H$ is catalyzed by $aa_3$ and *co*-type terminal oxidases as described above for *P. denitrificans*. Which of the terminal oxidases dominates in respiration is governed by the oxygen and the carbon supply, the $aa_3$-type oxidase being the preferred pathway to $O_2$ at high oxygen tension and under methanol-limitation. The $aa_3$-type oxidases found in various methylotrophic bacteria may differ in their capacity to pump protons.

The further oxidation of formaldehyde may occur in two principal ways. In a linear pathway, formaldehyde is oxidized to $CO_2$ by the $NAD^+$-dependent formaldehyde and formate dehydrogenases. Whereas $NAD^+$ serves as the principal electron acceptor in formate oxidation some methylotrophs link the oxidation of formaldehyde to unknown electron acceptors. The use of electron acceptors other than $NAD^+$ may have energetic implications, because lower energy yields can be expected for non-$NAD^+$-dependent dehydrogenases. Some methylotrophs such as *Methylophilus methylotrophus* oxidize formaldehyde by way of the dissimilatory hexulose monophosphate cycle (Fig. 9) which brings about the oxidation of formaldehyde to $CO_2$ according to the equation:

$$HCHO + 2NAD(P)^+ + H_2O \leftrightarrows CO_2 + 2NAD(P)H + 2H^+$$

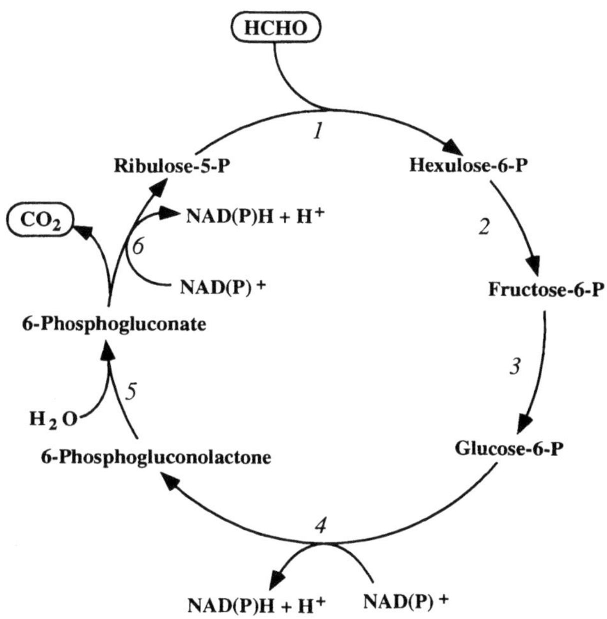


Fig. 9. The dissimilatory hexulose phosphate cycle. The italicized numbers represent the following enzymes: *1. hexulose phosphate synthase. 2. hexulose phosphate isomerase. 3. glucose phosphate isomerase. 4. glucose-6-phosphate dehydrogenase. 5. gluconolactonase. 6. 6-phosphogluconate dehydrogenase.*

Which of the formaldehyde oxidation pathways is used depends on the methylotrophic organism. The produced NADH is either used in the methane mono-oxygenase reaction or it is fed into the respiratory chain via NADH dehydrogenase (Fig. 8).

Although obligate methylotrophs represent very specialized bacteria, the principal setup of their electron transport chain shows a high degree of similarity with that of other aerobic chemotrophic bacteria. The differences encountered in detail reflect adaptations to the utilization of a particular substrate or simply evolutionary variations.

*3.1.8. Nitrate respiration:* So far, we have discussed the energetics of two groups of aerobic chemoorganotrophs: one represented by *P. denitrificans,* is versatile with regard to the electron donor utilized, whereas the other, represented by obligate methylotrophic bacteria, is very restricted. Many aerobic chemoorganotrophs are not only versatile with respect to the electron donor, but also with respect to the electron acceptor. Many of them are capable of utilizing nitrate as electron acceptor when $O_2$ is not available. In principal, nitrate respiration allows the complete oxidation of an organic substrate to $CO_2$. The degradation of a substrate to $CO_2$ is brought about by the same pathways operating under aerobic conditions. The reducing equivalents produced in the oxidation process are then fed into an alternative electron transport chain. This anaerobic respiratory chain is designed to transfer the electrons to nitrate instead of oxygen and to couple this electron transfer with the extrusion of protons. The upper part of the respiratory chain remains unchanged: NADH dehydrogenase and $bc_1$ complex are used in both aerobic and anaerobic electron transport. Enzymes particularly designed for the denitrification process are nitrate reductase, nitrite reductase, NO reductase, and $N_2O$ reductase. Denitrification occurs in four steps, $N_2$ being the final product (Fig. 10).

Nitrate reductase is a membrane-bound molybdoenzyme which catalyzes the reduction of nitrate to nitrite. The molybdenum is bound to a

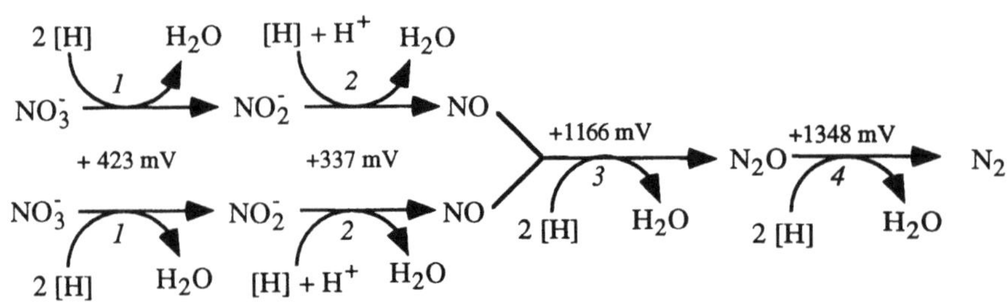


Fig. 10. The process of denitrification. The mV values represent the standard redox potentials of the respective redox pairs. The italicized numbers represent the following enzymes: *1. nitrate reductase. 2. nitrite reductase. 3. nitric oxide reductase. 4. nitrous oxide reductase.*

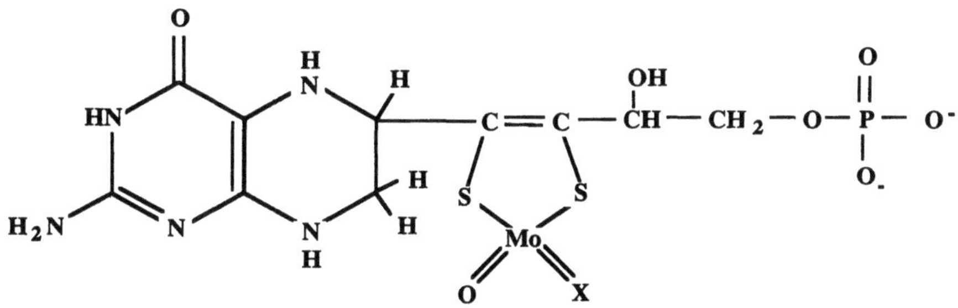

Fig. 11. Chemical structure of the molybdenum cofactor. X stands for S or Se or O.

pterin derivative also found in other enzymes. Its structure is shown in Fig. 11. Furthermore, nitrate reductase contains three or four 4Fe-4S centers and one 3Fe-3S center. The nitrate reductase is composed of three subunits ($\alpha$, $\beta$, $\gamma$), the smallest one ($\gamma$) being a *b*-type cytochrome with two heme *b* centers ($E_{m,7}$ = +95 and +210 mV). The oxidation of NADH is catalyzed by complex I and results, just like under aerobic conditions, in the reduction of ubiquinone to ubiquinol (Fig. 12). The $bc_1$ complex, however, is not involved in electron transfer to nitrate reductase. The electrons flow from ubiquinol by way of the heme *b* groups of the $\gamma$-subunit to the Fe-S clusters contained in the $\beta$-subunit. The $\alpha$-subunit which binds the molybdenum cofactor contains the catalytic site. Nitrate reduction occurs on the cytoplasmic side of the membrane. Nitrate enters the cell either via a nitrate/nitrite antiporter or via a proton-motive force-driven uptake system. The following three reduction steps from nitrite via nitric oxide and nitrous oxide to dinitrogen involve the passage of electrons by way of the $bc_1$ complex to cytochrome *c*. The electron pathway splits up at the level of cytochrome *c*. Cytochrome *c* reoxidation is catalyzed by $NO_2^-$, NO, or $N_2O$ reductase.

The reduction of nitrite to nitric oxide (NO) is catalyzed by nitrite reductase. Two types of nitrite reductase are present in *P. denitrificans.* One type, called cytochrome *cd,* is composed of two identical subunits. Each contains one heme *c* and one modified heme *d.* Although the main function of cytochrome *cd* is nitrite reduction it exhibits oxidase activity; this is reflected by a tenfold higher affinity towards nitrite than to $O_2$. Cytochrome *cd* is located in the periplasm where it takes over the electrons from the "normal" cytochrome *c*. The second type of nitrite reductase is a copper-containing protein.

The reduction of nitric oxide (NO) to nitrous oxide ($N_2O$) occurs in the periplasm. The nitrous oxide reductase which contains *b*- and *c*-type hemes catalyzes this reaction under formation of the N-N bond in $N_2O$. The nitrous oxide ($N_2O$) reductase is a periplasmic copper-containing protein being composed of two identical subunits. The enzyme catalyzes the formation of $N_2$ from $N_2O$.

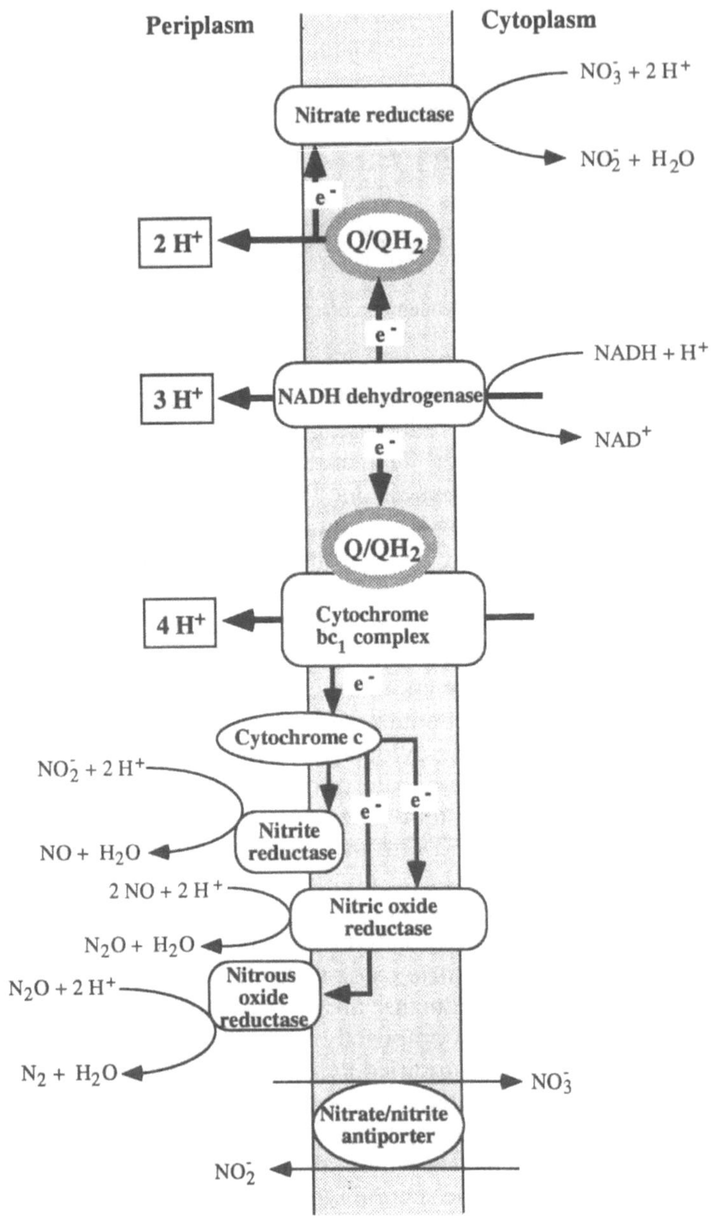


Fig. 12. Organization of the components involved in nitrate reduction in *Paracoccus denitrificans.* The stoichiometries of the electron transport steps have been disregarded. Further explanation are given in the text.

*3.1.9. Aerobic chemolithotrophic bacteria:* Aerobic chemolithotrophic bacteria oxidize a number of simple inorganic substrates using oxygen or nitrate as the terminal electron acceptor. Chemolithotrophic organisms can be distinguished on the basis of the electron donor. Typical electron donors are $H_2$, CO, $H_2S$, S, $Fe^{2+}$, $NH_4^+$, $NO_2^-$. The amount of energy available by the oxidation of these substrates and the ensuing electron transport to $O_2$ depends on the redox potential of the electron donor. Table 2 lists the redox potentials of a number of compounds that serve as electron donors for lithotrophic organisms. Chemolithotrophs have a conventional respiratory chain comparable to that of *P. denitrificans.* The enzymes catalyzing the oxidation of the lithotrophic substrates are either cytoplasmic or arranged in such a way that the electrons can enter the respiratory chain at the appropriate level. The predominant carbon source of chemolithotrophic bacteria is $CO_2$ although quite a number of species may additionally assimilate organic carbon sources. $CO_2$ fixation proceeds in most cases via the Calvin cycle which is also employed by phototrophic bacteria.

*3.1.9.1. Reverse electron transfer.* It is evident that all electron donors except $H_2$, CO, and possibly $SO_3^{2-}$ have a redox potential too positive to reduce $NAD(P)^+$ (Table 2). Since chemolithotrophic organisms utilizing these unfavorable reductants are not capable of lifting the electrons by light energy to more favorable energy level (sufficient to reduce $NAD(P)^+$) they have to use a different mechanism to generate reducing power. The mechanism employed is the so-called "reverse electron transfer". In principal, this mechanism can be envisaged as the reversal of normal electron transport. In normal electron transport the electrons are passed from carriers with a more negative potential to carriers with increasing redox potentials thereby generating $\Delta\tilde{\mu}_{H^+}$ which may subsequently be used for the synthesis of ATP. In reverse electron transport, $\Delta\tilde{\mu}_{H^+}$ is used to drive the transfer of electrons from an electron donor with

Table 2. Redox potential of lithotrophic substrates

| Redox reaction | | | $E_{m,7}$ |
|---|---|---|---|
| $CO_2 + 2\,e^- + 2\,H^+$ | → | $CO + H_2O$ | −524 mV |
| $SO_4^{2-} + 2\,e^- + 2\,H^+$ | → | $SO_3^{2-} + H_2O$ | −516 mV |
| $2\,e^- + 2\,H^+$ | → | $H_2$ | −420 mV |
| $S + 2\,e^- + 2\,H^+$ | → | $H_2S$ | −260 mV |
| $SO_3^{2-} + 4\,e^- + 6\,H^+$ | → | $S + 3\,H_2O$ | −45 mV |
| $NO_2^- + 6\,e^- + 8\,H^+$ | → | $NH_4^+ + 2\,H_2O$ | +340 mV |
| $NO_3^- + 2\,e^- + 2\,H^+$ | → | $NO_2^- + H_2O$ | +430 mV |
| $Fe^{3+} + e^-$ | → | $Fe^{2+}$ | +770 mV |

more positive redox potential to an acceptor with a more negative redox potential. In other words, $\Delta\tilde{\mu}_{H^+}$ is the driving force for this process. It is accompanied by proton movement from the periplasmic to the cytoplasmic side of the bacterial cell. Reverse electron transport has to occur whenever the redox potential of the electron donor is too positive to reduce $NAD(P)^+$. A schematic view of reverse electron transfer is given in Fig. 14 as exemplified by *Nitrobacter*, an organism that oxidizes nitrite to nitrate.

*3.1.9.2. Hydrogen-oxidizing bacteria.* Hydrogen-oxidizing bacteria are generally facultative chemolithotrophs. This means that they are also capable of oxidizing organic substrates. Hydrogen-oxidizing bacteria are taxonomically very diverse and belong to genera such as *Pseudomonas, Alcaligenes, Nocardia, Xanthobacter,* and *Paracoccus*. The enzyme responsible for the uptake of $H_2$ is hydrogenase. Two types of hydrogenases are found in hydrogen-oxidizing bacteria: a membrane-bound and a cytoplasmic hydrogenase. Most of these bacteria possess the membrane-bound type which feeds the electrons directly into the electron transport chain at the level of ubiquinone. In contrast, *Nocardia opaca* contains only a soluble hydrogenase. This enzyme transfers the electrons to $NAD^+$. The exception among the hydrogen-oxidizing bacteria is *Alcaligenes eutrophus* in that it contains both types of hydrogenase. Both hydrogenases contain nickel and iron sulfur centers. However, the $NAD^+$-dependent enzyme exhibits a more complex subunit structure and FMN serves as an additional prosthetic group.

*3.1.9.3. Carbon monoxide-oxidizing bacteria.* Carboxydotrophic bacteria are characterized by their ability to utilize carbon monoxide besides $H_2$ as a reductant. The oxidation of CO is catalyzed by CO dehydrogenase, a molybdenum cofactor-containing protein. The electrons derived from CO or $H_2$ are transferred to a branched electron transport chain. One of the two electron transport branches in the carboxydotrophic bacterium *Pseudomonas carboxydovorans* acts as ubiquinol oxidase transferring the electrons via cytochrome $b_{561}$, ubiquinone, and cytochrome $b_{563}$ to $O_2$. Cytochrome $b_{563}$ acts as terminal oxidase and is characterized by its insensitivity towards CO. The CO-sensitive terminal oxidase is a cytochrome *c* oxidase which contains cytochrome $a_1$. It is preferentially used during organotrophic growth. Both electron transport chains serve the organism to translocate protons and to generate a $\Delta\tilde{\mu}_{H^+}$.

*3.1.9.4. Bacterial ammonia oxidation.* The oxidation of ammonia to nitrate occurs in two steps catalyzed by two different organisms. In the first step *Nitrosomonas* oxidizes ammonia to nitrite. The second step, the oxidation of nitrite to nitrate, is carried out by *Nitrobacter*.

*Nitrosomonas* species are very specialized organisms: they are unable to oxidize electron donors other than ammonia. The ammonia oxidation cannot occur directly ($NH_3 + H_2O \leftrightarrows NH_2OH + 2\ H^+ + 2\ e^-$; $E_{m,7} = +900$ mV) because the redox potential of this reaction is to be coupled with the reduction of $O_2$ to $H_2O$. Instead, the oxidation of ammonia by *Nitrosomonas* is catalyzed by the membrane-bound ammonia mono-oxygenase according to the following equation:

$$NH_3 + O_2 + 2H^+ + 2e^- \leftrightarrows NH_2OH + H_2O$$
$$(E_{m,7} = +800\ mV)$$

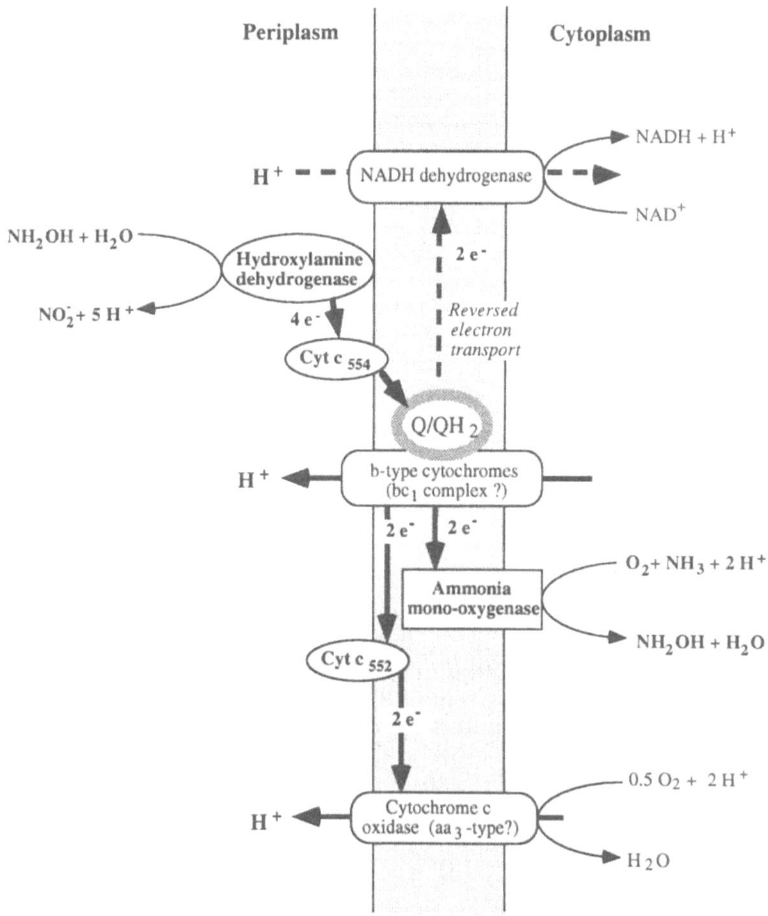


Fig. 13. Organization of the components involved in ammonia oxidation in *Nitrosomonas*. Reverse electron transport and the concomitant proton influx are indicated by dashed arrows. The stoichiometries of the proton translocation steps have been disregarded. Ammonia oxidation is shown to occur on the cytoplasmic side of the membrane. However, the exact location of the ammonia oxidation is not yet known. Further explanations are given in the text.

The introduction of one oxygen atom into $NH_3$ is accompanied by the reduction of the other oxygen atom to $H_2O$. $NH_3$ rather than $NH_4^+$ is the substrate of the ammonia mono-oxygenase. The electrons needed in the mono-oxygenase reaction are probably derived from the ubiquinone/cytochrome *b* region of the *Nitrosomonas* electron transport chain (Fig. 13). Ammonia oxidation is not coupled to proton translocation. Hydroxylamine is further oxidized to nitrite by hydoxylamine dehydrogenase

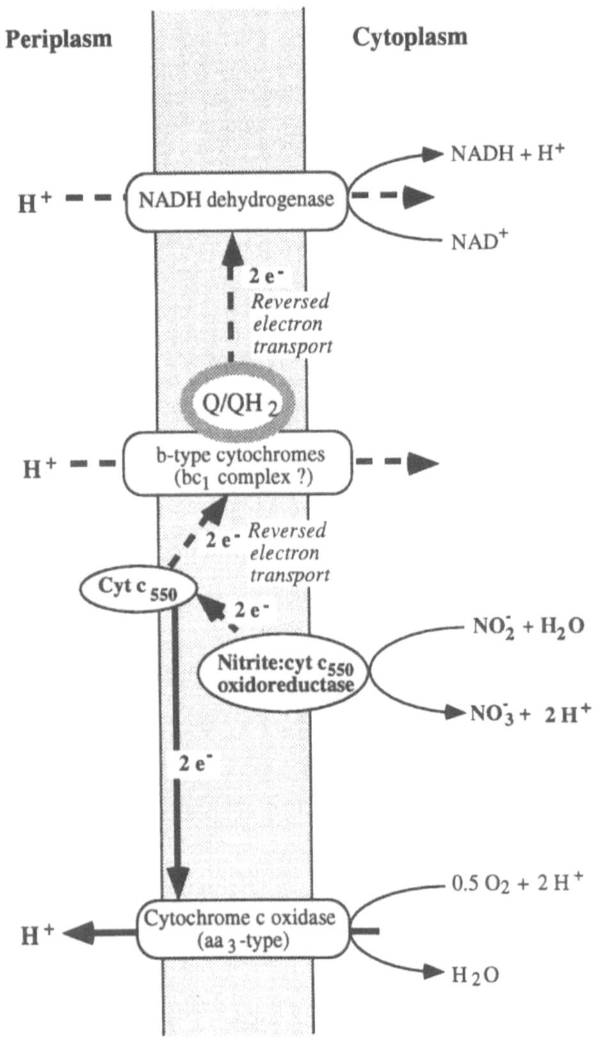


Fig. 14. Organization of the components involved in nitrite oxidation in *Nitrobacter.* Reverse electron transport and the concomitant proton influx are indicated by dashed arrows. The stoichiometries of the proton translocation steps have been disregarded. Further explanations are given in the text.

which is located in the periplasm. It has a complex structure containing 21 *c*-type hemes and 3 P-460; the latter are unusual covalently bound hemes. Since the redox potential of the $NO_2^-/NH_2OH$ couple is +60 mV, reverse electron transport has to occur in to order to reduce $NAD^+$ (Fig. 13). The electrons derived from the hydroxylamine dehydrogenase are transferred to the ubiquinone/cytochrome *b* region by way of tetraheme cytochrome $c_{554}$ and finally passed to the terminal oxidase containing cytochrome *a*. This electron transfer is accompained by $H^+$ translocation of uncertain stoichiometry.

In *Nitrobacter,* nitrite oxidation is reversibly catalyzed by nitrite: cytochrome $c_{550}$ oxidoreductase. The electrons are transferred to the terminal cytochrome $aa_3$-containing cytochrome oxidase by way of cytochrome $c_{550}$ which is located on the periplasmic face of the membrane (Fig. 14). The location of the nitrite-oxidizing enzyme (cytoplasm or periplasm) is still unclear. Since the redox potential ($E_{m,7}$) of the $NO_3^-/NO_2^-$ couple is +430 mV, the reduction of cytochrome $c_{550}$ ($E_{m,7}$ = +270 mV) is unfavorable and has therefore to be coupled to the release of $2H^+$ into the cytoplasm. In the ensuing cytochrome oxidase reaction $4H^+/2e^-$ are removed from the cytoplasm, two of which are translocated to the periplasmic side of the membrane, the other two protons are consumed during the reduction of $O_2$:

$$2\ \text{cyt}\ c_{red} + 4\ H^+_{in} + 0.5\ O_2 \leftrightarrows 2\ \text{cyt}\ c_{ox} + 2\ H^+_{out} + H_2O$$

*3.1.9.5. Bacterial oxidation of inorganic sulfur compounds.* The principal forms of sulfur utilized as electron donors by chemolithotrophic bacteria are sulfide ($S^{2-}$), sulfur (S), thiosulfate ($S_2O_3^{2-}$), and sulfite ($SO_3^{2-}$). The possible pathways of the oxidation of these compounds are outlined in Fig. 15. Which of these pathways is utilized depends on the organism.

Sulfite oxidation to sulfate as performed by *Thiobacillus thioparus* or *T. denitrificans* is initiated by the following reaction:

$$SO_3^{2-} + \text{AMP} \leftrightarrows \text{adenosine -5'-phosphosulfate (APS)} + 2e^-$$
$$(E_{m,7} = -60\ \text{mV})$$

The reaction is catalyzed by APS reductase, a cytoplasmic enzyme. Although the direct electron acceptor of sulfite oxidation is unknown, it is clear that the electrons enter the electron transport chain at the level of ubiquinone/cytochrome *b*. The product adenosine-5'-phosphosulfate (APS) may undergo the following reactions leading to ATP synthesis by substrate level phosphorylation:

$$\text{APS} + P_i \leftrightarrows \text{ADP} + SO_4^{2-}\ \text{(ADP sulfurylase) followed by:}$$
$$2\text{ADP} \leftrightarrows \text{ATP} + \text{AMP (adenylate kinase)}$$

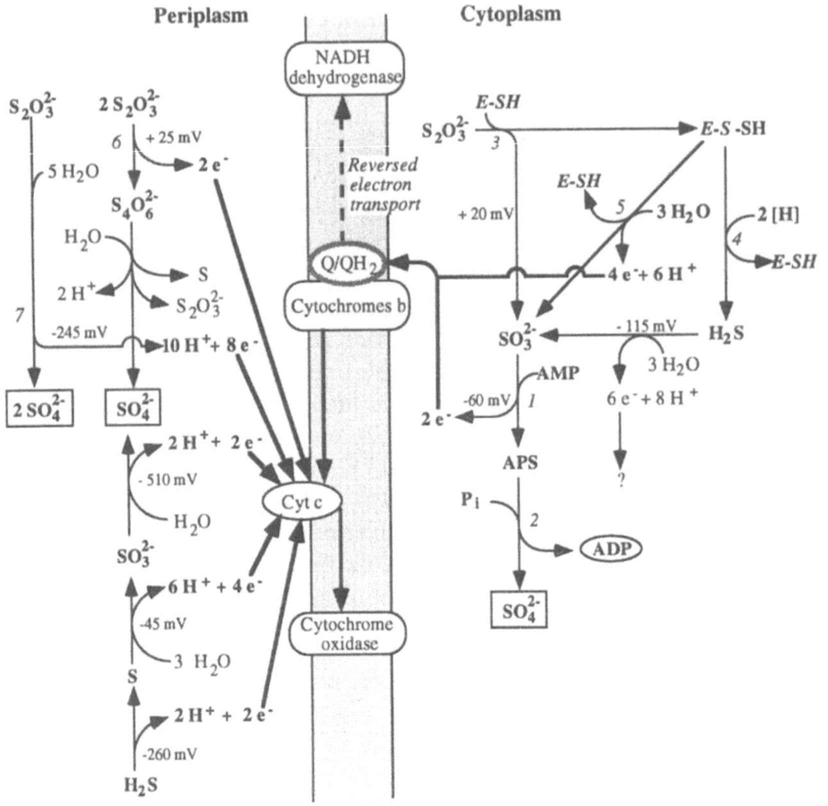


Fig. 15. Possible electron transport pathways involved in the oxidation of sulfur compounds. Electron transport pathways to the respiratory chain are indicated by bold arrows. Reverse electron transport is indicated by dashed arrows. For the sake of clarity, the NADH dehydrogenase reaction, the cytochrome oxidase reaction, and the sites of proton translocation are not shown. Further explanations are given in the text. The italicized numbers represent the following enzymes: *1. APS reductase. 2. ATP sulfurylase. 3. rhodanese. 4. thiosulfate reductase. 5. "reverse" sulfite reductase. 6. thiosulfate-combining enzyme. 7. thiosulfate-oxidizing enzyme.*

In other species such as *T. novellus* sulfite is directly oxidized to sulfate by sulfite oxidase. The electrons are then transferred to the periplasmic cytochrome $c_{551}$ and eventually to cytochrome oxidase, oxygen being the terminal electron acceptor. Although elemental sulfur is an important electron source for both chemolithotrophic and photolithotrophic bacteria, relatively little is known about the reactions catalyzing the oxidation of sulfur. One reason for this lack of knowledge is due to the complexity of sulfur chemistry. The product of sulfur oxidation is sulfite which is further oxidized as described above. Sulfide is either oxidized to elemental sulfur or directly to sulfite. However, in neither case are the electron acceptors or the location of the respective enzymes known.

Thiosulfate may be used in three ways:

1. Utilization starts with the cleavage of thiosulfate leading to the liberation of sulfite and the concomitant binding of S to the catalyzing enzyme rhodanese (*E*):

$$E\text{-SH} + S_2O_3^{2-} \leftrightarrows E\text{-S-S-H} + SO_3^{2-}$$

The enzyme-bound S is either reductively cleaved off by thiosulfate reductase under formation of $H_2S$ or it is oxidized to sulfite reductase operating in the reverse direction:

$$E\text{-S-S-H} + 3\ H_2O \leftrightarrows E\text{-SH} + SO_3^{2-} + 6H^+ + 4e^-$$

2. The so-called thiosulfate-combining enzyme catalyzes the oxidation of thiosulfate to tetrathionate:

$$2S_2O_3^{2-} \leftrightarrows S_4O_6^{2-} + 2e^-$$

The physiological electron acceptor is probably cytochrome *c*, rendering the periplasmic location of the enzyme very likely. Tetrathionate undergoes a non-enzymatical conversion yielding thiosulfate and sulfur.

3. In *T. versutus,* thiosulfate is directly oxidized to sulfate in an eight-electron process by a periplasmic thiosulfate-oxidizing enzyme complex consisting of four proteins: protein A, protein B, cytochrome $c_{551}$, and cytochrome $c_{552.5}$ exhibiting a stoichiometry of 1:2:6:2. Cytochrome $c_{551}$ contains four hemes: two with an $E_{m,7}$ = +240 mV and two with an $E_{m,7}$ = -115 mV. Cytochrome $c_{552.5}$ contains two hemes with redox potentials ($E_{m,7}$) of +220 mV and -215 mV. The electrons generated by thiosulfate oxidation are fed into the electron transport chain at the level of cytochrome *c*. Two protons are translocated during the passage of two electrons to $O_2$ and two protons are utilized for the production of $H_2O$. Little is known about the relative importance of the three pathways in the individual species of thiosulfate-oxidizing bacteria.

*3.1.9.6. Iron-oxidizing bacteria.* Representatives of iron-oxidizing bacteria belong to the genera *Gallionella, Sphaerotilus,* and *Leptothrix.* Their habitat is characterized by a zone of low oxygen tension and slightly acidic pH. The bioenergetics of $Fe^{2+}$ oxidation is difficult to study with the above organisms because of Fe(II) autooxidation and the insolubility of the oxidation product $Fe(OH)_3$ at neutral pH. More information is available on the $Fe^{2+}$ oxidation occuring at pH 2. The best studied example of an $Fe^{2+}$-oxidizing acidophile is *Thiobacillus ferrooxidans*. $Fe^{2+}$ is oxidized at the periplasmic side of the membrane involving rusticyanin, a copper-containing $Fe^{2+}$-binding protein. The electrons derived from $Fe^{2+}$ oxidation are transferred to cytochrome *c* ($E_{m,3.2}$ = +640 mV) or directly to the cytochrome oxidase which contains cytochrome $a_1$ (Fig. 16). At an external pH of 2, *T. ferrooxidans* maintains a large ΔpH across the cytoplasmic membrane, the intracellular pH being around 6.5, whereas the transmembrane electrical potential (Δϕ) is negligible and the protonmotive force (p) is −260 mV. A simple way to envisage energy conversion in *T. ferrooxidans* is to consider the cytochrome oxidase as a proton-consuming reaction by which protons are removed from the cytoplasm.

*3.1.10. Facultatively anaerobic chemoorganotrophs: P. denitrificans* has been taken as an example for a versatile aerobic organism. Enteric bacteria such as *E. coli* exhibit an even greater metabolic versatility. *E. coli* is a facultative anaerobe meaning that it is able to grow aerobically or anaerobically. The anaerobic growth of *E. coli* is in contrast to that of *P. denitrificans,* however, not restricted to nitrate as electron acceptor. The organism may take advantage of alternative electron acceptors such as fumarate,

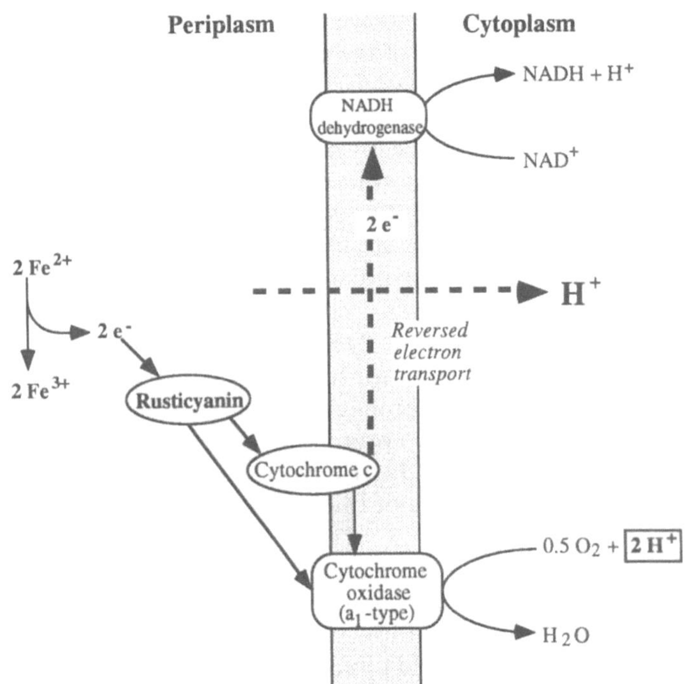


Fig. 16. Organization of the components involved in $Fe^{2+}$ oxidation in *Thiobacillus ferrooxidans.* Reverse electron transport and the concomitant proton influx are indicated by dashed arrows. The stoichiometry of the proton influx is not given. Further explanations are given in the text.

dimethylsulfoxide (DMSO) or trimethylamine-N-oxide (TMAO). *E. coli* is even capable of fermentative growth provided a fermentable substrate is present.

Whereas oxygen (0.5 $O_2/H_2O$; $E_{m,7}$ = +820 mV) and nitrate ($NO_3^-/NO_2^-$, $E_{m,7}$ = +425 mV) are strong oxidants, fumarate (fumarate/succinate; $E_{m,7}$ = +33 mV), dimethylsulfoxide (DMSO/DMS; $E_{m,7}$ = +160 mV), or trimethylamine-N-oxide (TMAO/TMA; $E_{m,7}$ = +130 mV) are much weaker oxidants. The use of various potential electron acceptors is therefore efficiently regulated in *E. coli.* The simple rule of this type of regulation is easily assessed: *E. coli* selects the electron acceptor that allows maximal energy generation. The sequence of their use is governed by their redox potential. Simply spoken, the presence of oxygen represses the expression of genes required for the use of all the other electron acceptors; nitrate represses fumarate reductase and DMSO/TMAO reductase: hence fumarate or DMSO or TMAO may only serve as electron acceptor under anaerobic conditions and in the absence of nitrate. Electron transport to these electron acceptors is accompanied by the generation of $\Delta\tilde{\mu}_{H^+}$ and by ATP synthesis. In the absence of any of these electron acceptors *E. coli* has to regenerate $NAD^+$ from NADH by dumping the reducing equivalents into intermediates of a degradative pathway (3.1.12.5.). Under these conditions, ATP may only be gained by substrate level phosphorylation.

*3.1.10.1. The aerobic electron transport chain of E. coli.* The aerobic degradation of sugars by *E. coli* follows the EMP pathway yielding two ATP by substrate level phosphorylation and two NADH per molecule of glucose. The reducing equivalents enter the respiratory chain mainly by way of NADH dehydrogenase (Fig. 17). Besides NADH dehydrogenase there are several other membrane-bound dehydrogenases which catalyze the oxidation of a number of substrates and feed the electrons directly into the respiratory chain. Examples are succinate dehydrogenase, D-lactate dehydrogenase, L-lactate dehydrogenase, *sn*-glycerol-3-phosphate dehydrogenase, and D-amino acid dehydrogenase. Ubiquinone may be considered as the central electron acceptor for all of these dehydrogenases including NADH dehydrogenase.

One important difference in comparison to *P. denitrificans* is the absence of a $bc_1$ complex and of cytochrome c from *E. coli*'s respiratory chain. Hence, electron transport from ubiquinol to oxygen is brought about by two ubiquinol oxidases. Depending on the conditions of growth one of the two oxidases may predominate.

Only two sites of *E. coli*'s respiratory chain are coupled to proton translocation: the NADH dehydrogenase and the terminal oxidase. Whereas in *P. denitrificans* transfer of two electrons from ubiquinol to $O_2$ via $bc_1$ complex and $aa_3$-type cytochrome *c* oxidase results in the translocation of six protons, only two protons are translocated in *E. coli* during the pas-

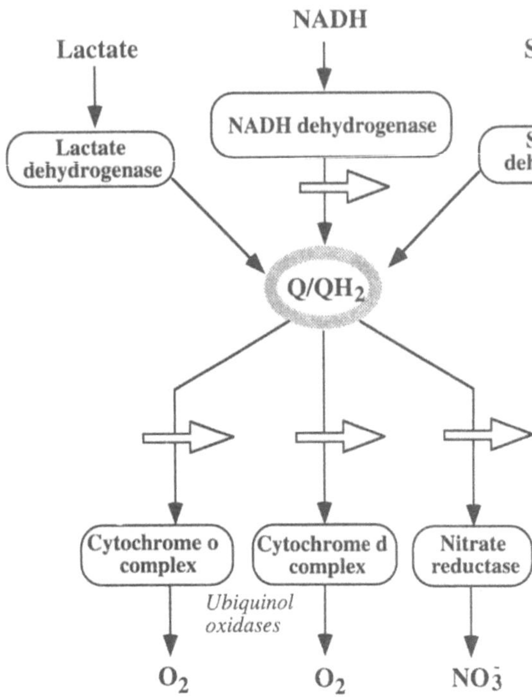


Fig. 17. Schematic representation of alternative electron transport pathways in *Escherichia coli*. The scheme depicts the oxidation of lactate, NADH, and succinate, and the ensuing electron transport to either of two alternate ubiquinol oxidases (under aerobic conditions) or to nitrate reductase (under anaerobic conditions in the presence of nitrate). The open arrows indicate those electron transport steps that are coupled with proton translocation.

sage of two electrons from ubiquinol to $O_2$. Ubiquinol oxidation in *E. coli* is catalyzed by the cytochrome *o* and/or by the cytochrome *d* complex. Both complexes are expressed under normal growth conditions, but the cytochrome *d* complex has a higher affinity for $O_2$ than the cytochrome *o* complex and is mainly produced when $O_2$ becomes limited. Both complexes couple the ubiquinol-dependent reduction of $O_2$ with the translocation of protons across the cytoplasmic membrane and the generation of a protonmotive force.

The cytochrome *o* complex is composed of four subunits. It contains two hemes *b* (cytochrome $b_{555}$ [*o*] and $b_{562}$) and two copper atoms (Fig. 18). Due to the presence of two hemes and two $Cu^{2+}$ the cytochrome *o* complex shares a certain similarity with the $aa_3$ complex from *P. denitrificans*. The cytochrome $b_{562}$ is analogous to cytochrome *a* in that it does not bind $O_2$, whereas the cytochrome $b_{555}$ component is analogous to the $O_2$-binding cytochrome $a_3$ and qualifies therefore as a cytochrome *o*. Proton translocation by the cytochrome *o* complex is accomplished in principally the same manner as described below for the cytochrome *d* complex.

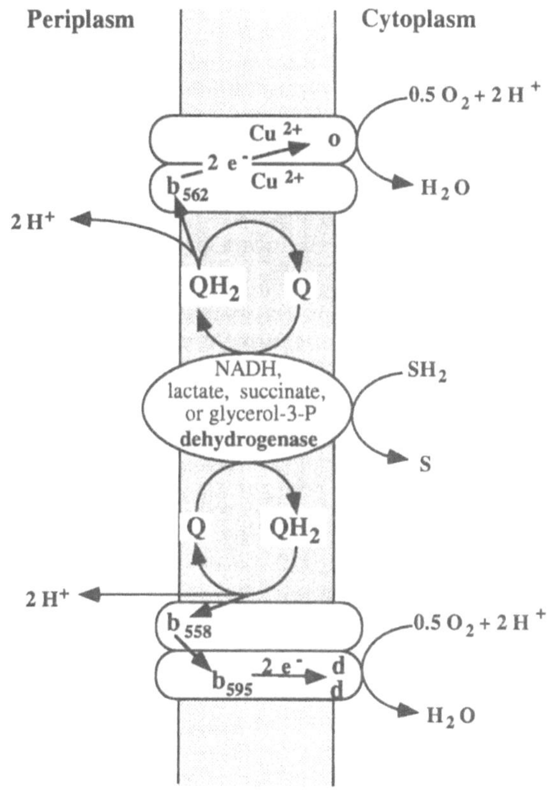


Fig. 18. Organization of the two alternative ubiquinol oxidases in *E. coli.* Further explanations are given in the text.

The cytochrome *d* complex is composed of two polypeptides which contain two *b*-type hemes (cytochrome $b_{558}$ and $b_{595}$) and two hemes of the *d*-type (chlorin). Erroneously, cytochrome $b_{595}$ had previously been designed cytochrome $a_1$ due to its unusual $\alpha$ absorption band. Heme *d* is derived from heme *b* by a reduction in pyrrol ring C followed by dihydroxylation. Heme *d* is directly involved in the binding of $O_2$ and its reduction to $H_2O$. Ubiquinol oxidation occurs near the periplasmic side of the membrane and involves cytochrome $b_{558}$. It is associated with the release of two protons into the periplasm and the transfer of electrons by way of cytochrome $b_{595}$ to cytochrome *d.* The latter catalyzes the reduction of oxygen to water at the cytoplasmic side of the membrane and the concomitant consumption of two protons. In this model the cytochrome *d* complex does not act as a proton pump such as the cytochrome *c* oxidase of *P. denitrificans,* but proton translocation is brought about indirectly by the concomitant release of 1 $H^+/e^-$ on the outside and the uptake of 1 $H^+/e^-$ on the inside of the cytoplasmic membrane (Fig. 18).

*3.1.10.2. Nitrate reduction by E. coli. E. coli* is also able to utilize nitrate as terminal electron acceptor, provided $O_2$ is absent. In contrast to *P. denitrificans,* however, *E. coli* is not able to perform denitrification, i.e. reduction to $N_2$, but nitrate is reduced to nitrite followed by the reduction to ammonia. Nitrate reduction occurs in *E.coli*—as in *P. denitrificans*—on the cytoplasmic side of the membrane and it is catalyzed by nitrate reductase which has the same composition as that of *P. denitrificans.* Electron transport from one of several membrane-bound dehydrogenases (see above) to this complex is associated with the formation of a $\Delta\tilde{\mu}_{H^+}$. One notable difference between the nitrate reductase from *P. denitrificans* and that from *E. coli* is related to the reductant: Whereas cytochrome *c* serves as electron donor for nitrate reduction in the former organism, it is ubiquinol in the latter organism (Fig. 17).

The further reduction of nitrite to ammonia may be catalyzed by any of three enzymes: two of these are soluble nitrite reductases. One of these soluble enzymes is a NADPH-dependent sulfite reductase which also unspecifically catalyzes nitrite reduction. Siroheme serves as a cofactor in this enzyme. Of more importance, however, is the second soluble enzyme: the NADH-dependent nitrite reductase. The enzyme is composed of two identical subunits: each of them contains one Fe-S cluster, one FAD and one siroheme. The major function of these soluble nitrite reductases is the reoxidation of NAD(P)H in order to regenerate $NAD(P)^+$. Only a small proportion of nitrite is reduced by a membrane-bound nitrite reductase. Nitrite reduction as catalyzed by this enzyme involves electron transport by way of a periplasmic cytochrome $c_{552}$ and is associated with the generation of $\Delta\tilde{\mu}_{H^+}$.

In *P. denitrificans,* nitrate is as efficient an electron acceptor as $O_2$ and organic substrate may be completely oxidized to $CO_2$. The same is true for other primarily aerobic bacteria such as *Bacillus licheniformis, Pseudomonas aeruginosa,* and *Alcaligenes faecalis.* Enteric bacteria, although capable of aerobic growth, have a somewhat different relationship towards $O_2$: even in the presence of $O_2$ the oxidation of an organic substrate is hardly ever complete and small amounts of fermentative products are formed. Under anaerobic conditions with nitrate as terminal electron acceptor, the proportion of substrate appearing as $CO_2$ is even smaller, for the TCA cycle does not operate under these conditions. As a consequence, the proportion of ATP that is synthesized by electron transport phosphorylation is smaller with nitrate than with oxygen as electron acceptor. This is in accordance with the observation that nitrite reduction is first in line for the regeneration of $NAD(P)^+$. Certain strict anaerobic bacteria such as some *Clostridium* species are not able to link nitrate reduction to energy generation at all. These organisms utilize nitrate or nitrite exclusively as an electron sink in order to regenerate $NAD(P)^+$.

*3.1.10.3. Fumarate, trimethylamine N-oxide (TMAO) and dimethylsulfoxide (DMSO) reduction by E. coli.* Another alternative terminal electron acceptor utilized by *E. coli* is fumarate. Fumarate reduction is catalyzed by the membrane-bound fumarate reductase. The enzyme is only expressed under anaerobic conditions in the absence ot nitrate. Whereas electron acceptors such as nitrate have to be added externally to the medium, fumarate may be produced to some extent by the organism during substrate degradation. The fumarate reductase complex consists of four polypeptides (Frd A, B, C, and D). Frd C and D are devoid of prosthetic groups and serve as membrane anchors for the membrane-extrinsic catalytic subunits Frd A and Frd B. An FAD moiety is covalently attached to a histidine residue of Frd A, whereas Frd B contains three Fe-S clusters.

Fumarate reduction may be coupled to the oxidation of molecular hydrogen, formate, NADH, glycerol-3-phosphate and lactate as catalyzed by hydrogenase and the respective dehydrogenases (Fig. 19). The electrons from these membrane-bound dehydrogenases are transferred to the fumarate reductase complex by way of menaquinone. Electron transport cannot be supported by ubiquinone. This is probably due to the fact that the redox potential of the ubiquinol/ubiquinone couple is too positive ($E_{m,7}$ = +113 mV) for a coupling with fumarate reduction ($E_{m,7}$ = +30 mV). The redox potential of the menaquinol/menaquinone couple ($E_{m,7}$ = −74 mV), however, is sufficiently negative to donate electrons to the fumarate reductase.

Fumarate-dependent electron transport is coupled to the generation of an electrochemical proton gradient. An energy-conserving fumarate reductase system is not only found in enterobacteria, but also in anaerobes such as propionibacteria. However, the composition of the fumarate reductase complex may differ from that of *E. coli.* The fumarate reductase from the strictly anaerobic bacterium *Wolinella succinogenes,* which may grow with formate and fumarate as sole energy source, is composed of three instead of four subunits: Whereas the catalytic polypeptides of both organisms (Frd A and B) show a high degree of similarity with respect to the prosthetic groups and the amino acid sequence there are differences as to the polypeptides anchoring these subunits in the membrane. In *W. succinogenes* the cofactor-free subunits Frd C and Frd D of *E. coli* are replaced by a diheme cytochrome *b* (one polypeptide containing two molecules of heme *b*).

Another anaerobic energy-conserving electron transport system which is relatively widespread in bacteria is the molybdenum-containing trimethylamine N-oxide (TMAO)/dimethylsulfoxide (DMSO) reductase (Fig. 19). The enzyme catalyzes the reduction of both TMAO to trimethylamine (TMA) ($E_{m,7}$ = +130 mV) and of DMSO to dimethylsulfide (DMS) ($E_{m,7}$ = +160 mV) in reactions that are analogous to nitrate reduction.

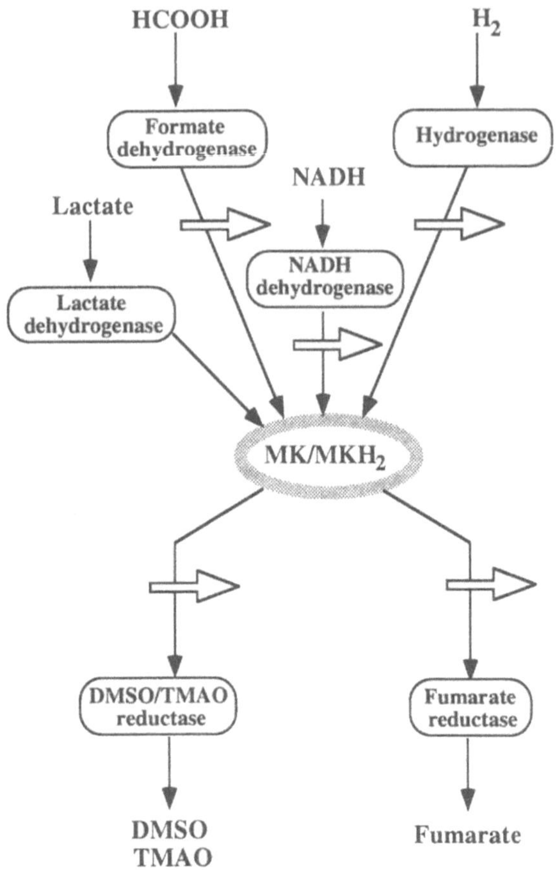


Fig. 19. Schematic representation of alternative anaerobic electron transport pathways in *E. coli.* The scheme depicts the oxidation of lactate, formate, NADH, and $H_2$, and the ensuing anaerobic electron transport pathways to either dimethylsulfoxide (DMSO)/tetramethylamine N-oxide (TMAO) or to fumarate as terminal electron acceptors. The open arrows indicate those electron transport steps that are coupled with proton translocation. MK, menaquinone.

*3.1.11. Strictly anaerobic chemotrophs:* The anaerobic electron transport chains discussed so far are not restricted to facultative anaerobes such as *E. coli,* but are also found in strict anaerobes. However, whereas facultative anaerobes have evolved the ability to respond efficiently to the availability of various electron acceptors and to changes in the redox potential of the medium, strict anaerobes are not able to grow in the presence of oxygen. Moreover, they are restricted to a certain biotope which is characterized by a low redox potential. They contribute to the biological transformations of biomass occuring in these anaerobic habitats and the mineralization of organic material. The toxic effect of oxygen on these organisms is due to two factors:

1. Strictly anaerobic bacteria are usually devoid of superoxide dismutase and/or catalase. Superoxide dismutase and catalase are commonly found in all aerobic bacteria, these enzymes detoxify the deleterious superoxide radical ($O_2^-$) and hydrogen peroxide ($H_2O_2$), respectively. $O_2^-$ and $H_2O_2$ are always by-products of aerobic respiration.
2. The metabolism of strict anaerobes is inhibited by $O_2$.

Strictly anaerobic chemoorganotrophic bacteria can be distinguished on the basis of their ability to conserve energy by electron transport phosphorylation or exclusively by substrate level phosphorylation (fermentation). The majority of the strictly anaerobic bacteria cover their ATP demand at least partly by electron transport phosphorylation. However, the ratio of ATP gained by substrate level phosphorylation to ATP gained by electron transport phosphorylation is usually higher in chemoorganotrophic anaerobes than in chemoorganotrophic aerobes because the electron transport chains of strict anaerobes have generally smaller redox spans. Whereas the anaerobic terminal reductases discussed above for *E. coli* are also found in many strict anaerobes, there are electron transport systems that are confined to strictly anaerobic bacteria. Such an example are the sulfate-reducing bacteria.

*3.1.11.1. Sulfate-reducing bacteria.* Sulfate reducers are characterized by their ability to couple the oxidation of organic and inorganic substrates to the reduction of sulfate, thereby producing the typical end product of their metabolism: $H_2S$; they may therefore also be called sulfidogens. The process of dissimilatory sulfate reduction is part of the global sulfur cycle. Sulfate reducers are found in anaerobic sulfate-rich habitats where they play in concert with other groups of anaerobic bacteria an important role in the anaerobic mineralization process of organic material (dead biomass). Typical representatives of this group belong to genera such as *Desulfovibrio, Desulfobacterium, Desulfomonas, Desulfobulbus,* and *Desulfotomaculum.*

The electron donors utilized by the sulfate-reducing bacteria are typical end products of fermentative bacteria: lactate, hydrogen, formate, acetate, propionate, butyrate, higher fatty acids, alcohols. A number of sulfate reducers are capable of oxidizing dicarboxylic acids or aromatic compounds. The ability of several sulfate-reducing bacteria to grow with $H_2$ and sulfate as the sole source of energy indicates that ATP is synthesized by electron transport phosphorylation.

*Sulfate reduction.* Sulfate itself is a poor electron acceptor due to its unfavorable redox potential ($SO_3^{2-}/SO_4^{2-}$: $E_{m,7} = -510$ mV). In order to make it a useful electron acceptor sulfate must first be activated to adenosine-5'-phosphosulfate (APS) by ATP sulfurylase (Fig. 20):

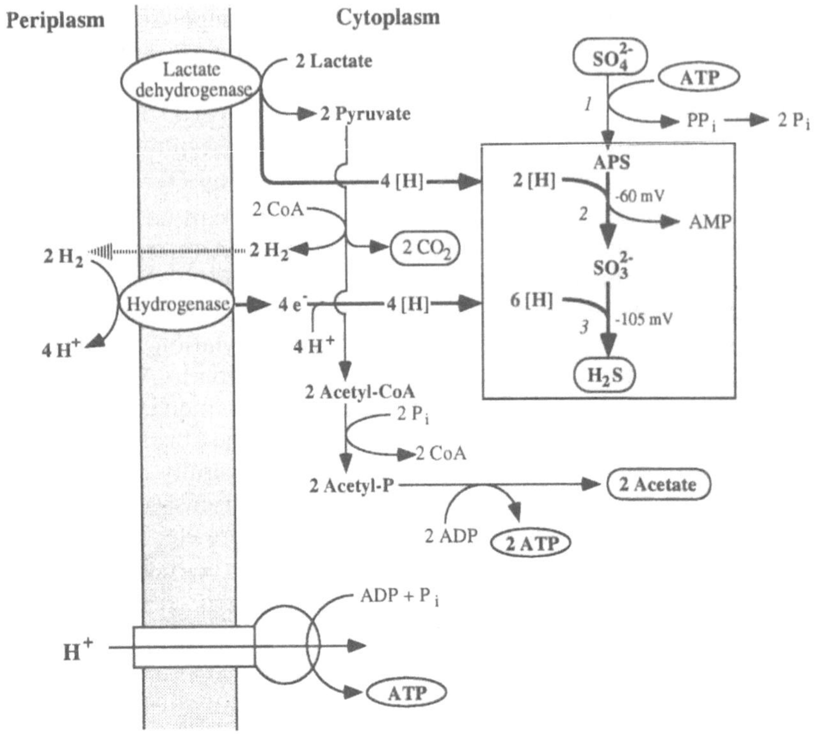


Fig. 20. Model for $\Delta\tilde{\mu}_{H^+}$ generation in sulfate-reducing bacteria based on "hydrogen cycling". The redox reactions of the sulfate reduction process are boxed. Further explanations are given in the text. The italicized numbers represent the following enzymes: *1. ATP sulfurylase. 2. APS reductase. 3. sulfate reductase.*

$$SO_4^{2-} + 2H^+ + ATP \leftrightarrows APS + PP_i$$
$$\Delta G^{0'} = +46 \text{ kJ/mol}$$

The equilibrium of this reaction lies under standard conditions on the left side. However, the equilibrium is somewhat shifted to APS formation by the ensuing hydrolysis of pyrophosphate ($PP_i$) as catalyzed by pyrophosphatase:

$$PP_i + H_2O \leftrightarrows 2\ P_i$$
$$\Delta G^{0'} = -21 \text{ kJ/mol}$$

The overall reaction remains endergonic under standard conditions, in spite of the hydrolysis of two energy-rich bonds ($\Delta G^{0'} = +25$ kJ/mol).

APS is the first electron acceptor in the sulfate reduction process. APS reduction to AMP and sulfite ($SO_3^{2-}$) is catalyzed by APS reductase, a Fe-S-flavoprotein. The redox potential of the $SO_3^{2-}$ + AMP/APS couple ($E_{m,7}$

= −60 mV) is sufficiently positive to couple APS reduction to oxidation of all relevant substrates. Sulfite reduction to sulfide is achieved by the transfer of six-electrons as catalyzed by sulfite reductase. Siroheme is the characteristic prosthetic group of sulfite reductase. Depending on the organism, the enzyme contains in addition one or four Fe-S centers. Sulfite reduction to sulfide may involve thiosulfate and trithionate as free intermediates. The site of sulfite reduction to sulfide appears to be the cytoplasm. Consequently, proton translocation cannot be directly coupled to this process. It seems plausible to assume that energy-related electron transport occurring at the membrane takes place between a membrane-bound dehydrogenase or hydrogenase as electron-donating system and an unknown membrane-bound electron-accepting system which is capable of reacting with the sulfate-reducing enzymes either directly or by way of unidentified soluble carriers. Menaquinones and various cytochromes have been detected in the membrane and/or periplasm of various sulfate-reducing bacteria, some of which might be involved in such an electron transport chain. However, more information is needed to propose an electron transport scheme.

*Complete and incomplete oxidation of organic substrate by sulfate-reducing bacteria.* The standard Gibbs free energy change associated with sulfate reduction depends on the electron donor utilized. Electrons may either be derived from $H_2$ or from organic substrates. When organic substrates are utilized, two groups of sulfate-reducing bacteria may be distinguished based on the mode of substrate oxidation: organisms completely oxidizing organic substrates to $CO_2$ are referred to as complete oxidizers; those organisms not able to do so are referred to as incomplete oxidizers, their main products being acetate and $CO_2$. Typical representatives of incomplete oxidizers are: *Desulfovibrio desulfuricans* and *D. vulgaris*, *Desulfobulbus* species, *Desulfotomaculum nigrificans*, and *Desulfomonas pigra*. Typical complete oxidizers are *Desulfobacter* species, *Desulfotomaculum acetoxidans*, *Desulfosarcina variabilis*, and *Desulfovibrio barsii*.

Interestingly, complete oxidizers exhibit slower growth than incomplete oxidizers. This difference will be discussed taking lactate as an example.

$$2 \text{ lactate} + SO_4^{2-} + 2H^+ \rightarrow 2 \text{ acetate} + 2\ CO_2 + H_2S + 2\ H_2O$$
$$\Delta G^{0'} = -175.6 \text{ kJ/mol}$$
$$2/3 \text{ lactate} + 1\ SO_4^{2-} + 2\ 2/3\ H^+ \rightarrow 2\ CO_2 + 1\ H_2S + 2\ H_2O$$
$$\Delta G^{0'} = -100.5 \text{ kJ/mol}$$

Lactate oxidation to pyruvate ($E_{m,7} = -190$ mV) is catalyzed by a membrane-associated, $NAD^+$-independent lactate dehydrogenase (Fig. 20). The electron carrier between lactate and sulfate has not yet been identified, but $H_2$ ($E_{m,7}$ = −420 mV) has definitely been ruled out. Pyruvate oxidation to acetyl-CoA and $CO_2$ ($E_{m,7}$ = −500 mV) is coupled to the reduction of ferredoxin. Ferredoxins are small Fe-S proteins which are often employed as low potential electron carriers in anaerobes. Reduced ferredoxin is reoxidized by a cytoplasmic hydrogenase leading to the release of $H_2$. The $H_2$ diffuses into the periplasm where it is oxidized by a periplasmic hydrogenase thereby releasing 2 $H^+$. The electrons are transferred to the cytoplasmic side of the membrane and used for sulfate reduction. This process is called hydrogen cycling and leads to the generation of a $\Delta\tilde{\mu}_{H^+}$ (Fig. 20). Sulfate-reducing bacteria being devoid of hydrogenase must, however, employ a different mechanism. The acetyl-CoA being produced from pyruvate is either further oxidized to $CO_2$ (complete oxidation) or it is converted to acetate (incomplete oxidation). The conversion of acetyl-CoA to acetate proceeds via acetyl phosphate and is catalyzed by phosphotransacetylase. Acetate kinase couples the formation of acetate from acetyl phosphate with the phosphorylation of ADP. This reaction is of great importance for many anaerobes because it effects ATP synthesis by substrate level phosphorylation.

*Acetate oxidation by sulfate-reducing bacteria.* Quite a number of sulfate-reducing bacteria are capable of growth with acetate as sole electron donor for sulfate reduction.

$$\text{acetate} + SO_4^{2-} + 3\,H^+ \leftrightarrows 2\,CO_2 + H_2S + 2\,H_2O$$
$$\Delta G^{0'} = -63\ \text{kJ/mol}$$

The complete oxidation of acetyl-CoA may either be brought about by a modified TCA cycle or by a linear pathway, referred to as carbon monoxide dehydrogenase pathway or acetyl-CoA pathway. The modified TCA cycle operates in organisms such as *Desulfobacter postgatei* and *Desulfobacter hydrogenophilus* (Fig. 21). In contrast to the normal TCA cycle, acetate is activated by a CoA transferase which transfers CoA from succinyl-CoA to acetate. The second modification has to do with the condensation of acetyl-CoA and oxaloacetate to citrate. This reaction is not catalyzed by citrate synthase, but by ATP citrate-lyase which couples this step with ATP synthesis by substrate level phosphorylation. The reducing equivalents generated in two out of four oxidation steps of this cycle are sufficiently negative to be coupled to the reduction of APS ($E_{m,7}$ = −60 mV) or sulfite ($E_{m,7}$ = −105 mV) and with proton translocation: these are NADPH ($E_{m,7}$ = −320 mV) and reduced ferredoxin ($E_{m,7}$ = −400 mV) which are formed during the oxidation of isocitrate and 2-oxoglutarate, respectively. The electron carrier employed for the oxidation of malate ($E_{m,7}$ = −166 mV) is unknown, but its reoxidation by APS or sulfite is not coupled with energy-conservation. The redox potential of the electrons

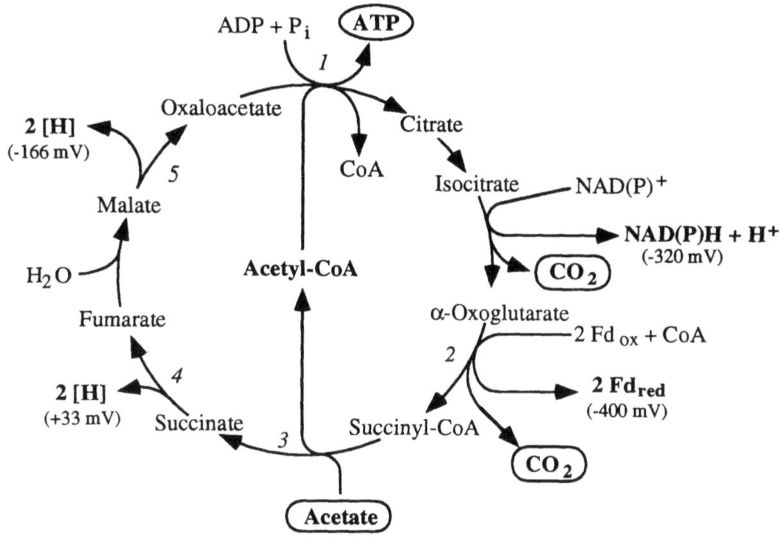


Fig. 21. Acetate oxidation via a modified tricarboxylic acid cycle in *Desulfobacter postgatei*. The mV values indicate the standard redox potential of the electrons generated in the respective redox reaction. Further explanations are given in the text. The italicized numbers represent the following enzymes: *1. ATP-citrate lyase 2. α-oxoglutarate:ferredoxin oxidoreductase. 3. succinyl-CoA: acetate CoA transferase. 4. succinate-oxidizing enzyme; physiological electron acceptor unknown. 5. malate-oxidizing enzyme, physiological electron acceptor unknown.*

generated during the oxidation of succinate to fumarate is so positive under standard conditions ($E_{m,7}$ = +33 mV) that reverse electron transport has to take place in order to allow the reduction of sulfite or APS.

Many other sulfate-reducing bacteria which are capable of complete acetate oxidation use the carbon monoxide dehydrogenase pathway (Fig. 22). Typical representatives of this group are *Desulfobacterium vacuolatum, Desulfosarcina variabilis, Desulfotomaculum acetoxidans.* Acetate oxidation by way of this pathway starts with the activation of acetate via acetate kinase and phosphotransacetylase which is associated with the consumption of one ATP. The key enzyme of this pathway, carbon monoxide dehydrogenase, cleaves the C-C bond of acetyl-CoA giving rise to methyl-tetrahydrofolate and enzyme-bound carbon monoxide. The latter is subsequently oxidized by CO dehydrogenase to $CO_2$ ($E_{m,7}$ = -524 mV). Transfer of these electrons to sulfite or APS may be coupled to energy conservation. Methyl-tetrahydrofolate is further oxidized by way of methylene-tetrahydrofolate, methenyl-tetrahydrofolate, formyl-tetrahydrofolate, and formate to $CO_2$. The formation of formate from formyl-tetrahydrofolate as catalyzed by formyl-tetrahydrofolate synthetase is associated with ATP synthesis by substrate level phosphorylation. The reducing equivalents produced during the oxidation of methylene-tetrahydrofolate

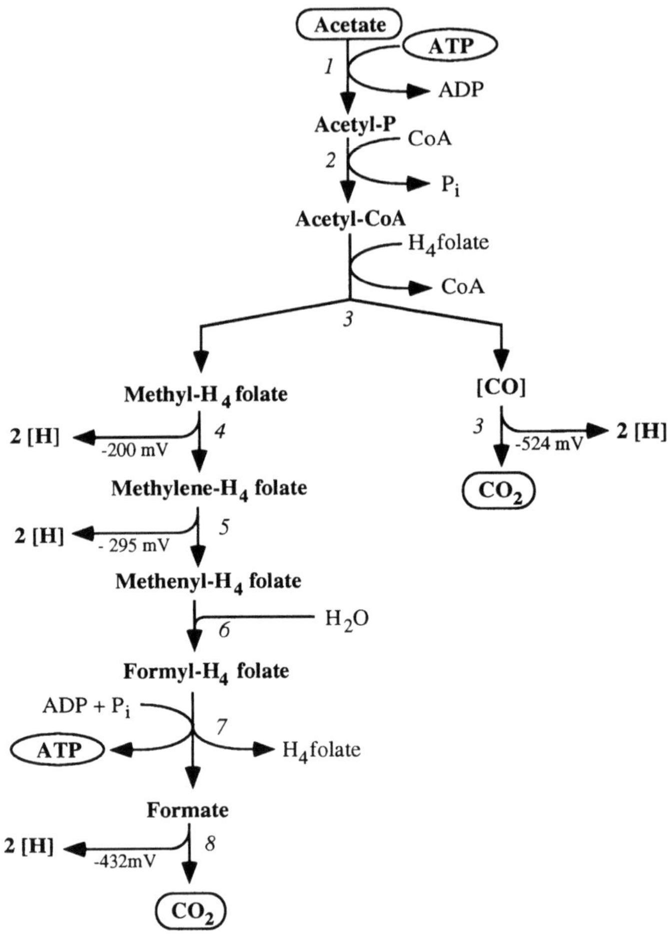


Fig. 22. Acetate oxidation via the linear acetyl-CoA pathway in *Desulfobacterium vacuolatum*. Further explanations are given in the text. The mV values indicate the standard redox potential of the corresponding redox pairs. $H_4$ folate: tetrahydrofolate. [CO], enzyme-bound CO. The italicized numbers represent the following enzymes: *1. acetate kinase. 2. phosphotransacetylase. 3. CO dehydrogenase. 4. methylene-tetrahydrofolate reductase. 5. methylene-tetrahydrofolate dehydrogenase. 6. cyclohydrolase. 7. formyl-tetrahydrofolate synthetase. 8. formate dehydrogenase.*

($E_{m,7}$ = −295 mV) and formate ($E_{m,7}$ = −432 mV) can be coupled with APS or sulfite reduction and lead to electron transport-driven ATP generation. The electrons from methyl-tetrahydrofolate oxidation ($E_{m,7}$ = −200 mV) can be used for APS or sulfite reduction, but cannot be coupled with energy conservation.

Sulfate-reducing bacteria perform electron transport-driven ATP synthesis, but it is obvious that the electron acceptor APS and sulfite have comparatively unfavorable redox potentials and are therefore poor electron acceptors for many electron donors. It is important to note that sulfate-reducing bacteria have to some extent the ability to gain energy exclusively by substrate level phosphorylation when sulfate is absent. Some *Desulfovibrio* species are for instance able to gain energy from the degradation of pyruvate to acetate, carbon dioxide and $H_2$. In this fermentation, acetate kinase is the only reaction linked to the phosphorylation of ADP. *Desulfobulbus propionicus* performs in the absence of sulfate a classical propionic acid fermentation according to the following equation (see section 3.1.12.4.):

$$3/2 \text{ lactate} \rightarrow 1 \text{ propionate} + 1/2 \text{ acetate} + 1/2\ HCO_3^- + 1/2\ H^+$$
$$\Delta G^{0'} = -82 \text{ kJ/mol}$$

Another way of gaining energy from lactate conversion without sulfate is used by a number of *Desulfovibrio* species. They convert lactate to acetate, $CO_2$ and 2 $H_2$ in syntrophic association with $H_2$-consuming organisms such as methanogenic bacteria. Whereas hydrogen evolution from pyruvate oxidation is exergonic under standard conditions, hydrogen evolution from lactate is thermodynamically only feasible if energy in the form of the electrochemical proton potential is invested to lift the electrons to a more negative redox potential by reverse electron transfer, and if the ambient redox potential of the $H^+/H_2$ couple is increased by constant removal of $H_2$ through a $H_2$ utilizer: from –420 mV under standard conditions ($10^5$ Pa) to –300 mV at 10 Pa $H_2$ partial pressure. Removal of reducing equivalents in the form of $H_2$ by a $H_2$-utilizer is the prerequisite for the anaerobic oxidation of substrates such as lactate, ethanol, and butyric acid in the absence of an external electron acceptor. The obligate transfer of $H_2$ from a fermenting organism to $H_2$-utilizing organism has been termed interspecies hydrogen transfer. This process plays a very important role in the mineralization of organic matter under anaerobic conditions (see also 3.1.13.).

*3.1.11.2. Carbon dioxide-reducing bacteria.* Two groups of anaerobic bacteria are capable of utilizing carbon dioxide as electron acceptor. Methanogenic and acetogenic bacteria couple the oxidation of a number of substrates to methane and acetate formation, respectively. They grow either chemolithotrophically or chemoorganotrophically. Both groups of organisms play a pivotal role in the anaerobic biodegradative microbial food chain.

*3.1.11.2.1. Acetogenic bacteria.* Acetogenic bacteria such as *Clostridium thermoaceticum* and *Acetobacterium woodii* convert sugars (hexoses and pentoses), $H_2$ and $CO_2$, methanol, and carbon monoxide to acetate according to the following equation:

| | |
|---|---|
| 1/3 fructose → acetate + $H^+$ | $\Delta G^{0\prime}$ = −104 kJ/mol |
| 2/5 xylose → acetate + $H^+$ | $\Delta G^{0\prime}$ = −106 kJ/mol |
| 2 $CO_2$ + 4 $H_2$ → acetate + 2 $H_2O$ + $H^+$ | $\Delta G^{0\prime}$ = −95 kJ/mol |
| 4/3 $CH_3OH$ + 2/3 $CO_2$ → acetate + 2/3 $H_2O$ + $H^+$ | $\Delta G^{0\prime}$ = −71 kJ/mol |
| 4 CO + 2 $H_2O$ → acetate + 2 $CO_2$ + $H^+$ | $\Delta G^{0\prime}$ = −175 kJ/mol |

During chemoorganotrophic growth, hexoses are degraded to pyruvate via the EMP pathway leading (per mol of hexose) to the synthesis of two mol ATP by substrate level phosphorylation and the production of two NADH. Pyruvate is converted by pyruvate: ferredoxin oxidoreductase to acetyl-CoA, $CO_2$ and reduced ferredoxin. Acetyl-CoA conversion to acetate is associated with ATP generation via acetate kinase. The reduced ferredoxin and the NADH being produced during glycolysis serve subsequently as electron donors for the reduction of two $CO_2$ to acetate as outlined below.

Chemolithotrophic growth on $H_2$ and $CO_2$ starts with the reduction of $CO_2$ to formate (Fig. 23). The latter is transferred to tetrahydrofolate in an ATP-dependent reaction to give rise to formyl-tetrahydrofolate which

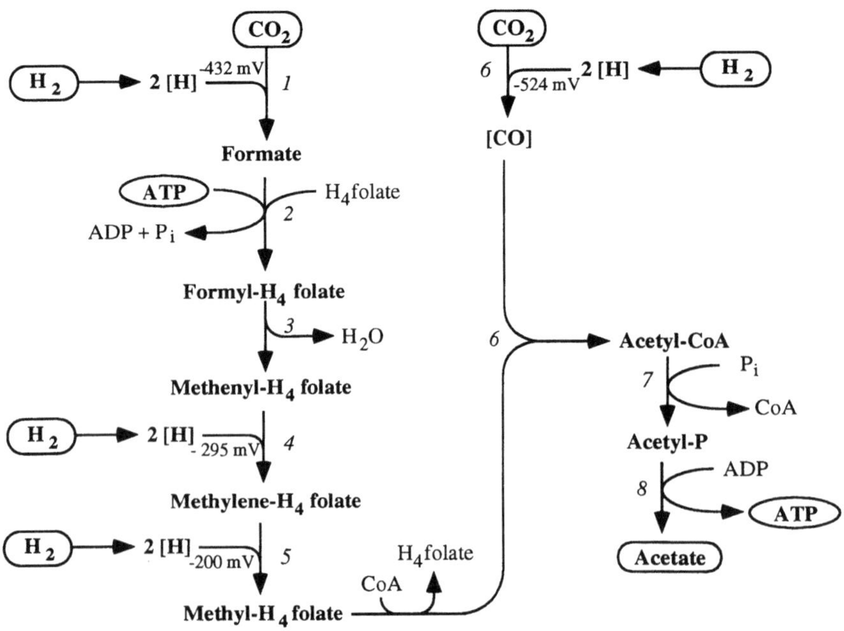


Fig. 23. Autotrophic acetate formation by acetogenic bacteria. Further explanation are given in the text. The mV values indicate the standard redox potentials of the corresponding redox couples. $H_4$ folate: tetrahydrofolate. [CO], enzyme-bound CO. The italicized numbers represent the following enzymes: *1. formate dehydrogenase. 2. formyl-tetrahydrofolate synthetase. 3. cyclohydrolase. 4. methylene-tetrahydrofolate dehydrogenase. 5. methylene- tetrahydrofolate reductase. 6. CO dehydrogenase. 7. phosphotransacetylase. 8. acetate kinase.*

is subsequently transformed to methenyl-tetrahydrofolate. Two $H_2$-dependent reduction steps yield methylene-tetrahydrofolate and then methyl-tetrahydrofolate. A second molecule of $CO_2$ is reduced by CO dehydrogenase to enzyme-bound CO with $H_2$ acting as electron donor. This reaction is endergonic and requires the coupling with the electrochemical proton potential (reverse electron transfer). After binding of coenzyme A, CO, and the methyl moiety from methyl-tetrahydrofolate to the catalytic site of CO dehydrogenase acetyl-CoA is formed and released from the enzyme. The conversion of acetyl-CoA to acetate as catalyzed by phosphotransacetylase and acetate kinase generates one ATP.

The acetyl-CoA pathway is not only employed as a catabolic route, but it is simultaneously used for $CO_2$ fixation. Furthermore, it operates in many sulfate-reducing bacteria in the opposite direction oxidizing acetyl-CoA to two $CO_2$. (Fig. 22).

At this point, it is difficult to see how net ATP can be gained in the acetyl-CoA pathway: one ATP is consumed in the formyl-tetrahydrofolate synthetase reaction and one is produced in the acetate kinase reaction. Hence, electron transport-driven ATP synthesis has to be the basis for net ATP generation. From the standard redox potentials of the intermediates involved it is evident that only the $H_2$-dependent reduction of methylene-tetrahydrofolate to methyl-tetrahydrofolate is sufficiently exergonic to be coupled with ATP formation. However, experimental evidence is still lacking. Furthermore, the energy output must be sufficiently high to drive both ATP synthesis and reverse electron transfer for the reduction of $CO_2$ to CO (see above). There are indications in *Acetobacterium woodii* that the reduction of methylene-tetrahydrofolate is coupled to the extrusion of sodium ions from the cytoplasm resulting in the generation of a transmembrane electrochemical sodium-ion gradient. The latter might drive ATP synthesis by means of a $Na^+$-dependent ATP synthase (see also 3.1.12.1.).

*3.1.11.2.2. Methanogenic bacteria.* Methanogenic bacteria represent the second group of strict anaerobic chemotrophs that are capable of utilizing $CO_2$ as electron acceptor. Most methanogens grow lithotrophically with $H_2$ as electron donor, but many of them are additionally able to utilize formate.

$$CO_2 + 4\ H_2 \rightarrow CH_4 + 2\ H_2O$$
$$\Delta G^{0'} = -130\ \mathrm{kJ/mol}$$
$$4\ HCOOH \rightarrow 3\ CO_2 + CH_4 + 2\ H_2O$$
$$\Delta G^{0'} = -120\ \mathrm{kJ/mol}$$

*The methanogenic $CO_2$ reduction pathway.* The methanogenic carbon reduction pathway differs from the acetogenic pathway in that it does not involve the consumption of ATP for the reduction of carbon dioxide to the formyl-level. Similarities exist, however, in the structure and function of the $C_1$ carriers tetrahydrofolate and tetrahydromethanopterin.

The methanogenic pathway of $CO_2$ reduction by $H_2$ to $CH_4$ involves four reduction steps and a number of unique coenzymes (Fig. 24). The

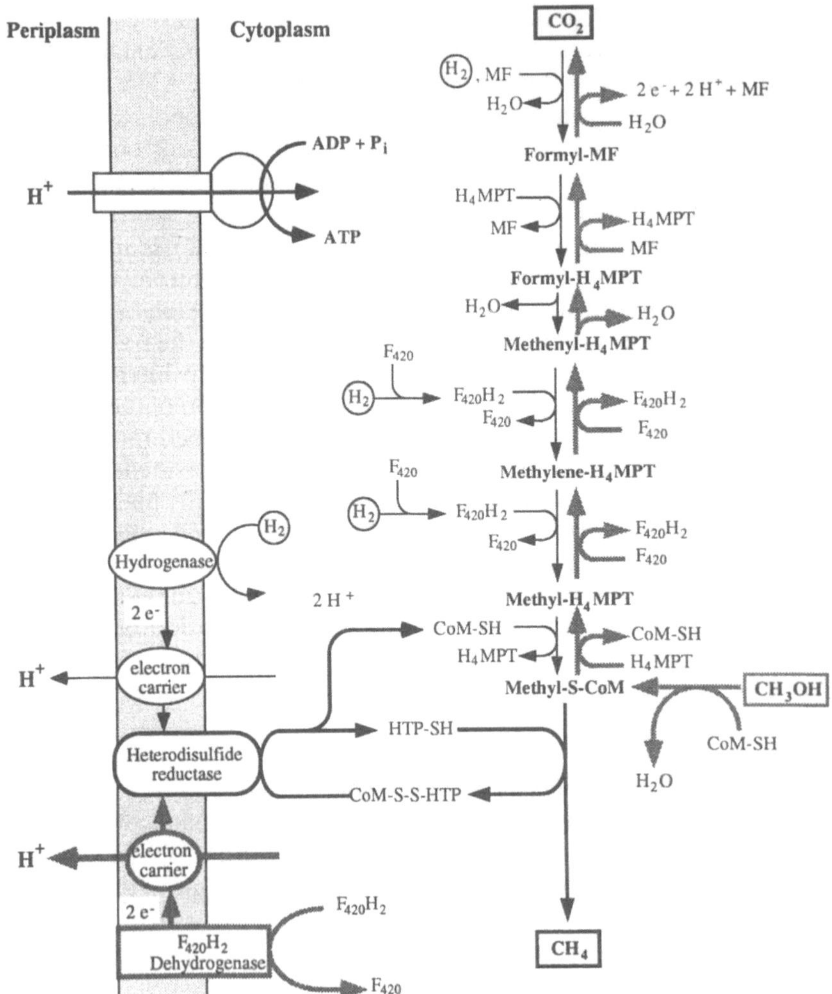


Fig. 24. Energy conservation during methanogenesis from $H_2 + CO_2$ or methanol. Arrows or boxes drawn with wide lines indicate reactions or proteins that are employed during the conversion of both, $H_2 + CO_2$ and methanol. Arrows or boxes in normal width indicate reactions or proteins that play a role during methanogenesis from $H_2 + CO_2$, whereas arrows or boxes in half tone mark reactions or proteins that are involved in methanogenesis from methanol. MF, methanofuran; $H_4MPT$, tetrahydromethanopterin. Further explanations are given in the text.

product of the first reduction step is not formate as in acetogens, but formyl-methanofuran which is formed from $CO_2$, $H_2$, and methanofuran. The physiological electron donor in this step (formyl-tetrahydromethanopterin dehydrogenase) is still unknown. After transfer of the formyl moiety to tetrahydromethanopterin the resulting formyl-tetrahydromethanopterin is transformed to methenyl-tetrahydromethanopterin which is subsequently reduced to methyl-tetrahydromethanopterin via methylene-tetrahydromethanopterin. The reduction of methenyl-tetrahydromethanopterin to the methyl-level is catalyzed by two factor $F_{420}$-dependent enzymes. The soluble deazaflavin $F_{420}$ acts as hydrogen carrier in these and in a number of other redox reactions including hydrogenase. Transfer of the methyl moiety to coenzyme M gives rise to methyl-CoM (mercaptoethanesulfonic acid) which is reductively demethylated leading to the formation of $CH_4$.

*Energy conservation in methanogens.* Methanogenesis from methyl-CoM and $H_2$ is accompanied by a standard Gibbs free energy change of ($\Delta G^{0'}$ = −85 kJ/mol). The reduction of methyl-CoM with $H_2$ occurs in two steps: methyl-CoM becomes reduced by 7-mercaptoheptanoylthreonine phosphate (HS-HTP) yielding $CH_4$ and a mixed disulfide:

$$CH_3\text{–S–CoM} + \text{HS–HTP} \rightarrow CH_4 + \text{CoM–S–S–HTP}$$
$$\Delta G^{0'} = -43 \text{ kJ/mol}$$

The sulfhydryl compounds are subsequently recovered by reductive cleavage of the heterodisulfide:

$$\text{CoM–S–S–HTP} + H_2 \rightarrow \text{CoM–SH} + \text{HS–HTP}$$
$$\Delta G^{0'} = -42 \text{ kJ/mol}$$

The latter reaction is accompanied by the generation of a transmembrane $\Delta\tilde{\mu}_{H^+}$ which is competent in driving ATP synthesis via an ATP synthase that differs somewhat from the classical $F_0F_1$-ATPase. The exact mechanism of proton translocation is not yet known, but it appears that the electrons derived from $H_2$ flow via a membrane-bound non-$F_{420}$-reducing hydrogenase along unknown carriers to the heterodisulfide reductase which transfers the electrons to the heterodisulfide.

The standard redox potential of the CoM-SH, HTP-SH/CoM-S-S-HTP couple is in the order of $E_{m,7}$ = −200 mV so that electron transport-driven phosphorylation with $H_2$ as electron donor ($E_{m,7}$ = −420 mV) is possible. However, it seems very likely that the actual redox potential ($E_h$) is more positive under cellular steady state conditions. This would be the case if the [CoM-S-S-HTP]/[CoM-SH][HTP-SH] ratio exceeds 1. At ratios of 10 and 100 the actual redox potential would increase to $E_h$ = −140 mV and $E_h$ = −80 mV, respectively. This would enlarge the redox span ($\Delta E$) between $H_2$ and the heterodisulfide considerably. Consequently, $\Delta G'$ for heterodisulfide reduction would increase to −54 kJ/mol and −65 kJ/mol, respectively. The [CoM-S-S-HTP]/[CoM-SH][HTP-SH] ratio is highly influenced by the methyl-CoM reductase reaction. Since the formation of CoM-S-S-HTP and $CH_4$ from $CH_3$-S-CoM and HS-HTP is thermodynamically highly favored ($K_{eq} = 34.5 \times 10^6$) a $\Delta G'$ considerably more negative than $\Delta G^{0'}$ can be envisaged for the $H_2$-dependent heterodisulfide reduction.

*Methylotrophic growth of methanogens.* Chemoorganotrophic growth is restricted to members of one methanogenic family, the Methanosarcinaceae. Members of this group are capable of utilizing in addition to $H_2$ and $CO_2$ substrates such as methanol, methylamines, and acetate. The conversion of methanol (and analogously of methylamines) occurs as follows:

$$4/3\ CH_3OH \rightarrow 1/3\ CO_2 + CH_4 + 2/3\ H_2O$$
$$\Delta G^{0'} = -105.5\ \text{kJ/mol}$$

One methanol is oxidized to $CO_2$ in order to generate six reducing equivalents which are in turn used to reduce three methanol to methane. The oxidation of methanol follows essentially the reverse route outlined above for $CO_2$ reduction (Fig. 24). The methyl group is transferred to tetrahydromethanopterin and the resulting methyl-tetrahydromethanopterin is oxidized to $CO_2$. The oxidation of both methyl-tetrahydromethanopterin and methylene-tetrahydromethanopterin is coupled with $F_{420}$ reduction. Methanol reduction is initiated by methyl group transfer to coenzyme M giving rise to methyl-CoM. The reduction of this central intermediate of methanogenesis is associated with electron transport-driven ATP formation and it proceeds as outlined above except that the reducing equivalents are not provided by $H_2$, but by the reducing equivalents gained during methanol oxidation. The reduced $F_{420}$ ($F_{420}H_2$) is reoxidized by a membrane-bound $F_{420}H_2$-oxidizing enzyme and the electrons are transferred to the heterodisulfide reductase. Membrane-bound *b*- and *c*-type cytochromes that were detected in all members of the Methanosarcinaceae are probably involved in this electron transfer.

Methanogenesis from acetate starts with an activation of acetate to acetyl-CoA. In *Methanosarcina* species acetyl-CoA formation is catalyzed by the concerted action of acetate kinase and phosphotransacetylase, whereas *Methanothrix* species employ acetyl-CoA synthetase for activation:

$$\text{acetate} + \text{CoA} + \text{ATP} \leftrightarrows \text{acetyl-CoA} + \text{AMP} + PP_i$$

Acetyl-CoA is subsequently cleaved by CO dehydrogenase giving rise to methyl-tetrahydromethanopterin and enzyme-bound CO. The methyl moiety is then transferred to coenzyme M yielding methyl-CoM which is reductively demethylated as described above. The oxidation of CO and the electron transfer to the heterodisulfide is linked to proton translocation and ATP synthesis. The amount of ATP synthesized in electron transport must exceed the one ATP that is consumed for the activation of acetate.

*3.1.12. Energy generation by chemoorganotrophic bacteria in the absence of external electron acceptors:* Sulfate-reducing bacteria, acetogenic bacteria, and methanogenic bacteria have managed to take advantage of electron transport phosphorylation even though the redox spans between the electron-donating and the electron-accepting systems are considerably smaller than in most chemoorganotrophic aerobes. Other groups of anaerobes rely completely on substrate level phosphorylation.

*3.1.12.1. Lactic acid fermentation. The homofermentative pathway.* The homofermentative lactic acid bacteria degrade glucose to lactate. This fermentation is carried out by *Sporolactobacillus inulinus* and by *Streptococcus* and *Lactobacillus* species. Glucose degradation to pyruvate proceeds along the EMP pathway (Fig. 25). The NADH produced in the glyceraldehyde-3-phosphate dehydrogenase reaction cannot be taken advantage of for electron transport-linked phosphorylation because of the

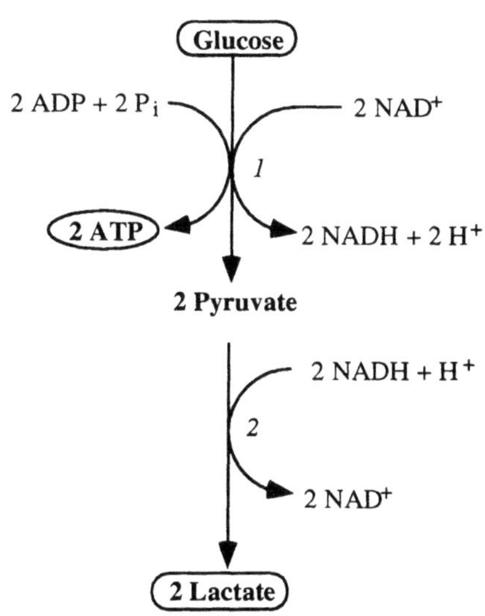


Fig. 25. The homofermentative pathway of lactic acid bacteria. The italicized numbers represent the following enzymes: *1. enzymes of the EMP pathway. 2. lactate dehydrogenase.*

unavailability of an external electron acceptor and because of the lack of an electron transport chain. In order to regenerate $NAD^+$, the reducing equivalents from NADH are transferred to pyruvate yielding lactate as the typical and sole end product. The net ATP gain per mol hexose being

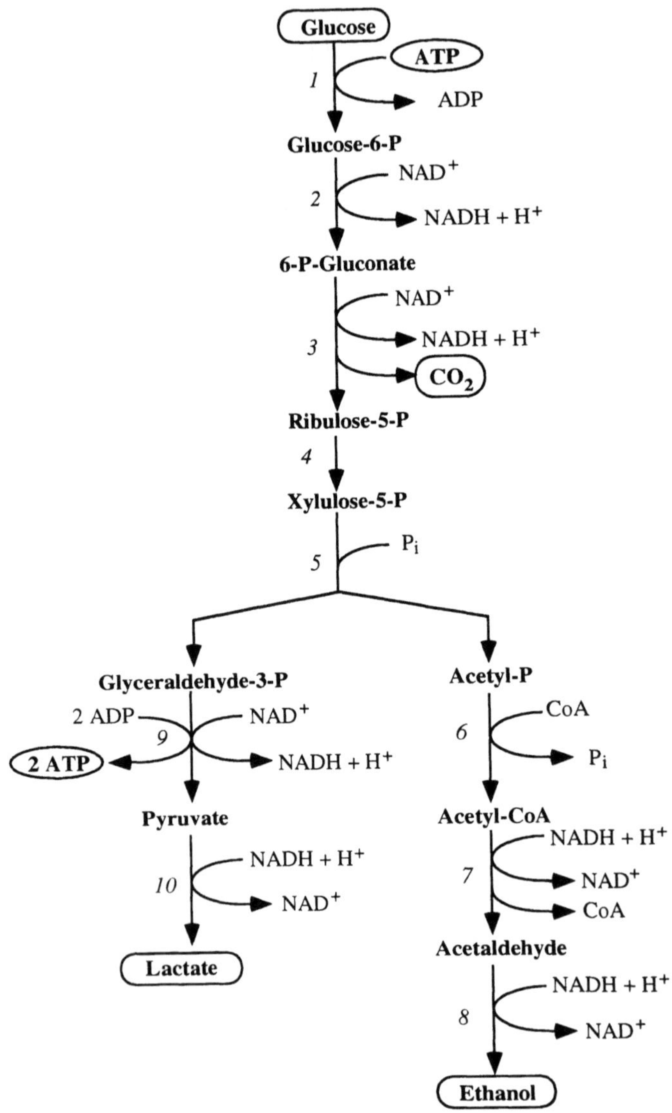


Fig. 26. The heterofermentative pathway of lactic acid bacteria. The italicized numbers represent the following enzymes: *1. hexokinase. 2. glucose-6-phosphate dehydrogenase. 3. 6-phosphogluconate dehydrogenase. 4. ribulose-5-phosphate 3-epimerase. 5. phosphoketolase. 6. phosphotransacetylase. 7. acetaldehyde dehydrogenase. 8. ethanol dehydrogenase. 9. enzymes of the EMP pathway. 10. lactate dehydrogenase.*

degraded is two: two ATP/hexose are invested for the conversion of glucose to fructose-1,6-bisphosphate and 4 ATP/hexose are produced during the conversion of glyceraldehyde-3-phosphate to pyruvate. The ATP-yielding reaction are 3-phosphoglycerate kinase and pyruvate kinase.

*The heterofermentative pathway.* Heterofermentative lactic acid bacteria such as *Leuconostoc* species are devoid of the glycolytic enzyme fructose bisphosphate aldolase and triosephosphate isomerase. As a result they degrade glucose differently (Fig. 26). Hexokinase catalyzes the activation of glucose to glucose-6-phosphate which is subsequently oxidized to 6-phosphogluconate and further to ribulose-5-phosphate and $CO_2$. These two oxidation steps are accompanied by the formation of 2 NADH. After epimerization of ribulose-5-phosphate to xylulose-5-phosphate the latter is split into glyceraldehyde-3-phosphate and acetyl phosphate in a thiamine pyrophosphate-dependent reaction. The latter reaction is catalyzed by phosphoketolase. Glyceraldehyde-3-phosphate conversion to pyruvate via glycolytic reactions is accompanied by the synthesis of two ATP. Acetyl phosphate is transformed into acetyl-CoA by the action of phosphotransacetylase. Acetyl-CoA is then reduced to acetaldehyde and further to ethanol whereby the NADH produced during glucose-6-phosphate oxidation becomes reoxidized. This fermentation is called heterofermentative because ethanol and $CO_2$ are produced besides lactate. The net ATP gain of this fermentation is one ATP/hexose.

*The Bifidobacterium pathway. Bifidobacterium bifidum* is a heterofermentative lactic acid bacterium that degrades glucose by way of yet another pathway (Fig. 27). After the transformation of 2 glucose to 2 fructose-6-phosphate, one of the latter molecules is split by fructose-6-phosphate phosphoketolase into erythrose-4-phosphate and acetyl phosphate. Acetate formation from acetyl phosphate is associated with the generation of one ATP. Erythrose-4-phosphate and the second molecule of fructose-6-phosphate are converted to xylulose-5-phosphate and ribose-5-phosphate by the concerted action of transaldolase and transketolase. Isomerization and epimerization of ribose-5-phosphate yields a second molecule of xylulose-5-phosphate. The two molecules of xylulose-5-phosphate are transformed to 2 glyceraldehyde-3-phosphate and 2 acetyl phosphate by xylulose-5-phosphate phosphoketolase. The conversion of glyceraldehyde-3-phosphate to lactate proceeds as in homofermentative bacteria and yields 2 ATP/glyceraldehyde-3-phosphate. Acetate kinase couples the conversion of acetyl phosphate to acetate with the phosphorylation of ADP. The energy yield of this pathway is 2.5 ATP/glucose.

*3.1.12.2. Ethanol fermentation.* Ethanol is a well known fermentation end product. It is produced by both yeasts, such as *Saccharomyces cerevisiae,* and by bacteria. Two principal pathways can be distinguished. The path-

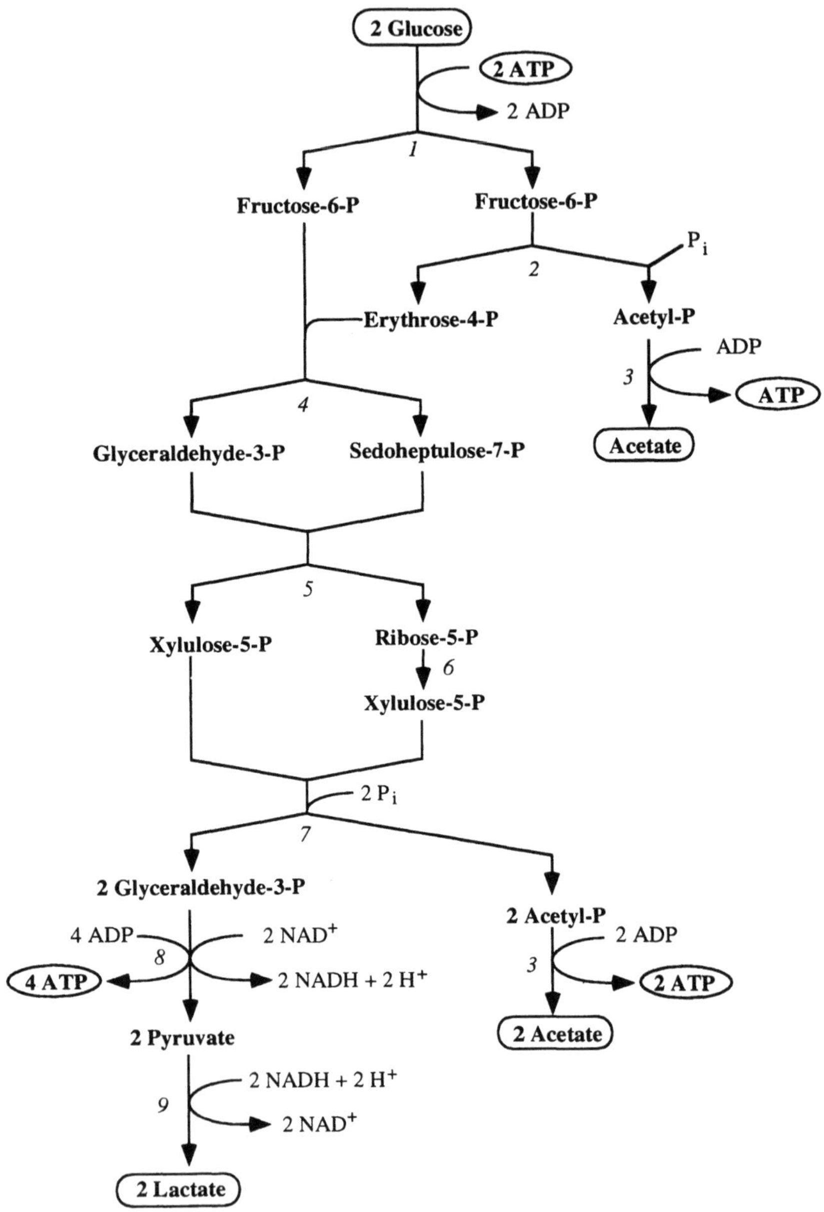


Fig. 27. The formation of lactate and acetate by *Bifidobacterium bifidum*. The italicized numbers represent the following enzymes: *1. hexokinase and glucose-6-phosphate isomerase. 2. fructose-6-phosphate phosphoketolase. 3. acetate kinase. 4. transaldolase. 5. transketolase. 6. ribose-5-phosphate isomerase and ribulose-5-phosphate 3-epimerase. 7. xylulose-5-phosphate phosphoketolase. 8. enzymes of the EMP pathway. 10. lactate dehydrogenase.*

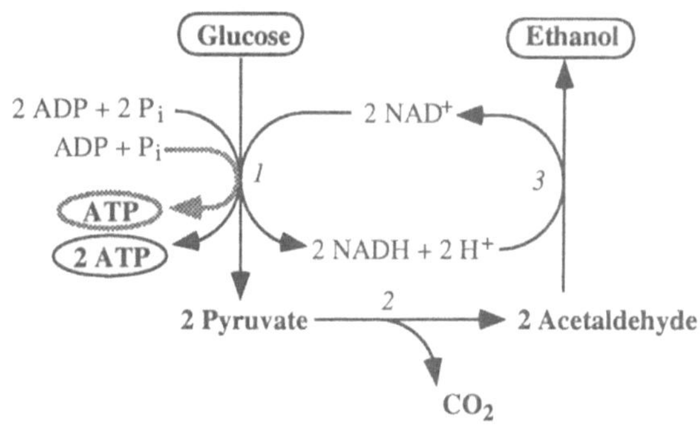


Fig. 28. Ethanol fermentations from glucose. See text for more details. The italicized numbers represent the following enzymes: *1. enzymes of the EMP pathway or enzymes of the ED pathway. 2. alcohol dehydrogenase.* Two molecules of ATP are generated during glucose breakdown in the EMP pathway (solid arrow) in contrast to one ATP in the ED pathway (half tone arrow).

way employed by *S. cerevisiae* is also used by certain bacteria such as *Sarcina ventriculi* and *Erwinia amylovora* and is based on the EMP pathway (Fig. 28). *S. cerevisiae* converts glucose to 2 pyruvate thereby generating 2 ATP and 2 NADH. In order to regenerate $NAD^+$, pyruvate is split by pyruvate decarboxylase into 2 $CO_2$ and 2 acetaldehyde. The latter is then reduced to ethanol in a NADH-dependent reaction.

In contrast, *Zymomonas mobilis* degrades hexoses via the Entner-Doudoroff pathway to pyruvate. The route from pyruvate to ethanol in *Zymomonas mobilis* is identical to that encountered in *S. cerevisiae.*

*3.1.12.3. Butyrate fermentation.* Butyric acid is the characteristic end product of several strictly anaerobic bacteria belonging to the genera *Clostridium, Butyrivibrio, Eubacterium,* and *Fusobacterium.* These organisms couple the conversion of glucose to 1 butyrate and 2 $CO_2$ with the phosphorylation of 3 ADP:

$$\text{Glucose} + 2\ H_2O \rightarrow \text{butyrate} + 2\ HCO_3^- + 2\ H_2 + 3\ H^+$$
$$\Delta G^{0'} = -255\ \text{kJ/mol}$$

Glucose enters the cell by way of the phosphotransferase system which brings about the uptake of glucose and simultaneously its phosphorylation to glucose-6-phosphate. The latter is further degraded to pyruvate via the glycolytic pathway (Fig. 29). Pyruvate is oxidized to acetyl-CoA, $CO_2$ and reduced ferredoxin as catalyzed by pyruvate: ferredoxin oxidoreductase, an enzyme that is widespread among strict and facultative anaerobes. The midpoint potential of various ferredoxins varies between −225 mV *(Azotobacter vinelandii)* and −480 mV *(Chromatium vinosum).* The redox potential of the clostridial ferredoxin ($E_{m,7}$ = −390 mV) is suffi-

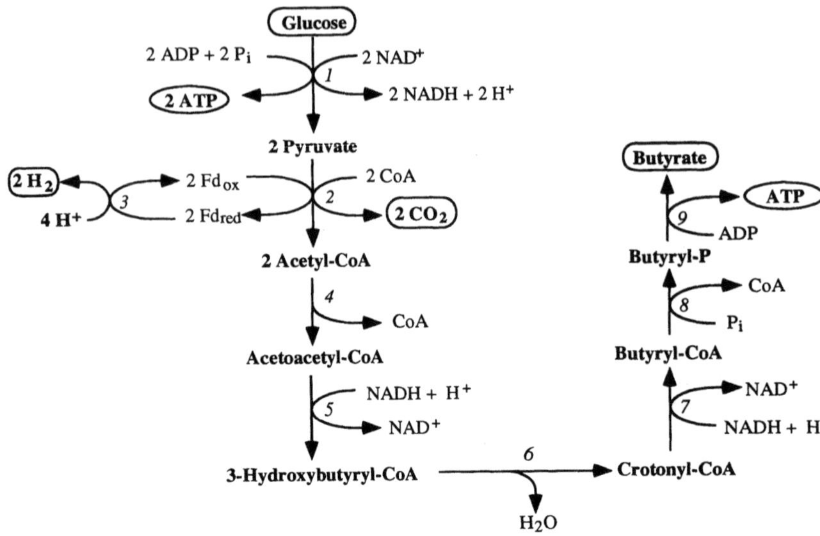


Fig. 29. Butyrate fermentation pathway from glucose. See text for more details. The italicized numbers represent the following enzymes. *1. enzymes of the EMP pathway or enzymes of the ED pathway. 2. pyruvate:ferredoxin oxidoreductase. 3. hydrogenase. 4. acetyl-CoA acetyltransferase. 5. hydroxybutyryl-CoA dehydrogenase. 6. crotonase. 7. butyryl-CoA dehydrogenase. 8. phosphotransbutyrylase. 9. butyrate kinase.*

ciently negative to be coupled with the reduction of protons and the release of hydrogen gas via hydrogenase.

Two molecules of acetyl-CoA are condensed giving rise to acetoacetyl-CoA. The latter serves now as electron acceptor for the reducing equivalents produced in the glycolytic pathway and is thereby stepwise reduced to butyryl-CoA. Phosphotransbutyrylase catalyzes the exchange of the CoA moiety for a phosphate group yielding butyryl phosphate. Conversion of the latter to butyrate by butyrate kinase is coupled with ATP synthesis. Phosphotransbutyrylase and butyrate kinase function analogously to phosphotransacetylase and acetate kinase in other fermentation pathways.

*3.1.12.4. Propionate fermentation.* Strictly defined, the conversion of glucose or lactate to propionic acid as performed by many anaerobic bacteria belonging to the genera *Propionibacterium, Bacteroides, Veillonella,* and *Desulfovibrio* is not a fermentation in the sense that the ATP produced in this pathway is exclusively gained by substrate level phosphorylation. Propionate formation involves electron transport to fumarate which generates a transmembrane proton potential. The latter is subsequently taken advantage of for ATP synthesis. Fumarate reduction as a source of energy has already been discussed above (3.1.10.3.). For historical reasons, however, propionic acid fermentation is considered as a fermentation. This may also be justified by the fact that fumarate is not externally added, but produced during substrate degradation. An important difference to the fumarate reduction as performed by *Wolinella succinogenes,* for instance, lies furthermore in the fact that not succinate but propionate is the end product. During propionate formation, additional ATP is formed by substrate level phosphorylation.

Lactate is a typical substrate of propionic acid bacteria and is converted according to the following equation:

$$3/2 \text{ lactate} \rightarrow \text{propionate} + 1/2 \text{ acetate} + 1/2\, HCO_3^- + 1\, H^+$$
$$\Delta G^{0'} = -82 \text{ kJ/mol}$$

The route of propionic acid formation follows the succinate-propionate pathway which is outlined in Fig. 30. The pathway starts with the oxidation of three lactate to three pyruvate by lactate dehydrogenase, a flavoenzyme. One pyruvate is further oxidized to acetyl-CoA and $CO_2$. The transformation of acetyl-CoA to acetate by way of acetyl phosphate is accompanied by substrate level phosphorylation. The two remaining molecules of pyruvate undergo a transcarboxylation reaction with methylmalonyl-CoA as $CO_2$ donor and biotin as $CO_2$-carrier resulting in the formation of oxaloacetate. The latter is subsequently reduced to malate which is turned into fumarate by the removal of water. As already mentioned, the subsequent reduction of fumarate to succinate with the electrons derived

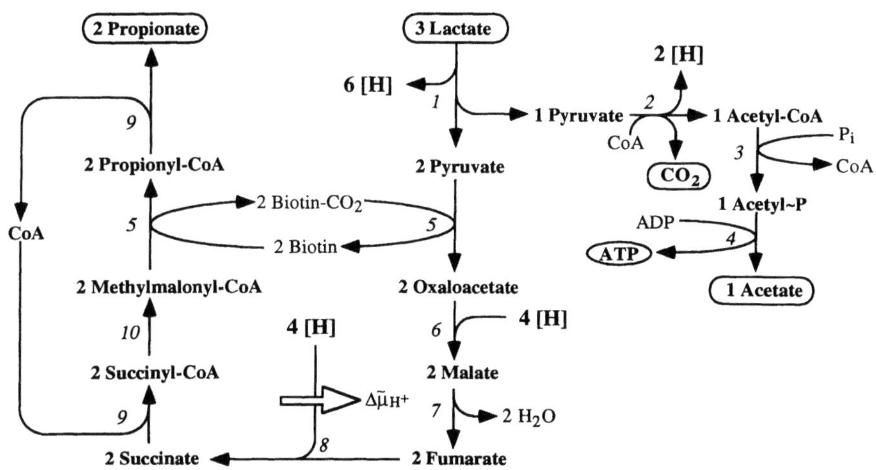


Fig. 30. Propionate fermentation pathway from lactate. See text for more details. The open arrow indicates a coupling with proton translocation. The italicized numbers represent the following enzymes: *1. lactate dehydrogenase. 2. pyruvate:ferredoxin oxidoreductase. 3. phosphotransacetylase. 4. acetate kinase. 5. methylmalonyl-CoA pyruvate transcarboxylase. 6. malate dehydrogenase. 7. fumarase. 8. fumarate reductase. 9. CoA transferase. 10. methylmalonyl-CoA mutase.*

from lactate or pyruvate oxidation is coupled to ATP formation by electron transport phosphorylation. Electrons derived from lactate oxidation flow through menaquinone and cytochrome *b* to fumarate reductase. After the activation of succinate by a CoA transferase to succinyl-CoA the latter undergoes a coenzyme $B_{12}$-dependent rearrangement reaction yielding methylmalonyl-CoA. Transcarboxylation of methylmalonyl-CoA yields propionyl-CoA. The subsequent CoA transfer to succinate gives rise to propionate.

Many propionic acid bacteria convert glucose as well:

$$3/4 \text{ glucose} \rightarrow \text{propionate} + 1/2 \text{ acetate} + 1/2 \; HCO_3^- + 2 \; H^+$$
$$\Delta G^{0'} = -231 \text{ kJ/mol}$$

*3.1.12.5. Mixed acid fermentation.* Bacteria belonging to the genera *Escherichia, Salmonella,* and *Shigella* convert sugars to $CO_2$, $H_2$, ethanol, as well as lactic, acetic, succinic and formic acid. Glucose degradation follows the EMP pathway (Fig. 31). Phosphoenolpyruvate is the first branching point in this fermentation. Phosphoenolpyruvate is either converted to pyruvate or carboxylated to yield oxaloacetate. Malate dehydro-

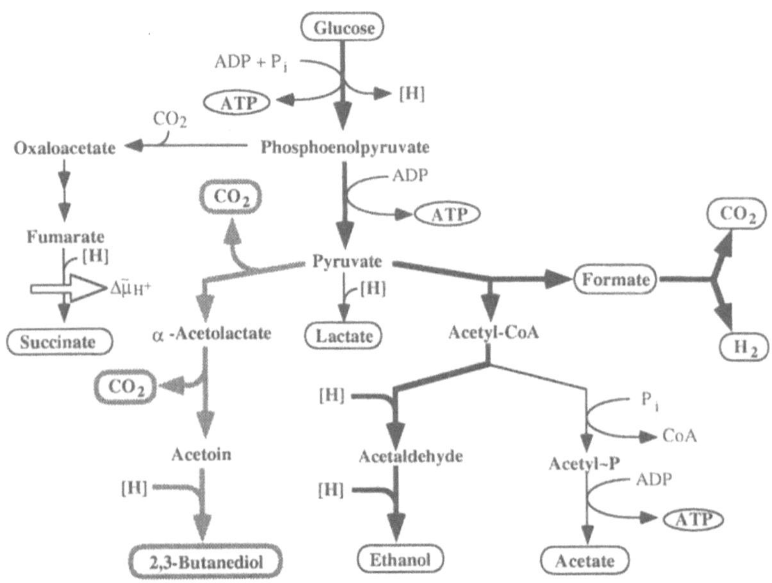


Fig. 31. Mixed acid and butanediol fermentations from glucose. The open arrow indicates a coupling with proton translocation. Wide arrows indicate reactions that are employed for both fermentations. Thin arrows indicate reactions of the mixed acid fermentation, whereas arrows in half tone mark reactions that are involved in the butanediol fermentation. The scheme does not take into account the stoichiometries of the fermentations. Further explanations are given in the text.

genase and fumerase catalyse the conversion of the latter to fumarate. The reduction of fumarate to succinate is accompanied by electron transport phosphorylation. Pyruvate is the second branching point: It may serve as electron acceptor giving rise to lactate or it is oxidized. The oxidation of pyruvate is catalyzed by pyruvate formate lyase which is only expressed under anaerobic conditions yielding formate and acetyl-CoA. Acetyl-CoA may be reduced via acetaldehyde to ethanol, or it may be converted to acetate by the concerted action of phosphotransacetylase and acetate kinase. The latter conversion is accompanied by the formation of ATP. Formate is either excreted or split to $CO_2$ and $H_2$ by formate hydrogen lyase. In contrast to *E. coli, Shigella* and *Erwinia* species do not contain formate hydrogen lyase and excrete therefore considerable amounts of formate.

*E. coli's* capability to grow not only with a variety of electron acceptors (oxygen, nitrate, fumarate, dimethylsulfoxide, and trimethylamine-N-oxide), but also in the absence of an external electron donor underlines once more the organism's astonishing metabolic versatility.

*3.1.12.6. Butanediol fermentation.* Enteric bacteria belonging to the genera *Enterobacter* and *Serratia* carry out a fermentation very similar to the one described above with the exception that succinate is not formed, excluding ATP gain by electron transport phosphorylation linked to fumarate reduction (Fig. 31). The second difference is the replacement of acetate by 2,3-butanediol as an end product. The enzyme responsible for the formation of $\alpha$-acetolactate, a precursor of 2,3-butanediol, is $\alpha$-acetolactate synthase which contains thiamine pyrophosphate: it converts two molecules of pyruvate to $\alpha$-acetolactate and $CO_2$. Decarboxylation of $\alpha$-acetolactate gives rise to acetoin which is subsequently reduced to 2,3-butanediol.

The ATP-yielding reactions in this fermentation are identical to those encountered in homofermentative lactic acid bacteria. In both fermentations, glucose degradation to pyruvate follows the EMP pathway. However, the reactions employed for the regeneration of $NAD^+$ differ significantly: Whereas the homofermentative lactic acid bacteria utilize only pyruvate as electron acceptor, which leads to the formation of their sole end product lactate, enterobacteria take advantage of a variety of degradation intermediates as electron acceptors.

The fermentations discussed so far are not the only ones playing a role in the anaerobic breakdown of organic material. The selection presented is restricted to well known pathways and to the utilization of sugars or lactate. There are of course other substrates which are subject to degradation by fermentative bacteria.

*3.1.13. Obligately syntrophic associations of bacteria:* Substrates such as ethanol, butyrate, and propionate cannot be degraded anaerobically to acetate by a single bacterial species, but only by bacterial consortia. This is due to the fact that the electrons generated from the oxidation of these compounds have to be transferred to $H^+$ leading to the formation of $H_2$. However, $H_2$ production from these compounds is endergonic under standard conditions which correspond to a $H_2$ partial pressure of 1 atmosphere (1 atm = 101 kPa):

$$\text{ethanol} + H_2O \leftrightarrows \text{acetate} + H^+ + H_2$$
$$\Delta G^{0'} = +9.6 \text{ kJ/mol}$$
$$1/2 \text{ butyrate} + H_2O \leftrightarrows \text{acetate} + 1/2\ H^+ + H_2$$
$$\Delta G^{0'} = +24 \text{ kJ/mol}$$
$$\text{propionate} + 3\ H_2O \leftrightarrows \text{acetate} + HCO_3^- + H^+ + 3\ H_2$$
$$\Delta G^{0'} = +25.5 \text{ kJ/mol}$$

Consequently, oxidation can only occur if the hydrogen partial pressure is kept low in the culture. This is accomplished by hydrogen-consuming organisms such as sulfate-reducing, methanogenic or acetogenic bacteria. In other words, the hydrogen-producing bacteria can only exist in syntrophic association with hydrogen consumers. The dependence of the actual Gibbs free energy change (ΔG') on the hydrogen partial pressure is to be illustrated by the conversion of ethanol to acetate and $H_2$. If the $H_2$ partial pressure is varied from $10^5$ to $10^4$, $10^3$, $10^2$, and 10 Pa (keeping all other parameters at standard conditions) ΔG' changes from +9.6 to −1.8, −13.2, −24.6 and −36.1 kJ/mol, respectively.

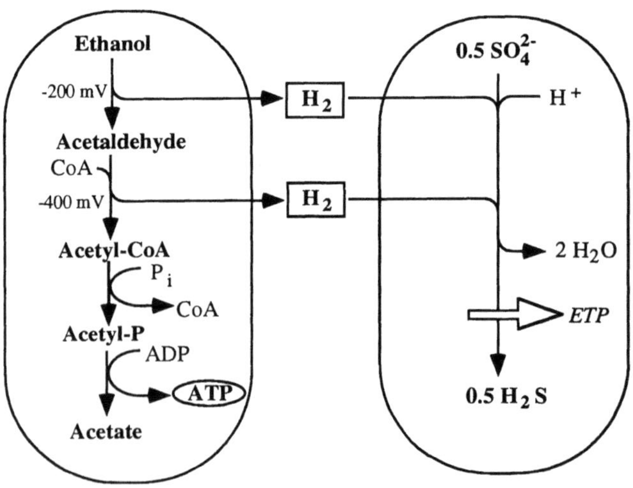


Fig. 32. Interspecies hydrogen transfer. The scheme shows an association of an anaerobic ethanol-oxidizing bacterium and a $H_2$-oxidizing sulfate-reducing bacterium. The open arrow indicates that sulfate reduction is coupled to ATP formation by electron transport phosphorylation (ETP). The mV values indicate the standard redox potential of the electrons generated in the respective redox reaction. Further explanation are given in the text.

The conversion of ethanol to acetate by a syntrophic culture consisting of an ethanol-oxidizing and sulfate-reducing bacterium is illustrated in Fig. 32. Ethanol is oxidized to acetaldehyde which upon oxidation yields acetyl-CoA. The conversion of the latter to acetate via acetyl phosphate generates one ATP by substrate level phosphorylation. The reducing equivalents generated in these reactions are released as $H_2$ which is immediately taken up by a sulfate-reducing bacterium. The ensuing $H_2$-dependent sulfate reduction is linked to ATP formation by electron transport phosphorylation. This process has been termed interspecies hydrogen transfer.

*3.1.14. Fermentation of amino acids:* Besides sugars, alcohols, and organic acids, nitrogenous compounds are also fermented by a variety of bacteria. The most important nitrogenous compounds are amino acids which are produced during protein breakdown, and purines and pyrimidines which may arise from DNA or RNA. Only two examples illustrating the degradation of amino acids will be discussed.

*3.1.14.1. Glycine degradation.* A number of strictly anaerobic bacteria such as *Clostridium sticklandii, Peptostreptococcus glycinophylus,* and *Eubacterium acidaminophilum* ferment glycine according to:

$$\text{glycine} + H_2O \leftrightarrows NH_4^+ + 1/2\ HCO_3^- + 3/4\ \text{acetate} + 1/4\ H^+$$
$$\Delta G^{0'} = -52\ \text{kJ/mol}$$

The pathway of glycine degradation is indicated in Fig.33. In an $NAD^+$-linked and tetrahydrofolate-dependent reaction one molecule of glycine is oxidized to $CO_2$, $NH_3$ and methylene-tetrahydrofolate. The latter is subsequently oxidized to $CO_2$ in a reaction sequence already discussed above (3.1.11.1.). This is accompanied by the formation of one ATP in the formyl-tetrahydrofolate synthetase reaction. The three pairs of reducing equivalents formed are used to reductively cleave three molecules of glycine to three molecules each of acetate and ammonia. Glycine reduction as catalyzed by glycine reductase is accompanied by the formation of acetyl phosphate, the conversion of which to acetate yields one ATP. All the ATP formed during this fermentation is gained by substrate level phosphorylation.

*3.1.14.2. Arginine degradation.* An unusual intermediate which leads to ATP formation by substrate level phosphorylation is formed during the degradation of arginine by species belonging to the genera *Clostridium* and *Eubacterium:* carbamoyl phosphate. The pathway of arginine breakdown starts with a deimination reaction yielding citrulline (Fig. 34). The latter is converted to carbamoyl phosphate and ornithine carbamoyl-transferase. The cleavage of carbamoyl phosphate to carbon dioxide and

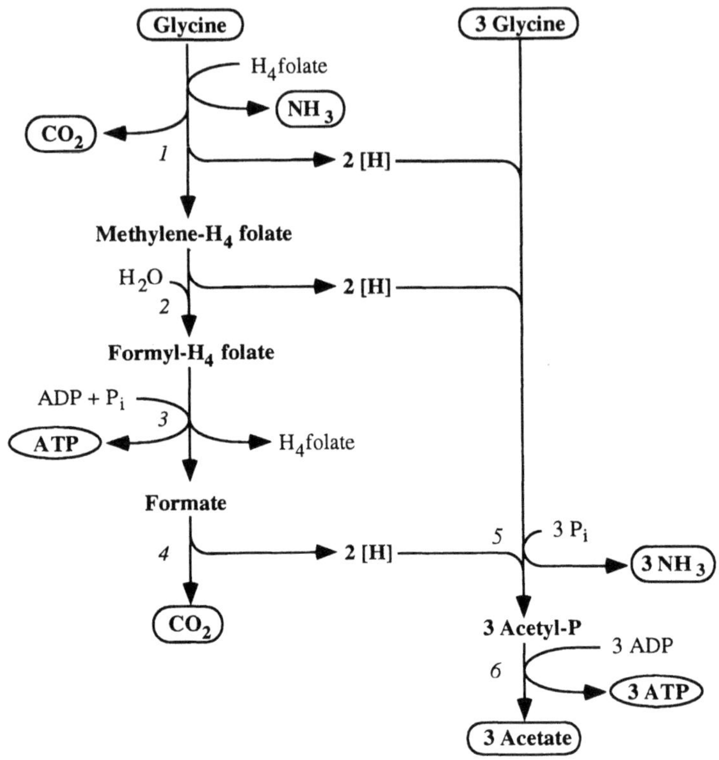


Fig. 33. Degradation of glycine by *Peptostreptococcus*. $H_4$ folate: tetrahydrofolate. The italicized numbers represent the following enzymes: *1. glycine decarboxylase. 2. methylene-tetrahydrofolate dehydrogenase and cyclohydrolase. 3. formyl-tetrahydrofolate synthetase. 4. formate dehydrogenase. 5. glycine reductase. 6. acetate kinase.* Further explanations are given in the text.

ammonia as catalyzed by carbamate kinase is accompanied by the formation of ATP. Depending on the organism ornithine may be further degraded or excreted. The examples discussed are only a small selection of a wealth of fermentative degradation pathways encountered in anaerobic chemoorganoheterotrophic bacteria. Despite this variety of pathways the underlying principle of ATP generation is quite simple: the Gibbs free energy change associated with the conversion of a substrate is stoichiometrically coupled with the formation of a phosphoryl donor intermediate (1,3-diphosphoglycerate, phosphoenolpyruvate, acetyl phosphate, or carbamoyl phosphate). The activated phosphoryl group is subsequently transferred to ADP.

*3.1.15. Unusual modes of ATP generation in bacteria: 3.1.15.1. Decarboxylation phosphorylation.* A number of strictly anaerobic bacteria gain ATP from decarboxylation reactions which are coupled to membrane ener-

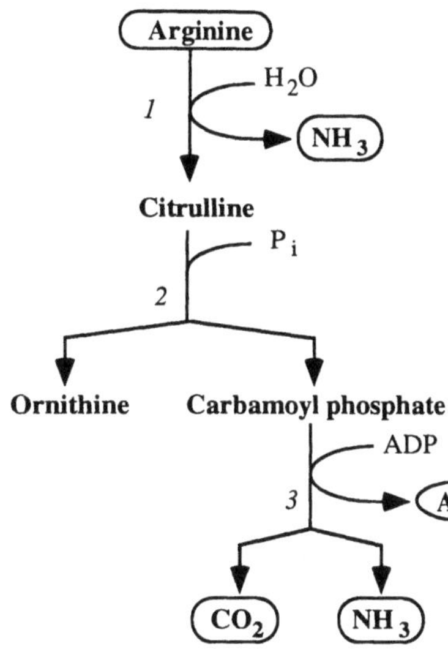


Fig. 34. The anaerobic degradation of arginine. The italicized numbers represent the following enzymes: *1. arginine deiminase. 2. ornithine carbamoyltransferase. 3. carbamate kinase.* Further explanations are given in the text.

gization. This kind of energy conservation is illustrated by the degradation of succinate to propionate and $CO_2$ as carried out by the strict anaerobe *Propionigenium modestum* (Fig. 35).

$$\text{succinate} + H_2O \rightarrow \text{propionate} + HCO_3^-$$
$$\Delta G^{0'} = -20 \text{ kJ/mol}$$

The conversion of succinate by this organism partly proceeds like the propionate fermentation pathway discussed above (3.1.12.4.): Succinate activation to succinyl-CoA is followed by a rearrangement reaction yielding methylmalonyl-CoA. The latter undergoes a decarboxylation yielding propionyl-CoA which serves as CoA donor in the activation of succinate. (In contrast, propionibacteria form propionyl-CoA from methylmalonyl-CoA by a transcarboxylation reaction).

The reaction sequence does not allow ATP synthesis by either substrate level or electron transport phosphorylation. Instead, the Gibbs free energy change of the decarboxylation reaction is converted into a transmembrane electrochemical potential difference of sodium ions ($\Delta\tilde{\mu}_{Na^+}$) which is defined in analogy to $\Delta\tilde{\mu}_{H^+}$:

$$\Delta\tilde{\mu}_{Na^+} = F\,\Delta\phi - 2.303\ RT\ \Delta pNa$$

Thus, $\Delta\tilde{\mu}_{Na^+}$ is composed of the transmembrane electrical potential difference ($\Delta\phi$) and the transmembrane concentration difference of sodium ions (2.303 RT/F ΔpNa), ΔpNa being defined as ($\lg[Na^+]_{out} - \lg[Na^+]_{in}$).

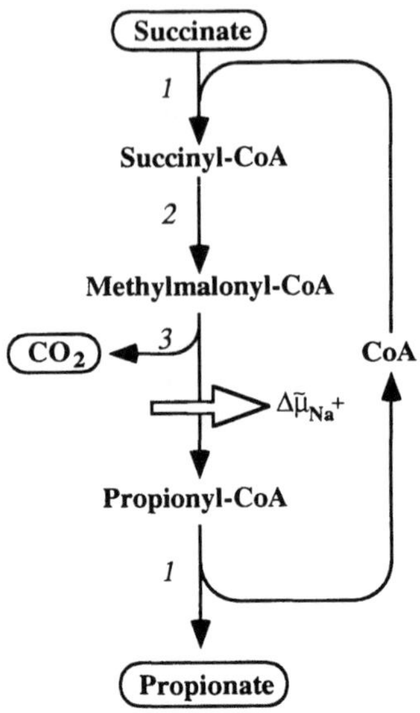


Fig. 35. The conversion of succinate to propionate by *Propionigenium modestum*. The open arrow indicates the coupling of the decarboxylase reaction with the formation of a $\Delta\tilde{\mu}_{Na^+}$. More details are given in the text. The italicized numbers represent the following enzymes: *1. CoA transferase. 2. methylmalonyl-CoA mutase. 3. methylmalonyl-CoA decarboxylase.*

In other words, the methylmalonyl-CoA decarboxylase couples the decarboxylation of methylmalonyl-CoA with the extrusion of $Na^+$ from the interior of the cell to the exterior. The $Na^+$ gradient generated is taken advantage of by a $Na^+$-specific ATP synthase which couples the influx of $Na^+$ with ATP synthesis. Interestingly, this ATP synthase displays the same structure and composition as a conventional, $H^+$-dependent, so-called $F_0F_1$-type ATPase, and is therefore classified as such. The specificity for the respective ion is determined by the $F_0$ part which may either prefer $H^+$ or $Na^+$ as a coupling ion. This way of energy conservation may be termed decarboxylation phosphorylation.

Other decarboxylation reactions coupled to $Na^+$ translocation are the oxaloacetate decarboxylase which plays a role in the anaerobic citrate degradation by enterobacteria, and the glutaconyl-CoA decarboxylase which is involved in the anaerobic breakdown of glutamate by *Acidaminococcus fermentans* and *Clostridium symbiosum* to ammonia, carbon dioxide, acetate, butyrate, and $H_2$:

$$\text{oxaloacetate} + 2\,Na^+_{in} \rightarrow \text{pyruvate} + CO_2 + 2\,Na^+_{out}$$
$$\text{glutaconyl-CoA} + 2\,Na^+_{in} \rightarrow \text{crotonyl-CoA} + CO_2 + 2\,Na^+_{out}$$

All of these $Na^+$-translocating decarboxylases are characterized by being membrane-bound proteins and by containing biotin as a prosthetic group. The oxaloacetate decarboxylase for instance is composed of three subunits ($\alpha$, $\beta$, $\gamma$): The membrane-peripheral, biotin-containing $\alpha$-subunit catalyzes the decarboxylation reaction, whereas the $\beta$- and $\gamma$-subunits which make up the membrane-intrinsic part of the complex are involved in $Na^+$ translocation.

*3.1.15.2. Respiratory $Na^+$ pump.* The $Na^+$-translocating decarboxylases are not the only primary $Na^+$ pumps encountered in bacteria. The aerobic respiratory chain of the marine bacterium *Vibrio alginolyticus* forms an electrochemical sodium gradient as a direct result of respiration when growing at alkaline pH. At neutral pH, however, protons instead of sodium ions are pumped and $\Delta\tilde{\mu}_{Na^+}$ is formed secondarily by the activity of a $Na^+$-$H^+$ antiporter which exchanges $H^+$ for $Na^+$. The $Na^+$ dependence of respiration in *V. alginolyticus* at alkaline pH could be assigned to the NADH: quinone oxidoreductase which translocates $Na^+$ from the interior to the exterior of the cell during the catalytic cycle. Similar $Na^+$-motive NADH dehydrogenases have also been found in other marine bacteria. It is not yet clear whether these organisms possess a $Na^+$-dependent ATP synthase like *P. modestum* or not.

Although *V. alginolyticus* utilizes $Na^+$ as a coupling ion the nutritional mode can be clearly characterized as being chemoorganoheterotrophic. In contrast, the metabolism of *P. modestum* can neither be termed organo- nor lithotrophic because energy generation is not brought about by a redox reaction, but by a decarboxylation reaction. Hence, the chemoheterotrophic growth mode of *P. modestum* cannot be further characterized within these categories.

*3.1.15.3. The role of $Na^+$ in methanogens.* In methanogens, electron flow from $H_2$ or $F_{420}H_2$ to CoM-S-S-HTP generates a $\Delta\tilde{\mu}_{H^+}$ (see section 3.1.11.2.2.). In contrast, the conversion of methylene-tetrahydromethanopterin to methyl-CoM is linked to the primary extrusion of $Na^+$, the reverse reaction being associated with $Na^+$ influx. There are indications that the $\Delta\tilde{\mu}_{Na^+}$ thereby established can be used for ATP synthesis via a $Na^+$-dependent ATP synthase which is present in addition to the $H^+$-dependent enzyme.

### *3.2. Phototrophic bacteria*

Whereas chemotrophic organisms exploit organic or inorganic substrates for the generation of metabolism energy, phototrophic organisms generate energy at the expense of light. The mechanism of ATP formation employed is photophosphorylation (Fig. 2A). Phototrophic prokaryotes can be divided into two groups. Group one comprises the cyanobacteria. They are autotrophs and the only phototrophic prokaryotes that are capable of utilizing $H_2O$ as electron donor. Group two comprises those photosynthetic prokaryotes that are not able to utilize $H_2O$ as electron donor. These are the purple and the green bacteria. Their mode of nutrition is either lithoautotrophic or organoheterotrophic.

*3.2.1. Cyanobacteria:* Unicellular representatives of the cyanobacteria are *Synechococcus* and *Gloeocapsa* species; *Oscillatoria, Anabaena,* and *Spirulina* are typical filamentous cyanobacterial genera. The photosynthetic apparatus of cyanobacteria resembles that of chloroplasts in green plants. It is composed of two photosystems and the evolution of $O_2$ is a characteristic feature (oxygenic photosynthesis). Flattened closed membrane vesicles called thylakoids are the site of photosynthetic energy transduction in cyanobacteria. The thylakoid membranes contain the light-absorbing pigments (chlorophyll a and carotenoids), various electron carriers and an $F_0F_1$-type ATPase. The typical blue-greenish color of the cyanobacteria is due to their special antenna pigments, the pycobilins. These are linear tetrapyrroles covalently attached to light-harvesting proteins called phycobiliproteins. They make up special structures known as phycobilisomes which are attached to the outer surfaces of the thylakoids and absorb light with a wavelenght around 600 nm. The excitation energy is funneled toward the reaction center of a photosynthetic unit which contains chlorophyll a. The reaction mechanism is essentially the same as in thylakoid membranes from higher plants (see Chapters 5, 8, and 9).

*3.2.2. Bacteria performing anoxygenic photosynthesis:* Purple and green bacteria are unable to perform oxygenic photosynthesis. They have only one photosystem and the organic or inorganic electron donors utilized for photosynthetic electron transport are $H_2S$, $H_2$, or organic substrates. Phototrophic growth of these bacteria is only possible under anaerobic conditions. In contrast to the green bacteria, the purple bacteria are not able to carry out a light-driven $NAD(P)^+$ reduction.

*3.2.2.1. Purple bacteria.* The photosynthetic apparatus of purple bacteria is embedded in photosynthetic membranes called chromatophores. These are formed by invaginations of the cytoplasmic membrane (Fig. 36). The photosystem of purple bacteria is restricted to cyclic electron transport which serves to generate a $\Delta\tilde{\mu}_{H^+}$. The light is trapped by antenna pigments (carotenoids and bacteriochlorophyll a or b) which make up the so-called light harvesting complex (LHC) and transferred to the membrane-spanning reaction center. A representative purple bacterial reaction center is composed of three protein subunits (L, M, H), four molecules of bacteriochlorophyll a or b, two molecules of bacteriopheophytin (BPhe), 2 ubiquinones ($Q_A$ and $Q_B$). Two of the four bacteriochlorophyll (BChl) molecules make up the so-called special pair ($[BChl]_2$), also re-

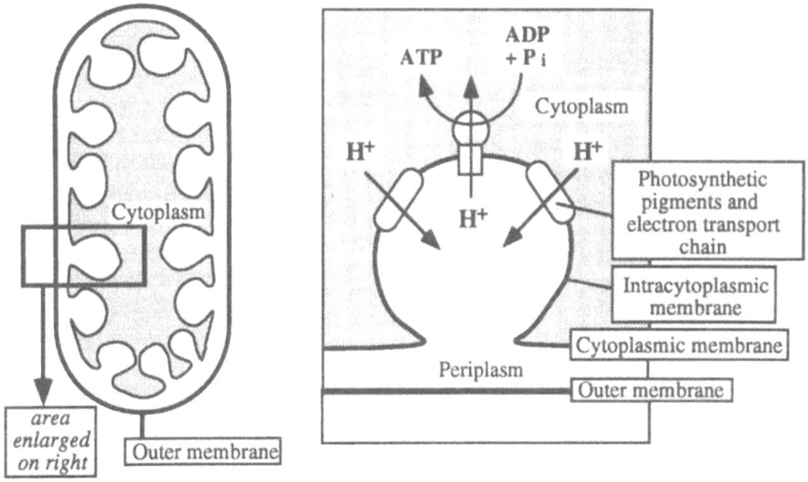


Fig. 36. Schematic diagram of a purple bacterial cell and location of the photosynthetic apparatus.

ferred to as P870. (For details see Chapter 7). Upon excitation, one electron is transferred from $[BChl]_2^*$ by way of BPhe to $Q_A$ giving rise to the semiquinone $Q_A^-$ and leaving behind one unpaired electron that is shared between the two BChl molecules ($[BChl]_2^+$) (Fig. 37). The further electron flow is by way of $Q_B$ to the ubiquinone pool. Ubiquinol is subsequently oxidized by the $bc_1$ complex which couples the transfer of two electrons to cytochrome $c_2$ with the extrusion of 4 $H^+/2e^-$ (Fig. 37 and 38). The rereduction of $[BChl]_2^+$ by cytochrome $c_2$ in the reaction center completes the cyclic electron transport.

The $\Delta\tilde{\mu}_{H^+}$ generated by cyclic electron transport is used for ATP synthesis via a $F_0F_1$-type ATP synthase and for reverse electron transfer to $NAD(P)^+$. Both NAD(P)H and ATP are needed for autotrophic growth. There are, however, differences within the purple bacteria as to their mode of nutrition. Two groups can be distinguished: the purple sulfur and the purple nonsulfur bacteria. Typical representatives of the latter group are *Rhodospirillum rubrum, Rhodobacter sphaeroides,* and *Rhodopseudomonas viridis.* The former group is represented by *Chromatium, Thiospirillum,* and *Thiopedia* species. The purple sulfur bacteria are characterized by the transient accumulation of elemental sulfur during the oxidation of $H_2S$. They have a preferentially photoautotrophic mode of growth, whereas the purple nonsulfur bacteria show a remarkable versatility with respect to their mode of metabolism. *Rhodobacter capsulatus* for instance may grow photoautotrophically, photoheterotrophically, chemoautotrophically, chemoheterotrophically with oxygen or other electron acceptors, and fermentatively. The electron transport pathways in *Rb. capsulatus* are therefore very complex and have to be highly regulated.

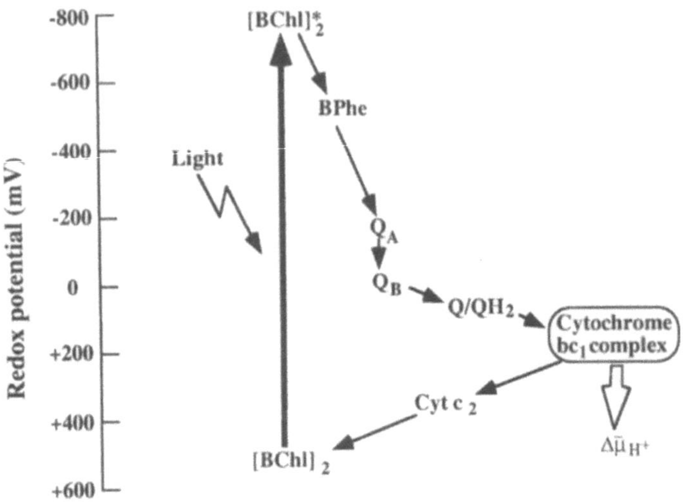


Fig. 37. Light-driven cyclic electron transport in purple bacteria. All abbreviations are explained in the text.

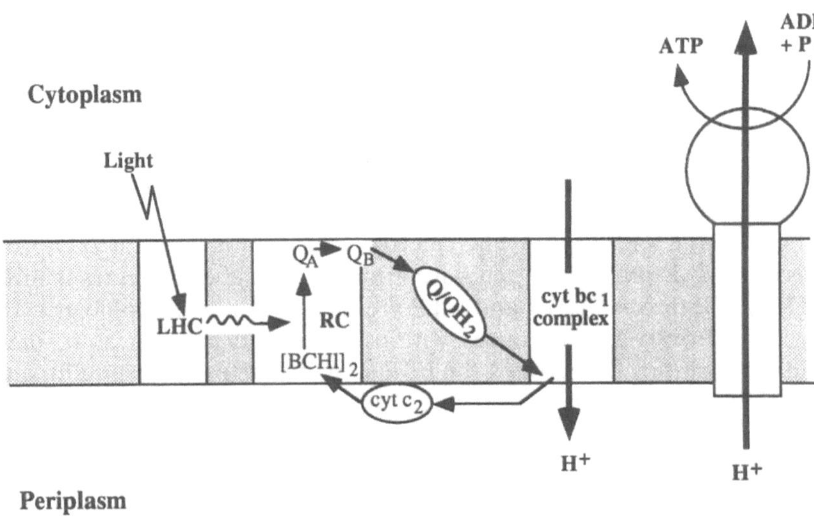


Fig. 38. Arrangement of the components of the light-driven electron transport in membranes of purple bacteria. Light harvesting complex (LHC), reaction center (RC).

*3.2.2.2. The green bacteria.* The two groups of green bacteria are represented by the genera *Chlorobium* (green sulfur bacteria) and *Chloroflexus* (gliding filamentous bacteria). *Chlorobium* grows preferentially photolithoautotrophically using $H_2S$ and $H_2$ as electron donors. The organism may however photoassimilate simple organic molecules such as acetate, but only in addition to $CO_2$. In contrast, *Chloroflexus* is a typical photoorganoheterotroph and chemoorganoheterotroph but unable to ferment.

All photosynthetic bacteria except the green sulfur bacteria employ the Calvin cycle for $CO_2$ fixation. The green sulfur bacteria fix $CO_2$ by the reductive tricarboxylic acid cycle which operates in the reverse direction as compared to the normal TCA cycle. Only three enzymes catalyzing irreversible or unfavorable reactions are replaced by other enzymes: succinate dehydrogenase by NADH-dependent fumarate reductase, $NAD^+$-dependent $\alpha$-oxoglutarate dehydrogenase by $\alpha$-oxoglutarate: ferredoxin (Fd) oxidoreductase, and citrate lyase by ATP citrate-lyase. In summary, the reductive TCA cycle catalyzes the formation of acetyl-CoA from two molecules of $CO_2$ at the expense of the hydrolysis of two ATP.

$$2\ CO_2 + 3\ NAD(P)H + 3\ H^+ + 2\ Fd_{red} + CoA + 2\ ATP$$
$$\leftrightarrows \text{acetyl-CoA} + 2\ ADP + 2\ P_i + 3\ NAD(P)^+ + 2\ Fd_{ox}$$

Acetyl-CoA may serve as building block for the synthesis of cell constituents.

The organization of the photosynthetic apparatus is, in contrast to differences in the carbon metabolism, similar in both group of the green bacteria. The light harvesting antenna pigments are localized in long bodies called chlorosomes. These are in direct contact with the inner face of the cytoplasmic membrane which contains the photosynthetic reaction centers. The chlorosomes harbor mainly carotenoids and bacteriochlorophyll c, d, or e, while the membrane contains chlorophyll a and bacteriopheophytin. Light is absorbed by the chlorosomes and funnelled to the reaction center in the cytoplasmic membrane. The photochemically active bacteriochlorophyll a (P840) expels an electron upon excitation (P840*). A bacteriochlorophyll a molecule ($A_0$) acts as an early electron acceptor (Fig. 39). The electron is subsequently transferred by way of $A_1$ (a quinone molecule) and iron sulfur clusters ($F_A$, $F_B$, $F_X$) to ferredoxin. In contrast to purple bacteria, $NAD^+$ may be reduced. $NAD^+$ reduction is catalyzed by ferredoxin-$NAD^+$ oxidoreductase. Alternatively, cyclic electron transport may occur. In this case, electrons flow from ferredoxin via menaquinone to the cytochrome $bc_1$ complex. Electron flow through the $bc_1$ complex is associated with proton translocation (2 $H^+/e^-$) across the cytoplasmic membrane. A mobile cytochrome $c_{555}$ transfers the electrons from the $bc_1$ complex to the oxidized reaction center ($P840^+$).

The electrons used for $NAD^+$ reduction have to be replaced. This is

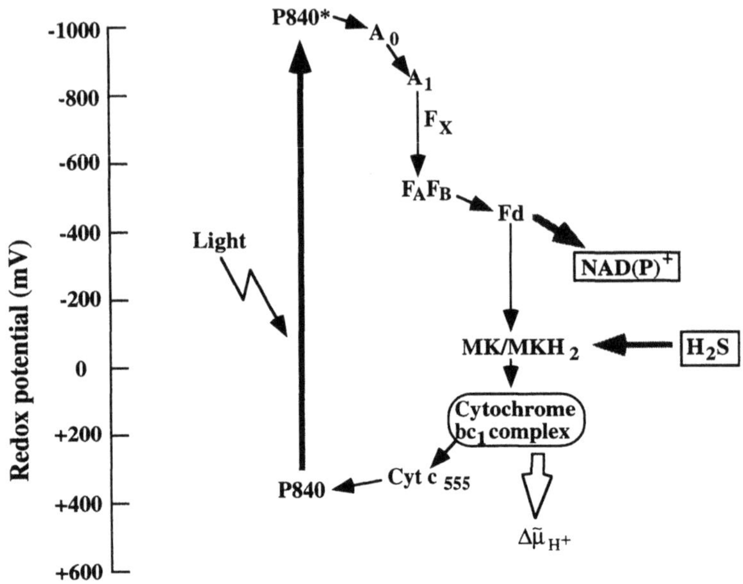


Fig. 39. Light-driven cyclic or non-cyclic electron transport in green bacteria. The hollow arrow indicates the step that leads to the generation of $\Delta\tilde{\mu}_{H^+}$. Noncyclic electron transport is indicated by half tone arrows. Abbreviations are explained in the text.

achieved by the oxidation of an external electron donor such as $H_2S$; the electrons enter the electron transport pathway probably by way of menaquinone.

*3.2.3. Light-dependent energy generation in halobacteria:* A special kind of light-driven $\Delta\tilde{\mu}_{H^+}$ generation is found in certain halobacteria, organisms that are adapted to high salt concentrations (2.5-5 M NaCl). The mechanism differs completely from those discussed so far. The light-driven proton translocation in halobacteria is independent of electron transport and achieved by the proton-pumping activity of bacteriorhodopsin. Bacteriorhodopsin is a protein to which the chromophore retinal is covalently bound via a lysine residue forming thereby a Schiff base. Light absorbed by bacteriorhodopsin leads to the deprotonation of the Schiff base as a result of an isomerization of the retinal from the *all-trans* to the *13-cis* form. The proton is released to the outside. This is followed by a light-independent reisomerization from the *13-cis* to the *all-trans* form which is accompanied by a protonation, the proton being taken from the cytoplasm. As a result a $\Delta\tilde{\mu}_{H^+}$ is formed which is competent in driving ATP synthesis via an ATP synthase. For details see Chapter 13.

## 4. Energy-requiring processes

The processes described so far have dealt with the various ways that are employed by bacteria to generate the ATP which is required for many biosynthetic reactions and transport processes. A variety of processes such as active transport and flagellar movement directly take advantage of the transmembrane electrochemical potential difference of protons ($\Delta\tilde{\mu}_{H^+}$) or in some cases of an electrochemical potential difference of sodium ions ($\Delta\tilde{\mu}_{Na^+}$). Both forms of energy, ATP and $\Delta\tilde{\mu}_{H^+}$ are interconvertible by means of a reversible ATPase /ATP synthase and both forms of energy have to be generated: Fermenting organisms form ATP by substrate level phosphorylation, i.e. without the intermediacy of $\Delta\tilde{\mu}_{H^+}$. The $\Delta\tilde{\mu}_{H^+}$ which is also required by the fermenting cell is generated secondarily: an ATPase pumps protons from the cell's interior to its outside at the expense of ATP hydrolysis (Fig. 2C). In contrast, respiring and phototrophic bacteria form primarily a $\Delta\tilde{\mu}_{H^+}$ in response to a redox or light-driven electron transport which may be secondarily interconverted to ATP by means of an ATP synthase.

### *4.1. Processes requiring ATP*

Every living cell needs ATP for synthesizing its cellular constituents: protein, lipid, DNA, RNA, and polysaccharides. The majority of the ATP generated is used for the synthesis of macromolecules such as protein, DNA, and RNA. This will not be further discussed here. However, a cell's ATP demand for biosyntheses depends also on whether the macromolecules are synthesized from preformed monomers such as sugars, amino acids, fatty acids, and nucleic acid bases, or from inorganic molecules such as $CO_2$, $H_2$, $NH_4^+$. The ATP requirement for the formation of cellular material from inorganic precursors is considerably higher than from preformed monomers. The difference becomes evident when we look at the ATP requirement of the *de novo* synthesis of glucose-6-phosphate from 6 molecules of $CO_2$ and 12 molecules of NAD(P)H. The Calvin cycle brings about the formation of one glyceraldehyde-3-phosphate from 3 $CO_2$ and 6 NAD(P)H; glyceraldehyde-3-phosphate may be further converted to glucose-6-phosphate. Eighteen molecules of ATP are altogether needed for the formation of one hexose from $CO_2$:

$$6\ CO_2 + 18\ ATP + 12\ NADH + 12\ H^+ \rightarrow \text{glucose-6-phosphate} + 18\ ADP + 17P_i + 12\ NAD^+$$

This example shows that autotrophic organisms have a much higher ATP demand than heterotrophs.

Autotrophic organisms are usually also lithotrophic. If the redox potential of the electron donor used by a certain bacterium is too positive to allow the reduction of $NAD(P)^+$, the energy demand increases even more because energy has to be invested for reverse electron transfer to reduce $NAD(P)^+$. This is for instance the case with $Fe^{2+}$-oxidizing bacteria (see 3.1.9.6.). Hence, the remarkable capability of bacteria to synthesize cell constituents from simple inorganic molecules has its price, namely a high ATP demand.

*4.1.1. Nitrogen fixation:* Another energetically costly process is nitrogen fixation. Nitrogen fixation is restricted to prokaryotic microorganisms enabling them to synthesize reduced nitrogen compounds from $N_2$ when ammonia or nitrate is unavailable. The ability to fix molecular nitrogen is relatively widespread among bacteria. Examples are *Anabaena azollae* (cyanobacterium), *Rhodospirillum rubrum* (phototrophic bacterium), *Clostridium pasteurianum* (strict anaerobe), *Klebsiella pneumoniae* (facultative aerobe), *Rhizobium japonicum* (strict aerobe), *Methanosarcina barkeri* (methanogenic bacterium). The organisms may either be free-living or exist in partnership with certain plants.

The reduction of $N_2$ to $NH_3$ is catalyzed by nitrogenase, an extremely oxygen-sensitive, molybdenum-containing Fe-S protein. The formation of $NH_3$ from $N_2$ and $H_2$ is an exergonic reaction:

$$N_2 + 3\ H_2 + 2\ H^+ \rightarrow 2\ NH_4^+$$
$$\Delta G^{0'} = -80\ \text{kJ/mol}$$

However, since $N_2$ is an extremely stable molecule, the activation energy required by nitrogenase is very high and has been estimated to be overcome by hydrolysis of 16 ATP per molecule of $N_2$ fixed:

$$N_2 + 6\ [H] + 16\ ATP \rightarrow 2\ NH_3 + 16\ ADP + 16\ P_i$$

The reducing equivalents are supplied in the form of reduced ferredoxin or flavodoxin.

*4.1.2. ATP-dependent transport:* The accumulation and extrusion of metabolites against a concentration gradient as carried out by a bacterial cell are energy-dependent processes which are catalyzed by membrane-bound transport proteins. Whereas one class of these carriers is energized by $\Delta\tilde{\mu}_{H^+}$, another class obtains its energy from ATP or from a phosphoryl donor derived from ATP. Whether the active transport of a metabolite is ATP- or $\Delta\tilde{\mu}_{H^+}$-driven depends on the organism and on the metabolite being transported. In *E. coli* and other Gram-negative bacteria ATP-dependent carriers are involved in the uptake of amino acids, some sugars as well as inorganic ions.

These ATP-dependent carriers are composed of two functional units: a binding protein located in the periplasm recognizes and binds the substrate; an ATP-requiring transport unit embedded in the membrane interacts with the substrate-binding complex and transports the substrate across the membrane at the expense of the hydrolysis of ATP or of another phosphoryl donor. Such ATP-driven porters may establish concentration differences of more than $10^6$-fold.

Not only metabolites are actively transported across the bacterial membrane, but also cations such as $Na^+$ and $K^+$. All living cells accumulate $K^+$ and expel $Na^+$. These cations play an essential role in the regulation of the cell turgor and the cytoplasmic pH. Two $K^+$ uptake systems of *E. coli* have been studied in detail: The constitutive TrK system which is characterized by a modest affinity for $K^+$ and a high transport rate depends on both ATP and $\Delta\tilde{\mu}_{H^+}$. The second $K^+$ uptake system (Kdp) which is inducible by high external osmolarity has a high affinity for $K^+$, but displays a modest uptake rate. Uptake of $K^+$ by the Kdp system is exclusively energized by ATP hydrolysis. The uptake mechanism involves the formation of a phosphorylated enzyme intermediate. Ion pumps with this characteristic are classified as P-type ATPases, in contrast to the $F_0F_1$-type ATPases which do not form such an intermediate.

### *4.2. Processes requiring* $\Delta\tilde{\mu}_{H^+}$

*4.2.1.* $\Delta\tilde{\mu}_{H^+}$*-driven transport:* Quite a number of transport systems depend directly on the protonic potential. Examples are the $Na^+$-$H^+$ antiporter and the lactose/$H^+$ symporter. The former catalyzes the extrusion of $Na^+$ in response to the influx of $H^+$. As already mentioned, the extrusion of $Na^+$ is essential for every living cell, and the concomitant influx of $H^+$ coupled to this $Na^+$ extrusion is important to maintain a neutral intracellular pH at alkaline external pH. The $\Delta\tilde{\mu}_{Na^+}$ being established through the $Na^+$-$H^+$ antiport may drive the uptake of certain substrates such as melibiose or glutamate. Cotransport of a substrate with $Na^+$ into the cell functions in analogy to the $\Delta\tilde{\mu}_{H^+}$-driven solute transport.

A well studied example of a $\Delta\tilde{\mu}_{H^+}$-driven uptake system is the β-galactoside porter of *E. coli,* also called lac permease. The accumulation of lactose and other β-galactosides is proportional to the size of $\Delta\tilde{\mu}_{H^+}$. The respective β-galactose is transported in symport with one $H^+$. This means that the uptake of the sugar is accompanied by the movement of a $H^+$ in the same direction. Either component of $\Delta\tilde{\mu}_{H^+}$ (ΔpH or Δϕ) can support the uptake. As an example, a $\Delta\tilde{\mu}_{H^+}$ of –180 mV would be sufficient to establish a 1000-fold accumulation of β-galactosides. Many other uptake systems function in a similar fashion.

*4.2.2. ATP synthesis:* One of the most prominent roles being played by $\Delta\tilde{\mu}_{H^+}$ in a respiring bacterium is its involvement in the synthesis of ATP from ADP and $P_i$. $\Delta\tilde{\mu}_{H^+}$-driven ATP synthesis is catalyzed by the membrane-bound $F_0F_1$-type ATP synthase which consists of a membrane intrinsic proton-conducting part ($F_0$) and of a catalytic membrane-peripheral part ($F_1$). $F_0F_1$-type ATPases comprise a distinct family of enzymes found in mitochondria, chloroplasts, and bacteria. They differ from one another only in certain details. The enzyme from *E. coli* represents a typical bacterial $F_0F_1$-ATPase. The $F_0$-part of the *E. coli* enzyme is composed of three different subunits, and the $F_1$-part displays an $\alpha_3\ \beta_3\ \gamma\ \delta\ \varepsilon$- stoichiometry. For details see Chapter 12.

*4.2.3. Flagellar movement:* One of the most fascinating $\Delta\tilde{\mu}_{H^+}$-dependent processes observed in motile bacteria is the rotational movement of the flagellar motor which enables these bacteria to actively swim in response to environmental stimuli (chemotaxis). The flagellum represents a device which transforms energy from $\Delta\tilde{\mu}_{H^+}$ into mechanical energy. However, it is not yet clear how the movement of protons down their electrochemical potential difference drives the rotation of the flagellum.

*4.2.4. Reverse electron transfer:* Reverse electron transport is another $\Delta\tilde{\mu}_{H^+}$-driven process already mentioned above (3.1.9.1.). Many photosynthetic and lithotrophic bacteria utilize electron donors with a relatively positive redox potential. The electrons generated can therefore not be directly used for the reduction of $NAD(P)^+$. Energy in the form of $\Delta\tilde{\mu}_{H^+}$ has to be invested in order to allow $NAD(P)^+$ reduction.

## 5. Concluding remarks

The simple organization, the fast growth, and the ease of molecular cloning have made bacteria a preferred object of bioenergetic research. Since the principles of energy transduction are the same in all living cells, bacteria are widely used as model systems to explore the molecular events that make available the Gibbs free energy change of a reaction in the form of ATP. Whereas much progress on the molecular level has been made in our understanding of the energy-transducing systems of many phototrophic and chemotrophic bacteria, there are still major gaps in our basic knowledge about the electron transport pathways in bacterial groups such as the acetogenic, methanogenic, and the sulfate-reducing bacteria (to name only a few). Many of these fascinating and ecologically important organisms are difficult to study because of experimental limitations. The application of molecular biological methods to these bacterial groups has just begun and will hopefully lead to an enhancement of our understanding of the underlying anaerobic energy-conserving electron transport pathways.

## Acknowledgements

We would like to thank Beate Kamlage and David Westenberg for critically reading the manuscript.

## References

*General interest and background*

FM Harold, *The vital force: a study of bioenergetics.* W. H. Freeman and Co., New York, 1986.

CW Jones in *Bacterial energy transduction,* C. Anthony (ed.), Academic Press, London, 1988, pp 1–82.

*Aerobic or facultatively anaerobic bacteria*

AH Stouthamer, J. Bioenerg. Biomembranes 23 (1991) 163–185.

Y Anraku, Ann. Rev. Biochem. 57 (1988) 101–132.

C Anthony and C.W. Jones in *Microbial growth on $C_1$ compounds,* HW van Verselfeld and JA Duine (eds.), Martinus Nijhoff Publishers, Dordrecht, 1987, pp 195–202.

PM Wood in *Bacterial energy transduction,* C Anthony (ed.), Academic Press, London, 1988, pp. 183–230.

DP Kelly in *Autotrophic bacteria,* HG Schlegel and B Bowien (eds.), Science Tech Publishers, Madison, 1989, pp. 193–217.

*Strictly anaerobic bacteria*

RK Thauer, K Jungermann and K Decker, Bacteriol. Rev. 41 (1977) 100–180.

W Hamilton in *Bacterial energy transduction,* C Anthony (ed.), Academic Press, London, 1988, pp. 83–149.

RK Thauer in *Autotrophic bacteria,* HG Schlegel and B Bowien (eds.), Science Tech Publishers, Madison, 1989, pp. 397–413.

M Blaut, V Müller and G Gottschalk, J. Bioenerg. Biomembranes 24 (1992) 529.

G Fuchs, FEMS Microbiol. Rev, 39 (1986) 181–213.

MJ Wolin and TJ Miller, ASM News 48 (1982) 561–565.

*Sodium ion-dependent energy conservation*

P Dimroth, Microbiol. Rev. 51 (1987) 320–340.

T Unemoto, H Tokuda and M Hayashi in *Bacterial energetics,* TA Krulwich (ed.), Academic Press, San Diego, 1990, pp. 33–54.

*Phototrophic bacteria*

W Nitschke and AW Rutherford, Trends Biochem. Sci. 16 (1991) 241–245.

RC Prince in *Bacterial energetics,* TA Krulwich (ed), Academic Press, San Diego, 1990, pp. 111–136.

JB Jackson in *Bacterial energy transduction,* C Anthony (ed.), Academic Press, London, 1988, pp. 317–375.

*Energy-requiring processes*

IR Booth in *Bacterial energy transduction,* C Anthony (ed.), Academic Press, London, 1988, pp. 377–428.

RH Fillingame in *Bacterial energetics,* TA Krulwich (ed.), Academic Press, San Diego, 1990, pp. 345–391.

S Khan in *Bacterial energetics,* TA Krulwich (ed.), Academic Press, San Diego, 1990, pp. 301–343.

Bioenergetics
ed. by P. Gräber and G. Milazzo†


# *CHAPTER 5*
# Chloroplasts

Elisabeth Haag and Gernot Renger

*Max Volmer Institute for Physical and Biophysical Chemistry, Technical University of Berlin, Berlin, Germany*

## 1. Introduction

Chloroplasts are the cell organelles enabling eukaryotic algae and plants to perform oxygenic photosynthesis, one of the essential processes in nature on which our present-day form of the life depends. By the use of solar radiation as energy source, water is cleaved into oxygen and chemically bound hydrogen which is subsequently used to reduce carbon dioxide.

The generation of an oxygenic atmosphere three billion years ago was thus tightly linked to the appearance of the first oxygen evolving photosynthetic organisms that were probably very similar to the still existing prokaryotic cyanobacteria. Until today cyanobacteria and chloroplasts are the only living structures capable of oxygenic photosynthesis. There is a strong resemblance in the overall organization of the photosynthetic reaction sequence between cyanobacteria and chloroplasts. They both exhibit an extended inner membrane system in which homologous pigment-protein complexes are embedded that are involved in the light-driven oxidation of water and generation of reducing power. Furthermore, chloroplasts contain their own bacterial-like genome as well as their own "prokaryotic" transcription and translocation system. This close relationship in genetic and metabolic organization gave rise to the hypothesis that chloroplasts evolved from cyanobacteria which were incorporated into a host cell and continued to live in endosymbiosis thereby conferring photoautotrophic capacity to the host cell.

However, the evolutionary period that has passed since the presumed event of endocytosis is long and despite the existing similarities between cyanobacteria and chloroplasts a number of important differences are also obvious. Therefore, before discussing the structure and function of chloroplasts, it seems worthwhile to briefly summarize analogous and different properties of prokaryotic cyanobacteria and chloroplasts in eukaryotic cells, and to mention some general features of chloroplasts in eukaryotic algae and higher plants. First of all, it has to be emphasized that cyanobacteria are fully autonomous cells whereas chloroplasts are semiautonomuos cell organelles with their genetic information and regulation being divided between the plastid genome and the nucleus. Second, in contrast to cyanobacteria, chloroplasts are metabolically neither omnipotent nor independent. For example, chloroplasts are not able to fix atmospheric nitrogen, and many metabolic pathways assigned to chloroplasts actually require the interaction of different cell compartments which, in turn, evokes the necessity of exchange mehanisms for metabolites. Third, chloroplasts are surrounded by a double lipid membrane and are missing a structure comparable to the cell wall of cynobacteria which renders them susceptible to osmotic stress. Fourth, there are striking differences in the pigment composition of the light absorbing antenna system. These differences exist not only between cyanobacteria and chloroplasts, but also between chloroplasts of different eukaryotic algae. The differences in pigment composition are responsible for many of the colloquial names, e.g., bluegreen, red, brown, or green algae, as well as for the classification of eukaryotic algae into Rhodophyta, Chromophyta and Chlorophyta. From the analysis of pigment composition it seems conceivable that the diversity of chloroplasts evolved not from one single endocytic event, but rather from several parallel events of endocytosis that established a variety of endosymbiotic relationships. However, this is still an open question and matter of current debate. A comparison of the pigment composition of Cyanobacteria, Rhodophyta, Chromophyta and Chlorophyta is given in Table 1.

In general, chloroplasts of green algae and land plants have a similar pigment composition and seem to be very closely related. However, chloroplasts of algae exhibit an enormous diversity in shape, size and number

Table 1. Pigment composition

| | Cyanobacteria | Rhodophyta | Chromophyta | Chlorophyta |
|---|---|---|---|---|
| chlorophyll a | + | + | + | + |
| chlorophyll b | – | – | – | + |
| chlorophyll c | – | – | + | – |
| chlorophyll d | – | (+) | – | – |
| phycobilins | + | + | – | – |

(+): present in all species; (-): absent in all species; ((+)): present in some species

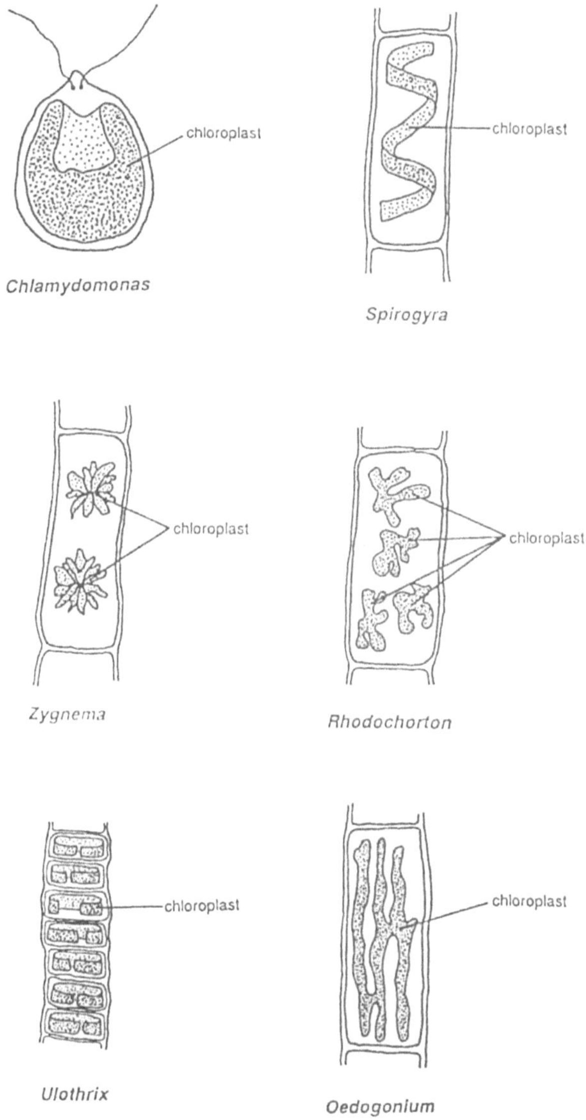


Fig. 1. Chloroplasts of different eukaryotic algae.

per cell. The variety of different shapes ranges from helical, reticular, stellate, libulate, lenticular, plate- or cup-like (Fig. 1). In contrast, chloroplasts of land plants are less variable in size and shape and will be described in more detail in this section.

In higher plants, chloroplasts develop in a light-dependent differentiation process from smaller proplastids that lack pigmentation and an inner membrane structure. Depending on seasonal changes, tissue differentia-

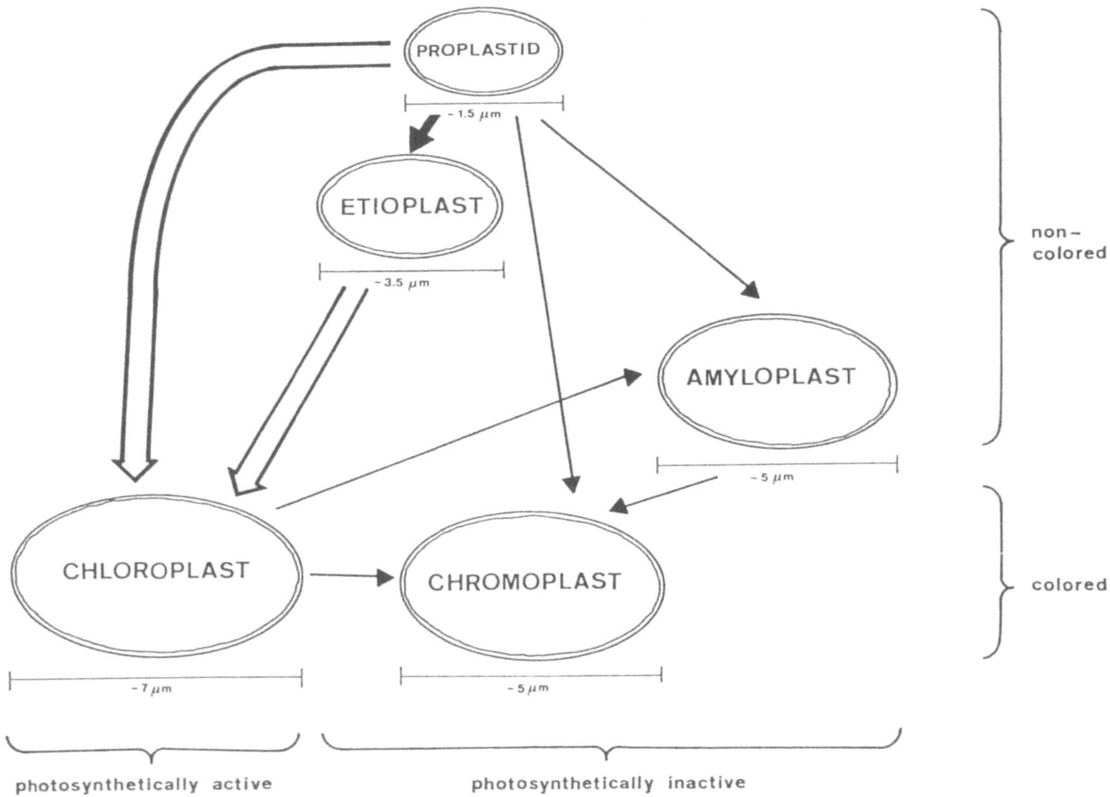


Fig. 2. Various types of plastids. The white arrows indicate light-dependent development, the bold dark arrow indicates dark-dependent development, and the thin dark arrows indicate light/dark-independent development and convertibility. Size specifications are given on a rough average.

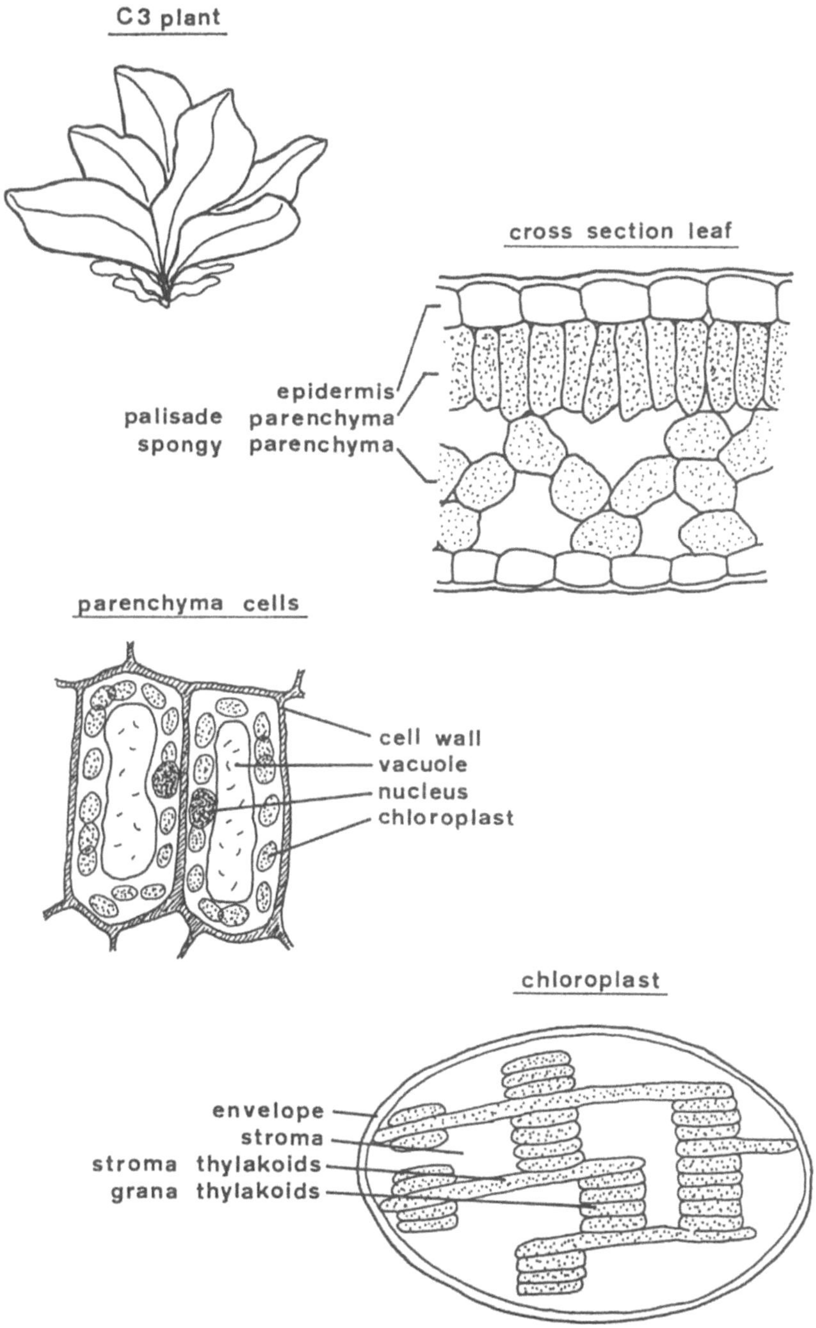


Fig. 3. Distribution and location of chloroplasts in higher plants.

tion or environmental constraints, proplastids can also differentiate into photosynthetically inactive plastid forms. A rough classification of the different members of the plastid family is given in Fig. 2. Most plastid forms are interconvertible, responding to changes in age, growth or metabolism of the cell.

Chloroplasts of higher plants are predominantly located in the assimilatory mesophyll cells of the palisade and spongy tissue of leaves. They have a lenticular shape with a length of approximately 5–10 μm and a height of 1–2 μm. Depending on the light conditions the number of chloroplasts per cell can be as large as 400. A cross-section through the leaf of a typical angiosperm C3-plant (see section 3) and the characteristic distribution and intraorganellar structure of its chloroplasts are depicted in Fig. 3. Each chloroplast is separated from the cytosol by a double lipid bilayer (envelope) and contains a proteinaceous matrix (stroma) and an extended inner membrane system (thylakoids). The thylakoids exhibit partitioning into stacked and unstacked membrane regions. The unstacked stroma thylakoids form single, non-appressed membranes whereas the stacked grana thylakoids are closely contacted. Nevertheless, the entire membrane system is arranged continuously and encloses a separate space which is distinct from the stroma by its chemical composition and is called thylakoid lumen. Taken together, a chloroplast can roughly be subdivided into four different structural units: the envelope, the stroma, the thylakoid membrane and the thylakoid lumen.

Each of these structural subdivisions is related to specific reactions and distinct functions in the overall process of photosynthesis (Fig. 4). The thylakoid membranes carry the pigment-protein complexes involved in light absorption and light-driven electron transport, leading to the oxidation of water in the thylakoid lumen and the generation of NADPH and ATP at the stroma exposed side. The thylakoid membranes can therefore be viewed as the energy trapping and converting system. However, as shown by a simple thermodynamic consideration, it would be very inconvenient, or even impossible, to accumulate and store photosynthetically converted solar energy in the form of NADPH and ATP for a longer time and on a larger scale. The free energy stored per mole (and mass) is significantly less for ATP as compared to reduced carbohydrates, e.g., hydrolysis of the terminal phosphoanhydro-bond of ATP yields 32 kJ/mol, corresponding to 0.064 kJ/g, whereas dissimilation of glucose delivers 2880 kJ/mol, equivalent to 16 kJ/g. Thus, the subsequent biochemical reactions of carbon reduction which take place in the stroma are an inevitable part of photosynthesis. They are often, but misleadingly, referred to as "dark reactions". Although photosynthetic carbon reduction using NADPH and ATP is an exergonic process and does not require light as driving force, the enzyme-catalyzed reaction sequence comprises a number of regulatory mechanisms which require light activation. The products of this light-regulated cycle of carbon reduction are either hexose-

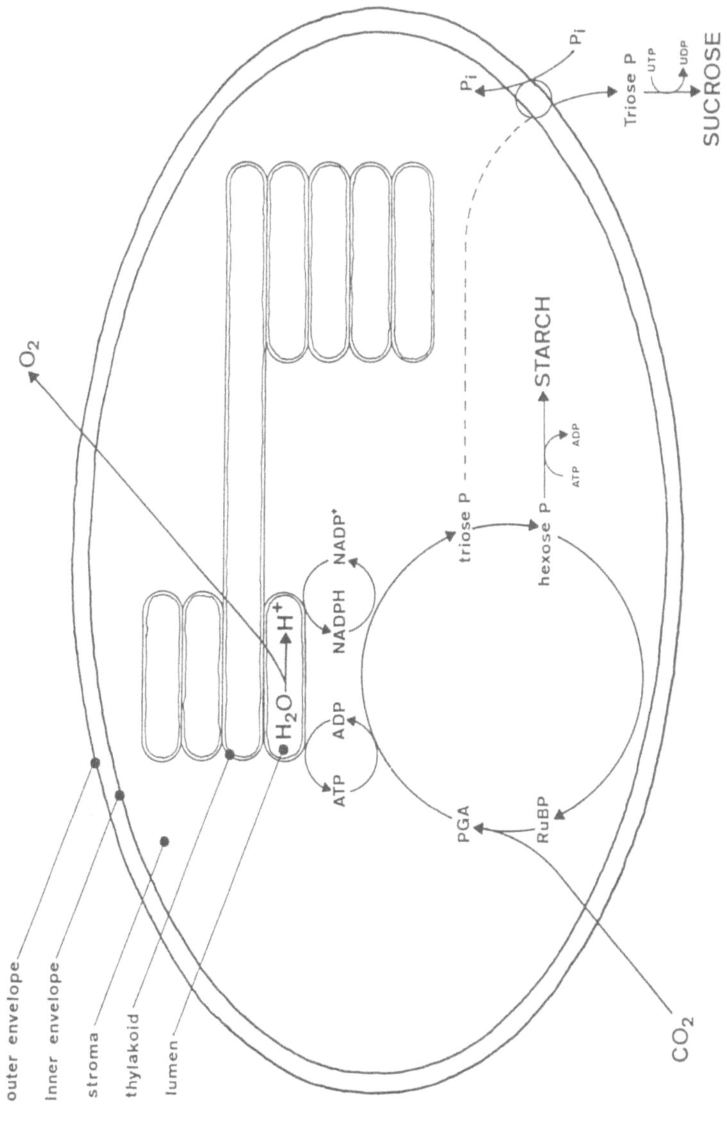


Fig. 4. Functional and structural organization of higher plant chloroplasts.

phosphates which are further assimilated to starch or triose-phosphates which are exported to the cystol. An accumulation of osmotically active assimilates during a longer photosynthesizing period would be deleterious for chloroplasts. Therefore, one of the essential functions of the envelope is to manage the metabolite transport from the chloroplast to the cytosol. The chloroplast itself can only store osmotically inactive assimilation products such as starch.

In summary, chloroplasts are cell organelles exhibiting a complex and highly organized structure which is necessary for the coordinated process of light-driven and light-regulated reaction sequences in photosynthesis.

The light-driven electron transport and proton translocation across the thylakoid membrane is discussed in section 2 of this chapter. Sections 3 and 4 describe biochemical pathways which take place in the stroma. Because of the predominant role of carbon metabolism, section 3 exclusively considers different ways of carbon reduction (and oxidation) in chloroplasts, whereas section 4 very roughly outlines other metabolic pathways. Section 5 deals with transport processes across the envelope. And last, section 6 catches on a few aspects of light-mediated regulatory mechanisms in chloroplasts.

## 2. Light-driven electron transport and proton translocation

The energy provided by the sunlight that reaches the surface of the earth amounts to roughly $5 \times 10^{21}$ kJ per year, the overwhelming majority of which is reflected or lost in the form of heat. Only 0.5–1% of the sunlight is absorbed and can be exploited photosynthetically by driving the electron transport chain from water to $NADP^+$ in chloroplasts (see Chapter 2).

### *2.1. Functional organization of the thylakoid membrane*

Light-driven electron transport and proton translocation across the thylakoid membrane is accomplished by the concerted operation of different membrane-integral protein complexes [see also refs. 1 and 2]. A simplified scheme of the photosynthetic electron transport chain and the participating protein complexes is given in Fig. 5A. Due to anisotropic orientation of the two membrane-spanning pigment-protein complexes involved in the primary photochemical reactions, phtosystem I (PS I) and photosystem II (PS II); the direction of light-induced electron transport is perpendicular with respect to the membrane plane. Within both photosystems the arrangement of functional redox groups is such that the photooxidizable chl *a* molecules are located near the lumenal side whereas the electron accepting redox components are incorporated near the stromal side. Thus,

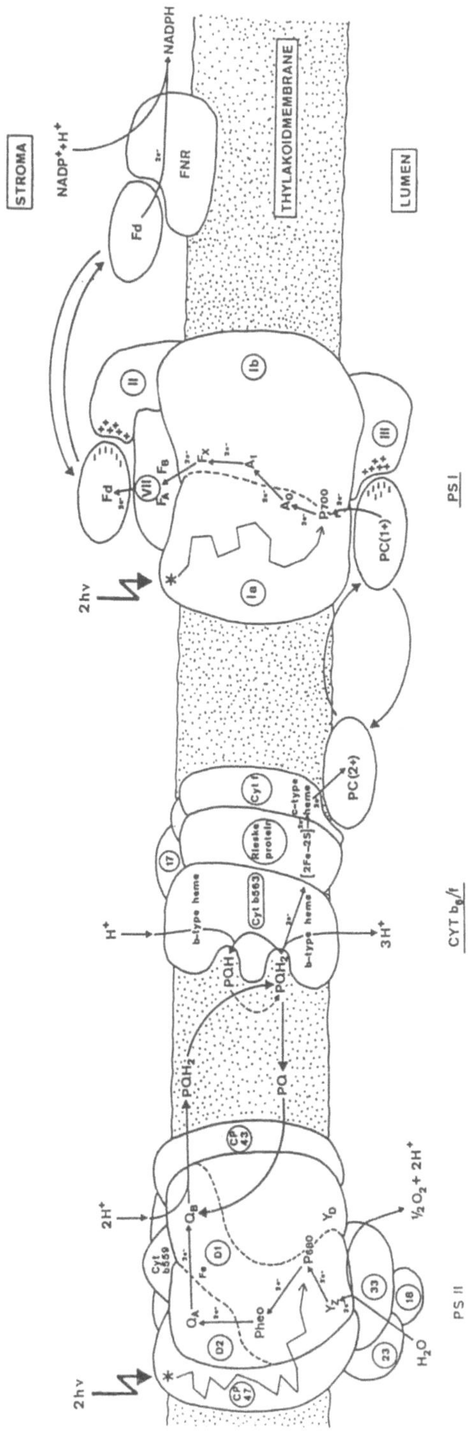
STROMA
NADP⁺+H⁺
NADPH
FNR
Fd
THYLAKOIDMEMBRANE
LUMEN
2hv
PC(1+)
PC(2+)
P700
PS I
CYT b6/f
PQH2
PQ
H2O
½O2 + 2H⁺
PS II

Fig. 5A.

photosynthetic electron transport is a vectorial process, i.e. upon illumination electrons are transported from the lumenal to the stromal side in PS I and PS II, respectively, thereby creating a transmembrane electrical potential difference. In the linear electron transport chain from water to $NADP^+$, PS I and PS II are connected via the cytochrome b6f complex.

The linear electron transport chain starts with the light-induced oxidation of the photoactive chl *a* molecule in the reaction center of PS II (P680) and concomitant electron transport to the primary quinone acceptor ($Q_A$) with pheophytin as intermediate. From there the electron proceeds to the secondary quinone acceptor ($Q_B$) which, upon reduction by two electrons and protonation, is released from its binding niche and migrates as plastoquinol ($PQH_2$) through the thylakoid membrane to the Cyt b6f complex. The Cyt b6f complex reoxidizes reduced plastoquinol and successively transfers the electrons to plastocyanin, thereby mediating between the two-electron carrier plastoquinol and the one-electron carrier plastocyanin. In turn, plastocyanin rereduces the photooxidized chl *a* molecule (P700) in the reaction center of PS I. In PS I, photochemistry leads to light-induced oxidation of P700, coincident reduction of a second chl *a* molecule ($A_0$), and subsequent electron transfer to a second-

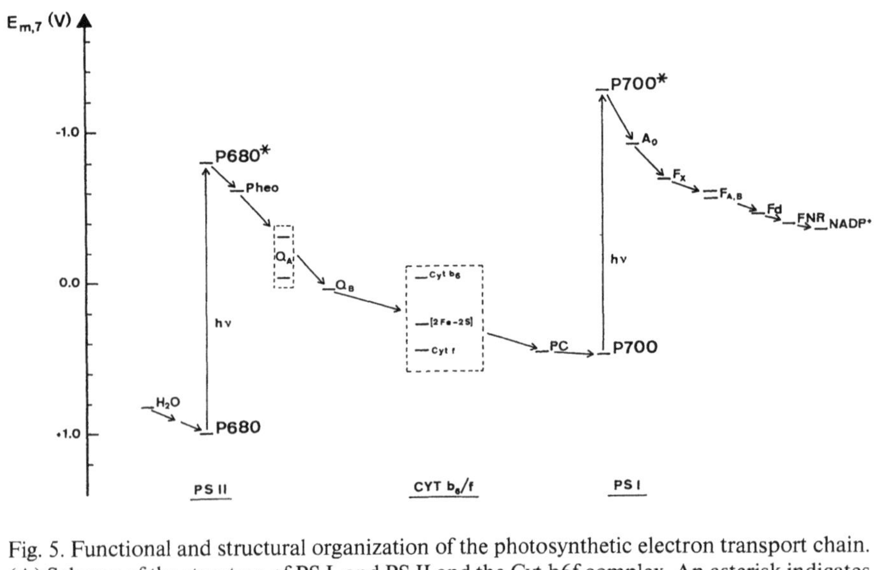


Fig. 5. Functional and structural organization of the photosynthetic electron transport chain. (A) Scheme of the structure of PS I, and PS II and the Cyt b6f complex. An asterisk indicates antenna chlorophyll molecules in their excited state. For further details see text. (B) Energetics of the photosynthetic electron transport chain. The redox potential indicated refer to the standard midpoint potential of oxidized versus reduced form. PS II exhibits a heterogeneity with respect to the redox potential of $Q_A$. The majority of PS II centers exhibits a midpoint potential in the range of 0 V for $Q_A/Q_A^-$, however, for a minor fraction, midpoint potentials of about −0.3 to −0.25 V are reported in the literature. (Midpoint potentials for redox active components of PS I are specified according to ref. 43.)

ary acceptor ($A_1$) which is supposed to be a phylloquinone (vitamine $K_1$). The electron then passes a series of consecutive iron-sulfur clusters, starting with $F_x$ and continuing to the clusters $F_A$ and $F_B$ (sometimes also referred to as P430) which are finally reoxidized by ferredoxin, a diffusible component delivering the electron to the ferredoxin: NADP oxidoreductase (FNR). The essential step to regenerate the system of linear electron transport is the rereduction of the oxidized $P680^+$ in the reaction center of PS II. This is provided by the abstraction of electrons from water at a manganese-containg catalytic site closely associated with the lumenal side of PS II.

A second consequence of the anisotropic architecture of the thylakoid membrane is the spatial segregation of deprotonating and protonating reactions, leading to proton release into the lumen ($H_l^+$) and proton uptake from the stroma ($H_s^+$). This can be demonstrated by dividing the overall reaction into the following partial reactions steps that are spatially segregated:

$$\text{PS II (lumen): } 2\ H_2O \rightarrow 4\ H_l^+ + 4\ e^- + O_2 \qquad (1a)$$

$$\text{PS II (stroma): } 4\ H_s^+ + 4\ e^- + 2\ Q_B \rightarrow 2\ PQH_2 \qquad (1b)$$

$$\text{Cyt b/f (lumen): } 2\ PQH_2 + 4\ PC^{ox} \rightarrow 4\ H_l^+ + 2\ PQ + 4\ PC^{red} \qquad (1c)$$

$$\text{PS I (stroma): } 4\ H_s^+ + 4\ e^- + 2\ NADP^+ \rightarrow 2\ NADPH + 2\ H^+ \qquad (1d)$$

In total, oxidation of two water molecules and reduction of two $NADP^+$ molecules give rise to a net translocation of eight protons from the stroma to the lumen. In addition, a cyclic electron transport around PS I is possible. In this process reduced ferredoxin is reoxidized by the Cyt b6f complex and the electron return to plastocyanin. The cyclic electron transport through PS I and the Cyt b6f complex contributes only to proton translocation without generating reducing power in form of NADPH. Taken together, a concentration difference of protons between the lumen and the stroma is eventually established upon illumination which is in the order of 3 pH-units (pH lumen ≈ 5; pH stroma ≈ 8) and is used to generate ATP in a reaction catalyzed by the ATP synthase. The synthesis of one molecule of ATP requires the reverse translocation of n protons (presumably n = 4, see Chapter 12) from the lumen to the stroma:

$$nH_l^+ + ADP^{3-} + HPO_4^{2-} + H_{cs}^+ \rightarrow nH_s^+ + ATP^{4-} + H_2O \qquad (2)$$

($H_{cs}^+$ is the proton which is chemically consumed at the stroma side.)

In summary, the photosynthetic electron transport chain is a membrane-integral array of anisotropically arranged redox components which are connected by diffusible electron carriers. The serial arrangement of PS II and PS I thereby provides sufficient stabilization of the difference in redox potential required for the subsequent reactions of water oxidation at the lumenal side of PS II and $NADP^+$ reduction at the stromal side of PS I. In addition, the accompanying deprotonating and protonating reactions create an electrochemical potential difference of protons which is used to drive ATP synthesis:

$$2\ H_2O + 2\ NADP^+ \rightarrow 2\ NADPH + 2\ H^+ + O_2 \tag{3a}$$

$$8\ H_l^+ + 2\ ADP^{3-} + 2\ HPO_4^{2-} + 2\ H_{cs}^+ \rightarrow 8\ H_s^+ + 2\ ATP^{4-} + 2\ H_2O \tag{3b}$$

Since generation of one oxygen requires abstraction of four electrons from two molecules of water, and taking into account the serial arrangement of PS I and PS II, a theoretical minimum of eight photons is required to perform the above reactions.

Each of the two photosystems is associated with antenna pigment-protein complexes to enlarge the light absorbing cross-section by about two orders of magnitude and to ensure efficient excitation energy transfer to the photoactive chl *a* molecules (P680 and P700) in the reaction centers of PS II and PS I, respectively (for further details see Chapters 8 and 9). The light-harvesting complexes (LHC I and LHC II) are chl *a*/*b*-binding protein complexes, where the pigment molecules are noncovalently bound at defined positions and orientation.

### *2.2. Bioenergetics of the light-driven reactions*

Primary charge separation following formation of the electronically excited singlet states of P680 and P700 can be either achieved by direct absorption or indirectly by excitation energy transfer from neighboring antenna molecules. The long wave absorption bands of the photoactive chl *a* molecules in the reaction centers of PS I and PS II peak at 700 nm and 680 nm, respectively. Therefore, the states $^1$P680 and $^1$P700 are about 1.8 eV above the ground state ($E = hc/\lambda$). This is the minimum photon energy required for charge separation.

The midpoint potentials ($E_m$) depicted in the energetic Z-scheme (see Fig. 5B) reveal that in the reactions of Eqn. 1a–1d four electrons are shifted from the redox level of water to that of $NADP^+$, spanning a difference in redox potential of 1.14 eV. These reactions concomitantly generate a transmembrane electrochemical potential difference of protons, $\Delta\tilde{\mu}_{H^+}$, which provides the driving force for the formation of the phosphoanhydro-bond of ATP.

$$\Delta\tilde{\mu}_{H^+} = RT \ln \frac{[H_l^+]}{[H_s^+]} + F(\phi_l - \phi_s) = -RT\,\Delta pH + F\Delta\phi \quad (4)$$

The proton motive force is defined as $\Delta\tilde{\mu}_{H^+}/F$ and it results:

$$\Delta\tilde{\mu}_{H^+}/F = -0.059 \text{ mV} + \Delta\phi \quad (5)$$

In chloroplasts, the electrical term becomes negligibly small under steady state illumination, because the proton translocation is accompanied by a compensatory transfer of ions across the thylakoid membrane. Assuming a ΔpH of 3, the proton motive force is in the order of 0.18 V per $H^+$ translocated.

Taken together, the energy stored in the products of the light-driven electron transport and the coupled formation of a transmembrane electrochemical potential difference of protons amounts to:

$$2 \text{ NADPH: } 4 \times 1.14 \text{ eV} = 4.6 \text{ eV} \quad (6a)$$

$$2 \text{ ATP: } 8 \times 0.18 \text{ eV} = 1.4 \text{ eV} \quad (6b)$$

Total energy: 6.0 eV

In this context it is interesting to note that the quantum efficiency for photooxidation of the reaction center chl *a* molecules is nearly 100% (see Chapters 7, 8, and 9). Therefore, in terms of energy, the abstraction of four electrons from water and subsequent transfer to $NADP^+$ could, in principle, be achieved by absorption of four red photons ($4\text{ h}\nu = 4 \times 1.8\text{ eV} = 7.2\text{ eV}$) in only one photosystem, instead of the coupled operation of PS II and PS I, leading to a minimum requirement of eight red photons ($8\text{ h}\nu = 8 \times 1.8\text{ eV} = 14.4\text{ eV}$). However, due to indispensable loss of energy caused by dissipative processes and non-equilibrium conditions (see Chapter 1), the serial arrangement of two photosystems is required to sufficiently stabilize the redox potential difference between water oxidation and $NADP^+$ reduction. The maximum efficiency of light energy exploitation is therefore under optimal conditions approximately 42% (ratio of energy output to input: 6.0 eV/14.4 eV). However, this calculation considers only the quantum yield of red photons. Taking the spectrum of solar radiation into account, the value markedly drops down. In addition, the theoretical minimum quantum requirement of eight photons per $O_2$ molecule evolved cannot be achieved in practice. The actual quantum yield of photochemistry under normal illumination is markedly lower and strongly depends on the physiological and environmental conditions. (For a more detailed discussion of the efficiency see Chapter 2).

Table 2. Lipid composition of chloroplast membranes from spinach

| | structure | side groups side chains fatty acids[b] | Thylakoids % molar ratio total lipids | Thylakoids % molar ratio acyl lipids | Inner envelope % molar ratio acyl lipids | Outer envelope % molar ratio acyl lipids |
|---|---|---|---|---|---|---|
| MGDG[a] | 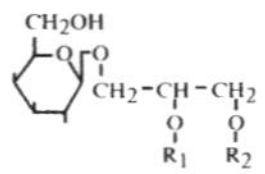 | 18:3/18:3<br>18:3/16:(0 or 3) | 27.2[c] | 38[c]/57[d] | 55[d] | 17[d] |
| DGDG | 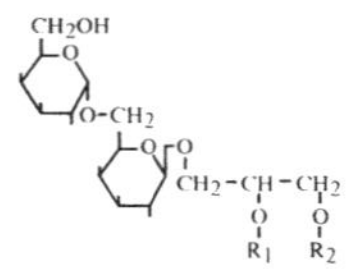 | 18:3/18:3<br>18:3/16:0 | 20.8[c] | 29[c]/27[d] | 29[d] | 29[d] |
| SQDG | 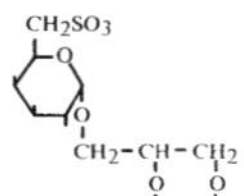 | 18:(3 or 2)/16:0<br>16:0/16/0 | 8.6[c] | 12[c]/7[d] | 5[d] | 6[d] |
| PG | 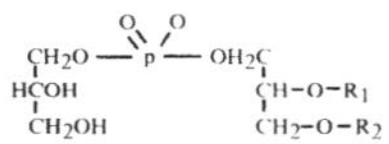 | 18:(3,2 or 1)/16:0<br>18:x/16:1t<br>16:0/16:1t | 10.0[c] | 14[c]/7[d] | 9[d] | 10[d] |
| PC | 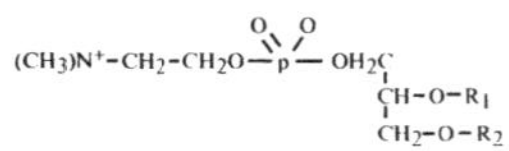 | 18:3/18:2<br>18:(3,2, or 1)/16:0 | 4.3[c] | 6[c]/0[d] | 0[d] | 32[d] |
| PQ | 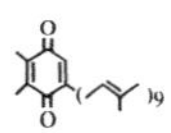 | | 0.7[c] | | | |
| α-CAR | 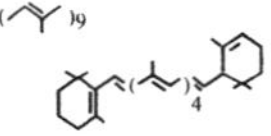 | derivatives:<br>Lutein | 2.2[c] | | | |
| β-CAR | 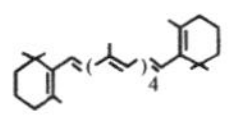 | Neo-, Zea-, Viola- } xanthin | 2.6[c] | | | |
| CHL a | 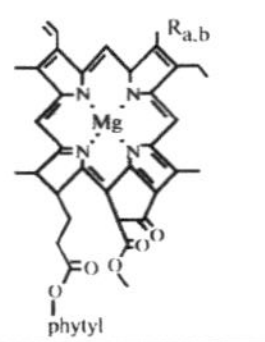 | $R_a = -CH_3$ | 17.3[c] | | | |
| CHL b | | $R_b = -CHO$ | 5.6[c] | | | |

([a]) Abbreviations see text; ([b]) 18:0, stearic acid; 18:1, oleic acid; 18:2, linoleic acid; 18:3, linoleic acid; 18:X, any C18 fatty acid; 16:0, palmitic acid; 16:1t, trans-D3-hexadecenoic acid; 16:3, palmitolenic acid. ([c]) Data according to Murphy, D. J., 1986, Biochim. Biophys. Acta 864, 33–94. ([d]) Data according to Douce, R. and Joyard, J., 1990, Annu. Rev. Cell Biol. 6, 173–216.

### *2.3. Structure and composition of the thylakoid membrane and its integral protein complexes*

*2.3.1. Lipid composition:* Like other energy-transducing membrane systems thylakoids exhibit an exceptionally high protein content. On a rough average they are composed to 60% (w/w) of proteins and to 40% (w/w) of lipids; lipophilic pigments and quinones thereby contributing approximately 25% to the total lipid content. However, compared to other biological membranes the thylakoid membrane exhibits a unique composition of lipids (see Table 2). (For further reading see also refs. 2, 3 and 4).

The vast majority of the thylakoid acyl lipids is comprised of uncharged glycolipids, i.e. mono- and digalactosyldiacylglycerol (MDGD and DGDG, respectively). In contrast, charged lipids only constitute a minority and are mainly made up of an unusual sulpholipid, sulfoquinovosylglycerol (SQDG), and the common phospholipid phosphatidylglycerol (PG). Whether or not phosphatidylcholine (PC) is present and to which extent still needs to be clarified conclusively. In addition to the preference for uncharged galactolipids, thylakoid acyl lipids also differ from those of most other eukaryotic membrane systems in the high degree of acyl unsaturation which is discussed to play a role for membrane fluidity. However, the exact mechanism to maintain membrane fluidity is still unknown. There is some evidence indicating that the lipid/protein ratio is primarily important, and that acyl unsaturation only exerts a disintegrative effect on the ordering structure of the high protein content.

Another striking feature of the thylakoid lipid composition is the high percentage of monogalactosyldiacylglycerol (MGDG) as compared to other acyl lipids. This is especially puzzling as it is known that unsaturated MGDG does not adopt a lamellar bilayer structure when purely dispersed in an aequeous medium. Therefore, a bilayer structure must either be stabilized by the presence of intrinsic proteins or other acyl lipids, e.g., the anionic sulpho- and phospholipids. In addition, MGDG is distinguished from all other thylakoid acyl lipids by its cone- or wedge-like shape resulting from a small headgroup and a rather bulky acyl moiety. This implies a possible role as structure-determining element. However, since no unambiguous pattern of lateral acyl lipid distribution can be assigned which corresponds to structurally different membrane regions, e.g., the highly curved thylakoid margin, a different role for MGDG is likely. It is conceivable, for instance, that MGDG is especially suited to seal the thylakoid membrane around intrinsic proteins. All other thylakoid lipids have larger polar headgroups and less bulky hydrophobic tails, leaving them with a cylindrical shape. Unfortunately, there is only little information on the precise function and role of acyl lipids in the thylakoid membrane.

In terms of the overall organization, a lateral segregation of thylakoid membranes into appressed and non-appressed membrane regions and the lateral distribution of the membrane integral protein complexes is well established. With respect to protein content and distribution, appressed thylakoids are enriched in protein compared to non-appressed thylakoids and incorporate the predominant part of PS II and LHC II. In contrast, PS I, LHC I and the ATP synthase complex are detected in non-appressed thylakoids. The Cyt b6f complex is evenly distributed, reflecting its role as mediator in the electron transport chain between PS II and PS I.

*2.3.2. The light-harvesting complexes:* With respect to their protein and pigment composition the light-harvesting complexes of PS I (LHC I) and PS II (LHC II) are fairly similar. Both harbor two types of chlorophyll, chl *a* and chl *b* and are composed of homologous protein subunits belonging to a family of chl *a*/*b*-binding polypeptides (CAB proteins) which exhibit molecular masses in the range of 25 to 30 kDa. The structure of CAB proteins as determined from LHC II studies reveals three membrane-spanning α-helices to which 15 chlorophyll molecules are ligated, i.e. 8 chl *a* and 7 chl *b*. The chlorophyll molecules are arranged on two levels with their porphyrin ring being oriented approximately perpendicular to the membrane plane [5]. The CAB monomers associate to trimers and several trimers are assembled within one light-harvesting complex. Functionally and structurally, LHC II can be separated into an inner part which is tightly associated with PS II and an outer part that can be phosphorylated. Upon phosphorylation negative charges are introduced, promoting a separation of the outer LHC II and a subsequent diffusion to the non-appressed thylakoid regions. This mechanism is suggested to be involved in adaptation to different light conditions (see section 6.1.1.). Furthermore, it seems very likely that surface exposed protein sections of LHC II, and maybe PS II, are involved in membrane stacking.

*2.3.3. Photosystem II and Photosystem I:* In contrast to the pigment composition of light-harvesting complexes, the core of PS II and PS I only binds one type of chlorophyll, i.e. chl *a*. In Fig. 5A a very simple picture of PS II is given: the reaction center is surrounded by a core antenna comprised of the chl *a*-binding proteins CP43 and CP47, and the lumenal side is covered by a regulatory protein cap associated with the water oxidizing system. The reaction center itself is comprised of two protein subunits with an apparent molecular mass of 32 and 34 kDa. According to the current model, D1 and D2 both exhibit five transmembrane α-helices and incorporate all redox groups involved in the light induced electron transport from P680 to $Q_B$, as well as a high-spin iron located between $Q_A$ and $Q_B$ whose role in electron transport is still unclarified. In addition, a tyrosin residue of D1 has been identified as the intermediate electron carrier $Y_Z$ involved in electron transfer from the manganese-containing site of water oxidation to P680. The homologous tyrosin residue on D2, $Y_D$, is not directely involved in the light-driven electron transport but is capable of interfering with the redox states of water oxidation in the dark. The lumen-exposed regulatory cap associated with water oxidation is comprised of three extrinsic, hydrophilic proteins with apparent molecular masses of 33, 24 and 18 kDa, respectively. It is known that the 24 and 18 kDa polypeptides are involved in the regulation of calcium and chloride requirement, whereas the 33 kDa is required to stabilize functional manganese in its binding site.

The reaction center of PS I is similarly comprised of a heterodimer of two polypeptides which each exhibit a molecular mass of 83 kDa as deduced from the amino acid sequence. According to hydrophobicity plots 11 transmembrane helices are predicted for subunit Ia and Ib, respectively. The heterodimer incorporates all functional redox groups from P700 to $F_X$ and, in addition, approximately 40 chl *a* molecules are ligated to each of the subunits. Apart from the reaction center heterodimer, PS I contains at least two other polypeptides, i.e. subunit II and subunit VII (see Fig. 5A). Subunit VII is closely associated with the stroma side of PS I and is most probably the apoprotein of the iron-sulfur clusters $F_A$ and $F_B$. Subunit II is presumably also located on the stroma side where it is supposed to function as the ferredoxin docking protein. One further extrinsic protein, subunit III, is located at the lumenal side and according to crosslinking studies is believed to provide the binding domain for plastocyanin. The copper containing plastocyanin and the iron-sulfur containing ferredoxin constitute the diffusible linkers between the lumenal side of PS I and the Cyt b6f complex and the stromal side of PS I and ferredoxin:NADP oxidoreductase (FNR), respectively.

*2.3.4. The cytochrome b6f complex:* The Cyt b6f complex (see Chapter 10) exerts a triple function in photosynthetic electron transport: (1) it serves as plastoquinol: plastocyanin oxidoreductase thereby acting as a $2e^-:1e^-$ gate, (2) it incidentally translocates protons from the stroma to the lumen, and (3) it catalyzes the Q-cycle.

The entire complex contains one c-type and two b-type cytochromes, one Rieske iron-sulfur cluster and two quinone binding sites. In this respect it is homologous to the Cyt b/$c_1$ complexes of respiratory membranes. Regarding the protein composition, the Cyt b6f complex is composed of five intrinsic subunits, i.e. cyt f, cyt b563, the Rieske protein, a 17 kDa structural polypeptide and a low molecular mass polypeptide of unknown function.

*2.3.5. The $H^+$-ATPase:* The typical "ball and stalk structure" of membrane associated ATP synthesizing enzymes is also inherent to the thylakoid ATPase complex (see Chapter 12). According to its dual function as "proton translocating reversible ATPase" the intrinsic stalk-like coupling factor 0 (CF0) is involved in proton translocation and the extrinsic ball-like coupling factor 1 (CF1) is responsible for nucleotide binding. CF0 and CF1 are attracted by electrostatical forces via negative surface charges that are compensated by divalent cations. Incubation with chelators promotes the reversible dissociation of CF1, giving rise to a situation where proton translocation is uncoupled from ATP synthesis. As a consequence, no potential difference of protons is generated across the thylakoid membrane which results in an increased photosynthetic electron transport rate and a decreased rate of photophosphorylation.

Regarding the molecular architecture, CF0 is assembled from four types of subunits (I, II, III and IV). The inner wall of the transmembrane proton channel is created by a dodekamer of subunit III, a 8 kDa proteolipid. The 12.5 kDa subunit II and subunit I, a 15 kDa polypeptide, serve as binding site for CF1. The ball-like CF1 consists of five different protein subunits with a presumed stoichiometry of $\alpha_3\beta_3\gamma\delta\varepsilon$.

In summary, thylakoids exhibit a very unusual chemical composition promoting a clearly ordered structure which is detectable as a pattern of appressed and non-appressed membrane regions. Nonetheless, thylakoids are a contiguous and dynamic membrane system since grana and stroma thylakoids are interconvertible due to lateral diffusion of different thylakoid components.

## 3. Carbon metabolism

Photosynthetic carbon reduction is the process responsible for the tremendous annual production of organic carbohydrates and provides not only the nutritional basis for all heterotrophic organisms but also gives rise to the accumulation of combustible resources, e.g., wood and fossil energy resources like coal and mineral oil. In total, about 200 billion tons of organic carbohydrates are assimilated on earth per year, two-thirds of which are due to the productivity of land plants. The pathway of $CO_2$-fixation and the identity of the first assimilation product was the topic of intense investigations until the mid-1950s. These efforts led to elucidation of the C3 photosynthetic carbon reduction cycle in 1956 by Calvin and Benson (Calvin-Benson cycle), and in 1961 M. Calvin was awarded the Nobel Prize in chemistry for this discovery [6].

### *3.1. C3 Photosynthetic carbon reduction (C3-PCR)*

The majority of land plants growing in the temperate zone synthesize a C3 sugar as the first detectable product of $CO_2^-$ fixation. For this reason the Calvin-Benson cycle is also often referred to as C3 photosynthetic carbon reduction cycle.

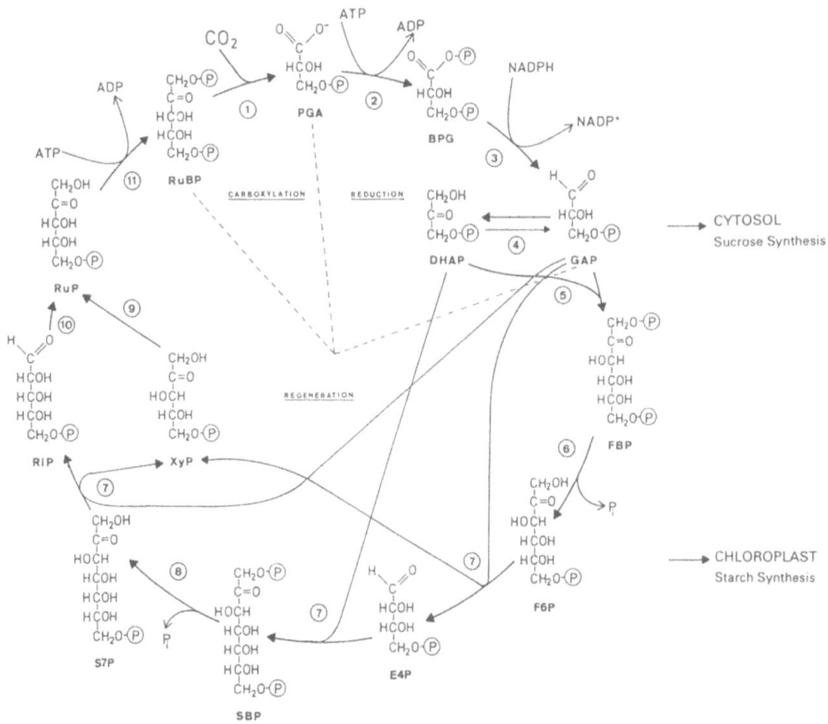


Fig. 6. Schematic diagram of the C3 photosynthetic carbon reduction cycle. Metabolites: RuBP, ribulose-1,5-bisphosphate; 3-phosphoglycerate; BPG, 1,3-bisphosphoglycerate; GAP, glyceraldehyde-3-phosphate; DHAP, dihydroxyacetonephosphate; FBP, fructose-1,6-bisphosphate; F6P, fructose-6-phosphate; E4P, erythrose-4-phosphate; SBP, sedoheptulose-1,7-bisphosphate; S7P, sedoheptulose-7-phosphate; RiP, ribose-5-phosphate; XyP, xylulose-5-phosphate; RuP, ribulose-5-phosphate. Enzymes: (1) Ribulose-1,5-bisphosphate carboxylase-oxygenase; (2) Phosphoglycerate kinase; (3) Glyceraldehyde phosphate dehydrogenase; (4) Triose phosphate isomerase; (5) Aldolase; (6) Fructose-1,6-bisphosphatase; (7) Transketolase; (8) Sedoheptulose-1,7-bisphosphatase, (9) Ribulose phosphate epimerase; (10) Ribose phosphate isomerase; (11) Ribulose-5-phosphate kinase.

*3.1.1. Biochemistry of the C3-PCR cycle:* The complete C3-PCR cycle is a reaction sequence comprising a carboxylating, reducing, and regenerating phase (see Fig. 6). In the carboxylation reaction one molecule of $CO_2$ becomes chemically bound to one molecule of ribulose-1,5-biophosphate (RuBP) resulting in an instable C6-sugar that is immediately cleaved into two molecules of 3-phosphoglyceric acid (PGA).

The initial reaction of the reducing phase is the ATP-dependent activation of PGA generating 1,3-bisphosphoglycerate (BPG) which is subsequently reduced by NADPH to glyceraldehyde-3-P (GAP). Most of the reduced glyceraldehyde-3-P reversibly isomerizes to dihydroxyacetonephosphate (DHAP). Under equilibrium conditions, 90% of reduced GAP is converted to DHAP. In the following reaction one molecule of GAP and DHAP are linked together to form one molecule of fructose-1,6-biphosphate (FBP) which is subsequently dephosphorylated to fructose-6-P (F6P). Hence, the reducing phase generates either triose-phosphates or hexose-phosphates. In general, triose-phosphates represent the metabolite form that can be exported to the cytosol whereas the hexose-phosphates remain in the chloroplast and are the offshoot for chloroplast starch synthesis (see section 3.5.). However, most of the triose-phosphates are required to regenerate RuBP in order to prevent a depletion of the primary $CO_2$-acceptor molecule.

The acceptor regenerating phase is a complex pathway of entangled reactions restoring three molecules of RuBP from five triose-phosphates:

$$\text{GAP} + \text{DHAP} \rightarrow \text{Fru-1,6-BP} \rightarrow \text{Fru-6-P} \tag{7a}$$

$$\text{Fru-6-P} + \text{GAP} \rightarrow \text{Ery-4-P} + \text{xylulose-5-P} \tag{7b}$$

$$\text{Ery-4-P} + \text{DHAP} \rightarrow \text{Sed-1,7-BP} \rightarrow \text{Sed-7-P} \tag{7c}$$

$$\text{Sed-7-P} + \text{GAP} \rightarrow \text{ribose-5-P} + \text{xylulose-5-P} \tag{7d}$$

The generated pentose-phosphate are converted to ribulose-5-P by an epimerase and isomerase, respectively:

$$2\ \text{xylulose-5-P} \rightarrow 2\ \text{ribulose-5-P} \tag{7e}$$

$$1\ \text{ribose-5-P} \rightarrow 1\ \text{ribulose-5-P} \tag{7f}$$

Ribulose-P is then phosphorylated to RuBP:

$$3\ \text{ribulose-5-P} + 3\ \text{ATP} \rightarrow 3\ \text{ribulose-1,5-BP} + 3\ \text{ADP} + 3\ \text{P}_i \tag{7g}$$

To assimilate one hexose-P the PCR cycle has to turn over six times in total, leading to fixation of six $CO_2$ to six molecules of RuBP. From the 12 triose-P molecules generated primarily, ten are required for regeneration of six RuBP leaving only two as net products for sucrose or starch synthesis.

*3.1.2. Bioenergetics of the C3-PCR cycle:* An energetic scheme of the C3-PCR cycle is given in Fig. 7 in the form of a $\Delta G^0$-diagram summarizing all reactions involved in the production of 1 mole glucose. Energetically, the crucial step of the PCR cycle is the activation of PGA and the subsequent reduction to glyceraldehyde-3-P. For every $CO_2$ fixed to RuBP, 2 PGA's are generated requiring 2 ATP and 2 NADPH for reduction. Per RuBP, one more molecule of ATP has to be taken into account with respect to the last regeneration step. In summary, the production of one mole glucose can be described in the following equation:

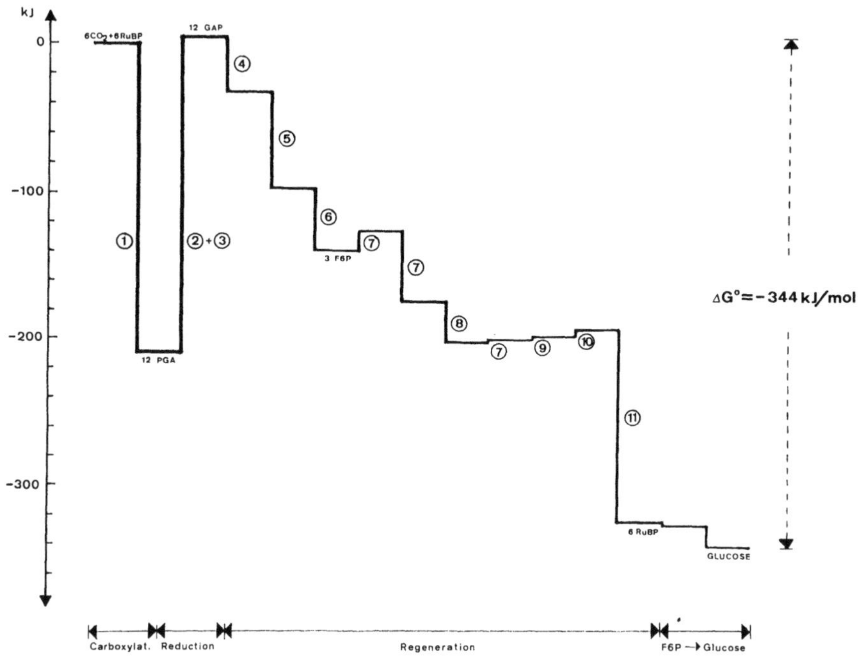


Fig. 7. Energetics of the C3 photosynthetic carbon reduction cycle. ($\Delta G^0$ values according to ref. 44.)

$$6\ CO_2 + 18\ ATP + 12\ NADPH \rightarrow C_6H_{12}O_6 + 18\ P_i + 12\ NADP^+ \quad \Delta G^0 = -344\ kJ/mol \qquad (8)$$

or on the basis of one mole $CO_2$ reduced:

$$CO_2 + 3\ ATP + 2\ NADPH \rightarrow [CH_2O] + 3\ ADP + 3\ P_i + 2\ NADP^+ \quad \Delta G^0 = -57\ kJ/mol \qquad (8a)$$

Considering overall photosynthesis, the exergonic process of carbon reduction producing one $[CH_2O]$-unit requires the light-driven generation of 3 ATP and 2 NADPH which, in turn, requires a minimum absorption of eight red photons with a free energy of 175 kJ/mol. Thus, the energy required to reduce one $CO_2$ is given by $8 \times 175$ kJ/mol = 1400 kJ/mol.

For an evaluation of the overall efficiency of the photosynthetic machinery the amount of free energy regained by respiratory oxidation of one $[CH_2O]$-unit to that being required to reduce one $CO_2$ to a $[CH_2O]$-unit is compared. The former reaction is given by:

$$[CH_2O] + O_2 \rightarrow CO_2 + H_2O \quad \Delta G^0 = -477\ kJ/mol \qquad (9)$$

Hence, the available free energy stored per mole [$CH_2O$]-units is 477 kJ. The theoretical overall efficiency for conversion of absorbed light energy is therefore 477/1400 resulting approximately in 30–35% efficiency. However, as outlined above, various modes of energy loss severely reduce the actual efficiency of photosynthesis and on a rough average only 5% of the absorbed radiation is actually transformed into chemical free energy.

*3.1.3. Enzymes involved in the C3-PCR cycle:* Rubisco, the enzyme catalizing the unique reaction of RuBP-carboxylation, constitutes roughly 50% of the total protein content of leaves and, consequently, is the most abundant protein on earth. For convenience, "Rubisco" is the synonym generally used instead of the long-winded term ribulose-1,5-bisphosphate carboxylase-oxygenase. Unfortunately, the abbreviation is less informative. As indicated by the correct name, Rubisco possesses twofold enzymatic activity, catalyzing both the carboxylation and oxygenation of RuBP. The competitive oxygenation reaction leads to a substantial loss of assimilates (see section 3.2.: photorespiratory carbon oxidation). The enzymatic activity of Rubisco therefore decides between photosynthesis and photorespiration, an annoying circumstance with respect to crop productivity [7]. For this reason, Rubisco is a topic of unabated scientific interest. The enzyme structure and biochemical properties have been analyzed extensively, and currently numerous attempts are made to manipulate the enzyme genetically.

The chloroplast type enzyme reveals a hexadecameric structure comprised of eight large subunits of 55 kDa (L-subunit) and eight small subunits of 15 kDa (S-subunit) giving the holoenzyme a molecular mass of 560 kDa. The $L_8S_8$ configuration can be structurally divided into two $L_4S_4$ halves with four-fold symmetry (see Fig. 8). However, each L-subunit on one-half is interlocked with one L-subunit on the other half and the catalytic activity is entirely inherent to the dimetric structure of two interlocked large subunits. Functionally, the $L_8$ configuration therefore has to be segregated into a tetramer of interlocked $L_2$ units. Considering the molecular structure of the large subunit, significant progress has been achieved by investigating the enzyme of the purple bacterium *Rhodospirillum rubrum* which simply contains one $L_2$ dimer. Although the overall homology between the large subunit of purple bacteria and that of higher plants is only ≈ 30%, the conservation is markedly higher in the domains of the active site. According to the three-dimensional structure determined for the $L_2$ dimer of purple bacteria the C-terminal protein domain of the large subunit folds into an alpha/beta barrel structure. It is also this domain that is the site of interaction between the dimerizing L-subunits. The active site is constituted by the mouth of the C-terminal barrel of one L-subunit and a portion of the N-terminal domain of the adjacent L-subunit. As in purple bacteria, the active site of chloroplasts-type Rubisco is solely constituted by large subunits. Nevertheless, the presence of the small subunits

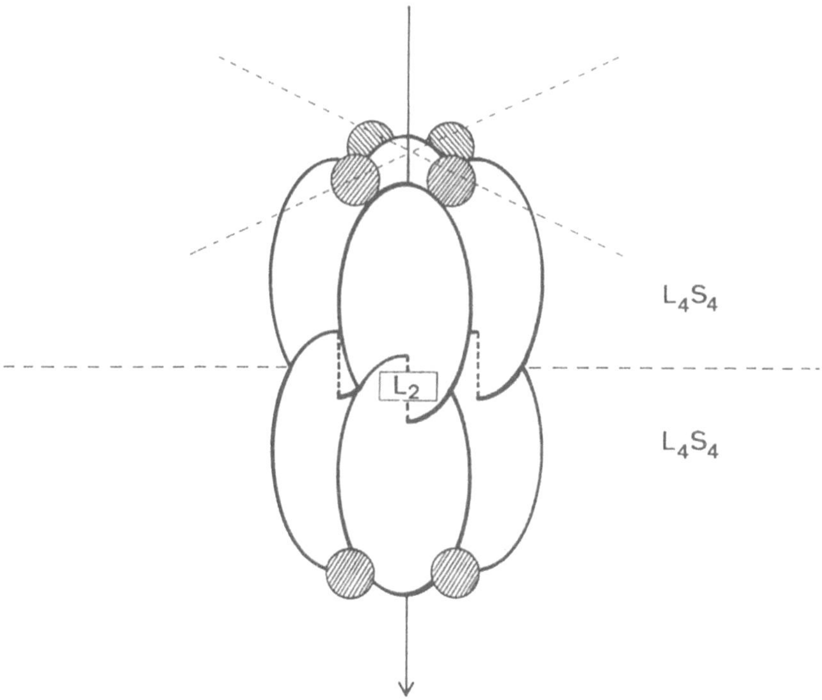


Fig. 8. Schematic drawing of the subunit configuration and structure of Rubisco from higher plants. Shaded areas represent small subunits, non-shaded areas large subunits. For further details see text.

is necessary for full catalytic activity. This finding gives rise to questions about the role of the small subunit. It is conceivable that the small subunit is important to maintain an active conformation of the catalytic site. However, this is speculative and other functional or structural roles are also possible. The functional difference of the two subunit types is in line with the observation that the amino acid sequence of the chloroplast-encoded large subunit is highly conserved (≈90% among higher plants) whereas that of the nuclear encoded small subunit is less conserved (≈70% among higher plants). Both, the chloroplast and the nuclear gene coding for the large and the small subunit, respectively, have been cloned in *E. coli*. But due to missing self-assembly of large and/or large and small subunits expression of these genes in *E. coli* does not yield an active holoenzyme. It now seems well established that the correct assembly into the $L_8S_8$ configuration of active Rubisco requires the additional presence of an assembly-mediating protein, the so-called plastid chaperonin. (For further reading see also refs. 8 and 9).

The enzymatic activity of Rubisco is regulated in a diurnal pattern, i.e. the enzyme is inhibited in the dark and activated in the light. As a consequence, not only photosynthetic electron transport but also carbon reduction is dependent on the availability of light. However, it has to be stressed that in the former reaction sequence light is the indispensable driving force whereas in the latter case it exerts only a regulatory function. This light regulation is of physiological relevance in order to prevent a circulation of $CO_2$, ATP and reducing power between dissimilating and assimilating reactions in the dark, and hence to prohibit an idling of metabolism. The primary mechanism of dark inhibition of Rubisco activity is due to the presence of a competitive inhibitor [10]. In contrast, the primary mechanisms of light-induced activation involve not only the dissociation of the noctural inhibitor but also a particular activation reaction which is dependent on the light-driven increase in $Mg^{2+}$ and $H^+$ concentration of the stroma (for further details of regulation see section 6.1.2.). Activation requires the consecutive reaction of an activator $CO_2$ molecule ($^ACO_2$) and a divalent metal cation, usually $Mg^{2+}$, with Rubisco (R):

$$^{inact}E + {}^ACO_2 \rightarrow [^{inact}E \times {}^ACO_2] + Mg^{2+} \rightarrow [^{act}E \times {}^ACO_2 \times Mg^{2+}] \quad (10)$$

On the molecular basis, a lysine residue in the active site reacts with $^ACO_2$ generating a carbamate which is stabilized by $Mg^{2+}$. It is very likely that the $Mg^{2+}$ is directed towards the introduced negative charge of the carbamate and, in turn, is involved in binding and aligning the substrate $CO_2$. In this context it should be mentioned that the interaction between $^ACO_2$ and Rubisco is kinetically and spectroscopically distinct from the binding of the substrate $CO_2$ molecule ($^SCO_2$):

$$\text{Lys-NH}_2 \xrightarrow{^ACO_2} \text{Lys-NH-COO}^- \xrightarrow{Mg^{2+}} \text{Lys-NH-COO}^- \times Mg^{2+} \xrightarrow{^SCO_2}$$
$$\rightarrow \text{Lys-NH-COO}^- \times Mg^{2+} \times SO_2C \quad (11)$$

It is interesting to note that for activation as well as for catalysis, $CO_2$ and not $HCO_3^-$ is the active species. Since the permeability of the chloroplast envelope is high for $CO_2$ and low for $HCO_3^-$, $CO_2$ is the imported species into the chloroplast. However, at the alkaline pH of the stroma the equilibrium for

$$CO_2 + H_2O \rightleftarrows HCO_3^- + H^+ \quad (12)$$

is clearly on the side of bicarbonate formation. Nevertheless, due to substrate binding Rubisco steadily removes $CO_2$ thereby causing a redissociation of bicarbonate. A sufficient supply of $CO_2$ to Rubisco is provided by the activity of carboanhydrase, the enzyme accelerating the readjustment of the equilibrium by a factor of $10^7$. Under stromal conditions (10 μM $CO_2$, 5–10 mM $Mg^{2+}$, pH 8–8.5) the rate of carbon fixation is approximately 150–180 μmol $CO_2$/mg chl·h at 20°C, and is mainly limited by the atmospheric $CO_2$ concentration of 0.033%. For a typical C3-plant the substrate affinity of Rubisco is in the order of 10 μM for $CO_2$ and approximately 30 mM for RuBP. The surprisingly low substrate affinity of Rubisco for $CO_2$ is partly overcome by the extraordinary abundance of this protein.

Regarding the carboxylation reaction in more detail, the addition of $CO_2$ to the C2-atom of RuBP requires the creation of a nucleophilic center at the C2-atom [see ref. 7]. The simplest possibility to accomplish this is the formation of an enediol by removal of a proton from the C3-atom of RuBP. This was already postulated by M. Calvin. The proposed mechanism of carboxylation is shown in Fig. 9 and involves the following four steps:

Step 1: removal of proton from the C3-atom of RuBP through interaction with an alkaline component (possibly a lysine residue in the active site) leading to enediol formation.

Step 2: binding of substrate $CO_2$ molecule ($^{S}CO_2$) and RuBP to Rubisco. This converts the activated ternary complex into a quintenary complex [RuBP × $^{S}CO_2$ × $Mg^{2+}$ × $^{A}CO_2$ × $E^{active}$].

Step 3: electrophilic atack of $^{S}CO_2$ on the C2-atom leading to the formation of a transition state (2-carboxy-3-keto-D-arabinitol 1,5-bisphosphate). It is this intermediate that is mimicked by the nocturnal inhibitor 2-carboxy-D-arabinitol 1-phosphate (CA1P) and competitively inactivates Rubisco in the dark.

Step 4: hydrolysis of the transition state and protonation of the top moiety generates two PGA molecules. At this point Rubisco must assure a precise stereochemical control since a non-enzymatic hydrolysis yields both, D- and L-type PGA, and Rubisco only produces PGA in the D-configuration.

Besides Rubisco, the key enzyme in the carbon fixation, ten other enzymes are involved in the subsequent reactions of the C3-PCR cycle [see also ref. 11]. Since many of the reaction steps of the reducing and regenerating phase of the C3-PCR cycle are also involved in cytosolic pathways, e.g., glycolysis, it is not surprising that most PCR enzymes have cytosolic counterparts. In this context it is interesting to note that all enzymes, with exception of the large subunit of Rubisco, are encoded by the nuclear genome and are transported from the cytosol into the chloroplast as precursor proteins containing a specific transit sequence. This implies that the nuclear genes encoding the PCR enzymes are distinct from those encoding the cytosolic isoenzymes. With respect to the existence of chloroplast enzymes and their cytosolic counterparts it is interesting, especially for evolutionary reasons, to consider the degree of amino acid sequence homology. In the case of GAPDH, for example, the enzyme of chloroplasts is present in two forms, either as homotetramer ($A_4$) or as heterotetramer ($A_2B_2$), whereas the cytosolic enzyme is always a homotetramer ($C_4$). Although the reason for the heterogeneity of the chloroplast enzyme

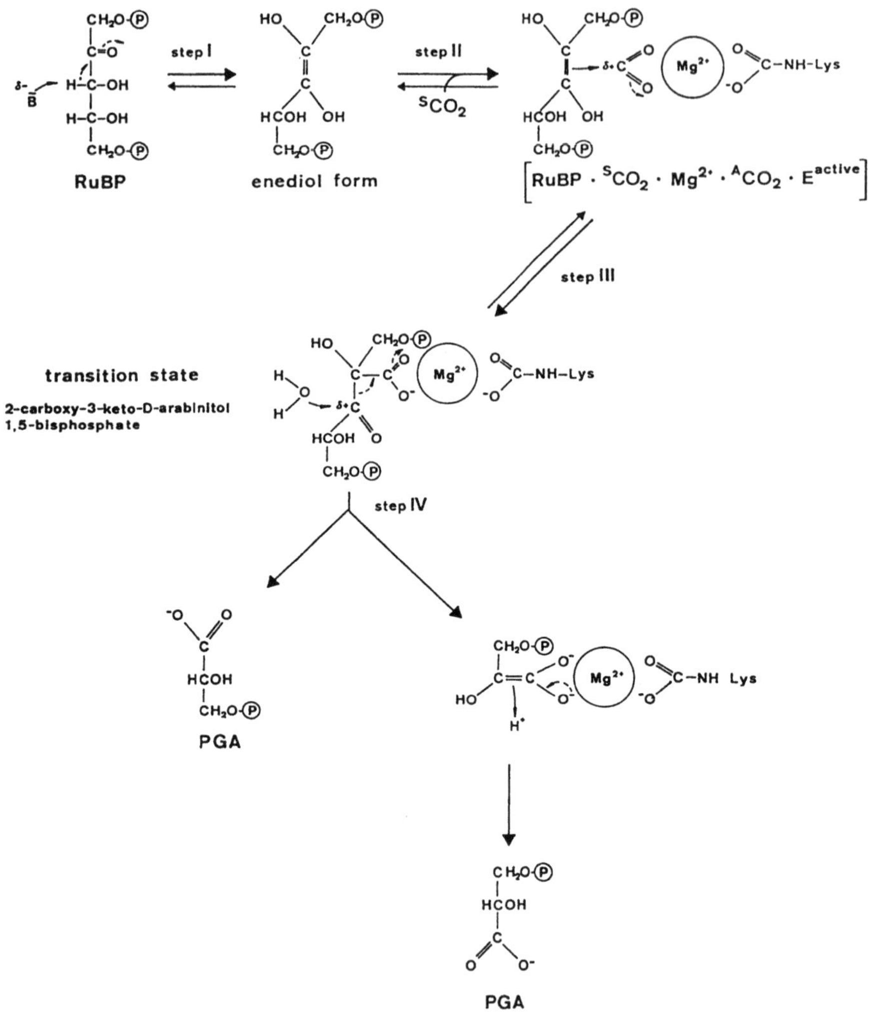


Fig. 9. Mechanism for the carboxylation of ribulose-1,5-bisphosphate. (According to ref. 7.)

is not completely clarified, it is suggested that stretches of amino acids present on subunit B and missing from subunit A are responsible for anchoring the enzyme to the inner membrane of the envelope. Despite these differences, the amino acid sequences of the two chloroplast subunits, A and B, exhibit a higher degree of homology among each other (77%) than compared to the cytosolic subunit (45%). In addition, chloroplast GAPDH also exhibits other pronounced prokaryotic features, which could indicate that the chloroplast and the cytosolic enzyme evolved from two different evolutionary lineages.

Finally, it should be mentioned that three of the enzymes involved in the PCR cycle exhibit the same structural motif in the active site. Actually, the $\alpha/\beta$ barrel structure, a motif also common to many other enzymes, was first identified in the triose-P isomerase. The barrel-like structure is constituted from eight parallel $\beta$-strands that are surrounded by eight $\alpha$-helices, probably pursuing the joint issue to place the site of catalysis into an anhydrous environment.

### *3.2. C2 photorespiratory carbon oxidation (C2-PCO)*

The term "Photorespiration" refers to the fact of light-induced respiratory gas exchange associated with this pathway, that is an uptake of $O_2$ and a release of $CO_2$.

*3.2.1. Biochemistry of the C2-PCO cycle:* The biochemical reactions involved in the C2 photorespiratory carbon oxidation cycle were to a large extent deciphered by Tolbert and co-workers. As illustrated in Fig. 10 the entire cycle requires the interaction of three different cell compartments: chloroplasts, peroxisomes and mitochondria.

The key reaction is the oxygenation of RuBP that, like RuBP carboxylation, is catalyzed by Rubisco and includes a C-C bond cleavage between carbon atoms 2 and 3 of RuBP. Consequently, one C2 and one C3 carbohydrate (P-glycolate and P-glycerate) are generated instead of the two molecules of P-glycerate usually gained from RuBP carboxylation. Thus, P-glycolate is produced at the expense of triose-P, implying a loss of assimilates. In the PCO cycle, two molecules of P-glycolate are transformed and decarboxylated to restore one molecule of P-glycerate, i.e. part of the lost carbon is regained.

In more detail the PCO cycle includes the following reactions: P-glycolate is dephosphorylated to glycolate which, after transfer to peroxisomes, is oxidized to glyoxylate. The glyoxylate is, in turn, transaminated to glycine either by serine or by glutamate and then exits the peroxisome. In the mitochondrion two molecules of glycine are oxidized and decarboxylized to serine, releasing $CO_2$ and $NH_3$. Serine returns to a peroxisome where it is converted into hydroxypyruvate which is subsequently reduced to glycerate. The glycerate is reimported into a chloroplast via the dicarboxylic acid transfer located in the envelope (see section 5.1.2.). In the reaction completing this cycle within the chloroplast, glycerate is phosphorylated to P-glycerate feeding the salvaged carbon back to carbon reduction.

*3.2.2. Bioenergetics of the C2-PCO cycle:* Photorespiration is generally viewed as being an energy-consuming process, since oxygenation of 2 RuBP only yields 3 P-glycerate and 1 $CO_2$ (instead of the four PGA generated through carboxylation) and requires additional ATP and reducing power:

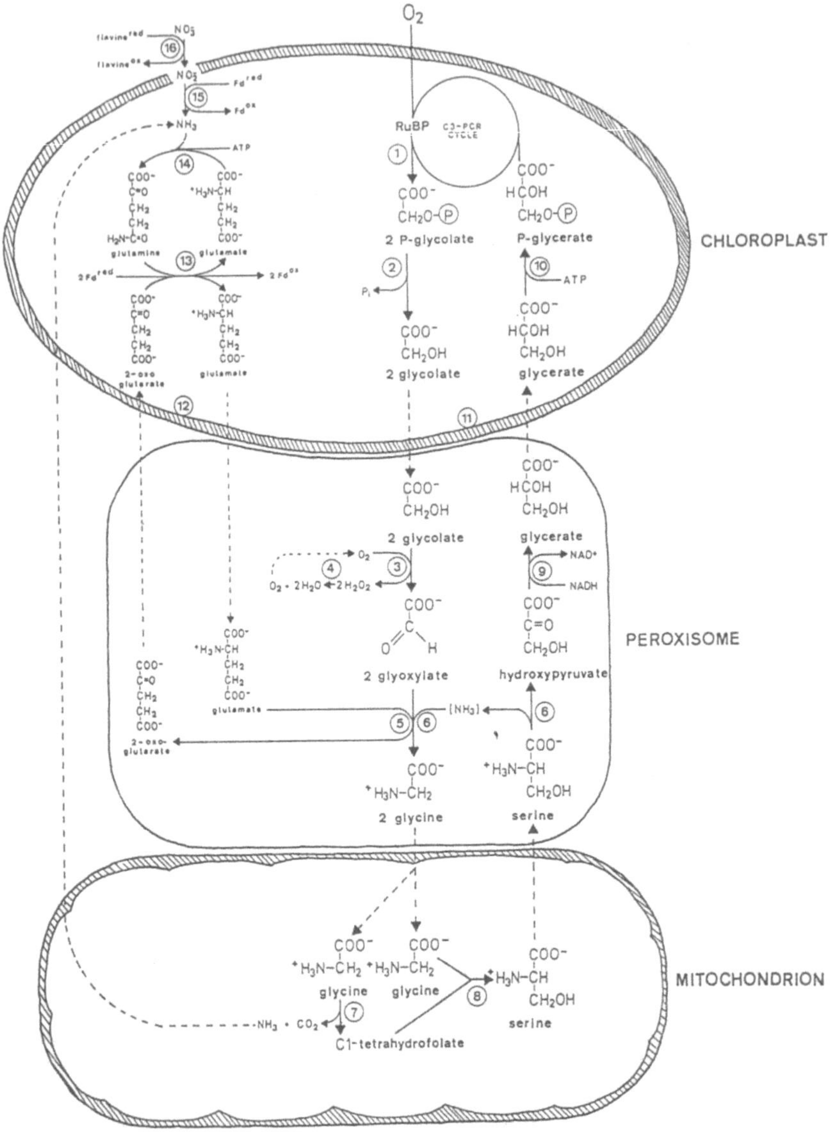


Fig. 10. Schematic diagram of the C2 photorespiratory carbon oxidation cycle. The diagram includes the "photorespiratory" cycling of nitrogen. Enzymes: (1) Ribulose-1,5-bisphosphate carboxylase-oxygenase; (2) Phosphoglycolate phosphatase; (3) Glycolate oxidase; (4) Catalase; (5) Glutamate-glyoxylate aminotransferase; (6) Serine-glyoxylate aminotransferase, (7) Glycine decarboxylase complex; (8) Serine hydroxymethyltransferase; (9) Hydroxypyruvate reductase; (10) Glycerate kinase; (11) Glycolate/glycerate translocator; (12) Dicarboxylate translocator; (13) Glutamine: oxoglutarate aminotransferase (GOGAT); (14) Glutamine synthase (GS); (15) Nitrite reductase; (16) Nitrate reductase.

$$2\ RuBP + 2\ O_2 \rightarrow 2\ PGA + 2\ PGL \quad (13a)$$

$$2\ PGL + 2\ O_2 + 2\ [NH_3] \rightarrow 2\ Glycine + O_2 + 2\ P_i \quad (13b)$$

$$2\ glycine \rightarrow CO_2 + NH_3 + Serine \quad (13c)$$

$$Serine + NADH + ATP \rightarrow PGA + [NH_3] + NAD^+ + ADP \quad (13d)$$

$$2\ RuBP + 3\ O_2 + NADH + ATP \rightarrow 3\ PGA + CO_2 + NAD^+ + ADP \quad (13e)$$

Taking into account that 2 mole NADPH and 3 mole ATP were previously required to assimilate the lost $CO_2$, the total balance for the evolution of 1 mole $CO_2$ results in 3 mole NAD(P)H and 4 mole ATP that seem to be wasted.

However, considering the PCO cycle as an isolated metabolic pathway does not really meet the actual situation (see also ref. 12). As outlined above, photosynthetic carbon reduction and photorespiratory carbon oxidation are due to a competitive catalytic activity of Rubisco. Whether the carboxylation or the oxygenation reaction takes place depends on the $CO_2$ and $O_2$ concentration in the stroma and on the affinity of Rubisco to $CO_2$ and $O_2$, respectively. The internal ratio of $CO_2/O_2$ can partly be altered by changes in stomatal conductance or, in some plants, by $CO_2$-concentrating mechanisms (see sections 3.3. and 3.4.). Likewise, the substrate affinity of Rubisco is strongly influenced by the temperature, i.e. an increase in temperature reduces the affinity of Rubisco for $CO_2$ whereas the affinity for $O_2$ remains more or less the same. Thus, the ratio of carboxylation/oxygenation is not a fixed parameter, and it decreases with increasing temperature. For a typical C3-plant growing in a moderate climate, the ratio is approximately 4:1, corresponding to 15–25% photorespiration. However, the range is rather wide and can also reach values up to 50%. This means that generally the C3-PCR and the C2-PCO cycle occur concomitantly.

*3.2.3. The physiological role of the C2-PCO cycle:* Although the question on the physiological role of photorespiration is being discussed for more than a decade, a conclusive answer is still lacking (see also ref. 12 and 13).

At first sight, the simplest role that can be assigned to photorespiration is the dissipation of excess light-energy, i.e. the consumption of excess NADPH and ATP generated in the light-driven reactions. Under certain conditions the cycling of $CO_2$ between the PCO and PCR pathway might be advantageous. For example under drought conditions the stomates have to be closed in order to prevent transpiratory loss of water and the normal $CO_2$ supply is interrupted causing the internal $CO_2$ concentration to drop below the $CO_2$ compensation point. (The $CO_2$ compensation point corresponds to the $CO_2$ concentration at which the rate of (photo) respiratory $CO_2$ evolution equals that of photosynthetic $CO_2$ consumption). Thus, in a situation where leaves are illuminated and the stomates are closed, photorespiration might allow $CO_2$ cycling, and therefore allow electron transport to continue. In addition, it would consume the generated NADPH and ATP with approximately the same ratio and stoichiometry as the PCR cycle and also eliminate oxygen produced in the light-driven reactions. However, whether or not photorespiration is a significant mechanism to protect leaves from photoinhibition under conditions of high light intensity and low water potential in the tissue is still a matter of debate.

Substantial evidence indicates that catalysis of RuBP oxygenation is just an inevitable accompaniment of catalysis, and hence, an inherent property to Rubisco. This is implied by the finding that oxygenation is catalyzed by Rubisco in all organisms investigated so far, e.g., RuBP oxygenation is also catalyzed by Rubisco from the anaerobic purple bacterium *Rhodospirillum rubrum.* Furthermore, catalysis of both oxygenation and carboxylation requires the same activation step (see section 3.1.2.). If it is true that RuBP oxygenation occurs more or less accidentally, the role of the subsequent C2-PCO cycle would simply be to regain part of the lost carbon, to decontaminate the intermediate glyoxylate and to prevent a depletion of amino nitrogen. The release of ammonia during oxidative decarboxylation of glycine in mitochondria would be deleterious if it were not returned to chloroplasts and reassimilated to glutamate. It has to be emphasized that apart from the recycling of carbon, the photorespiratory cycle also provides an important recapture of nitrogen [14].

### *3.3. C4 photosynthetic carbon assimilation (C4-PCA)*

Some plants, preferentially those growing in warm and very sunny areas where they are exposed to high light intensities, have evolved a pathway that almost completely overcomes photorespiration. This enables them to sustain a notable crop productivity, although high light intensities usually come along with elevated temperatures that favor RuBP oxygenation instead of carboxylation (see section 3.2.2.).

*3.3.1. Biochemistry of the C4-PCA cycle:* The mechanism by which the occurrence of photorespiration is nearly abolished relies on a local accumulation of $CO_2$ at the site of Rubisco activity. The $CO_2$ concentrating system is based on primary $CO_2$-fixation in cells that are distinct from those where the secondary $CO_2$-fixation, catalyzed by Rubisco, takes place. Plants exhibiting this pathway are referred to as C4-plants since the $CO_2$ prefixation yields a C4-dicarboxylic acid and therefore the pathway is also known as the C4-dicarboxylic acid pathway or as Hatch-Slack cycle. The crucial reaction of primary $CO_2$ fixation is the carboxylation of phosphoenolpyruvate (PEP) which is catalyzed by PEP carboxylase (PEP-case) and is not competitively inhibited by oxygen. The generated oxaloacetate is then converted to either aspartate or malate and the dicarboxylic acids are transported to and decarboxylated in the cells performing secondary $CO_2$ fixation and $CO_2$ reduction. Therefore, in spite of the fact that Rubisco isolated from these plants is indistinguishable from that of other plants, RuBP oxygenation is suppressed as a consequence of a "$CO_2$ pump" which concentrates $CO_2$ at the catalytic site.

C4-plants can be distinguished from C3-plants not only physiologically but also morphologically as shown in Fig. 11. A cross-section through a leaf of a typical C4-plant reveals bundles that are surrounded with a corona of dark green bundle sheath cells. Besides this characteristic leaf anatomy, in most C4-plants the chloroplasts of the bundle sheath cells differ in their ultrastructure from those of the adjacent mesophyll cells. The chloroplast dimorphism is a structural and physiological one, reflecting the spatial segregation of primary $CO_2$ fixation in the mesophyll cells and secondary $CO_2$ fixation and reduction in the bundle sheath cells. Accordingly, chloroplasts of the mesophyll cells are small and contain many grana thylakoids indicating a vivid turnover of the light-driven electron transport. However, they almost completely lack assimilated starch due to the absence of carbon reduction. In contrast, chloroplasts of the bundle sheath cells are large and deposit a considerable amount of assimilated starch resulting from a higher turnover of the PCR cycle. Likewise, these chloroplasts do not exhibit grana thylakoids which indicates that PS II is largely missing, presumably to avoid oxygen evolution associated with PS II activity.

Although the initial carboxylation reaction is always the same, C4-plants are generally classified into three different types according to the nature of the dicarboxylic acid generated in the mesophyll cell and the decarboxylating enzyme system used in the bundle sheath cell. The reaction common to all three types is the carboxylation of PEP which generates oxaloacetate and $P_i$ and occurs in the cytoplasm of the mesophyll cells. Oxaloacetate is then either imported into a mesophyll chloroplast where it is reduced to malate (type I) or remains in the cytoplasm where it is transaminated to aspartate (types II or II).

In the first type of plants malate directly enters a chloroplast in the adjacent bundle sheath cell where it is oxidized and decarboxylated to pyruvate by the NADP malic enzyme (NADP-ME). The advantage in this case is a concomitant import of reducing power and $CO_2$ into the chloroplast performing carbon reduction (see Fig. 11). In the second type of plants aspartate is transported to the cytoplasm of the bundle sheath cells where it is deaminated to oxaloacetate and then decarboxylated by the PEP carboxykinase (PEPCK). The third type of plants realizes more or less a mixture of the above described two ways in that aspartate first enters a mitochondrion of the bundle sheath cell where it is converted to

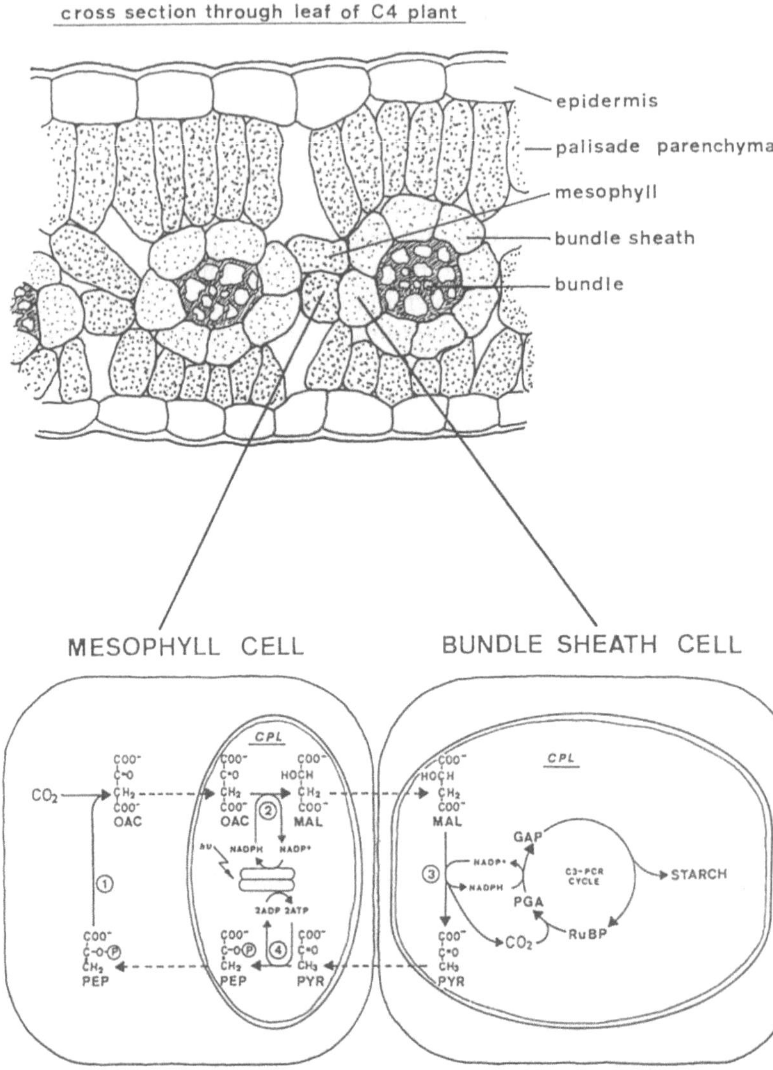


Fig. 11. Schematic diagram of the structural and functional organization of C4 metabolism. ATP consumption associated with the C3-PCR cycle and starch synthesis is not shown. (1) Phosphoenolpyruvate carboxylase; (2) NADP:malate dehydrogenase; (3) NADP malic enzyme (NADP-ME); (4) pyruvate:$P_i$ dikinase. *CPL:* chloroplast.

malate which, in turn, is decarboxylated by NAD malic enzyme (NAD-ME). However, neither in type II nor in type III is hydrogen imported into the chloroplasts of the bundle sheath cells along with $CO_2$.

*3.3.2. Bioenergetics of the C4-PCA cycle:* To complete the C4 pathway outlined above, pyruvate, the remnant of decarboxylation of either malate or aspartate in the bundle sheath cells, must return to the chloroplasts of mesophyll cells where it is regenerated to PEP. Essentially, this causes the additional energy-requirement of the cycle. The regeneration of PEP from pyruvate and $P_i$ is catalyzed by pyruvate:$P_i$ dikinase and involves an unusual activation step requiring the hydrolysis of two phosphoanhydro-bonds of ATP. In the preceding events of activation, the terminal phosphoryl group of ATP is transferred to $P_i$ yielding a diphosphate, and the β-phosphoryl group of ATP is coupled to a histidine residue in the catalytic site of the enzyme. During catalysis, pyruvate:$P_i$ dikinase transfers this phosphoryl group to pyruvate to regenerate PEP. Energetically, the reaction can be summarized as follows:

$$\text{pyruvate} + P_i + 2\,\text{ATP} \rightarrow \text{PEP} + PP_i + \text{AMP} + \text{ATP} \qquad (14a)$$

$$\text{AMP} + \text{ATP} \rightleftarrows \text{ADP} + \text{ADP} \qquad (14b)$$

Since primary fixation and transport consumes one PEP per $CO_2$, two additional ATP are required per reduced carbon in the C4 pathway as compared to C3 photosynthesis. Thus, the total balance of C4 carbon assimilation is 2 NADPH and 5 ATP per assimilated $CO_2$.

However, the additional energy requirement is of no significant importance provided that C4-plants are exposed to high light intensities enabling them to generate supplementary ATP from the light-driven reactions. Therefore, as long as light provides a sufficient energy supply, C4 photosynthesis is advantageous due to the suppression of photorespiration, especially at elevated temperatures, and C4-plants reveal high crop productivity. Among the best known C4-plants of practical use in agriculture are maize, sugar cane and sorghum.

*3.3.3. PEP carboxylase, the key enzyme of the C4-PCA cycle:* Comparing the carboxylation mechanisms catalyzed by PEPcase or Rubisco, the reaction patterns resemble each other in that an electrophilic carboxy-moiety adds to a nucleophilic center of the $CO_2$ acceptor molecule. However, important differences arise since PEPcase uses $HCO_3^-$ as substrate instead of $CO_2$. The advantage is that $HCO_3^-$ is more abundant in the cytoplasm than is $CO_2$. This, in addition to the high substrate affinity of PEPcase ($K_M$ ≈ 7 μM for $HCO_3^-$), brings about that carboxylation can occur even at reduced stomatal conductance and reduced $CO_2$ concentration in the intercellular spaces. Therefore, not $CO_2$ but light is the limiting factor in C4 photosynthesis. Nevertheless, a mechanistic disadvantage of using $HCO_3^-$ as substrate is that it is a rather poor electrophile compared to $CO_2$. Thus, PEPcase like all other carboxylases using $HCO_3^-$ requires activation of the bicarbonate prior to catalysis. It is interesting to note that among the many carboxylases using bicarbonate as substrate, PEPcase is the only one that does not depend on biotin, a mobile carrier of "activated $CO_2$", as a cofactor. This indicates that the reaction catalyzed by PEPcase is mechanistically unique. The only essential cofactor required for catalysis is a divalent cation, usually $Mg^{2+}$.

As shown in Fig. 12 (see also ref. 15), bicarbonate activation is accomplished in the first reaction step in which the phosphoanhydro-bond of PEP is hydrolized and the phosphoryl group is transferred to $HCO_3^-$ creating carboxyphosphate. In addition to this, the dephosphorylation of PEP also dismisses enolpyruvate with a nucleophilic center at the C3-atom which then undergoes nucleophilic attack on the carboxyphosphate in the second reaction step generating oxaloacetate and $P_i$.

### *3.4. Crassulacean acid metabolism (CAM)*

In terms of the biochemical reactions, CAM can be considered to be a variation of the C4-PCA cycle with the apparently small difference that the $CO_2$ prefixation and the carbon reduction cycle are not compartmentalized into different cells but occur at different times, i.e. day and night. However, the physiological and morphological differences between C4 and CAM-plants exceed the biochemical differences by far. A plant performing CAM can adapt to very arid environments and can better support water deficiency and great temperature differences between day and night. Generally, CAM-performing plants have a succulent appearance like species belonging to the Crassulaceae and Cactaceae. For some species of other plant families CAM is a facultative metabolic pathway that can be induced by environmental changes (salt or water stress). Since the phenomenon of CAM is of more interest considering overall cell metabolism and plant physiology, only a very brief description can be given in this context.

*3.4.1. Biochemistry of CAM:* Due to nocturnal $CO_2$ prefixation CAM is associated with a diurnal fluctuation of dicarboxylic acids, in this case almost exclusively malate. At night starch is dissimilated to PEP which is carboxylized by PEPcase and reduced to malate in the cytoplasm. The generated malic acid is then transported into the vacuoles where it is

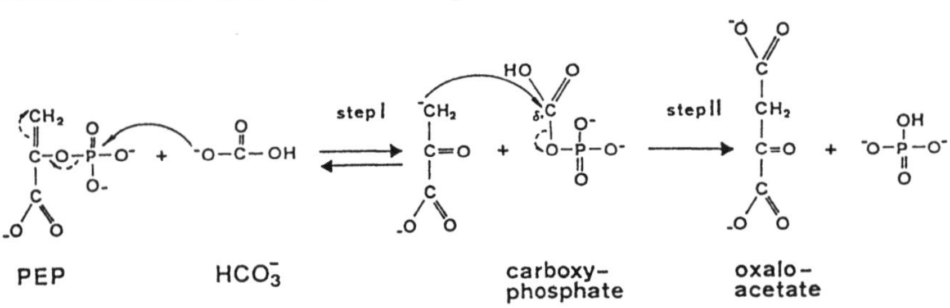


Fig. 12. Reaction mechanism for the carboxylation of PEP.

stored. During day time malic acid is released from the vacuoles and is decarboxylated by NADP-ME or by PEPCK (see also section 3.3.1.). The decarboxylation reaction produces a considerable increase of the $CO_2$ concentration and normal C3 photosynthesis can be performed in the chloroplasts. Depending on the decarboxylation enzyme, PEP is either the direct product of decarboxylation of oxaloacetate or has to be regenerated from pyruvate by pyruvate:$P_i$ dikinase catalysis. In any event, PEP is subsequently fed to gluconeogenesis or a related sequence in order to restore the deposit of carbohydrates. The obvious advantage of this cycle is that the stomates can be kept closed during day time which extremely diminishes the loss of water due to transpiration. (For further reading see ref. 16.)

A fundamental problem of CAM concerns the regulation of carboxylation activity, since PEPcase-catalyzed $CO_2$ prefixation and Rubisco-catalyzed $CO_2$ fixation occur, though in different compartments, in the same cell. Without going into detail, there seem to be at least two basic mechanisms involved in controlling PEPcase- and Rubisco-catalyzed carboxylation: (1) Rubisco is light regulated whereas PEPcase is not, and (2) PEPcase is inhibited by malate, the reduced end product of carboxylation. Therefore, in order to maintain nocturnal PEPcase activity, malic acid is constantly removed from the cytosol and stored in the vacuole. The reverse process during day time causes a massive increase of the malate concentration in the cytoplasm and thus an inhibition of PEPcase.

*3.4.2. Bioenergetics of CAM:* The characteristic feature of CAM metabolism is the drastic increase of malic acid concentration in the vacuoles during night when the stomates are open. This gives rise to the succulent appearance and is also the main energy-consuming process. During night time, malic acid is imported into the vacuoles against an increasing difference in concentration. This is very likely achieved by a proton pump-mediated transport of malate requiring the activity of specific ATPase in the tonoplast. It is estimated that approximately one ATP is consumed per imported malate. At the end of a longer dark period the vacuolic malate concentration can reach values in the order of 0.2 M whereas the cytosolic concentration has to be kept very low in order to prevent inhibition of PEPcase activity (see also ref. 16). Thus at the end of the night, the difference in concentration is extremly high and the energy requirement to sustain further import can hardly be covered by respiratory processes. In contrast, malate efflux during day time is a passive process. Essentially, the massive energy requirement for malate storage and the necessity to cover the additional ATP requirement at night through respiratory processes is the reason for the low overall efficiency of carbon assimilation in CAM-plants. The only CAM-plant of some agricultural importance is pineapple.

### *3.5. Partitioning of assimilated carbon between the chloroplast and the cytosol*

Although carbon partitioning between the chloroplast and the cytosol is not involved in $CO_2$ fixation and reduction, it influences carbon assimilation in a regulatory manner. Eventually, the pathway of carbon metabolism branches off at the level of assimilated triose-P. As already mentioned above, triose-phosphates can either be exported to the cytosol where they are further processed to sucrose or they remain within the chloroplast and are converted to a non-osmotic form, i.e. starch. The decision about which path to follow should meet the instantaneous metabolic situation and requirement. Therefore, in order to achieve a fine adjustment according to different needs, carbon partitioning must be regulated at several levels. In spite of many still unresolved questions and a limited space to discuss these in detail, a few basic principles controlling the distribution of carbon between starch and sucrose synthesis shall be briefly outlined in this section. Virtually common to all plants are the regulatory role of the PGA/$P_i$ ratio in chloroplasts as well as in the cytosol and the concentration of cytosolic fructose-2,6-bisphosphate (F-2,6-BP). (For further reading see also refs. 17 and 18.)

As shown in Fig. 13, chloroplast starch synthesis starts with the formation of F-1,6-BP which is then converted to glucose-1-P involving several intermediate steps. Subsequently, adenylation generates activated ADP-glucose which is the immediate precursor for starch synthesis. In contrast, triose-P exported to the cytosol is used to generate UDP-glucose. Addition of UDP-glucose and F-6-P produces phosphorylated sucrose.

The PGA/$P_i$ ratio exerts its regulatory control at various points: First, the formation of ADP-glucose within the chloroplast is stimulated by PGA and inhibited by inorganic phosphate. Thus, a high PGA/$P_i$ ratio, resulting from a rate of carbon reduction exceeding that of triose-P export to the cytosol, enhances starch synthesis in the chloroplast. Vice versa, an increased stromal $P_i$ level at night favors starch breakdown. Second, a high level of inorganic phosphate in the cytosol inhibits formation of sucrose-6-P and, in addition, positively affects generation of F-2,6-BP. In turn, the produced F-2,6-BP inhibits cytosolic F-1,6-BP formation which is an important precursor for sucrose synthesis. Furthermore, F-2,6-BP can induce cytosolic glycolysis. A high level of $P_i$ in the cytosol therefore effectively suppresses sucrose synthesis in a multiplicative manner. In contrast, a high concentration of PGA in the cytosol inhibits F-2,6-BP formation and thus favors sucrose synthesis.

In summary, inorganic phosphate inhibits starch and sucrose synthesis and induces glycolysis. The ratio of PGA to $P_i$ in either cell compartment is largely dependent on the exchange rate for triose-P and $P_i$ via the phosphate translocator (see section 5.1.1.). During the course of photosynthetic carbon reduction an export of assimilated triose-P from the chloroplast in exchange for cytosolic $P_i$ will on one hand supply and increase su-

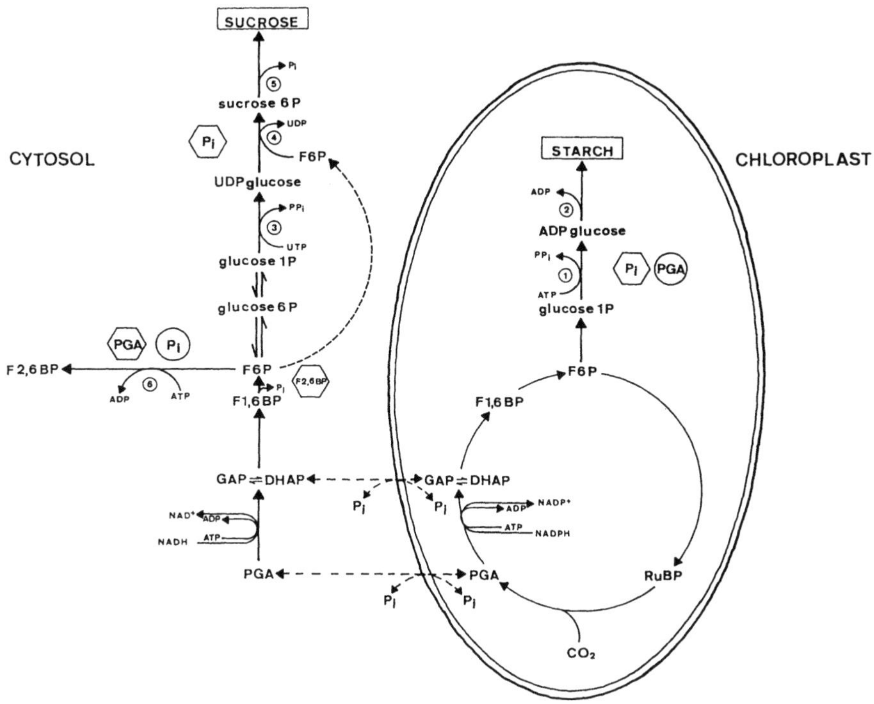


Fig. 13. Simplified scheme of some of the regulatory mechanisms involved in carbon partitioning between starch and sucrose synthesis. Effector molecules leading to activation are shown in circles, effector molecules leading to inhibition are shown in a hexagonal form. (1) ADP-glucose pyrophosphorylase; (2) Starch synthase; (3) UDP-glucose synthetase; (4) Sucrose phosphate synthase; (5) Sucrose phosphate phosphatase; (6) Fructose phosphate-2-kinase.

crose synthesis in the cytosol and on the other hand decrease starch synthesis in the chloroplast and provide $P_i$ for assimilation. Hence, the relative rates of carbon synthesis and export largely determine carbon partitioning between the chloroplast and the cytosol.

## 4. Other metabolic pathways

Besides their principal metabolic role for $CO_2$ reduction and carbohydrate assimilation, chloroplasts perform a number of other essential metabolic pathways. Some of them will be described very briefly in this section.

### *4.1. Reduction of $NO_3^-$ and $SO_4^{2-}$ and assimilation into amino acids*

In higher plants the majority of inorganic $NO_3^-$ and $SO_4^{2-}$ taken up by roots is initially transported to leaves where it is then reduced and assimilated to amino or sulfhydryl moieties of organic compounds. Essentially, the reactions involved occur in chloroplasts. However, the first step of $NO_3^-$ reduction takes place in the cytoplasm. The reduction of nitrate to nitrite is catalyzed by the cytosolic nitrate reductase containing flavine, cytochrom b and molybdenum as cofactors. $NO_2^-$ enters the chloroplast either as protonated neutral $HNO_2$ or via a specific carrier whose activity depends on the stromal pH-value. Within the chloroplast, nitrite is further reduced to $NH_4^+$ by the chloroplast nitrite reductase which catalyzes the transport of hydrogen from reduced ferredoxin to nitrite. The generated ammonia is subsequently assimilated to glutamate by the consecutive action of two enzymes, together representing the GOGAT cycle (see also Fig. 10). In the first reaction $NH_4^+$ is transferred to glutamate thereby yielding glutamine. This ATP-consuming reaction is catalyzed by glutamine synthetase (GS). In the second reaction one molecule of glutamine and one molecule of 2-oxoglutarate are transaminated and concomitantly reduced generating two molecules of glutamate. This reaction is catalyzed by glutamine:oxoglutarate aminotransferase (GOGAT) which also uses ferredoxin as reductant. A NADP- dependent glutamate dehydrogenase only plays a minor role in chloroplasts.

In the case of $SO_4^{2-}$, uptake into chloroplasts relies on the presence of a specific carrier in the envelope mediating a counterexchange of inorganic sulfate and phosphate. Prior to reduction $SO_4^{2-}$ is activated by a twofold reaction with ATP generating phosphoadenosine phosphosulfate. In the following, the "activated $SO_4^{2-}$" is coupled to a carrier protein of low molecular mass and is subsequently reduced to generate a protein bound sulfhydryl group. Transfer of the reduced SH-moiety to the acceptor molecule O-acetylserine then yields cysteine. The consecutive reduction steps involved in protein-bound $SO_4^{2-}$ assimilation require ferredoxin as reductant and do not release any intermediates.

### *4.2. Biosynthesis of proteins*

Chloroplasts possess their own organelle genome as well as their own transcription and translation machinery and thus are capable of own protein biosynthesis. Approximately three-fourths of the chloroplast genes code for products involved in the transcription/translation system, including genes for RNA polymerase subunits, ribosomal RNAs, transfer RNAs, and ribosomal proteins. The other group of genes codes for proteins involved in the photosynthetic electron transport system.

In essence, gene expression in chloroplasts resembles that of prokaryotes in many features: (1) chloroplast transcription units are somewhat similar to bacterial-like operons yielding polycistronic messenger RNAs, (2) chloroplast ribosomes (70S) resemble those of prokaryotes and many of the ribosomal proteins are homologous to those of bacteria, and (3) chloroplast and prokaryotic protein biosyntheses are usually susceptible to chloramphenicol and insensitive to cycloheximide, an inhibitor of eukaryotic protein synthesis [19]. Nevertheless, in spite of the prokaryotic nature several important differences should be mentioned. In contrast to prokaryotic genes, chloroplast genes can carry introns and, concerning the regulation of gene expression, they seem to reveal supplementary promoter sequences in addition to the bacterial-like -10 and -35 sequences. Furthermore, there are distinct differences between the chloroplast and prokaryotic RNA polymerase. Whereas the former requires supercoiled template for transcription initiation and is inhibited by novobiocin, the latter transcribes relaxed DNA and is susceptible to rifampicin [20]. In general, a resemblance to prokaryotic features seems to be more pronounced on the translational than on the transcriptional level. This is supported by the similarity of chloroplast rRNA genes to their bacterial counterparts, revealing only minor differences.

### *4.3. Biosynthesis of lipids*

Membrane lipids are composed of a nonpolar tail of two fatty acids, each esterified to a central glycerol which, in turn, also binds to a polar or charged headgroup. In case of the prevalent galactolipids, the polar headgroup contains either one or two galactose molecules. In contrast, the charged headgroups of phosphatidylglycerol and sulfolipid are comprised of phosphoglycerol and sulfoquinovose, respectively. In plants, medium-chain fatty acids are synthesized to a large extent or perhaps even exclusively in chloroplasts. In addition, it seems very likely that synthesis of the vast majority of plastid membrane lipids takes place in chloroplasts. Virtually all enzymes required for lipid biosynthesis are present in chloroplasts and are associated with the envelope, generally the inner membrane. (For further reading see ref. 3.)

Membrane lipid biosynthesis starts off with glycerol-3-P formation by reduction of DHAP. A twofold acylation of glycerol-3-P with fatty acids involves the consecutive operation of two acyltransferases and yields diacylglycerol-3-P (phosphatidate). Subsequent dephosphorylation which is catalyzed by phosphatidate phosphatase generates diacylglycerol, the central intermediate in chloroplast membrane lipid synthesis. Finaly, galacto-, sulfo- or phospholipids are produced by addition of the activated form of the various headgroups to diacylglycerol, e.g., addition of UDP-galactose to diacylglycerol generates MGDG and is catalyzed by a specific envelope-bound galactosyltransferase (MGDG synthetase). DGDG is formed either by galactosyl transfer from UDP-galactose to MGDG or by transfer of the galactosyl moiety from one molecule of MGDG to a second one. Likewise, UDP-sulfoquinovose plus diacylglycerol yields SQDG. In contrast, phosphoglycerol synthesis requires a preceding activation of diacylglycerol generating CDP-diacylglycerol which reacts with glycerol-3-P to produce PG. It is postulated that desaturation of fatty acids occurs subsequent to lipid synthesis. However, only very little is known on the nature and localization of the enzymes involved in fatty acid desaturation.

### *4.4. Biosynthesis of porphyrins*

The pathway of chlorophyll and heme synthesis includes generation of the same porphyrin frame, namely protoporphyrin IX. This common precursor molecule is synthesized from 5-aminolevulinic acid via porphobilinogen, the first pyrrolic intermediate, and uroporphyrinogen, the first cyclic tetrapyrrole. Chlorophyll synthesis branches off with insertion of $Mg^{2+}$ into protoporphyrin IX. All reaction steps from 5-aminolevulinic acid to protoporphyrin IX production occur in the stroma phase of chloroplasts whereas all subsequent reaction steps seem to be associated with membranes. Probably $Mg^{2+}$ chelation and transformation to protochlorophyllide occur in the envelope as does reduction to chlorophyllide. This is supported by the finding that envelope membranes contain small amounts of both, protochlorophyllide and chlorophyllide. Furthermore, $Mg^{2+}$ chelatase and protochlorophyllide reductase are presumably located in the envelope. Finally, chlorophyll is produced in thylakoids by addition of the phytyl moiety to chlorophyllide. (For further reading see ref. 21.)

### *4.5. Biosynthesis of terpenoids*

Terpenoids are generated by assembly of a varying number of equal C5-units (isopentenyl pyrophosphate). Accordingly, the length of their carbon chain is variable and largely determines the lipophylic character of the molecule. Several important lipophilic membrane components are terpenoids or contain a terpenoid moiety that anchors them in the membrane. Examples are carotinoids, the prenyl side chains of phylloquinone (vitamine $K_1$), plastoquinone, $\alpha$-tocopherol (vitamine E), and the phytyl side chain of chlorophyll. The chloroplast prenyltransferase and other enzymes involved in the biosynthesis of the above mentioned compounds are located in chloroplasts and are probably associated with the inner envelope membrane [3].

## 5. Transport across the envelope

Despite its important role for biosynthesis of lipids (see section 4.3.), the main function of the envelope is to compartmentalize the chloroplast from the surrounding cytosol and hence to control the exchange of material with the cytosol. Therefore, the envelope membrane system which is comprised of two lipid bilayers including an intermembrane space contains a set of proteins that is involved in specific and non-specific translocation processes. Due to the presence of a non-specific pore-forming protein (porine), molecules smaller than 10 kDA can easily penetrate the outer membrane, and it is mainly the inner membrane that limits the permeability of the envelope and carries specific translocator proteins [3]. Likewise, the protein to lipid ratio is much higher for the inner than for the outer envelope membrane. Thus, all three membrane types found in chloroplasts, i.e. outer and inner envelope membrane and thylakoid membrane, are clearly distinct from each other by a different amount and composition of proteins. In contrast, the lipid composition is somewhat similar in that the majority of acyl lipids is comprised of uncharged glycolipids containing a high degree of polyunsaturated fatty acids (see Table 2). Especially the inner envelope membrane and thylakoids exhibit a striking similarity in lipid composition whereas the outer membrane reveals some significant differences, especially a high content of PC. The unusual composition of the outer envelope membrane raises questions on its evolutionary origin since PC is absent in cyanobacterial membranes and glycolipids are absent in all other eukaryotic cell membranes. Functionally, however, the outer and inner envelope form a tight unit that can easily be distinguished from thylakoids by the complete lack of chlorophyll. Besides the acyl lipids the envelope only contains some carotenoids, quinones, and small amounts of protochlorophyllide and chlorophyllide probably due to its involvement in chlorophyll biosynthesis.

As outlined in sections 3 and 4, chloroplasts are involved in several essential metabolic pathways. Among the most important are reduction of $CO_2$, $NO_2^-$ and $SO_4^{2-}$ and their assimilation into organic matter such as carbohydrates and amino acids. In order to fulfill these multiple functions within cell metabolism, selective uptake of inorganic compounds and controlled exchange of metabolic intermediates with the cytoplasm has to be provided by the envelope. In general, the envelope has a high permeability for gases, e.g., $CO_2$ and $NH_3$, and a lower permeability for ions. However, the latter can be specifically enhanced by the presence of carrier mechanisms, which seems to be the case for $NO_2^-$ and $SO_4^{2-}$, and probably also for $Mg^{2+}$ and $Ca^{2+}$. Although the permeability for protons is generally low, it is higher than that of thylakoids and proton leakage across the envelope is presumably mediated by salts of weak acids. This leads to the problem that a passive flux of protons could partly counteract the light-driven proton translocation since the pH-value of the stroma (pH ≈ 8) is raised above that of the surrounding cytosol (pH ≈ 7) during illumination. Whether or not a passive proton influx is significant and by which mechanism it is compensated is still not fully understood. The existence of a light-dependent active proton extrusion and/or compensation via discrete connecting channels between the thylakoid lumen and the intermembrane space of the envelope are discussed. (For further reading see refs. 3 and 22.)

The following part of this section will focus on different carriers involved in exchange of metabolic intermediates between the chloroplast and the cytoplasm. In addition, a second important function of the envelope, i.e. its role in mediating the protein import of nuclear encoded chloroplast polypeptides, will be described in the last part of this section.

### *5.1. Translocation of metabolites*

The description of metabolic pathway in sections 3 and 4 readily shows that chloroplasts eventually attempt to specifically exchange metabolic intermediates with the cytoplasm. Of course the exchange of carbohydrates plays a central role and depending on the specific pathway of carbon metabolism (see section 3) different intermediates have to be translocated. In general, chloroplasts of C3-plants are $P_i$ importing and triose-P exporting. Moreover, due to the occurrence of photorespiration they also require a mechanism for glycolate and glycerate translocation. Chloroplasts in bundle sheath cells of C4-plants (type I) require malate import and pyruvate export. In contrast, chloroplasts in mesophyll cells of C4-plants import pyruvate, $P_i$, and in some cases also oxaloacetate, whereas they export PEP and possibly malic acid (C4-plants, type I). As will be described in the following, only a couple of carrier proteins exist that satisfy the diversity of demands. However, in different plants they might be present in different isoforms with regard to substrate affinity and activity. (For further reading see refs. 3, 23 and 24.)

*5.1.1. The phosphate translocator:* According to its function the correct name for this carrier protein is actually phosphate/triose-P/phosphoglycerate translocator. However, for convenience it is simply called phosphate translocator. The abundance and activity of this translocator is predominantly responsible for the export of assimilates to the cytosol. The phosphate translocator accounts for approximately 15–20% of the total protein content of the envelope and reveals a very high molecular activity of transport with a turnover rate of approximately 80 per second.

The phosphate translocator preferentially accepts phosphate or phosphorylated C3 carbohydrates at its binding site and the substrates are translocated on the basis of strict counterexchange. Therefore, a stable level of the total phosphate content (inorganic plus esterified) is maintained within the chloroplast. Carbohydrates with a deviating number of C-atoms or a phosphorylated site other than a terminal C-atom are poorly or not at all accepted.

Basically, two modes of exchange are mediated by the phosphate translocator (see Fig. 14). First, during illumination triose-P, mainly DHAP, is exported from the chloroplast and $P_i$ is imported in exchange. This triose-P/$P_i$ shuttle supplies the cytosol with assimilates where they are mainly used to synthesize sucrose, the main metabolite form for long distance transport between source and sink via the phloem. The second mode of exchange mediated by the phosphate translocator is an exchange of triose-P and P-glycerate. The triose-P/PGA shuttle predominantly works in one direction during illumination, i.e. triose-P export in exchange to PGA import. The unidirectionality during illumination prevents a depletion of fixed $CO_2$ in form of non-reduced PGA from the chloroplast. The importance of this shuttle is to establish an indirect transport of ATP and reducing power from the chloroplast to the cytosol in form of either NADPH or NADH, depending on the cytosolic dehydrogenase. In the dark, however, the unidirectionality of the shuttle is lost and transport of PGA in both ways is feasible. At a first glance it seems puzzling that export of PGA can be specifically suppressed in the light, especially since after a longer period of illumination its concentration is approximately ten times higher than that of triose-P. However, the apparent discrepancy can be explained taking into account that suppression of PGA export is closely related to establishment of the light-induced potential difference of protons. The phosphate translocator preferentially accepts double negatively charged anions as substrate, and P-glycerate is overwhelmingly present as trivalent anion at physiological pH, in contrast to the divalent triose-P and $P_i$. This implies that a counterexchange of triose-P and PGA is accompanied by an indirect proton translocation. A presumed export of divalent PGA would cause a concomitant export of protons against the already existing potential difference of protons. In addition to this powerful restriction mechanism, export of PGA during illumination also seems to be inhibited by the raised level of stromal $Mg^{2+}$ that is established.

The phosphate translocator is nuclear encoded and is synthesized as a larger precursor in the cytosol. It is posttranslationally imported into the chloroplast where it is processed to its mature form. The active carrier is comprised of two identical subunits with a molecular mass of 36 kDa as determined from the amino acid sequence. Presumably, each of the subunits spans the inner envelope membrane with 6 or 7 $\alpha$-helices, some of which have amphiphilic character and could therefore generate a hydrophilic lining of the putative translocation channel. It is conceivable that two cationic amino acid residues, one lysine and one arginine residue, that are present in the membrane integral part of the protein are involved in binding of the double negatively charged substrate. Furthermore, it is suggested that the 62 kDa dimer exhibits only one substrate binding site

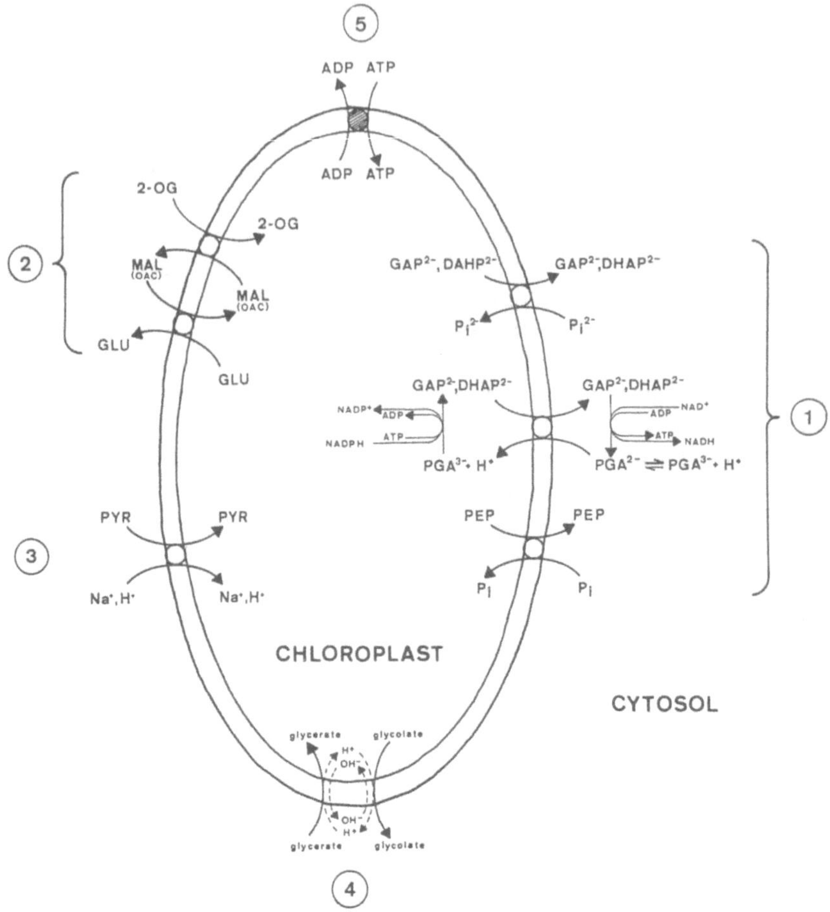


Fig. 14. Different types of translocators of the chloroplast envelope. (1) Phosphate translocator; (2) Dicarboxylate translocator(s); (3) Pyruvate translocator; (4) Glycolate/glycerate translocator, (5) Aldenylate transport. For details see text.

which is probably alternatively exposed to either side of the inner envelope membrane. This structural organization appears to be suited to assure strict counterexchange.

*5.1.2. The dicarboxylate translocator(s):* An exchange of a number of dicarboxylic acids between the chloroplast and the cytosol is either required for photorespiratory nitrogen cycling, for C4 carbon metabolism, or for indirect transport of reducing power. Accordingly, dicarboxylate exchange mainly involves a translocation of malate, oxaloacetate, 2-oxoglutarate, glutamate, and aspartate and is mediated by different carriers with overlapping specifities. Dicarboxylate translocation occurs essentially via counterexchange, however, less strictly than the phosphate translocator.

Exchange of dicarboxilic acids is a crucial step in the photorespiratory nitrogen cycle. In the course of recapturing released $NH_3$, chloroplasts support a constant inward flux of 2-oxoglutarate and an opposing efflux of glutamate. In essence, an exchange of 2-oxoglutarate and glutamate is provided by the concomitant action of two different shuttle mechanisms, each using malate (or oxaloacetate) as counterpart (see Fig. 14). Thus, the combined activity of a 2-oxoglutarate-importing/malate-exporting shuttle and a malate-importing/glutamate-exporting shuttle leads to a net exchange of 2-oxoglutarate and glutamate. Obviously, a net translocation of malate is also feasible as soon as the two translocators work with different efficiency. Therefore, the two translocators provide a system sufficient for transport of 2-oxoglutarate, glutamate, malate, and to some extent possibly also for oxaloacetate since this is bound with roughly the same affinity as malate. However, the contribution to oxaloacetate transport is probably very low because the intracellular concentration of malate is thought to be much higher than that of oxaloacetate, leading to competitive inhibition of oxaloacetate transport. In any case, a specific oxaloacetate translocator with a very high substrate specifity is also present and establishes an additional shuttle mechanism for oxaloacetate. Likewise, this translocator uses malate as counterpart. Its activity is crucial for C4 metabolism and in C3-plants it is involved in transport of reducing power across the chloroplast envelope.

Furthermore, a specific glutamine translocator that functions as glutamine-importing/glutamate-exporting shuttle has been identified in plants which are capable to rescue photorespiratory $NH_3$ in the cytosol via the cytosolic isoform of glutamine synthase.

*5.1.3. The pyruvate translocator:* Chloroplasts of typical C3-plants such as spinach support only a very slow transport of pyruvate across the envelope. In contrast, pyruvate translocation is markedly enhanced in C4 mesophyll chloroplasts. In all C4-plants the mesophyll cells are the site of $CO_2$ prefixation and accordingly their chloroplasts act as pyruvate-importing and PEP-exporting compartments. Likewise, a rapid transport of PEP seems to be unique to chloroplasts of mesophyll cells of C4-plants. However, pyruvate and PEP are not exchanged by the same translocator. PEP is probably exported by a modified phosphate translocator whereas pyruvate uptake is suggested to be mediated by a distinct translocator. Pyruvate uptake seems to work on the basis of a cation/pyruvate symport, in which case the cation can either be provided by a proton or by $Na^+$. In C4-plants of type I (NADP-ME) an active pyruvate uptake is apparently coupled to the light-induced potential difference for protons generated across the envelope. In C4-plants of type II and III (PEPCK and NAD-ME) pyruvate uptake is presumably driven by a potential difference for $Na^+$ across the envelope. Thus, the pyruvate translocator seems to be regulated by slightly different means among the different types of C4-plants.

*5.1.4. The glycolate/glycerate translocator:* Both, glycolate and glycerate, readily permeate across the envelope, a fact that is essential for photorespiratory carbon cycling. For a long time it was assumed that glycerate and glycolate unspecifically penetrate the envelope as neutral compounds in the form of the undissociated acid. However, this idea is not easy to reconcile with the finding that glycolate and glycerate transport is sensitive to specific inhibitors. Meanwhile evidence has substantiated that translocation of both compounds is mediated by a specific carriers. Whether glycolate and glycerate are translocated by the same or by two different carriers is still not clarified unambiguously, even though some evidence favors the existence of only one translocator. In any case, glycolate and glycerate translocation is stimulated by the light-induced electrochemical potential difference for protons generated across the envelope membrane and the mechanism is presumbly either a proton/carboxylate symport or a hydroxyl/carboxylate antiport. A direct counterexchange of glycolate with glycerate is also conceivable.

*5.1.5. Glucose transport:* During longer dark periods assimilated carbon that is stored in the chloroplast in the form of starch becomes mobilized. Starch breakdown can occur via two modes: (1) Phosphorolytic degradation yields glucose-1-P that is subsequently subjected to the oxidative pentose phosphate cycle generating triose-P or P-glycerate which in the dark can both be exported via the phosphate translocator. (2) Hydrolytic starch degradation produces glucose which seems to be exported from the chloroplast by a glucose translocator. With respect to specifity, substrate affinity, and activation energy this translocators exhibits striking similarities to glucose carriers of human erythrocytes and rat liver cells.

*5.1.6. ATP/ADP transport:* A specific translocator for ATP and ADP is present in the inner envelope membrane of chloroplasts. However, its activity is low and its physiological role remains obscure. This carrier mediates a counterexchange of adenylates and exhibits a high specificity for external ATP. Therefore, it presumably represents an ATP-importing/ADP-exporting system which supplies chloroplasts with respiratory ATP whenever needed, e.g., during longer dark periods or during biogenesis.

*5.1.7. Amino acid transport:* Generally, neutral amino acid can penetrate the envelope membrane by simple diffusion very slowly but nevertheless at a rate sufficient to supply protein synthesis. As outlined above (see section 5.1.2.) aspartate, glutamate, and glutamine are transported across the envelope membrane via specific translocators. Specific transport of other amino acids remains to be demonstrated conclusively. The presence of two translocators mediating the transport of glycine and serine could only be demonstrated in pea chloroplasts. There is also some evidence for leucine and isoleucine carriers in pea chloroplasts.

*5.2. Import of proteins*

Approximately three-fourths of the chloroplast proteins are nuclear encoded and are transcribed and translated in the cytoplasm yielding larger precursors that have to be imported into the chloroplast. The cytoplasmic precursor proteins are usually extended at the N-terminus and are processed to the mature protein by chloroplast peptidases. The precursor protein must contain all the information needed for import into the chloroplast and for precise intraorganellar targeting of the protein. According to the structural subdivisions of chloroplasts, generally four different types of protein targeting have to be distinguished: the imported protein can either be integrated into a membrane, i.e. envelope or thylakoid, respectively, or can be directed towards the stroma or thylakoid lumen. In the latter case, the imported protein has to be transported across a second membrane in order to reach its final location. (For further reading see also refs. 25-28.)

*5.2.1. Import and intraorganellar targeting of inner envelope membrane proteins:* The import and intraorganellar targeting of stroma proteins, thylakoid-integral proteins, and thylakoid lumen proteins have been studied with some emphasis during the past. Concerning the mechanism of correct insertion of nuclear-encoded proteins into the envelope, however, progress has been achieved only very recently. In fact, the only information available refers to the nuclear-encoded phosphate translocator of the inner envelope membrane. Like all other nuclear-encoded chloroplast proteins, the phosphate translocator is synthesized as a larger cytoplasmic precursor with a N-terminal extension of roughly 80 amino acids and is processed by a specific peptidase after import into the chloroplast. The transit sequences of the two phosphate translocators investigated so far (from pea and spinach, respectively) both exhibit an amphiphilic α-helix which is thought to be an essential structural feature for import and correct targeting. This is of special interest since an amphiphilic α-helical structure was already determined for mitochondrial transit sequences, and therefore had long before been postulated to play a crucial role for protein uptake into choroplasts. Unfortunately, in all other cases studied previously, it could not be identified unequivocally. However, these cases mainly dealt with cytoplasmic precursors of chloroplast proteins directed towards the stroma or the thylakoid membrane (see below).

*5.2.2. Import and intraorganellar targeting of stroma proteins:* The nuclear-encoded small subunits of Rubisco (SSU) was the first example studied of protein import into chloroplasts. Meanwhile the transit sequences of many nuclear-encoded chloroplast proteins are known and have been compared in order to determine common features guiding the post-translational import. Unfortunately, this attempt revealed a high variability in length and very low degree of sequence homology among different chloroplast transit peptides. However, linked to other proteins all transit peptides correctly mediate the import into the stroma, and therefore, they must contain sufficient information in some way. The only commonly shared properties that are recognizable are a depletion in acidic amino acids, a low content of basic amino acids, and an enrichment in hydroxylated amino acid residues (especially serine) as well as hydrophobic residues (especially alanine). Moreover, the search for commonly exposed secondary structures largely failed. However, there seems to be a tendency to form an amphiphilic β-strand on the C-terminal half of the transit sequence, but this is more likely to be important as recognition site for the stromal processing peptidase than for import since deletion of this part does not alter the import significantly but instead affects the processing of the precursor. Ultimately, the only obvious common feature of chloroplast transit peptides is a general amphiphilicity which is not necessarily related to a specific secondary structure. This might imply a possible role of membrane lipid interaction in the course of protein import since amphiphilic proteins are known to interact directly with biological membranes. In additon to this, substantial evidence indicates the involvement of receptor proteins located in the outer envelope membrane in binding and mediating protein import into chloroplasts. At present, it cannot be decided whether protein-protein or protein-lipid interactions or both are crucial for binding and importing the chloroplast precursor protein. In any case, protein uptake is an energy dependent process and it is now fairly well established that energy must be provided in the form of ATP and unlike the mitochondrial system not in the form of a proton motive force. Presumably, ATP is required at the outside of the envelope for binding to the receptor as well as at the inside for translocation of the protein. Once the precursor protein is imported into the chloroplast, the transit sequence is cleaved by a stromal processing peptidase generating the mature protein.

*5.2.3. Import and intraorganellar targeting of integral thylakoid proteins*: The best studied examples of nuclear-encoded thylakoid-integral proteins are the chl *a*/*b*-binding proteins belonging to the family of CAB proteins. The mechanism for precursor import is virtually the same as described above for nuclear encoded stromal proteins. Accordingly, the transit sequence is removed by a stromal processing peptidase subsequent to the translocation across the envelope. Chimeric CAB precursors in which the mature CAB protein has been fused to another chloroplast transit sequence are not only imported into the chloroplast but also correctly integrate into the thylakoid membrane with equal efficiency. This indicates that the feature directing the CAB protein to the thylakoid membrane does not rely on the transit sequence but is inherent to the mature protein. However, the integration does require the presence of a stromal protein, a so-called unfoldase, which is thought to aid the CAB protein to maintain an integration-competent conformation. This is in line with the finding that integration depends on the presence of ATP.

*5.2.4. Import and intraorganellar targeting of thylakoid lumen proteins:* Import and intraorganellar targeting across the thylakoid membrane into the lumen have been investigated mainly for three nuclear encoded hydrophilic proteins, i.e. plastocyanin, and the 33 kDa and 23 kDa proteins involved in water oxidation. A common principle to all three is the two-stage mechanism reflected by a bipartite presequence. The first part of this presequence is equivalent to the transit sequence of other chloroplast precursors and directs the transport across the envelope membrane. The second part of the presequence, the transport sequence, enables transport across the thylakoid membrane. The transfer peptide resembles bacterial-like signal sequence and reveals a positively charged N-terminal sequence, a hydrophobic core, and a rather hydrophilic C-terminal region. In total, it is more hydrophilic than the transit peptide. In contrast to the transit sequence which is cleaved by a stromal processing peptidase, the transfer sequence is degraded by a thylakoidal processing peptidase releasing the mature protein into the lumen.

The presence of transit and transfer sequences in all three precursor proteins implies a generality of the two-stage mechanism for lumen oriented proteins. However, there also seem to be subtle mechanistic differences. Efficient import and targeting of the 33 kDa and 23 kDa proteins requires a proton motive force while the transport of plastocyanin does not. The presence of ATP can only substitute for a missing proton motive force. Likewise, a dependence on the presence of stromal extract is different for the three hydrophilic lumen proteins. It seems possible that a stromal chaperone-type protein is required, at least for an efficient import of the 33 kDa protein.

In summary, the import into chloroplasts very likely relies on a feature common to all transported proteins regardless of their final location (stroma, thylakoid or thylakoid lumen) and this is provided by the transit peptide. The transit peptide could therefore be viewed as an eukaryotic sorting system mediating the import. In contrast, the correct targeting within the chloroplast rather depends on a prokaryotic (chloroplast) sorting system, e.g. the prokaryotic nature of the transfer peptide sequence. However, many questions concerning the import and location of nuclear encoded chloroplast proteins remain to be investigated.

## 6. Photoregulation of chloroplast function

Numerous regulatory mechanisms are involved in optimizing the function of chloroplasts. This requires various adaptation mechanisms to different parameters (e.g., light conditions, temperature, $CO_2$ supply, etc.). Generally, proteins provide an ideal material for the fine-tuning of each individual reaction step involved in photosynthesis. The regulatory mechanisms can be categorized in several ways, e.g. in terms of the structural organization level (entire plant, specific tissue, single cell or organelle), in terms of the regulatory level (regulation of gene expression at transcriptional or translational level, regulation of enzyme activity at post-translational level), in terms of the mediating signal (response to light, substrate concentration, etc.), or in terms of the considered time scale (long- or short-term). Considering the whole plant, optimization of chloroplast function and adaptation to different light conditions include mechanisms such as orientation of leaves depending on the direction of irradiance or induction of a facultative pathway of carbon fixation, e.g. CAM. On the cell level, regulatory sequences can determine the number, movement and orientation of chloroplasts. Moreover, a coordinated expression of nuclear and chloroplast encoded genes is required to optimize chloroplast function. In general, expression of nuclear genes coding for chloroplast proteins is regulated at the transcriptional level, whereas expression of chloroplast genes is usually regulated in the organelle at a post-transcriptional level [36]. In addition, an intraorganellar regulation of the rate of photosynthetic electron transport and of the enzyme activity involved in $CO_2$ fixation and reduction occurs. Regardless of all details of sometimes highly sophisticated regulatory mechanisms, it should be emphasized that light is the most important regulatory signal for chloroplast function and development.

This section will only consider light-mediated regulation of enzyme activity and enzyme abundance within the chloroplast. The former phenomenon can generally be viewed as a reversible short-term adaptation. In contrast, regulation of enzyme abundance is achieved by controlled gene expression and coordinated assembly of the gene products. Naturally, this is rather a long-term adaptation process. (For further reading see refs. 29–38.)

### *6.1. Short-term adaptation and photoregulation of activity*

In terms of the overall function, the role of virtually all regulatory mechanisms is to balance the rate of ATP and NADPH production with that of consumption, i.e. to couple the rate of light-driven electron transport and proton translocation to the rate of carbon fixation and reduction [30]. Since light conditions can change very rapidly, chloroplasts need to accomplish reversible short-term adaptation. Optimization of ATP and NADPH production includes mechanisms regulating the distribution of light-energy between PS I and PS II as well as processes involved in dissipation of excess light-energy. A coupling of ATP and NADPH consumption is achieved by enzyme regulation of the photosynthetic carbon reduction cycle.

*6.1.1. Light-dependent modulation of photosynthetic electron transport and proton translocation across the thylakoid membrane:* Maximum efficiency of linear electron transport from $H_2O$ to $NADP^+$ requires a well balanced excitation of the two photosystems. In higher plant chloroplasts, the structural segregation of thylakoids and partitioning of PS II into appressed membrane regions leads to a preferential excitation of PS I at very low light intensities (shade). Excitation of a given photosystem can, however, be enhanced by increasing the size of the absorbing cross-section provided by the associated light-harvesting antenna complex. Thus, the large light-absorbing cross-section of LHC II (accounting for 50% of the total chlorophyll) associated with PS II is helpful to ensure an efficient exploitation of excitation energy at very low light intensities. However, the situation changes at slightly elevated light intensities causing a preponderant excitation of PS II. Under these circumstances, reduced plastoquinol accumulates since the rate-limiting step in linear electron transport is the PS I-dependent reoxidation of reduced plastoquinol generated by PS II. Therefore, mechanisms that regulate the distribution of excitation energy funneled to PS II and PS I under light-limiting conditions are crucial in order to provide a maximum efficiency of linear electron transport. Photoinduced reversible phosphorylation of thylakoid membrane proteins, especially of LHC II and PS II proteins, plays a major role in the fine adjustment of the relative excitation of PS II and PS I [31]. Although several PS II proteins, e.g., CP43, D1, D2 and a 9 kDa protein, can be phosphorylated in the light, LHC II is the predominant target of photoinduced phosphorylation. As mentioned above (see section 2.3.2.), LHC II can be divided into an inner and outer part. While the inner part remains tightly associated with PS II at all times, phosphorylation of the outer part causes a detachment from PS II, subsequent migration from appressed to non-appressed membrane regions, and reattachment to PS I, thereby leading to a change in the relative absorption cross-section of PS II and PS I. In addition, phosphorylation causes a repulsive force between adjacent LHC II complexes on opposing grana membranes. Dissociation decreases energy transfer between opposing LHC II complexes, and therefore reduces cooperativity among different PS II centers in the grana partition. In total, the regulatory mechanism exerted by protein phosphorylation is referred to as state transition between state 1, representing a state of large PS II antenna size and high PS II cooperativity among opposing centers in the grana thylakoids, and state 2, characterized by low PS II cooperativity and reduced PS II antenna size. However, although the decreased interaction of phosphorylated LHC II with PS II certainly reduces excitation of PS II, there still is a lack of direct evidence that the increased interaction with PS I actually enhances transfer of excitation energy to PS I. Nevertheless, phosphorylation is induced by overexcitation of PS II as compared to PS I and is catalyzed by one or more redox-sensitive chloroplast kinase(s). Vice versa, dephosphorylation is catalyzed by (a) chloroplast phosphatase(s). Both, phosphorylation and dephosphorylation, are controlled by the redox state of an intermediate component. There is still some uncertainty concerning the nature of the communicating redox component which might be the plastoquinone pool itself, or instead some component of the Cyt b6f complex. In any event, the chloroplast kinase(s) catalyze(s) the phosphorylation of N-terminal threonine residues facing the stroma. Since phosphorylation introduces negative charges to the stroma exposed surface, the molecular basis of the mechanism is thought to be mediated either by direct electrostatic repulsion or by an induced local change in secondary protein structure or by both [31].

There is some evidence indicating that the catalytic activity of the kinase(s) may be regulated by an additional control system which depends on the ATP/NADPH ratio within the chloroplast. According to this, a decreased electrochemical potential difference of protons resulting from an increased demand for ATP as compared to NADPH stimulates LHC II phosphorylation. It is tempting to speculate that the presumed regulatory role of the Cyt b6f complex for kinase activation might be related to this mode of control since the Cyt b6f complex is also involved in cyclic electron transport. An unfavorable ATP/NADPH ratio can be overcome by cyclic electron flow around PS I which increases the $\Delta$pH and thereby promotes ATP production without generation of NADPH (see section 2.1.). Consequently, phosphorylation of thylakoid membrane proteins could be induced by and respond to an altered metabolic requirement for ATP and NADPH, and a changed distribution of excitation energy between PS II and PS I could be involved in the branching between linear and cyclic electron transport.

Regarding the role of PS I in modulating photosynthetic electron transport, two basic levels could obviously accommodate regulation, i.e. the rate of reoxidation of the reduced acceptor side and the rate of rereduction of the photooxidized donor side. The redox reactions at the reducing side are generally not rate limiting under continuous illumination provided that a sufficient supply of $CO_2$ is sustained. Over a wide range of light intensities the rate of photosynthetic carbon reduction can cope with PS I electron transport and therefore effectively circumvents an accumulation of reduced redox components at the acceptor side of PS I. However, this is different during an initial dark/light transition which is partly due to the light-dependence activation of the enzymes involved in carbon reduction (as will be discussed below), but also depends on the light-dependent activation of the ferredoxin:NADP oxidoreductase (FNR) [32]. FNR is inactive in the dark and is activated during dark/light transitions by light-induced alkalization of the stroma. Subsequently, FNR binds to the membrane and undergoes a conformational change enabling catalysis. Inhibition of FNR activity in the dark seems necessary in order to prevent a waste of light-generated NADPH since FNR can, in principle, also catalyze the reverse reaction of NADPH oxidation. Moreover, a second intriguing question with respect to the regulatory role of FNR for modulation of photosynthetic electron transport concerns, once more, the branching between linear and cyclic electron transport. FNR resides at the strategic point from which electrons are either returned to the Cyt b6f complex or are transferred to $NADP^+$. Although the mechanistic details are largely unknown, a modulation of the catalytic activity of FNR again appears to be regulated by either protein phosphorylation and/or by conformational changes depending on the electrochemical potential difference of protons across the thylakoid membrane.

The other conceivable level of regulation within PS I is a modulation of the kinetics of rereduction of photooxidized $P700^+$ by plastocyanin following different light conditions. Compared to plastoquinone diffusion within the membrane, diffusion of the hydrophilic plastocyanin within the thylakoid lumen is fast. The rate of rereduction of the PS I donor side is therefore determined by the rate of the redox reaction between plastoquinol and the Cyt b6f complex which depends on both plastoquinol concentration and pH-value of the thylakoid lumen. Since reoxidation of plastoquinole is coupled to a release of protons into the thylakoid lumen, the rate is naturally slowed by an increasing acidification of the thylakoid lumen. Therefore, the pH-value of the thylakoid lumen has to be balanced so as to drive ATP synthesis on one hand without restricting the rate of plastoquinol oxidation or $P700^+$ reduction, respectively, on the other hand. Likewise, the activity of the chloroplast ATPase is light-regulated via energetization of the thylakoid membrane on one hand and via the thioredoxin system (see section 6.1.2.) on the other hand. As is the case for FNR, inhibition of the ATPase in the dark is necessary in order to suppress a reverse catalysis of ATP hydrolysis.

Essentially, the above described control mechanisms optimize the system under conditions of limiting light. However, under saturating or oversaturating light conditions additional problems arise and supplementary regulating mechanisms are required. Oversaturating light intensities can cause photoinduced damage, e.g. photooxidation of pigments or inactivation of photosynthetic electron transport. Although the mechanistic details still remain to be elucidated, it is well documented that especially PS II is susceptible to photoinhibition. The strategies to prevent PS II from photoinduced damage at high light intensity include mechanisms to reduce the amount of absorbed energy and to dissipate excess energy. The former mechanism corresponds to a state 1-state 2 transition as described previously in this section. The latter mechanism involves non-photochemical dissipation processes within the antenna, e.g., fluorescent or non-radiative dissipation as well as photochemical pathways of dissipation that effectively remove electrons from the reduced acceptor side of PS II. In this context, mechanisms such as electron cycling around PS II, reduction of $O_2$ by PS I (Mehler reaction) or reduction of peroxide by ascorbate peroxidase are discussed. Reduction of peroxide is also part of the ascorbate- glutathione cycle which can in several ways be interpreted as a dissipative cycle between PS II and PS I [30]. The crucial reaction in this cycle is a de-epoxidation of LHC II carotenoid (formation of zeaxanthin from violaxanthin) thereby oxidizing ascorbate and removing oxygen radicals that could, for example, result from a Mehler-type reaction. Dihydroascorbate is rereduced by glutathione which itself is subsequently regenerated by NADPH. Altogether, the cycle maintains electron transport through PS II and PS I, consumes NADPH and produces zeaxanthin which, in turn, is suggested to be somehow involved in non-photochemical dissipation of excess energy. Although the signal-transducing mechanisms initiating the dissipative pathways are not always completely understood, it has been shown that all dissipative processes require a proton motive force or an acidic pH-value of the thylakoid lumen.

In summary, light-dependent modulation of the photosynthetic electron transport intends to maximize ATP and NADPH production at light limiting conditions. At saturating and oversaturating light conditions, dissipative reactions play an additional important role in order to prevent photoinduced damage. Generally, light-dependent modulation of electron transport is transmitted by feed-back mechanisms.

*6.1.2. Light-dependent modulation of the rate of carbon fixation and reduction:* In contrast to the regulatory mechanisms described in the former section, light-dependent modulation of the rate of carbon assimilation is, in general, subjected to feed-forward type of control. Apparently, feedback mechanisms triggered by the products of assimilation are rather involved in metabolite partitioning (starch versus sucrose synthesis, see section 3.5.) than in determining the rate of carbon fixation. Carbon assimilation requires $CO_2$, ATP and NADPH. Provided $CO_2$ is not a limiting factor, the enzymatic activity of the PCR cycle can be adjusted to cover the rates of ATP and NADPH production over a wide range of irradiances. Therefore, even in changing light conditions the ratio of ATP/ADP and NADPH/$NADP^+$ are kept fairly constant. Basically, adjustment of the activity relies on several intervening variables such as the $Mg^{2+}$ concentration, the pH-value or the redox state of the stroma [33]. Thus, the light-dependent regulation of the activity of several PCR enzymes is due to a feed-forward regulation via products of the light-driven reactions of photosynthetic electron transport.

According to its role in $CO_2$ fixation, regulation of Rubisco is exerted at multiple levels which differ mechanistically from the regulation of the other light-regulated PCR enzymes. For this reasons, regulation of Rubisco will be described in more detail. However, it is impossible to consider all levels of regulation, and the following description will only focus on the prevailing regulatory mechanisms, i.e. control by CA1P (2-carboxy-D-arabinitol 1-phosphate) and control of the activation step (see also refs. 7 and 10).

The activity of Rubisco is very effectively suppressed by the competitive inhibitor CA1P that mimics the instable C6 intermediate of $CO_2$ fixation and binds with high affinity to the activated catalytic site. Changes in the binding of CA1P are brought about by controlling the concentration of CA1P rather than by controlling its affinity. In the dark, the concentration of CA1P exceeds that of Rubisco whereas in the light it decreases to a steady state level that depends on the light intensity. The light-dependent level of CA1P is thought to be controlled by coupling the rate of CA1P degradation to the turnover of PS II; however, the mechanistic details are not understood. Although control via CA1P is a powerful regulation system, it is apparently not present in all plant species, e.g., it could not be detected in spinach. In any event, Rubisco is subjected to another level of light-dependent regulation which controls the activation reaction. As outlined in section 3.1.3., Rubisco must be activated prior to catalysis. In this reaction an activator $CO_2$ molecule ($^{A}CO_2$) binds to a lysine residue in the catalytic site and subsequently adds a $Mg^{2+}$ cation to form the activated ternary complex [$^{act}E \times {}^{A}CO_2 \times Mg^{2+}$] (see Eqn. 10 in section 3.1.3.).

In the presence of sufficient $CO_2$ and $Mg^{2+}$, activation is a function of the stroma pH-value. The pH-dependency is even more pronounced in the presence of RuBP, which could be due to formation of a Rubisco-RuBP complex involving a protonation reaction and probably takes place prior to the activation step. There is strong evidence that the addition of $^{A}CO_2$ and $Mg^{2+}$ to Rubisco (or the Rubisco-RuBP complex) is catalyzed by "Rubisco activase" in an ATP-consuming reaction. Presumably, "Rubisco activase" strongly participates in initiating and maintaining Rubisco activity. Thus, the active state of Rubisco is light-modulated by the $Mg^{2+}$, $H^+$ and ATP concentration of the stroma and therefore depends on the light-driven reactions.

In addition to Rubisco, several other PCR enzymes are light-regulated via the level of $Mg^{2+}$ cations in the stroma. However, the four other light-regulated enzymes, i.e. GAPDH, FBPase and RuPKase, are subjected to an additional and different control mechanism in which they are activated via the ferredoxin/thioredoxin system. This system represents a cascade of redox reactions in the course of which the redox state of ferredoxin determines the activity of a given enzyme [34]. The redox sequence starts with PS I-mediated reduction of ferredoxin. Reduced ferredoxin then reduces the ferredoxin:thioredoxin reductase (FTR) which, in turn, reduces thioredoxin which finally reduces the target enzyme thereby converting it from its inactive into its active form:

$$\begin{array}{lllllll} & \text{PS I} & & & & & \\ & \downarrow & & & & & \\ \text{Fd}^{\text{ox}} & \rightarrow & \text{Fd}^{\text{red}} & & & & (15a) \\ & & \downarrow & & & & \\ & \text{FTR}^{\text{ox}} & \rightarrow & \text{FTR}^{\text{red}} & & & (15b) \\ & & & \downarrow & & & \\ & & \text{Td}^{\text{ox}} & \rightarrow & \text{Td}^{\text{red}} & & (15c) \\ & & & & \downarrow & & \\ & & & \text{inactive E}^{\text{ox}} & \rightarrow & \text{active E}^{\text{red}} & (15d) \end{array}$$

Thioredoxins themselves are low molecular mass proteins and, like FTP and the target enzyme, contain two cysteine residues in close proximity which are either present as reduced dithiol or are oxidized to form a disulfide.

In summary, it is obvious that a multitude of different control levels is established to couple photosynthetic carbon reduction to the state of photosynthetic electron transport. Light-dependent modulation is regulated in a feed-forward cycle and is mainly communicated via the $Mg^{2+}$ concentration, the pH-value and the redox state of the stroma.

### *6.2. Long-term adaptation and photoregulation of gene expression*

In contrast to short-term adaptation which provides mechanisms that enable to respond to a sudden change of light conditions, long-term adaptation is related to more constant environmental light conditions. There are several remarkable differences between chloroplasts of sun or shade plants, respectively. In order to maximize light absorption, chloroplasts of shade plants exhibit more thylakoids, higher granal stacks, and more antenna pigments and proteins as compared to chloroplasts of sun plants. This indicates that, in contrast to short-term adaptation which is characterized by rapid rearrangements, long-term adaptation involves changes in the amount and composition of the thylakoid membrane and its protein components. Basically, this requires a modulation in the turnover and assembly of the involved components, especially chloroplast proteins and pigments, in response to changing light conditions. Likewise, synthesis of chlorophylls and carotenoids is light-modulated according to functional requirement, i.e. they are not needed in the dark, largely needed under shady conditions and less needed in bright sun. Chlorophyll synthesis is basically subjected to two light-dependent control levels: (1) the abundance of 5-aminolevulinic acid is indirectly regulated via the phytochrome system, and (2) the reduction of protochlorophyllide to chlorophyllide is light-dependent in angiosperm plants. Less information is available on regulation of carotenoid synthesis. It is suggested that regulation is also somehow mediated via phytochrome. In addition, a supplementary control level must exist for both chlorophyll and carotenoid synthesis, since neither is produced in excess to the pigment-binding apoproteins. Thus, pigment abundance can be indirectly controlled via availability of the apoproteins. This is probably the predominant means of control at high irradiance.

Turning to mechanisms regulating protein abundance, it should be mentioned beforehand that there are no rules applicable to all species and all times of development. Expression of genes coding for chloroplast proteins is controlled very species-, tissue- and time-specifically. Therefore, only general tendencies can be described according to the few examples examined in more detail. The abundance of nuclear- and chloroplast-encoded proteins can, in principle, be regulated at all possible levels, i.e. transcriptional, post-transcriptional and post-translational control. However, light-regulated gene expression occurs predominantly by transcriptional control for nuclear genome encoded proteins and, in contrast, by post-transcriptional or even post-translational control for chloroplast genome encoded proteins. (For further reading see also refs. 35–39.)

*6.2.1. Light-regulated expression of nuclear genes:* Light-regulated expression of the nuclear genes are best studied for the gene encoding the small subunit of Rubisco (*rbcS* gene) and for those encoding the light-harvesting chl *a*/*b* proteins (*cab* genes). Contrary to most of the light-regulated short-term adaptation mechanisms which indirectly communicate via the photosynthetic pigments, signal perception for light-regulated expression of nuclear genes is accomplished by at least two different receptors, phytochrome and a blue light receptor, respectively. Additionally, protochlorophyllide may be involved in some cases to some extent. Besides the light quality (red or blue light) the light quantity can also play a role in that red light-dependent mechanisms are triggered either by low fluence (LF) or by very low fluence (VLF). Since the LF response is reversible it resembles more the typical phytochrome response than the irreversible VLF response. Red light in the LF range enhances transcription of the nuclear genes for ferredoxin and Rubisco SSU and decreases transcription of the protochlorophyllide reductase gene. Light-regulation of *cab* gene expression is additionally controlled in the VLF range whereas *rbcS* gene expression is additionally controlled by blue light.

Subsequent to perception, signal transduction must occur. Unfortunately, the transducing mechanisms are only poorly understood. It seems clear, however, that they require the interaction of regulatory DNA sequences (cis-acting elements) with sequence-specific DNA-binding proteins (trans-acting elements). The cis-acting elements, in the case of light-regulation also called light regulatory elements (LRE), are usually located upstream from the promoter. For example, phytochrome controlled LREs of *rbcS* and *cab* genes are located within a region covering at maximum 350 bp upstream of the promoter and are presumably arranged as repeating boxes of very short GT-rich stretches that need to be spaced correctly in order to exert regulatory control (GT motif). The current notion is that each box interacts with one subunit of a proteinaceous regulatory factor which subsequently assembles to the holocomplex. The correct spacing of the boxes is probably not essential for subunit binding but for subunit assembly. In addition, binding of other regulatory factors to other LREs that are, e.g., regulated either by blue light and/or indirectly by phosphorylation are discussed [40]. In general, it is assumed that the trans-acting element is a nuclear-derived factor. However, there is strong evidence favoring an additional involvement of at least one factor originating from chloroplasts because inhibition of chloroplast protein biosynthesis influences the light-regulation of expression of *cab* genes.

In summary, light-regulated gene expression of nuclear genes is predominantly controlled at the transcriptional level, is mainly mediated via phytochrome or a blue light receptor, and involves the interaction of cis-acting DNA elements with trans-acting protein elements.

*6.2.2. Light-regulated expression of chloroplast genes:* The rate of plastid transcription is very low in the dark, is rapidly increased during chloroplast biogenesis following initial illumination, and is then reduced to a steady state level in mature chloroplasts. Thus, transcription is maximized during light-induced chloroplast development. However, an enhanced overall transcription activity during chloroplast biogenesis appears to result primarily from an increased amount of DNA templates (multiplication of the chloroplast genome). A differential transcription of chloroplast genes is more or less controlled in a light-independent manner by promoter strength and the stoichiometry of primary transcripts remains fairly constant. Furthermore, many chloroplast genes are transcribed polycistronically and therefore it seems easier to regulate a differential expression at a post-transcriptional level, i.e. control of RNA stability, RNA splicing and RNA translation, or at a post-translational level, e.g., protein degradation and assembly (see also refs. 36, 41 and 42).

For example, RNA turnover is suggested to be largely influenced by RNA-protein interaction leading to the formation of mRNP particles and thereby stabilizing the transcript. It is likely that such a mechanism is responsible for the increased stability of the *psb*A transcript (coding for PS II-D1 protein) relative to that of other primary transcripts following illumination. Moreover, transcription of the *psb*C/*psb*D operon (encoding PS II proteins CP43 and D2, respectively) yields a set of different polycistronic mRNAs in the dark and in the light. Those derived in the light are extended by approximately 150 bp downstream of the 5' end as compared to the dark transcripts. The transcripts synthesized in the light exhibit an increased stability which might be due to a different RNA processing or to protein-RNA interactions or to both [42].

For some chloroplast-encoded proteins light-dependent regulation is additionally (or primarily) controlled at the translational or post-translational level. In this case, accumulation of a certain protein exceeds accumulation of the corresponding stable mRNA transcript. Such a situation was first observed for chl *a*-binding protein of PS I in spinach (*psa*A and *psa*B gene product, i.e. subunit Ia and Ib, respectively). The light-dependent accumulation could either result from an increased protein stability in the light or from an enhanced translation. Protein stability is usually increased by correct assembly into the complete holocomplex and/or by binding of cofactors such as chlorophylls. In addition, it is also conceivable that the rate of translation is influenced, e.g., by coupling to light-dependent pigment synthesis.

In any event, it seems very likely that all regulatory levels of chloroplast gene expression are, at least partly, controlled by nuclear gene products. Interaction of nuclear-encoded proteins with the primary RNA transcripts in chloroplasts can alter RNA stability, processing and translation, and nuclear-encoded proteins are required for correct assembly of mixed holocomplexes. Light-regulation of the intervening nuclear gene product leads, in turn, indirectly to light-modulated chloroplast gene expression.

## References

1. L Bogorad, J. Cell Biology 91 (1981) 256s.
2. DJ Murphy, Biochim. Biophys. Acta 864 (1986) 33.
3. R Douce and J Joyard, Annu. Rev. Cell Biol. 6 (1990) 173.
4. MS Webb and BR Green, Biochim. Biophys. Acta 1060 (1990) 133.
5. W Kühlbrandt and DNWang, Nature 350 (1991) 130.
6. M Calvin, Photosynth. Res. 21 (1989) 3.
7. GH Lorimer, Annu. Rev. Plant Physiol.32 (1981) 349.
8. J Barber, Nature 325 (1987) 663.
9. AA Gatenby, Annu. Rev. Cell. Biol. 6 (1990) 125.
10. DB Knaff, Trends in Biochem. Science 14 (1989) 159.
11. CA Raines, JC Lloyd and TA Dyer, Photosynth. Res. 27 (1991) 1.
12. WL Ogren, Annu. Rev. Plant Physiol. 35 (1984) 415.
13. U Heber and GH Krause, Trends Biochem. Sci. (1980) 32.
14. CV Givan, KW Joy and LA Kleczkowski, Trends Biochem. Sci. 13 (1988) 433.
15. DH Gonzales and CS Andreo, Trends Biochem. Sci. 14 (1989) 24.
16. IP Ting, Annu. Rev. Plant Physiol. 36 (1985) 595.
17. J Preiss, Trends in Biochem. Sci. (1984) 24.
18. C Cseke, A Balogh, JH Wong, BB Buchanan, M Stitt, B Herzog et al. Trends Biochem. Sci. (1984) 533.
19. PR Whitfeld and W Bottomley, Annu. Rev. Plant Physiol. 34 (1983) 279.
20. L Hanley-Bowdoin and NH Chua, Trends Biochem. Sci. 12 (1987) 67.
21. PA Castelfranco and SI Beale, Annu. Rev. Plant Physiol. 34 (1983) 241.
22. U Heber and HW Heldt, Annu. Rev. Plant Physiol. 32 (1981) 139.
23. UI Flügge and HW Heldt, Annu. Rev. Plant Physiol. Plant. Mol. Biol. 42 (1991) 129.
24. UI Flügge and HW Heldt, Trends Biochem. Sci. (1984) 530.

25. S Smeekens, P Weisbeek and C Robinson, Trends Biochem. Sci. 15 (1990) 73.
26. K Keegstra and LJ Olsen, Annu. Rev. Plant Physiol. Plant Mol. Biol. 40 (1989) 471.
27. EK Archer and K Keegstra, J. Bioenergetics Biomembranes 22 (1990) 789.
28. RM Mould, JB Shackleton and C Robinson, J. Biol. Chem. 266 (1991) 17286.
29. JM Anderson and B Andersson, Trends in Biochem. Science 13 (1988) 351.
30. C Foyer, R Furbank, J Harbinson and P Horton, Photosynth. Res. 25 (1990) 83.
31. JF Allen, Biochim. Biophys. Acta 1098 (1992) 275.
32. N Carrillo and RH Vallejos, Trends Biochem. Sci. (1983) 52.
33. BB Buchanan, Annu. Rev. Plant Physiol. 31 (1980) 341.
34. DF Knaff, Trends Biochem. Sci. 14 (1989) 433.
35. JM Anderson, Annu. Rev. Plant Physiol. 37 (1986) 93.
36. WC Taylor, Annu. Rev. Plant Physiol. Plant Mol. Biol. 40 (1989) 211.
37. EM Tobin and J Silverthorne, Annu. Rev. Plant Physiol. 36 (1985) 569.
38. C Kuhlemeier, PJ Green and NH Chua, Annu. Rev. Plant Physiol. 38 (1987) 221.
39. JE Mullet, Annu. Rev. Plant Physiol. Plant Mol. Biol. 39 (1988) 475.
40. JF Allen, Photosynth. Res. 36 (1993) 95.
41. PE Gamble and JE Mullet, J. Biol. Chem. 264 (1989) 7236.
42. PE Gamble, T Berends Sexton and JE Mullet, EMBO J. 7 (1988) 1289.
43. JH Golbeck, Biochim. Biophys. Acta 895 (1987) 167.
44. JA Bassham and GH Krause, Biochim. Biophys. Acta 189 (1969) 207.

Bioenergetics
ed. by P. Gräber and G. Milazzo†


# *CHAPTER 6*
# The mammalian mitochondrial respiratory chain

G. von Jagow, B. M. Geier and T. A. Link

*Gustav Embden Zentrum der Biologischen Chemie, Klinikum der Uni Frankfurt, Frankfurt am Main, Germany*

## 1. Introduction

Approximately 90% of the ATP synthesized in the cells of animals is supplied by oxidative phosphorylation. This process takes place at the inner mitochondrial membrane. The respiratory chain builds up an electrochemical membrane potential across the inner mitochondrial membrane which is used by the $F_1F_0$ ATP sythase to synthesize ATP within the mitochondrial matrix. ATP is then exported into the cytosol; this export is also driven by the membrane potential. Hence, mitochondria are the "powerhouse of the cell".

Mitochondria are organelles containing four compartments (from inside to outside): The mitochondrial matrix; the inner membrane which is impermeable to most solutes including protons; the intermembrane space; the outer membrane, which is permeable to solutes up to a size of 5 kDa. The inner membrane has a high protein content (> 3 mg protein/mg phospholipid) and has invaginations into the matrix so that the surface area is enlarged. It contains the mitochondrial respiratory chain enzymes, the $F_1F_0$ ATP synthase, and multiple substrate carrier proteins. Since the outer membrane is essentially permeable to most solutes, it is often neglected and the outside of the inner mitochondrial membrane will be referred to as "cytosol" in schematic representations.

Mitochondria resemble structurally aerobic bacteria like *Paracoccus denitrificans*. They have a small genome and perform bacterial-type protein biosynthesis; this supports the endosymbiont hypothesis that mitochondria elvolved from autonomous bacteria which were incorporated into a host cell. The mitochondrial genome codes for 13 of the most hydrophobic membrane subunits of the oxidative phosphorylation system while the majority of the proteins is synthesized in the cytosol and imported into mitochondria by a specific import machinery.

The respiratory chain oxidizes substrates coming from a broad range of metabolic pathways. Some of them are located in the mitochondrial matrix (e.g., the citric acid cycle and the β-oxidation pathway for fatty acids) so that the reducing equivalents can directly enter the respiratory chain. Other substrates have to be transported into the mitochondria or can be oxidized at the outside of the mitochondria by specific dehydrogenases. Most of the chemical energy of the nutrient molecules is transformed into the group transfer potential of the hydrogen donors NADH and $FADH_2$.

Figure 1 gives a block scheme of the electron transfer chain. The respiratory chain represents a system consisting of four multisubunit enzyme complexes which are arranged in a side-specific orientation in the inner mitochondrial membrane [1]. According to their enzymatic function they are referred to as NADH:ubiquinone oxidoreductase (complex I), succinate:ubiquinone oxidoreductase (complex II); ubihydroquinone:cytochrome c oxidoreductase (complex III) and cytochrome oxidase (complex

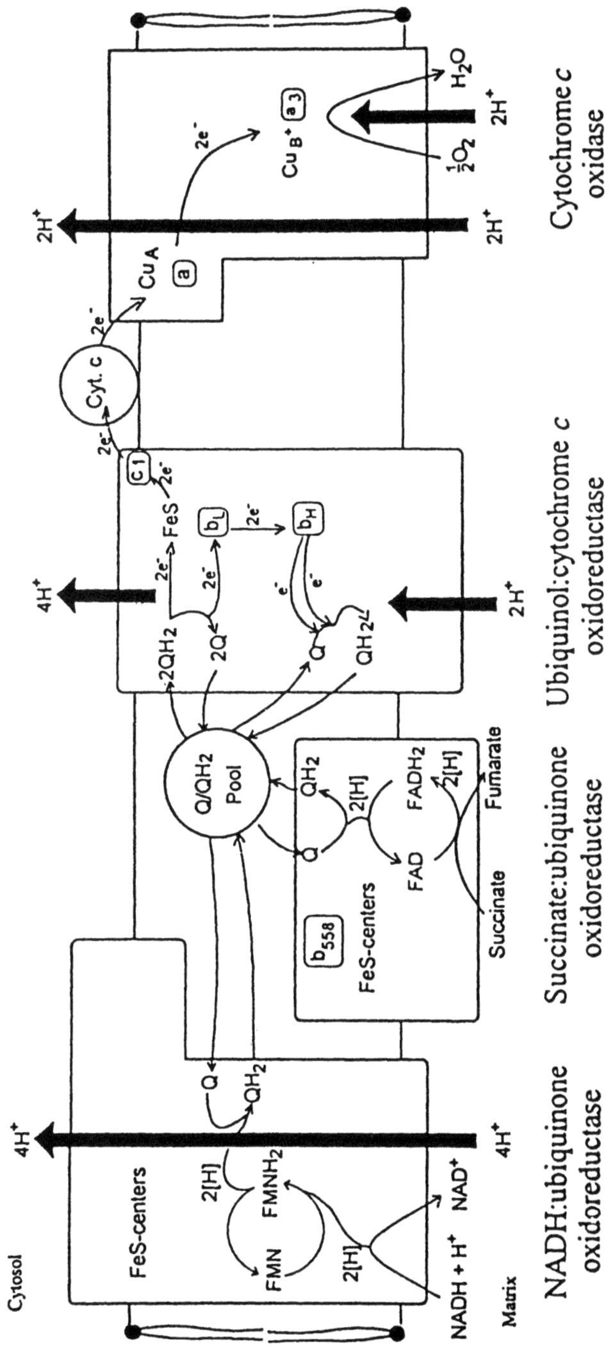


Fig. 1. Block scheme of the mitochondrial respiratory chain.

IV). The scheme disregards some complexes with side entries of redox equivalents into the ubiquinone pool like the electron-transferring protein complex (ETF) or α-glycerophosophate dehydrogenase.

Every complex is built up of numerous peptides. In mammals, in the case of complex I, there are some 40 different peptides. Complex II consists of 4–5 peptides, complex III of 11 peptides, and complex IV of 13 peptides. With the exception of complex II, the vast number of predominantly smaller proteins lacking any redox center is not presents in the homologous bacterial enzymes [2]. Deletion of the genes of some of them in yeast prevents correct assembly of the complexes; therefore, these subunits may be involved in assembly of the complexes or in its regulation.

Figure 2 shows how the electrons flow down the sequence of redox carriers with a steady increase of the redox potentials, hence losing redox energy. At the entry of the electron transfer chain at NADH, the electrons have a redox potential of about −320 mV. After leaving complex I on ubiquinone they have a potential of about +50 mV. On cytochrome c their potential is +250 mV; and the water/oxygen couple has a potential of +820 mV. The increment of the redox potentials is not always equivalent. Therefore, we meet electron transition steps between two redox centers

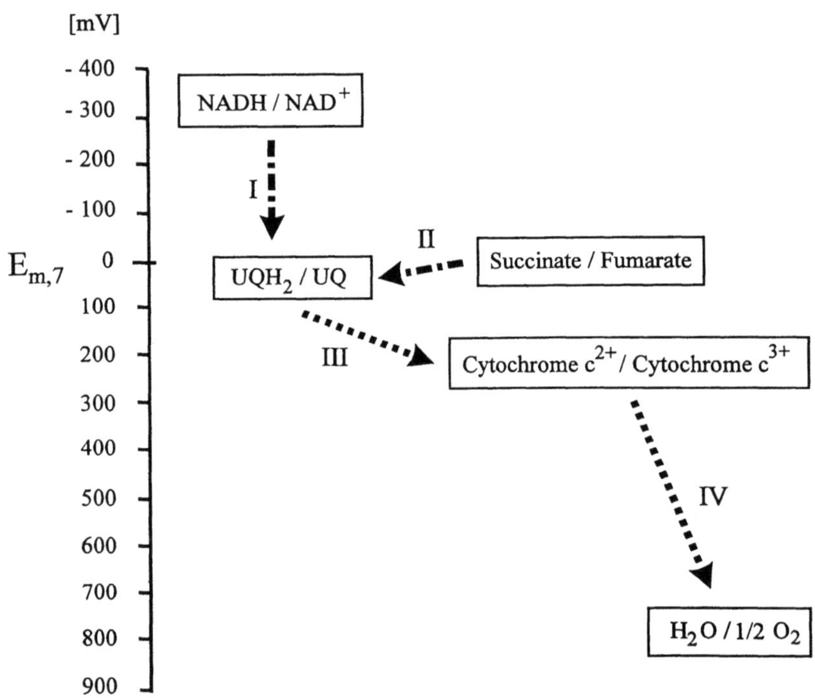


Fig. 2. Reactions of the mitochondrial respiratory chain. - · · -, hydrogen transfer; · · · , electron transfer.

with a large potential gap, and those with a small potential gap. In the first case the large energy change drives conversion of redox energy into a new form of energy, i.e., into a protonmotive force [3]. Electron transfer steps with a small potential gaps are not directly involved in energy conversion. A loss of redox energy and the production of heat might occur, for instance, under uncoupling conditions (cf. section 9). The release of the energy of the membrane potential as heat is used, for example, by the mitochondria of the brown adipose tissue of hibernators.

Two accessory redox components are essential for completion of the electron transfer chain. These are the small lipophilic organic molecule, ubiquinone and the relative small hydrophilic peptide, cytochrome c. Both are diffusible redox carriers. Ubiquinone connects electronically complexes I and II with complex III, whereas cytochrome c transfers the electrons from complex III to complex IV. Ubiquinone (see Fig. 8) is located in the lipid phase of the inner mitochondrial membrane while cytochrome c is attached to and moving on the cytosolic surface of the membrane.

When we look at the function of these two molecules, we see a recurrent motive in the transfer chain of redox equivalents (Figs. 1 and 2): Namely, a consecutive switching between hydrogen and electron transfer carriers. Hydrogen carriers are NADH, $FMNH_2$, $FADH_2$ and $UQH_2$. Electron carriers are the iron-sulfur centers (FeS-centers), the heme centers and the copper centers.

The hydrogen carriers have quinoide or quinone structure, the electron carriers represent transition metal complexes. NADH like ubiquinone is a mobile component, but unlike ubiquinone not lipophilic but rather water soluble. Therefore, it migrates in the matrix of the mitochondrion or in the cytosolic compartment, whereas ubiquinone is anchored with its isoprenoide side chain in the hydrocarbon core of the membrane. The flavin and iron-sulfur clusters are embedded in the peptide scaffold of the proteins. The heme centers are the catalytic centers of the cytochromes. Two copper centers reside besides heme a and $a_3$ in cytochrome oxidase.

The consequences which arise due to a change between a hydrogen and an electron carrier and back from an electron to a hydrogen carrier are obvious. The first swich will lead to a release of $2H^+$, the second switch to an uptake of $2H^+$. This might occur at the surface of or with access to the surface of the inner mitochondrial membrane either at the positively charged cytosolic side or at the negatively charged matrix side of the membrane.

## 2. Principles of electron transfer-linked proton translocation

During electron transfer the released redox energy is partially conserved by a proton transport creating a transmembrane difference of the electrochemical potential of protons, $\Delta\tilde{\mu}_{H^+}$ (protonmotive force) [2]. This reaction immediately introduces a new dimension, that of a spatial vector. Hence the scalar chemical reactions we are usually dealing with are brought to the dimension of vectorial reactions. The membrane gains sidedness and polarity which do not only determine the arrangment of the individual redox centers and proteins within the membrane, but also confer physical reaction parameters like osmolarity and electrogenicity.

Two principal mechanisms of electron transfer-linked proton translocation are under discussion. The first one are redox loops and redox cycles, the second mechanism includes all kinds of proton pumps.

In the case of a redox loop or a redox cycle protons are released or taken up by scalar, electroneutral reactions on opposite sides of the membrane. In mitochondria, the direction of the membrane potential is such that the cytosolic side (outside) is positive and the matrix side (inside) is negative. A redox loop requires three rections: (i) oxidation of a hydrogen carrier at the positive side of the membrane, (ii) reduction of a hydrogen carrier at the negative side of the membrane, and (iii) connection of these two processes by an electrogenic electron transfer which crosses the membrane from the positive to the negative side (Fig. 3A).

The release of the $2H^+$ occurs when the redox donor ($DH_2$) on the cytosolic side of the mitochondrial membrane donates successively two electrons to its next electron acceptor. The electron transfer across the membrane can include several electron carriers. The uptake of $2H^+$ takes place when the redox acceptor (A) accepts two electrons on the matrix side of the mitochondrial membrane. This coupled process leads to the formation of a membrane potential (cytosol positive) and a transmembrane proton concentration difference (higher proton concentration, i.e., lower pH in the cytoplasm than in the mitochondrial matrix space).

A prerequisite for a redox loop is that the redox centers involved are in contact with opposite sides of the membrane. This requires either a location of the centers close to the side where the reaction occurs or proton channels connecting the respective redox center with the bulk phase. Through this reaction mechanism proton translocation is stoicheiometrically linked to the redox rection. Characteristic for a redox loop is a constant $H^+/2e^-$ stoicheiometry of 2 ($2H^+$ translocated per two electrons transferred from $DH_2$ to A).

The redox cycle is a kind of a modified and refined redox loop mechanism. It relies on the same principles as the redox loop, but the flow of electrons endows a higher $H^+/2e^-$ stoicheiometry. It requires a bifurcation of the electron transfer chain at the $DH_2$ oxidation site. Such a redox cycle mechanism has been proposed by Mitchell to describe the electron and proton transport mechanism in complex III (Fig. 3B) [4].

On oxidation of the redox donor $DH_2$, $2H^+$ are released to the cytosolic side of the membrane, the first electron goes to the electron acceptor A, the second electron back to D. In order to reduce D at the matrix side, a second electron is required, a second molecule of $DH_2$ has to be oxidized,

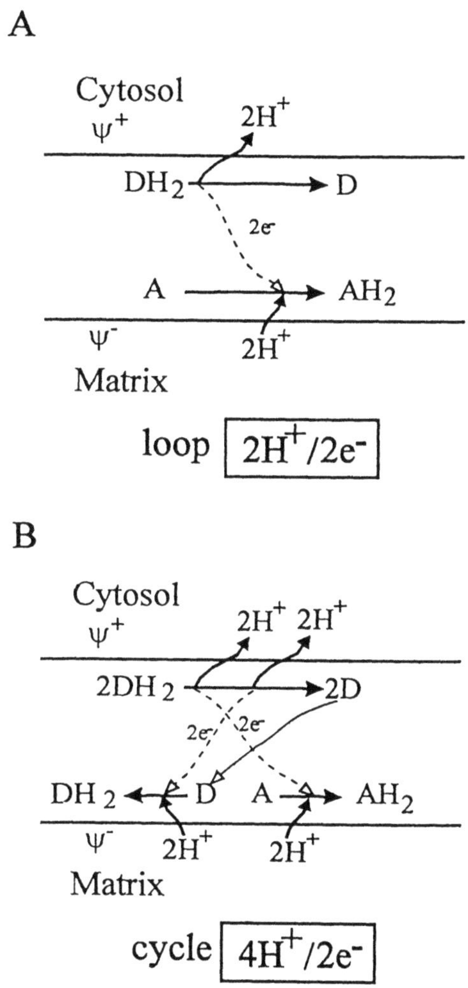


Fig. 3. Conceived mechanisms of proton translocation: (A) Redox loop and (B) Redox cycle.

and again $2H^+$ are released to the cytosolic side. One molecule D is reduced at the matrix side of the membrane with an uptake of $2H^+$ from the matrix. In order to complete the cycle, $2e^-$ have to cross the membrane from the cytosol to the matrix reducing A to $AH_2$. Drawing the balance of one catalytic cycle which consists of two turnovers, it turns out that $4H^+$ are released to cytosol when one molecule of $DH_2$ is oxidized. $2H^+$ are taken up from the matrix while one molecule of D is reduced and the further $2H^+$ are taken up from the matrix when one molecule of A is reduced. Further details of the Q cycle mechanism will be discussed in section 5.2.

Our knowledge about proton pump mechanisms, especially in the mammalian respiratory chain, is more vague than our knowledge about proton transport mediated by a redox loop or cycle. All our speculations are based on the knowledge of proton transport in bacteriorhodopsin in which light energy drives a cis-trans epimerisation of a Schiff base. The Schiff base represents the protonic group driving the proton transport. This group is arranged more in the middle of the membrane and is connected to both sides of the bacterial membrane by proton channels (see Chapter 11).

A few redox components of the mammalian respiratory chain reveal a pH-dependent change of their redox potential, i.e. they posses a redox-dependent change of the pK value of a protonic amino acid side chain. Such redox state/pK relations might contribute to a proton channel involved in a proton pump mechanism. To what extent short- or long-distance conformational changes of the protein scaffold are implicated in those pump mechanisms remains to be clarified. It has been suggested that a presumptive proton pump mechanism may operate in combination with a redox loop mechanism. Conceivable mechanisms have been discussed in detail by Mitchell [5].

## 3. NADH:ubiquinone oxidoreductase (complex I)

### *3.1. Subunit composition, genetic origin, content of redox centers and three-dimensional gross structure*

Complex I, by far the largest multisubunit complex of the respiratory chain, is a phylogenetically ancient enzyme [6]. Already in bacteria a piericidine-sensitive complex with iron-sulfur clusters very similar to those of the mitochondrial enzyme has been detected [7]. Weiss [8] has suggested that the complex phylogenetically emerges and is assembled from two preexisting subcomplexes. A bacterial $NAD^+$ hydrogenase resembles in subunit composition and in redox centers the hydrophilic head part of complex I (small complex); a bacterial formate hydrogenlyase resembles the membrane moiety of complex I. Pulse chase experiments have shown that the membrane part is synthesized first. Besides this, there exist various bacterial NADH dehydrogenases which are comparable to one or the other half of the complex.

Complex I consists of some 40 subunits in both mammals and *N. crassa*. The molecular masses are about 800 and about 700 kDa, respectively. In the following we will follow the concept of Weiss, who distinguishes between the small form, represented mainly by the extramembranous moiety, and the large form consisting of the head and the intramembranous further half (Fig. 4).

It was detected long ago by Hatefi [1] that bovine heart complex I can be cleaved by chaotropic agents into a water-soluble fraction and into a

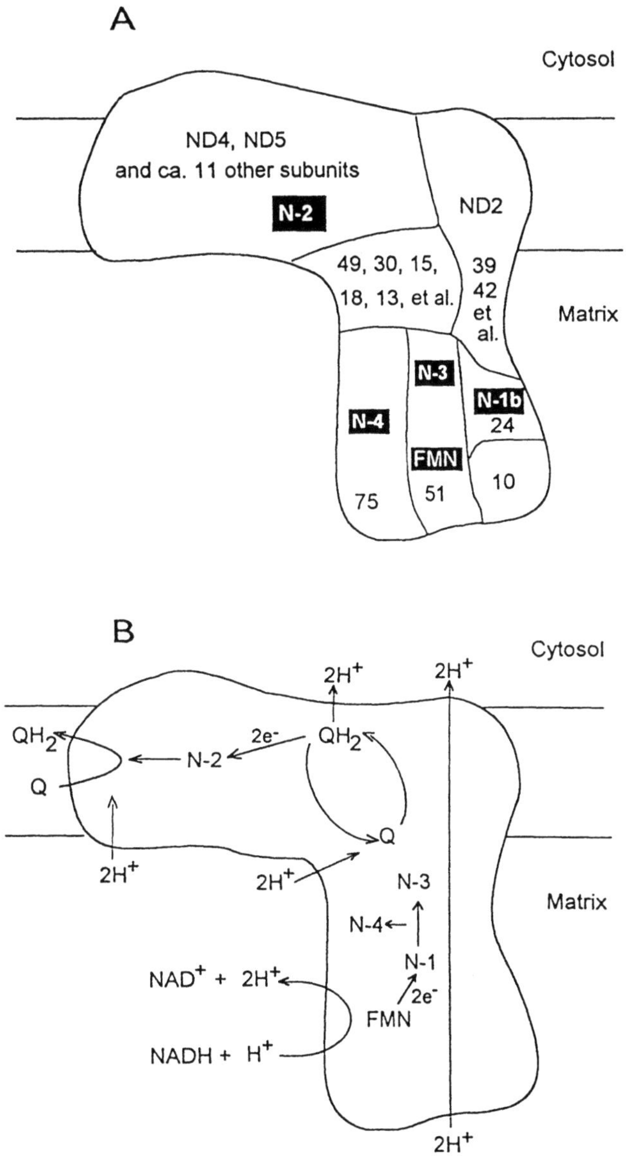


Fig. 4. Complex I: (A) Gross structure and subunit assignment; (B) Path of electron and proton flow.

hydrophobic fraction. The water-soluble fraction dissociated into a flavin containing subcomplex consisting of three polypeptides and into an iron-sulfur protein containing subcomplex consisting of five polypeptides.

The gross structure of complex I is known from image reconstruction pictures of electron microscopic patterns of two-dimensional crystals of the *N. crassa* complex. As shown in Fig. 4, the complex has an L-shaped structure with two arms of about similar size. The moiety which is arranged perpendicular to the plane of the membrane consists of about 13 subunits which are all nuclear encoded. The moiety arranged within the plane of the membrane consists of about 15 subunits, of which seven are mitochondrially encoded. In the head of the complex three subunits with molecular masses of 75, 51 and 24 kDa seem to constitute a structural and functional unit. Photoaffinity labeling experiments have shown that the 51-kDa subunit forms the NADH binding site. It is not quite clear which subunit binds the FMN. Three iron-sulfur clusters (N-1, N-3 and N-4) are arranged in this half of the complex, iron-sulfur cluster N-2 is situated close to the membranous part. There, the 36-kDa subunit encoded by the mitochondrial gene ND-1 seems to form the reaction center where the quinone molecules of the ubiquinone pool do react. The 36-kDa subunit is photolabeled by rotenone. Rotenone, a competitive inhibitor with regard to ubiquinone, blocks electron flow from NADH to ubiquinone.

When *N. crassa* is grown on chloramphenicol, an inhibitor of the mitochondrial protein synthesis, a NADH dehydrogenase consisting of 13 subunits is synthesized. The protein pattern of this complex is highly analogous to the small form and together with the hydrophobic half complementary to the total complex. The complex grown under chloramphenicol possesses the NAD binding site, the FMN and the iron-sulfur clusters N-1, N-3 and N-4, but is devoid of cluster N-2 and therefore inactive.

### *3.2. The redox centers of complex I*

An exact allocation of the redox centers to the individual subunits and to the amino acid ligands cannot be given so far. It is known, however, that the FMN center is not bound covalently. The redox potential of FMN ($E_{m,7}$) is about −320 mV. The redox potentials and character of the known iron-sulfur clusters are summarized in Table 1.

Table 1. Characteristics of the iron-sulfur centers of bovine complex I

| Cluster | Cluster structure | $E_{m,7}$ (mV) |
|---|---|---|
| N-1a | binuclear | −370 |
| N-1b | binuclear | −240 |
| N-2 | tetranuclear | −150 |
| N-3 | tetranuclear | −245 |
| N-4 | tetranuclear | −245 |
| N-5 | tetranuclear | −260 |

The clusters N-1, N-3 and N-4 are isopotential. Cluster N-2 has a more positive redox potential than the other iron-sulfur clusters. The existence of iron-sulfur cluster N-1a revealing a very negative redox potential is still controversial. So far it has only been detected by Ohnishi in bovine heart complex I.

Iron and acid-labile sulfur analyses indicate the presence of more iron-sulfur clusters which have not been identified unambiguously [1]. Additional to the flavin center and the iron-sulfur clusters, complex I seems to contain a tightly bound ubisemiquinone species which reacts at cluster N-2, as shown by EPR spectroscopy.

### *3.3. Path of electron and proton flow*

The sequence of reactions can be summarized by the following equation:

$$NADH + Q + 5\ H^+_m \rightarrow NAD^+ + QH_2 + 4\ H^+_c$$

The scheme of the path of electron and proton flow remains still highly hypothetical. Any reaction scheme, however, has to match the redox potentials as well as the well known $H^+/2e^-$ stoicheiometry of 4. Different flow schemes have been proposed; Fig. 4B shows a modification of a scheme proposed by Weiss et al.

NADH is oxidized at the flavin center in a reaction cleft which might be visible in the electron microscopic pictures of the *N. crassa* complex. Two electrons are transferred under release of $2H^+$ from the flavin center through the multitude of iron-sulfur clusters within the peripheral arm to a quinone binding site within the membrane part. Proton translocation could occur by a quinone cycle analogous to that discussed for the $bc_1$ complex (see below). The existence of multiple inhibitor binding sites has been taken as evidence for independent quinone binding sites; two classes of complex I inhibitors are represented by rotenone and piericidine, respectively.

Whether the flavin participates in proton pumping is still a matter of discussion; this would require a long proton channel from the peripheral arm out in the matrix to the cytosol for which there is no precedent.

## 4. Succinate:ubiquinone oxidoreductase (complex II)

### *4.1. Subunit composition and redox centers*

Complex II is by far the simplest multisubunit complex with respect to its number of subunits. It consists of only four or five subunits. Like complex I it is composed of a hydrophilic extramembranuos part, and a hydrophobic membranuos part, which serves as membrane anchor [9].

The membrane anchor is composed of at least two subunits with molecular masses of 15.5 kDa and 7.0 kDa, respectively. The head is assembled of subunits of higher molecular masses. The flavoprotein protein subunit (Fp) has a molecular mass of 70 kDa and the iron-sulfur protein subunit (lp) of 27 kDa. These two subunits form the so-called "succinate dehydrogenase" ($SDH_s$) which is capable of transferring electrons from succinate to various artificial electron acceptors, but not to ubiquinone. Soluble $SDH_s$ can be cleaved off the membrane anchor by addition of chaotropic salts. Full catalytic activity, i.e. electron flow form succinate to ubiquinone, requires the complete four to five subunit complex. All subunits are coded for by the nuclear DNA and synthesized in the cytoplasm.

The complex contains five redox centers, a covalently bound FAD center and three different iron-sulfur clusters. EPR measurements have shown that additionally it comprises tightly bound ubisemiquinone $Q_s$.

The (8a-(N(3)-histidyl) FAD center is covalently bound to the 70 kDa subunit. The three iron-sulfur clusters are coordinated by the 27 kDa subunit. Sequence analysis has shown that this subunit contains three conserved arrangements of cysteine residues which are significant for (2Fe-2S), (4Fe-4S) and (3Fe-4S) clusters. The three iron-sulfur clusters have the following characteristics:

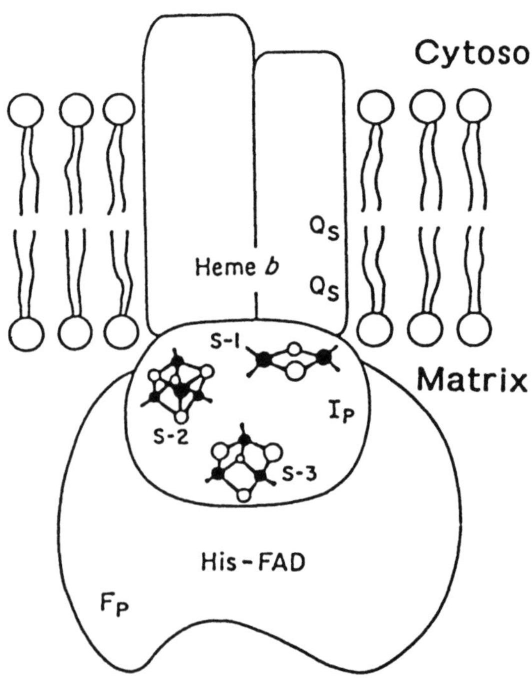


Fig. 5. Complex II: Gross structure and location of the redox centers. $Q_S$: bound semiquinone.

Table 2. Characteristics of the redox centers of complex II

| Cluster | | Cluster structure | Ligands | $E_{m,7}$ (mV) |
|---|---|---|---|---|
| FAD | (8a-(N(3)-histidyl) | - | | +50 |
| S-1 | (2Fe-2S) | binuclear | 3 Cys,1 ? | +70 |
| S-3 | (3Fe-4S) | trinuclear | 3 Cys | +10 |
| S-2 | (4Fe-4S) | tetranuclear | 4 Cys | -270 |
| Cyt. b-560 | heme b | | 2 His | -194 |

The complex contains three different iron-sulfur clusters, one of each type. The heme b of cytochrome b-560 is supposed to be coordinated by two histidine residues. The two histidines are ascribed to two transmembrane $\alpha$-helices, one contributed by the 15.5 kDa and one by the 7.0 kDa subunit. Each of the small subunits contains three hydrophobic sequences long enough to form three transmembrane $\alpha$-helices, and one conserved histidine residue.

### *4.2. Path of electron and proton flow*

The catalytic activity of complex II can be summarized by the following reaction equation:

$$\text{succinate} + Q \rightarrow \text{fumarate} + QH_2$$

From EPR measurements the distances between the FAD and the iron-sulfur centers are known (cf. Fig. 5). They are in electronic contact. The distance between FAD and S-1 is about 15 Å, between S-1 and S-3 about 10 Å. The distance between S-3 and heme b is unknown. It is striking that we are dealing with two groups of redox carriers, one with a potential intermediate between succinate and quinone consisting of FAD, cluster S-1 and S-3, and a low potential group consisting of cluster S-2 and cytochrome b-560. Cytochrome b cannot be reduced by succinate, but is oxidized by fumarate [5,8]. It has been speculated that two electron transfer chains function in parallel; one composed of S-1 and S-3, and the other of S-2 and cytochrome b-560. However, since the low potential chain has a more negative potential than both donor and acceptor, it is more likely that the low potential chain is functionally quiet and represents an evolutionary relic. Most of the bacterial succinate dehydrogenases do not have a heme bound to the hydrophobic membrane anchor subunits.

The electron flow is sensitive to thenoyltrifluoroacetone (TTFA) which destabilizes specifically the semiquinone radical ($Q_s^{\cdot}\cdot$). The electron flow is not coupled to a proton transport across the membrane. $2H^+$ are released into the matrix space during oxidation of succinate and $2H^+$ are taken up from the matrix space during reduction of ubiquinone.

## 5. Ubihydroquinone:cytochrome c oxidoreductase (complex III)

### *5.1. Subunit composition, genetic origin and tertiary structure*

Complex III, also referred to as cytochrome $bc_1$ complex according to its heme centers, is an oligomeric membrane protein complex. In mammals it has molecular mass of 250 kDa. It seems to be integrated in the membrane as a dimeric complex. Its complexity is comparable to cytochrome oxidase. The $bc_1$ complex is phylogenetically widely distributed [2]. It transduces energy not only in bacteria (including phototrophic bacteria, not in archebacteria), but also in chloroplasts, algae and mitochondria. To date $bc_1$ complexes of about ten species have been isolated. The redox subunits are highly conserved in their protein folding and coordination of the redox centers.

The complexity of the subunit composition has increased in the course of evolution. Bacterial $bc_1$ complexes consist of only three or four subunits [2], chloroplast complexes of four to five subunits, those of yeast of ten and those of mammals of 11 subunits [10]. The three subunit complexes consist of the peptides carrying the redox centers. These are the iron-sulfur protein, cytochrome $c_1$ and a b-type cytochrome containing the two heme b centers.

The mitochondrial $bc_1$ complexes consist either of ten or of 11 subunits. Again the three redox subunits contain the four redox centers, but additional protein subunits exist that lack prosthetic redox groups: two proteins with a molecular mass higher than that of the three redox-catalytic subunits and five or six peptides with smaller molecular masses.

In order to demonstrate briefly the degree of divergence in the course of evolution, a comparison of the subunits of the mitochondrial complexes from *S. cerevisiae* and bovine heart is given in Table 3. The amino acid sequences of the subunits from beef were obtained by protein sequencing with the exception of the two core proteins and cytochrome b; those of *S. cerevisiae* were deduced from the DNA sequences of the respective cDNA's.

The two core proteins reveal distinct species-specific differences in their molecular masses, whereas the three redox subunits are of comparable size. Molecular genetic studies, especially gene deletion studies in yeast, have unraveled the importance of the small subunits for the assembly of the oligometric complexes and for the integration of the redox centers into the larger "apo-redox peptides". Ten of the 11 subunits of the mammalian complex are nuclear encoded, only cytochrome b is a mitochondrial gene product.

Table 3. Peptide subunits of the $bc_1$ complexes of beef heart and of *S. cerevisiae*

| | Bovine | | | Homology | Yeast | | |
|---|---|---|---|---|---|---|---|
| | Subunit | $M_r$ | Helices | | Subunit | $M_r$ | Helices |
| I | Core I | 49.209 | 0 | 29% | Core I | 48.225 | 0 |
| II | Core II | 46.250 | 0 | 20% | Core II | 46.808 | 0 |
| III | Cytochrome *b* | 42.588 | 8 | 51% | Cytochrome *b* | 43.998 | 8 |
| IV | Cytochrome $c_1$ | 27.285 | 1 | 56% | Cytochrome $c_1$ | 27.966 | 1 |
| V | Iron-Sulphur Protein | 21.782 | 1/2? | 55% | Iron-Sulphur Protein | 21.804 | 1/2? |
| VI | Cyt *b*-associated | 13.389 | 0 | 29% | 14 kDa | 14.430 | 0 |
| VII | Core-associated | 9.588 | ? | 20% | 11 kDa | 12.269 | ? |
| VIII | Hinge protein | 9.174 | 0 | 36% | 17 kDa | 14.513 | 0 |
| IX | ISP targeting peptide | 7.955 | 0 | – | ? | | |
| X | Cyt $c_1$-associated | 7.326 | 1 | 30% | 7,3 kDa | 7.340 | 1 |
| XI | ISP-associated | 6.519 | 1 | 18% | 8.5 kDa | 8.461 | 1 |

Image reconstruction patterns of electron microscopic pictures of two-dimensional crystals of the *Neurospora* complex (consisting of ten subunits) have provided us with a gross picture of the whole complex [11]. Studies of the dissociation of the yeast and bovine complexes into subcomplexes and single subunits by higher salt and detergent concentrations also gave insight into the structure. The arrangement of the subunits within this structure has been studied using side-specific chemical and photoaffinity labels and limited protease digestion which probe the accessibility of proteins on both surfaces of the membrane.

In bovine complex the nucleus is formed by cytochrome b together with the 13.4 kDa (subunit VI) and the 8 kDa (subunit IX) proteins. On the cytosolic side of the complex the iron-sulfur protein (ISP, subunit V) is connected to the 6.4 kDa (subunit XI) protein. In bovine complex these two subunits are easily cleaved off under mild salt and detergent conditions. The second redox subunit at the cytosolic side is cytochrome $c_1$. Cytochrome $c_1$ is connected to the 9.2 kDa (subunit VIII) and the 7.2 kDa (subunit X) proteins. The two core proteins (subunit I and III) are located at the matrix side in close vicinity to the 9.5 kDa (subunit VIII) protein. According to x-ray studies of the *N. crassa* complex, the main part of the protein mass of the two core proteins extends into the matrix space.

EPR studies have revealed that the iron-sulfur center is located near the cytosolic surface, whereas the two heme b centers span the membrane. An intriguing question remains about the location of the reaction center for the electron donor ubihydroquinone ($QH_2$) at the cytosolic side of the membrane and location of the electron acceptor ubiquinone (Q) at the matrix side of the membrane. These questions will be discussed together with the description of the folding pattern of cytochrome b and the path of electron flow.

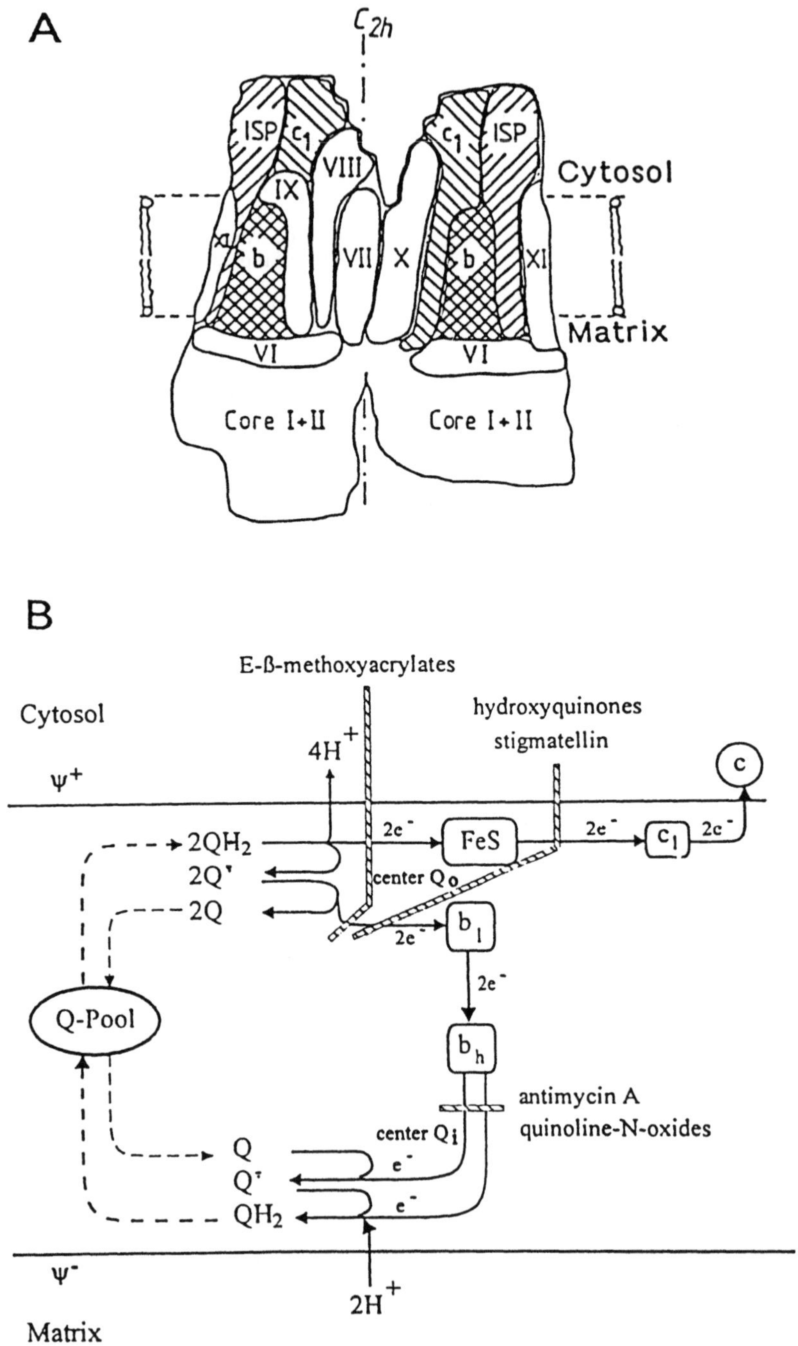


Fig. 6. Complex III: (A) Presumptive topographical arrangement of the eleven subunits in the mammalian cytochrome complex; (B) Protonmotive Q-cycle pathway.

### *5.2. Path of electron and proton flow*

As already mentioned during discussion of the principles of proton translocation, the path of electron flow in the $bc_1$ complex follows the scheme of a redox cycle [12]. Electrons of a redox potential of about +50 mV are transferred from ubihydroquinone to cytochrome c ($E_{m,7}$ = +250 mV) after passing the redox centers of the $bc_1$ complex.

A redox cycle requires a branching of the electron transfer chain which consist of a high potential pathway (FeS-cluster, heme $c_1$ and heme c) and a low potential pathway (heme $b_l$ and heme $b_h$). The sequential path of electron flow is shown in Fig. 6B. For a better understanding the most likely sequence of single electron transfer steps will be discussed. In order to channel two electrons onto two cytochrome c, two molecules of $QH_2$ have to be oxidized. This results in the recycling of two electrons across the membrane back to quinone Q. Therefore, one complete catalytic Q-cycle requires two "redox turnovers". The Q-cycle requires two reaction centers for quinone: one at the cytosolic side of the membrane where ubihydroquinone is oxidized (center *o*), and one at the matrix side of the membrane where ubiquinone is reduced (center *i*). The reactions can be described by the equations in Fig. 7:

(1) $$QH_2 + FeS^{ox} + b_l^{ox} \rightarrow Q + FeS^{red} + b_l^{red} + 2\,H_c^+$$

(2) $$FeS^{red} + b_l^{red} + b_h^{ox} + c_l^{ox} \rightarrow FeS^{ox} + b_l^{ox} + b_h^{red} + c_l^{red}$$

(3) $$b_h^{red} + c_l^{red} + Q + c^{ox} \rightarrow b_h^{ox} + c_l^{ox} + Q^{-\cdot} + c^{red}$$

(4) $$QH_2 + FeS^{ox} + b_l^{ox} \rightarrow Q + FeS^{red} + b_l^{red} + 2\,H_c^+$$

(5) $$FeS^{red} + b_l^{red} + b_h^{ox} + c_l^{ox} \rightarrow FeS^{ox} + b_l^{ox} + b_h^{red} + c_l^{red}$$

(6) $$b_h^{red} + c_l^{red} + Q^{-\cdot} + c^{ox} + 2\,H_m^+ \rightarrow b_h^{ox} + c_l^{ox} + QH_2 + c^{red}$$

---

Σ $$QH_2 + 2\,c^{ox} + 2\,H_m^+ \rightarrow Q + 2\,c^{red} + 4\,H_c^+$$

Fig. 7. (Equations 1 and 4) On oxidation of $QH_2$ at hydroquinone oxidation center *o* the first electron is transferred to the iron-sulfur cluster, the second electron to heme $b_l$. (Equations 2 and 5) The first electron on the iron-sulfur cluster is transferred to heme $c_1$, the second electron from $b_l$ to $b_h$. (Equation 3) During the first turnover $b_h$ reduces at the quinone reduction center *i* quinone to the semiquinone which is stabilized by tight binding, and at the high potential pathway the electron flows from heme $c_1$ to cytochrome c. (Equation 6) During the second turnover $b_h$ reduces at center *i* the semiquinone to ubihydroquinone, and a second molecule of cytochrome c is reduced by cytochrome $c_1$.

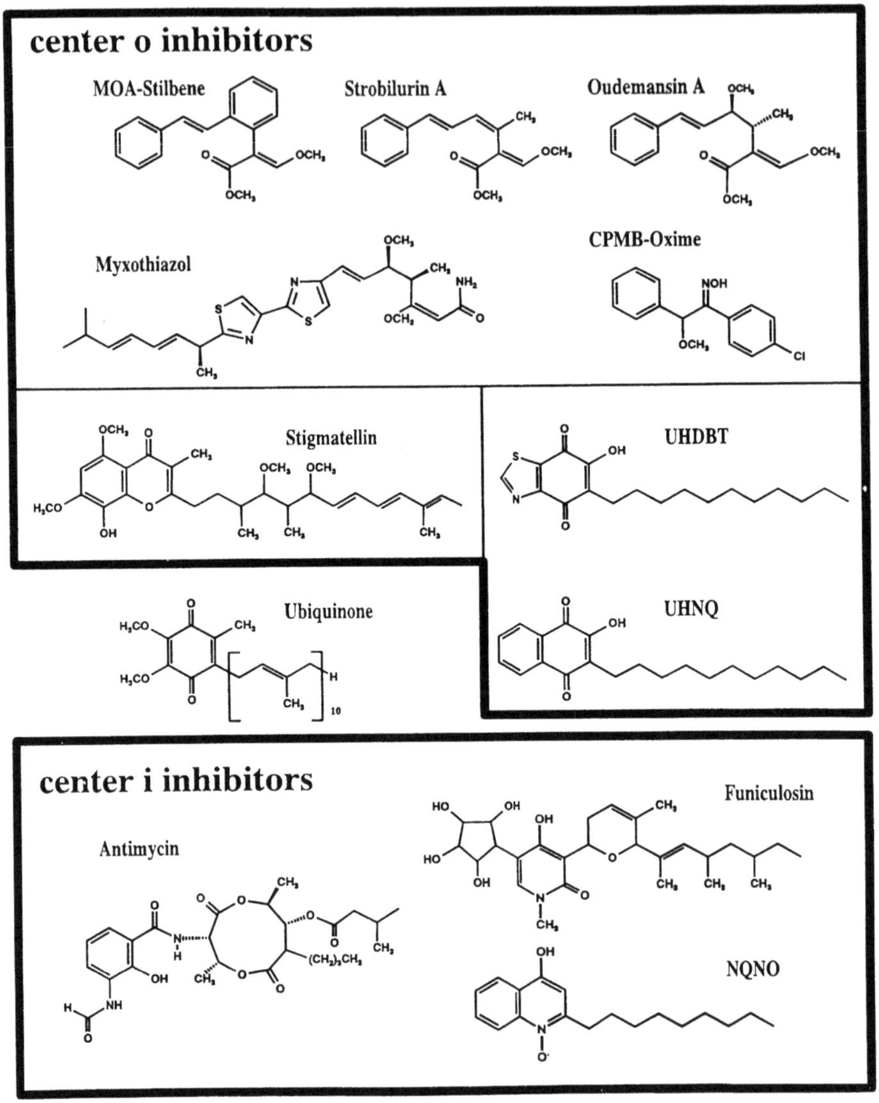


Fig 8. Classification and structural formulae of the inhibitors of center *o* and center *i* of the $bc_1$ complex.

During net oxidation of one molecule of ubihydroquinone $2H^+$ are taken up from the matrix side at center *i* and $4H^+$ are released to the cytosolic side of the membrane at center *o*. As already discussed in section 2, when we include in the account the uptake of $2H^+$ from the matrix side during oxidation of oxygen at complex IV (cf. Fig. 1), the Q-cycle is working with a stoicheiometry of $2H^+/e^-$, i.e. with an efficiency twice as much as that of a redox loop.

In the $bc_1$ complex from *Rb. capsulatus,* two quinone or hydroquinone molecules have been found to bind within center *o* [13]. This finding has led to the proposal of specific reaction schemes for the bifurcation of electron flow occuring during hydroquinone oxidation at this center.

Several inhibitors highly specific for center *o* or center *i* have been isolated from microorganisms and studied in detail [13]; for simplicity, we will refer to them as center *o* and center *i* inhibitors. By means of these inhibitors and their application to inhibitor-resistant yeast mutants, protein domains of different subunits involved in the formation of the two reaction centers have been characterized.

The center *o* inhibitors block the electron transfer from ubihydroquinone to cytochrome $c_1$ and to cytochrome b. They can be subdivided into inhibitors which bind predominantly to the cytochrome b domain of center *o* (methoxyacrylate (MOA) inhibitors) and those binding predominantly to the iron-sulfur domain of center *o*. The latter group is composed of the stigmatellins and of the hydroxyquinones.

The center *i* inhibitors block the electron transfer from heme $b_h$ to ubiquinone. The mutations conferring resistance to the MOA-inhibitors and to antimycin have all been mapped on the cytochrome b gene. The consequences of the interpretation of the folding of cytochrome in the membrane will be discussed in section 6.3.3.

### *5.3. Protein folding of the redox subunits and coordination of the redox centers*

*5.3.1. Iron-sulfur protein:* The iron-sulfur protein has a molecular mass of 21.5 kDa. The [2Fe-2S] cluster is coordinated within a C-terminal water soluble part which is bound to the rest of the complex through an N-terminal membrane anchor. The cluster binding domain forms part of center *o*; the [2Fe-2S] cluster is the primary acceptor of electrons from ubihydroquinone.

The reduced [2Fe-2S] cluster is in a $[Fe^{III}, Fe^{II}]$ mixed valence state with spin $S = 1/2$. This gives rise to a distinct EPR signal centered at $g = 1.91$ which is sensitive to the state of occupancy of center *o* with quinones.

Recently, the structure of the water soluble domain containing the intact [2Fe-2S] cluster has been determined by X-ray crystallography at 1.5 Å resolution (Fig. 9) [15]. The overall fold contains three layers of antiparallel β sheets. The cluster is coordinated by two cysteine and two histidine residues in agreement with previously proposed models. These ligands originate from two loops within a separate domain-like fold; two additional conserved cysteines form a disulfide bridge connecting the two loops. A third loop containing the conserved residue Pro 175 covers the cluster from the other side. The two cluster binding loops show structural similarity to the metal-binding loops of rubredoxin; two of the coordinating cysteine residues of rubredoxin superpose with the two coordinating cysteine residues of the ISF while the Sγ atoms of the other two coordinating cysteine residues superpose with the acid labile sulfur atoms of the [2Fe-2S] cluster.

The distance between the two iron atoms of the [2Fe-2S] cluster is 2.70 Å. The symmetry of the cluster is reduced around the bis-histidine coordinated iron atom (Fe-2) both by the presence of the histidine ligands as well as by a distortion of the tetrahedral coordination geometry. The sulfur atoms of the [2Fe-2S] cluster participate in multiple NH-S and OH-S hydrogen bonds to peptide nitrogen atoms and to hydroxyl groups of two sidechains (serine and tyrosine, respectively). In addition to the presence of electron withdrawing histidine ligands and to the neutral overall charge of the cluster, these hydrogen bonds are essential for maintaining the high redox potential of the cluster (approx. +300 mV).

The two histidine ligands coordinate the Fe-2 through their Nδ atoms while the Nε atoms are fully accessible to solvent and have been proposed to be the binding site for quinones and quinonoid inhibitors. From the pH

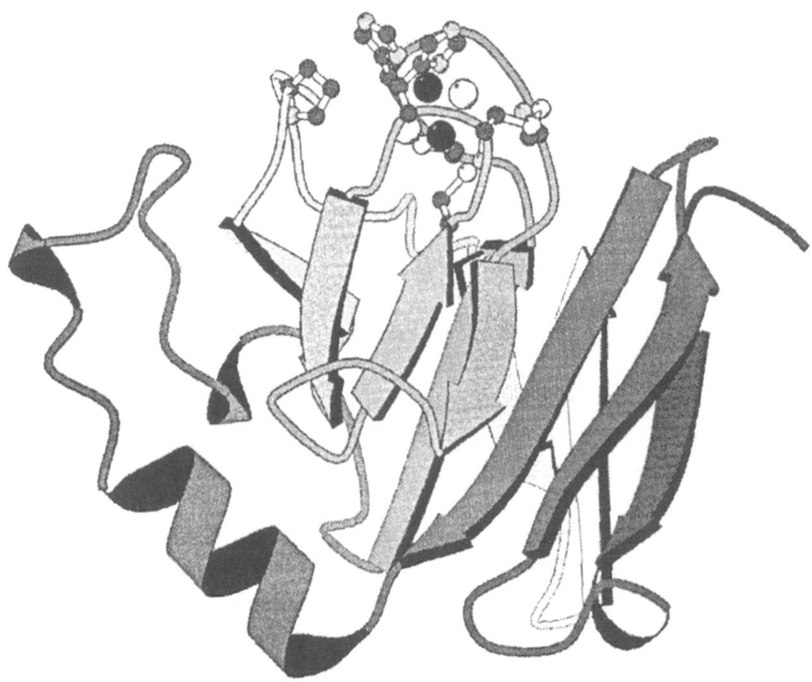

Fig. 9. Schematic ribbon diagram of the structure of the water soluble fragment of the "Rieske" iron-sulfur protein. The [2Fe-2S] cluster is shown as ball-and-stick-model at the top of the protein with the histidine ligands exposed. The disulfide bridge and residue Pro 175 are also indicated.

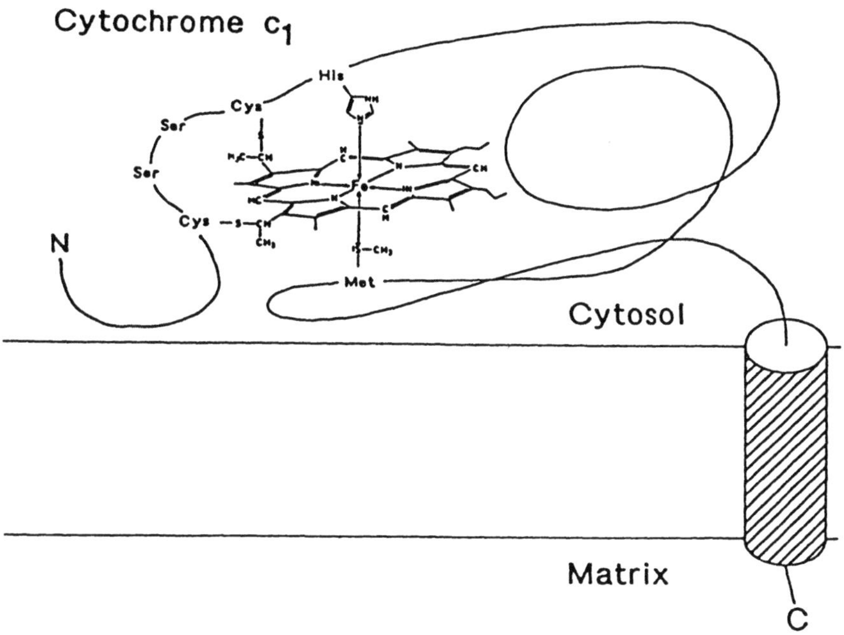


Fig. 10. Predicted folding pattern of bovine cytochrome $c_1$. The porphyrine is covalently bound by two conserved cysteins (Cys-37 and Cys-40) and by His-41 and Met-160.

dependence of the redox potential, the presence of two groups with redox dependent $pK_a$ values has been determined. These groups are most likely the histidine ligands of the [2Fe-2S] cluster. Therefore, the histidines will become protonated upon reduction of the cluster by bound hydroquinone. This reaction may be essential for hydroquinone oxidation at center *o* as this process requires an organized sequence of proton and electron transfer steps.

Binding of some center *o* inhibitors, for instance 5-n-undecyl-6-hydroxy-4,7-dioxobenzothiazol (UHDBT) or stigmatellin, to the iron-sulfur protein leads to an increase of its redox potential. This shows that these inhibitors bind much tighter to the reduced than to the oxidized iron-sulfur protein.

*5.3.2. Cytochrome $c_1$:* The heme center of cytochrome $c_1$ is accessible from the cytosolic surface of the membrane, since cytochrome c –its electron acceptor– diffuses laterally on the cytosolic surface of the mitochondrial membrane. The transient binding of cytochrome c is promoted by the acidic character of cytochrome $c_1$. Cytochrome c possesses basic residues at its binding domain for cytochrome $c_1$ (cf. Fig. 13). The formation of ionic bonds between cytochrome c and $c_1$ can be disturbed, for instance, by high salt concentrations.

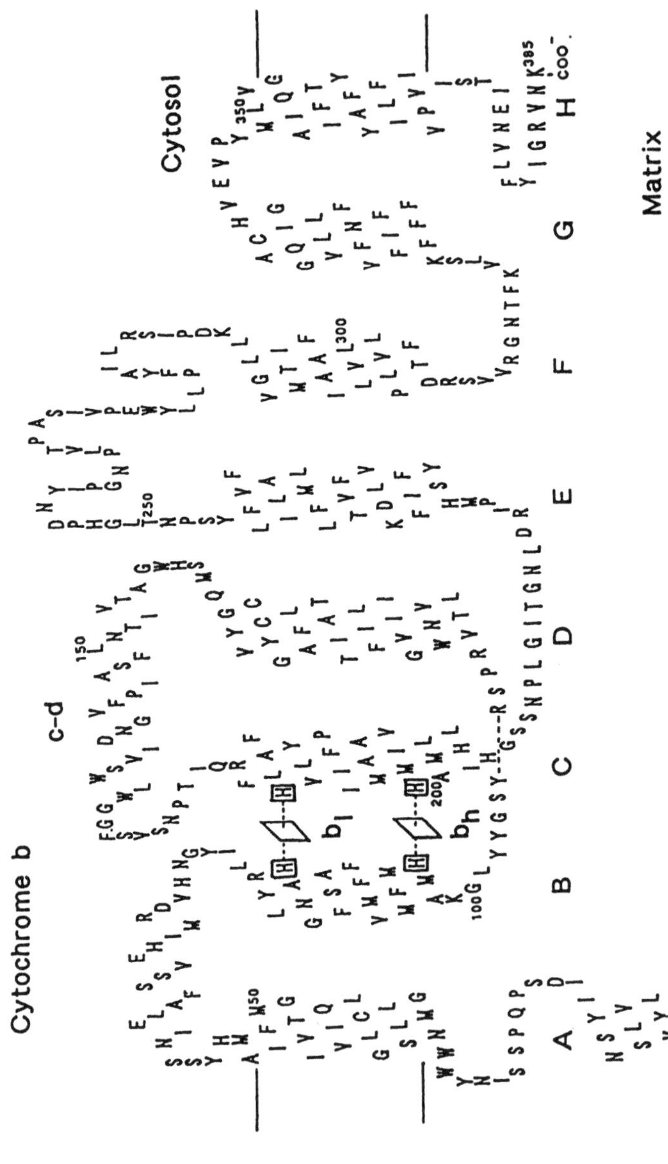


Fig. 11. Predicted secondary structure of yeast cytochrome b. The predicted structure shows eight transmembrane α-helices which are connected by extra-membranous loops. The four conserved histidines which are supposed to supply the fifth and sixth ligands of the porphyrins are boxed in black.

Bovine cytochrome $c_1$ has a molecular mass of 27.2 kDa. The protein can be divided into a hydrophobic membrane anchor and into a hydrophilic head which extends into the intermembrane phase. The hydrophilic domain accommodates the heme center. It transports one electron by reducing the central iron atom from the $Fe^{3+}$ to the $Fe^{2+}$ state.

The primary structure of cytochrome $c_1$ has no detectable homology to that of cytochrome c with the exception of a pentapeptide protein stretch: Cys-X-X-Cys-His. Cys-37 and Cys-40 form covalent thioether bonds with two vinyl groups of the protoporphyrin ring system. His-41 is the fifth ligand, and Met-160 the sixth ligand of the iron atom. The redox potential ($E_{m,7}$) of cytochrome $c_1$ is approximately +250 mV.

*5.3.3. Cytochrome b:* Cytochrome b is the most hydrophobic subunit of the $bc_1$ complex with a molecular mass of 42.5 kDa.

Cytochrome b contains nine hydrophobic helical domains. Eight α-helices are predicted to span the inner mitochondrial membrane. Former helix IV, now designated as the interhelical span c-d, is not transmembrane, but located on the outside of the membrane.

Cytochrome b contains two heme b centers. They are coordinated by four conserved histidine residues which are present in helices B and D. In both helices, the two pairs of histidines are 13 residues apart from each other which results in an iron-iron distance of about 20 Å. Thus the two heme groups allow electrons to traverse the inner mitochondrial membrane since they are electronically connected. Heme $b_l$ seems to be located at the cytosolic surface of the membrane whereas heme $b_h$ is more in the middle of the membrane bilayer. Conserved arginines at the end of the two helices may assist in the binding of the porphyrine ring systems by interacting with the propionic side chains.

Cytochrome b type hemes are iron protoporphyrin IX systems. The two b hemes, both bis-histidine coordinated, have different physical properties. They exhibit different redox potentials and different optical and EPR spectra. Heme $b_l$ has a redox potential of −50 mV, heme $b_h$ one of +50 mV. These differences depend on the specific protein enviroment conferring "steric strains" to the heme centers. Since all six ligand positions of the iron atom are occupied they do not react with oxygen, cyanide or carbon monoxide in their native state.

Various models for the packing of the transmembrane α-helices were suggested. All of the models have helices B and D in close proximity, which supply the four histidine ligands for the fifth and sixth coordination sites of the two atoms, surrounded by other helices. A four α-helical bundle was proposed. The nearest neighbor relations of the helices have been deducted mainly from the analysis of inhibitor-resistant mutants.

Figure 12 shows a model which includes helices A, B, C and D which form the core of cytochrome b [16]. Recently, Crofts gave a model of a

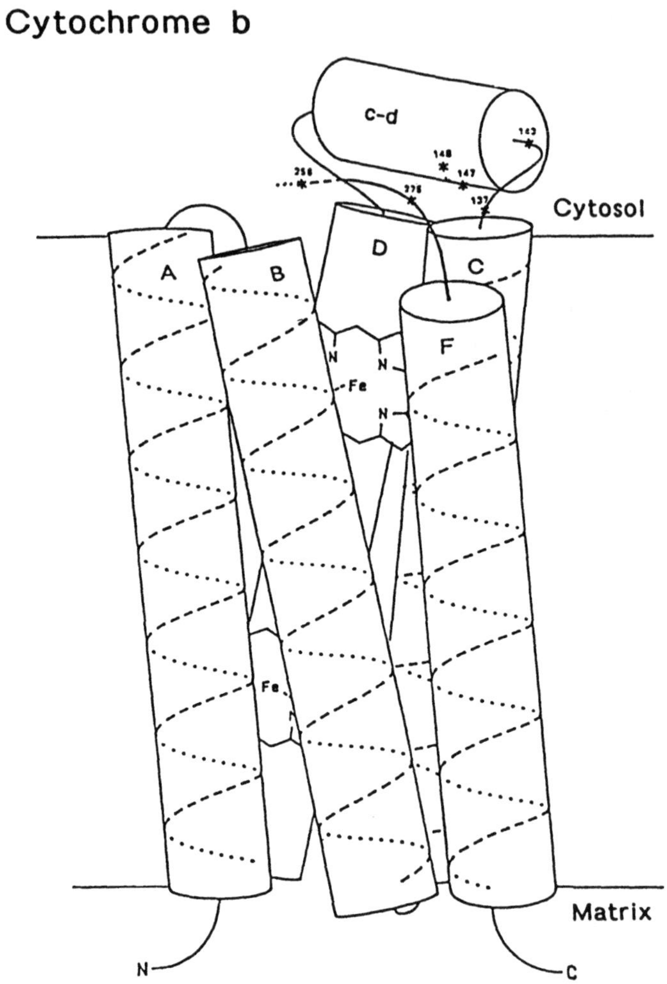


Fig. 12. Model of a 4-α-helical bundle of cytochrome b.

six-α-helical bundle obtained by adding helices E and F to the central core. This model takes into account mutability moments, packing needs at places where conserved glycines occur and conserved protonic residues. Moreover, an important feature leading to the supposed arrangement of the α-helices in the bundle are the mutations conferring inhibitor resistance. Mutations conferring resistance to center *o* inhibitors fall close to cytochrome $b_l$ on the cytosolic side of the membrane and those of center *i* inhibitors close to cytochrome $b_h$ on the matrix side of the membrane.

*5.4. The non-redox subunits*

As already discussed in Chapter 5.1, the number of the non-redox subunits depends on the species studied. The $bc_1$ of *P. denitrificans* is supposed to consist only of the three redox subunits. The complex from *Rhodobacter* contains one or two non-redox subunits. The mitochondrial complexes contain seven (yeast) or eight non-redox subunits (bovine) (cf. Table 3). In bovine mitochondria, more than half of the protein mass is contributed by non-redox subunits.

Following the low-resolution three-dimensional structure picture of the *N. crassa* complex [11], we have calculated that about one-third of the protein mass is embedded in region of the mitochondrial inner membrane. This corresponds to about 600 amino acid residues. Assuming a length of 30 residues for one transmembrane helix including tight loops between the helices which may lie in the membrane region, at least 20 helices can be predicted to be in the membrane. Twelve of these helices are contributed to the redox subunits. Therefore, in accordance with our structural predictions, we assume that the non-redox subunits are anchored within the complex and the membrane by one or two transmembrane helices each.

The knowledge about the function of the non-redox subunits is still meager. Subunit VI (13.4 kDa) of the bovine complex has been reported to bind ubiquinone, hence it is often referred to as the ubiquinone binding protein (QP-C). However, in later experiments it was found that ubiquinone rather binds to subunit VII (9.5 kDa). Presumably one of both subunits contributes structurally to the formation of one or the other of the two quinone reaction centers. Subunit VIII (9.2 kDa) is designated as hinge protein. It is assumed to be involved in the binding of cytochrome c; in the cow it contains a sequence of eight consecutive glutamic acid residues at its N-terminus. The core proteins belong to a protein family formed by matrix processing peptidase (MPP) and the protease enhancing peptide (PEP). It is not clear whether the core proteins are involved in the sorting and assembly of mitochondrial proteins or whether the homology is coincidential. Recency, in potato mitochondria a third core protein has been detected which is the MPP of this system.

Subunit IX was found to be the 78-amino acid presequence of the iron-sulfur protein [17]. The protein is nuclear coded and requires the presequence for mitochondrial targeting; the protein is processed and the presequence is cleaved in a single proteolytic step after it is inserted into the $bc_1$ complex. This is the first instance in which a cleaved presequence has been shown to be retained as a subunit of a mature complex. The functional relevance of this finding is still unclear.

## 6. The diffusible redox carriers ubiquinone and cytochrome c

### *6.1. Structure and function of ubiquinone*

Ubiquinone is a derivative of 2,3-dimethoxy-5-methyl-1,4-benzoquinone with an isoprenoid side chain attached to C-6. The length of the side chain depends on the species studied; in yeast, the length of the side chain is six isoprenoid units, in mammals nine or ten. Ubiquinone occurs in about tenfold excess over the $bc_1$ complex and forms a redox pool. This implies that ubiquinone diffuses freely in the inner mitochondrial membrane and transports redox equivalents between complexes I, II, and III (cf. Fig. 2).

Ubiquinone is highly hydrophobic; the isoprenoid tail diffuses in the membrane interphase. The head group is more hydrophilic and may be located closer to the interface between membrane and aqueous phase.

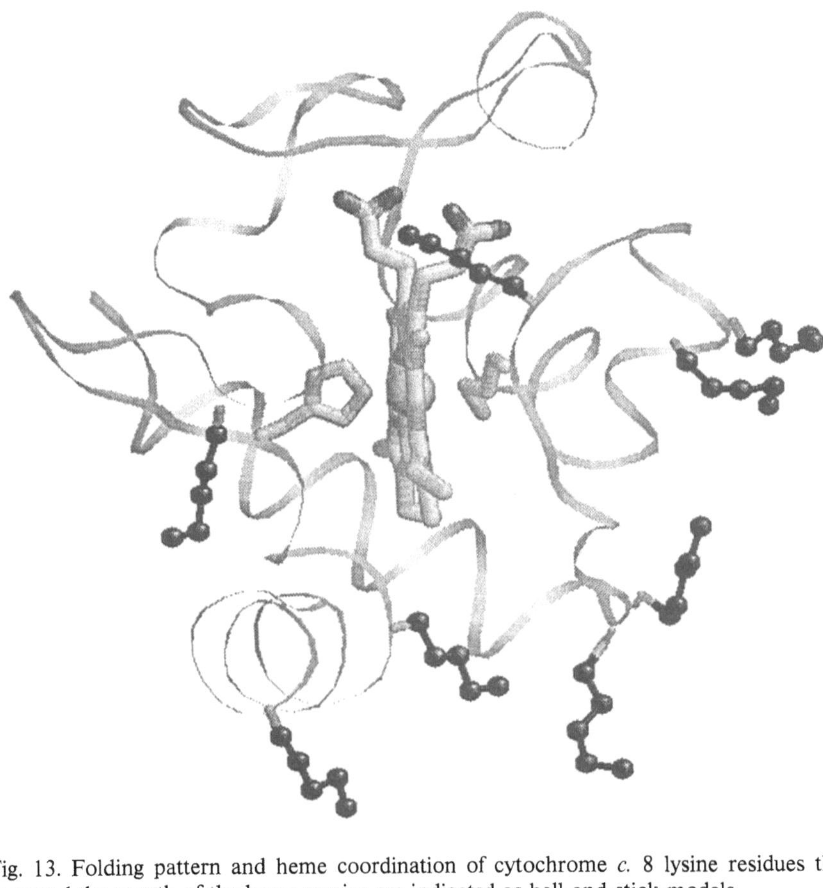

Fig. 13. Folding pattern and heme coordination of cytochrome *c*. 8 lysine residues that surround the mouth of the heme crevice are indicated as ball-and-stick-models.

The existence of tightly bound quinone has been proposed for complexes I and II. Protein subunits of these complexes have been designated as "ubiquinone-binding proteins" since they were labeled in photoaffinity labeling experiments using photoactivatable analogs of quinone. The $bc_1$ complex does not contain tightly bound ubiquinone; all quinone can be removed by detergent washing together with lipids, while the complex can still accept electrons from added hydroquinones.

### *6.2. Structure and function of cytochrome c*

Cytochrome c is a small water-soluble protein with a molecular mass of about 12 kDa. The heme is covalently attached to the apoprotein through two thioether bonds resulting from the addition of cysteine side chains to the two vinyl groups of the porphyrine ring. The heme ligands are one histidine and one methionine. A single electron is transported when the central iron atom is reduced from the $Fe^{3+}$ to the $Fe^{2+}$ state.

The heme group is buried within the hydrophobic core of cytochrome c, but the edge of porphyrin ring C and the adjacent thioether bridge are exposed to the aqueous environmemt. Several basic residues form a positively charged ring around the mouth of the heme crevice and control binding of redox patners. Recently, the crystal structure of a complex between cytochrome c and soluble cytochrome c peroxidase has been solved which is thought to serve as a model for electron transfer complexes of the respiratory chain [19]. The structure differs significantly from previously proposed models. The two heme rings are not coplanar but make an angle of about 60° to each other with an iron-iron distance of 26.5 Å.

## 7. Energy balance of electron transfer reactions

The table shows a summary of the charges $q$ and of the protons translocated across the inner mitochondrial membrane for each molecule of NADH oxidized (i.e., for $2e^-$ transferred from NADH to $1/2\ O_2$).

The numbers for the proton translocation of complexes III and IV are well established while there is still some uncertainty about the stoichiometry of proton translocation by complex I, $H^+/2e^- = 4$ being the most likely number; values up to $H^+/2e^- = 5$ have been proposed. Complexes III and IV both do not translocate an equal number of protons and charges. Two of the four protons appearing during oxidation of $QH_2$ by complex III are "scalar" protons released during the chemical reaction and not transported across the membrane bilayer. Complex IV translocates two charges when protons are taken up during oxygen reduction cycle; therefore, less protons appear on the cytosolic side of the membrane.

Table 4. Charge and proton transfer during electron flow

| | NADH | → | Q | → | cyt c | → | $O_2$ |
|---|---|---|---|---|---|---|---|
| Complex | | I | | III | | IV | |
| $q/2e^-$ | | 4? | | 2 | | 4 | |
| $H^+/2e^-$ | | 4? | | 4 | | 2 | |
| | NADH | | | → | | | $O_2$ |
| Complex | | | | I + III + IV | | | |
| $q/2e^-$ | | | | 10 | | | |
| $H^+/2e^-$ | | | | 10 | | | |
| | succinate | | | → | | | $O_2$ |
| Complex | | | | II + III + IV | | | |
| $q/2e^-$ | | | | 6 | | | |
| $H^+/2e^-$ | | | | 6 | | | |

The most likely proton stoichiometry of respiratory chain under ideal, fully coupled (state 3) conditions is 10 $H^+/2e^-$ for the oxidation of NADH and 6 $H^+/2e^-$ for the oxidation of succinate since complex II does not contribute to the proton translocation. The mitochondrial membrane potential is approximately 0.20 V. Proton translocation by the respiratory chain occurs close to equilibrium, except the last step, electron transfer from cytochrome c to $NADH^+$ by an ATP-induced membrane potential [20] while water can be oxidized by complex IV only to the peroxide level.

## 8. Carrier proteins involved in energy transduction

The synthesis of ATP occurs at the matrix side of the mitochondrial membrane. Therefore, the synthesis of ATP from ADP produced in the cytosol and the provision of ATP for cytosolic energy dependent processes requires the import of ADP and phosphate into the mitochondrion and the export of ATP. This transport requires two carrier membrane proteins: the ADP/ATP carrier (AAC) catalyzing the electrogenic antiport of $ADP^{3-}$ for $ATP^{4-}$ and the phosphate carrier (PiC) catalyzing the electroneutral symport of $H_2PO_4^-$ and $H^+$ (or the equivalent electroneutral antiport of $H_2PO_4^-$ and $OH^-$). Both carriers belong to the family of mitochondrial carrier proteins which includes the mitochondrial substrate carrier proteins and the uncoupling protein. All carriers sequenced so far have a conserved protein structure consisting of three repeats of approximately 100 amino acid residues each.

### *8.1. Protein structure*

The AAC and the PiC both have molecular masses of 33–35 kDa; in the membrane they are present as dimers. Each monomer has between 297

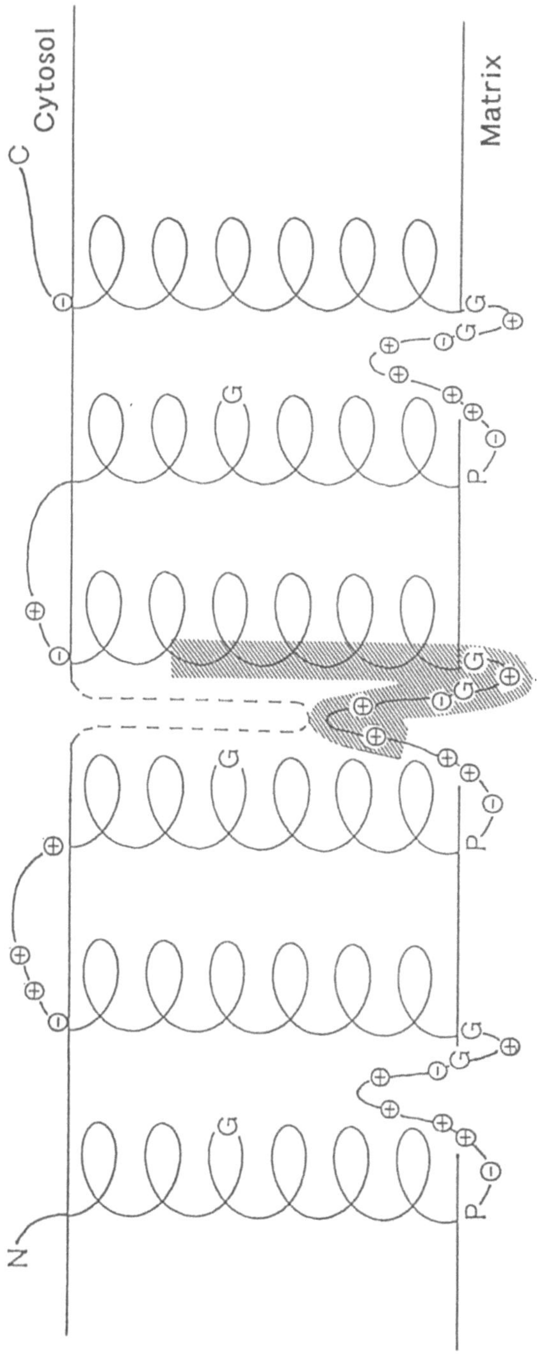


Fig. 14. Folding of the ADP/ATP carrier (AAC) and the phosphate carrier (PiC) monomers in the inner mitochondrial membrane. The shaded area indicates the part of the AAC labeled by the photolabile ATP derivate, 2-azido-ATP.

and 318 amino acid residues. The structure consists of a threefold repeated sequence of approximately 100 residues each. Comparison of the individual repeats shows between 13 and 22% conserved residues. No homology with the bacterial carrier proteins can be detected.

Each repeat contains two hydrophobic membrane-spanning domains; therefore, each peptide chain contains six and each dimer 12 transmembrane α-helices. A putative folding model for both the AAC and the PiC monomers indicating the residues conserved between the two proteins is shown in Fig. 14. Several clusters of conserved residues occur at the end of the transmembrane helices, particularly on the matrix side of the membrane.

Domains involved in substrate binding of the AAC have been identified by photoaffinity labeling with the photolabile ATP derivate, 2-azido-ATP. In bovine heart as well as in yeast AAC, the label was incorporated into the central loop on the matrix within the second repeat.

### *8.2. Catalytic activity*

The catalytic mechanism will be discussed for the AAC since this protein has been studied most extensively. The basic kinetic scheme is shown in Fig. 15 [21].

The sequential mechanism depicted is that of a "single site gated pore" [21] in which only a single substrate molecule is bound to the carrier protein dimer. The carrier has its substrate binding site exposed either to the side (c-state) or the matrix side of the membrane (m-state). In the AAC, these states can be distinguished by binding of two inhibitors, atractyloside from the side to the c-state or bongkrekic acid from the matrix side to the m-state. These inhibitors bind competitively to each other and to the substrates, ADP and ATP.

A full translocation cycle can be described as follows (starting with the free carrier protein in the c-state):

(1) One substrate molecule (e.g., ADP) binds from the cytosolic side.
(2) The protein changes its conformation to the m-state; thereby, the substrate is transported across the membrane.
(3) The bound substrate dissociates, leaving the free carrier in the m-state.
(4) One substrate molecule (e.g.,ATP) binds from the matrix side.
(5) The protein changes its conformation to the c-state.
(6) The bound substrates dissociates, leaving the free carrier in the c-state.

Now, the carrier is in the same state as in the beginning of the cycle and another cycle can start. For this mechanism, it is essential that the carrier cannot change from the c-state to the m-state without bound substrate; otherwise, the carrier would swich to a uniport mechanism and uncoup-

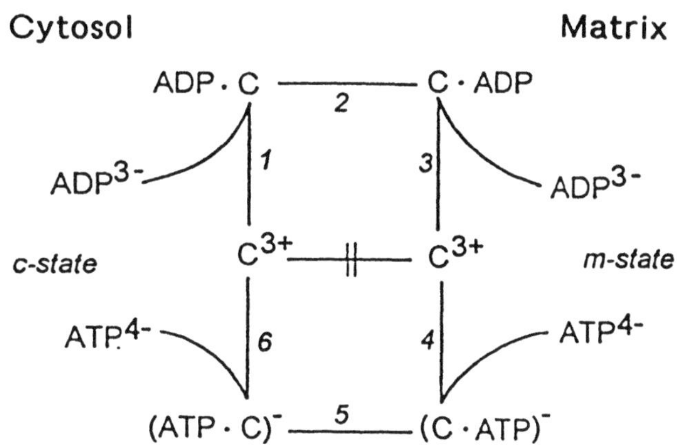


Fig. 15. Catalytic cycle of the ADP/ATP carrier (AAC).

ling would occur. In the case of the AAC, this would lead to the efflux of both ADP and ATP from the mitochondrion, driven by the membrane potential.

The transport cycle of the PiC is similar assuming an antiport of $H_2PO_4^-$ and $OH^-$. For a symport of $H_2PO_4^-$ and $H^+$, the sequence of binding steps must change and the carrier must be able to swich from the c-state to the m-state only in the free form and with both phosphate and the proton bound. During a symport mechanism, uncoupling would occur if the carrier could switch conformations with a single substrate bound.

For the mitochondrial carriers, alternative kinetic schemes have been discussed which involve two substrate molecules bound and transported in opposite directions simultaneously. Neither the structural nor the kinetic data allow to decide unambiguously between these models.

## 9. Interaction of electron transfer chain, ATP synthase, adenine nucleotide carrier, and phosphate carrier, resulting in the coupling of oxidative phosphorylation

According to Mitchell's chemiosmotic theory, the coupling between the exergonic redox processes of the respiratory chain and the endergonic synthesis of ATP is achieved through the membrane potential generated by the proton translocation across the membrane coupled to electron transfer [3]. In this chapter we will discuss the functional interaction between the exergonic and the endergonic processes and the energy balance of ATP synthesis.

*9.1. Coupling*

An experiment demonstrating the coupling of mitochondrial oxidative phosphorylation is shown in Fig. 16.

When isolated, tight mitochondria are suspended in buffer containing substrates and oxygen, only a low "state 4" (controlled state) respiratory activity is observed using, for example, a Clark-type oxygen electrode. When now ADP is added to the suspension, an onset of respiratory activity occurs and the rate of oxygen consumption increases by about a factor of 10 (depending on the quality of the mitochondria). This "state 3" (active state) activity continues until all added ADP is converted into ATP. At this moment, the respiration rate returns to the low "state 4" level which is determind basically by the leak proton conductance of the membrane under these conditions. From the amount of ADP added and the amount of oxygen consumed, P/O (or P/2e$^-$) coefficient can be calculated which indicate the number of ATP molecules generated for each 1/2 $O_2$ reduced to $H_2O$ (see below). These values have to be corrected for leakage, contaminants of non-phosphorylating broken mitochondria, etc.

During state 4, i.e., when no phosphorylation occurs due to the lack of the substrate ADP, the membrane potential is approximately 220 mV and the low respiratory activity balances the leak of protons and other ions through the membrane in order to maintain the membrane potential. After addition of ADP, ATP synthesis leads to a rapid decrease of the membrane potential to approximately 170 mV. The respiratory chain re-

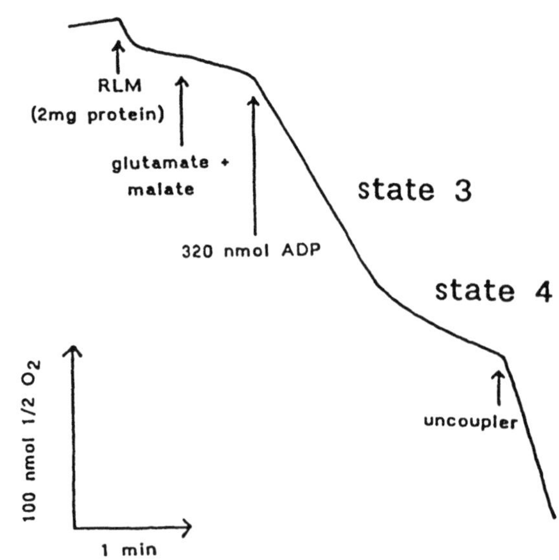


Fig. 16. Coupling of electron transfer and ATP synthesis.

sponds to the decrease of the membrane potential by increasing the respiration rate in order to restore the initial value of the membrane potential. Simultaneously, the leak rate drops to essentially zero due to the strongly non-linear (non-Ohmic) dependence of the leak rate on the potential.

An important tool for studying mitochondrial oxidative phosphorylation were chemical compounds that stop the process by acting at three different steps:

(i) Inhibitors of the electron transfer chain: Numerous compounds inhibit electron transfer, including rotenone and barbiturates (complex I); malonate and thenoyltrifluoroacetone (complex II); antimycin, myxothiazol and other MOA-inhibitors, and stigmatellin (complex III); cyanide and carbon monoxide (complex IV). By addition of these compounds, electron transfer is blocked and ATP synthesis stops since the driving membrane potential is no longer built up. The ATP synthase remains active and it still can hydrolyse ATP.

(ii) Inhibitors of the ATP synthase: The most well known compounds are oligomycin and dicyclohexyl-carbodiimide (DCCD). DCCD binds to the proton translocating $F_0$ part of the ATP synthase, oligomycin to the stalk region of the $F_1$ part. ATP synthesis stops but the electron transfer chain is still active; this resembles "state 4" conditions, however, the electron transfer cannot be activated by addition of ADP.

(iii) Uncouplers: These compounds uncouple electron transfer and ATP synthesis from each other by dissipating the membrane potential. Uncouplers are hydrophobic weak acids or bases; they dissolve in the membrane and shuttle protons across the membrane. Many hydrophobic phenolic compounds have uncoupling activity, including 2,4-dinitrophenol and the highly potent SF6847 (3,5-di-*tert*-butyl-4-hydroxybenzyliden-malodinitrile). In the uncoupled state, ATP synthesis no longer occurs since the proton circuit is short-circuited. Unlike in the electron transfer inhibited state, fast oxygen reduction is observed which is not controlled by the rate of ATP synthesis or the concentration of ADP and ATP. The explanation of the mode of action of the uncouplers was an important piece of evidence in favour of the chemiosmotic mechanism of mitochondrial oxidative phosphorylation.

### *9.2. P/O values and energy balance*

The ATP synthase drives ATP synthesis within the $F_1$ part by allowing protons to pass from the cytosol back to the mitochondrial matrix through the membranous $F_0$ part (see Chapter 12). The most likely stoichiometry of ATP synthesis is 3 $H^+$/ATP generated on the matrix side. For the synthesis of ATP which is used in the cytosol, the additional energy requirement of the transport processes has to be taken into account. The import of ADP and free phosphate into the mitochondrion and the export of ATP into the cytosol require in total 1 $H^+$/ATP. Therefore, the total energy requirement for ATP synthesis is:

| | |
|---|---|
| Cytosolic ATP | 4$H^+$/ATP |
| Mitochondrial ATP | 3$H^+$/ATP (no transport of ATP, ADP, or $P_i$ involved). |

The combination of the proton stoichiometry of both respiratory chain and ATP synthesis and transport gives mechanistic P/O values. Using the most likely values given above, we obtain:

| | P/O values | |
|---|---|---|
| Substrate | NADH | succinate |
| Cytosolic ATP | 2.5 | 1.5 |
| Mitochondrial ATP | 3.3 | 2.0 |

P/O values have been measured in different types of isolated mitochondria [22]. For a long time, values of 3 and 2 using NADH and succinate, respectively, as substrates have been used and these values are still found in most textbooks. However, it is important to understand that if we accept the chemiosmotic theory there is no reason to assume integer P/O values. Integer coupling constants are characteristic for directly coupled, substrate phosphorylation-type mechanisms, since during every single oxidation reaction one molecule of ATP is formed. In respiratory chain-type mechanisms, each oxidation step acts as a generator of membrane potential which is then used by potential consuming processes to drive ATP synthesis, transport of substrates and ions, flagellar motion in bacteria, and other endergonic processes.

Using the P/O values given above, we can calculate the total energy yield of oxidative phosphorylation under fully coupled conditions. From the concentrations of ATP, ADP, and phosphate determined in the cell, the free energy of the reaction

$$ATP + H_2O \rightarrow ADP + P_i$$

in the cytosol has been estimated as

$$\Delta G' = -51 \text{ kJ/mole.}$$

The redox potential for the oxidation of NADH is

$$\Delta E = \{0.82 - (-0.32)\}\ V + (RT\ \ln 10/F) \times \{0.25 \log [O_2] + 0.5 \log ([NADH)]/[NAD^+]) + 0.5\ (pH_{matrix} - 7)\}$$

Using $[O_2] = 0.24$ mM, $[NADH]/[NAD^+] = 0.25$ and $pH_{matrix} = 8$ for respiring mitochondria, we obtain

$$\Delta E = 1.01\ V$$

and

$$\Delta G = -2 \times F \times \Delta E = -195\ kJ/mole.$$

The efficiency for coupled processes is defined as

$$\eta = -\ (J_{ATP}/J_{ox}) \times (X_{ATP}/X_{ox}) = -\ (P/O) \times (X_{ATP}/X_{ox})$$

since the flux ratio $J_{ATP}/J_{ox}$ is the P/O coefficient.

The efficiency for the synthesis of cytosolic ATP from mitochondrial NADH as substrate (e.g., from the citric acid cycle) can now be calculated as

$$\eta = -2.5 \times 51/-159 = 0.65$$

In the cell, the efficiency might be lower due to slip or proton leakage; also, oxidation of NADH generated in the cytosol may have a lower efficiency if transport of the redox equivalents into the mitochondrion requires energy. For example, in insect flight muscle oxidation of cytosolic NADH proceeds largely through $\alpha$-glycerophosphate dehydrogenase where electrons are fed into the respiratory chain at the level of ubiquinone; in this case, the P/O for NADH is only 1.5.

## 10. Perspectives and summary

The current picture of the respiratory chain has emerged from more than 30 years of biochemical work, starting with the identification of the respiratory complexes I to IV. After purification and protein chemical characterisation, the prosthetic groups were identified and the basic reaction mechanisms could be established. During the last ten years, molecular biology has provided us with sequence information and with information on the role of individual residues from the analysis of mutants obtained either by random or site-directed mutagenesis. The availability of this information has allowed the construction of models for the catalytic centers. The next step forward in understanding the protein structure will require high resolution structural information such as has become available for the photoreaction center.

However, although the 3D-structure is a prerequisite, it will not be sufficient for establishing the structure-function relationship. The elucidation of the mechanism will require the application of modern sophisticated time-resolved biophysical, spectroscopic techniques. For example, in cytochrome oxidase Fourier-transform infrared (FT-IR) and Raman spectroscopy has provided insight into the reaction steps of oxygen reduction at the binuclear center. Photosynthetic research has demonstrated how the interplay between spectroscopy, crystallography, and molecular biology leads to a high level of understanding of the system and how the different approaches mutually benefit from each other. This research will require scientists with a deep physicochemical background who will be able to apply the most imaginative physical methods to study the most highly developed biological systems. From this research, we can hope to get fascinating insights at the frontier of nature and science.

## References

1. Y Hatefi, Annu. Rev. Biochem. 54 (1985) 1015–1069.
2. BL Trumpower, Microbiol. Rev. 54 (1990) 101–129.
3. P Mitchell Nature 191 (1961) 144–148.
4. P Mitchell J. Theor. Biol. 62 (1976) 327–367.
5. P Mitchell in Advances in *Membrane Biochemistry and Bioenergetics,* CH Kim, H Tedeschi, JJ Diwan, and JC Salerno (eds.) Plenum Press, New York/London, 1988, pp. 25–52.
6. JE Walker Q. Rev. Biophys. 25 (1992) 253–324.
7. T Yagi Biochim. Biophys. Acta 1141 (1993) 1–17.
8. H Weiss, T Friedrich, G Hofhaus and D Preis Eur. J. Biochem. 197 (1991) 563–576.
9. L Hederstedt and T Ohnishi in *Molecular Mechanisms in Bioenergetics,* L Ernster (ed.) Elsevier, Amsterdam, 1992, pp. 163–198.
10. H Schägger, U Brandt, S Gencic and G von Jagow Methods Enzymol. 260 (1995) 82–96.
11. K Leonard, P Wingfield, T Arad and H Weiss J. Mol. Biol. 149 (1981) 259–274.
12. G von Jagow, TA Link and T Ohnishi J. Bioenerg. Biomembr. 18 (1986) 157–179.
13. H Ding, DE Robertson, F Daldal and PL Dutton Biochemistry 31 (1992) 3144–3158.
14. G von Jagow and TA Link Methods Enzymol. 126 (1986) 253–271.
15. S Iwata, M Saynovits, TA Link and H Michel Structure 4 (1996) 567–579.
16. TA Link, H Schägger and G von Jagow FEBS Lett. 204 (1986) 9–15.
17. U Brandt, L Yu, CA Yu and BL Trumpower J. Biol. Chem. (1993) 8387–8390.
18. RJ Almassy and RE Dickerson Proc. Natl. Acad. Sci. USA 75. (1978) 2674–2678.
19. K Pelletier and J Kraut Science 258 (1992) 1748–1754.
20. M Klingenberg Angew. Chem. 75 (1963) 900–907.
21. M Klingenberg Ann. N. Y. Acad. Sci. 456 (1985) 279–288.
22. PC Hinkle, AM Kumar, A Resetar and DA Harris Biochemistry 30 (1991) 3576–3582.

## Appendix A. Glossary

| | |
|---|---|
| FAD | Flavin-adenine-dinucleotide |
| $FADH_2$ | reduced form |
| FMN | Flavin-mononucleotide |
| $FMNH_2$ | reduced form |
| $NAD^+$ | Nicotinamide-adenine-dinucleotide |
| NADH | reduced form |
| UQ | ubiquinone (2,3-dimethoxy-5-methyl-6-decyl-1,4-benzoquinone) |
| $UQH_2$ | ubihydroquinone (reduced form). |

Bioenergetics
ed. by P. Gräber and G. Milazzo†


# *CHAPTER 7*
# Photosystem II and water oxidation in cyanobacteria, algae and higher plants

Gernot Renger

*Max Volmer Institute for Physical and Biophysical Chemistry, Technical University of Berlin, Germany*

## 1. Introduction

The cornerstone of the evolution of highly organized forms of living matter was the "invention" of a biomolecular device in the early biosphere that was able to dissociate water into dioxygen and metabolically bound hydrogen by using visible light. This event, which occurred in photosynthesizing organisms about 2–3 billion years ago, had two consequences of paramount importance: (a) it allowed the huge water pool on the earth's surface to become available as a hydrogen source for the biosphere, and (b) the resulting formation of dioxygen as a photosynthetic "waste" product led to the present day aerobic atmosphere and as consequence to the formation of the protective ozone layer. The advent of $O_2$ in the atmosphere opened the road for the much more efficient exploitation of the free energy content of food through aerobic respiration of heterotrophic organisms. This immense bioenergetic advantage which can hardly be overestimated, however, had to be paid for by a markedly enhanced risk to the organism of damage due to the formation of very reactive oxygen radicals. Protective mechanisms are therefore indispensible in order to survive in an aerobic atmosphere (for further reading, see ref. 1).

In the following only the processes of the oxygenic ($O_2$-evolving) photosynthesis will be considered. The overall process for the water cleavage can be simply described by:

$$2\ H_2O + 2\ T \xrightarrow{h\nu} O_2 + 2\ H_2 \times T \qquad (1)$$

where T represents a molecule which chemically binds hydrogen. An overall scheme of the functional and structural organisation of photosynthesis in higher plants is shown in Fig. 1. of Chapter 5. Within the cell a special organelle, the chloroplast (see Chapter 5), is the locus of photosynthetic activity. Chloroplasts exhibit a characteristic subdivision into a structurally highly organized membrane system referred to as thylakoids and a less ordered gel-like phase, the stroma. Water cleavage by visible light is carried out through the cooperation of three integral membrane protein complexes that are anisotropically incorporated into the thylakoid membrane: (a) Photosystem II (PS II), (b) Photosystem I (PS I) (see Chapter 9), and (c) Cytochrome $b_6$/f-complex (see Chapter 10).

The key steps of this process take place within the PS II complex. This complex catalyzes the light-induced water cleavage into dioxygen and a moderately reducing form of bound hydrogen. In this part of the reaction sequence, plastoquinone (PQ) serves as the hydrogen acceptor molecule, i.e. the overall reaction of PS II can be described by:

$$2\ H_2O + 2\ PQ \underset{PS\ II}{\xrightarrow{h\nu}} O_2 + 2\ PQH_2 \qquad (2)$$

In the following section the basic principles of this reaction sequence and the structural organisation of the PS II complex will be described.

## 2. Overall organisation of water cleavage in photosystem II

Regardless of the details of its molecular mechanism, the overall process of water cleavage in PS II has to comprise the light-induced abstraction of four electrons and four protons from two water molecules, concomitantly with the formation of a dioxygen bond and an eventual transfer of these electrons to an appropriate acceptor followed by a proton uptake. This goal has been achieved by an ordered array of light-induced redox reactions within the PS II complex in which PQ reduction and $H_2O$ oxidation take place as spatially separated processes. Accordingly, three functionally distinguishable types of reaction sequences take place in PS II:

(1) the generation of an electronically excited state and its transformation into an electron-hole pair with sufficient stability and redox power:

$$PA \xrightarrow{h\nu} P^*A \rightarrow P^+A^- \qquad (3)$$

where P represents a photochemically active pigment and A an electron acceptor.

(2) the cooperative reactions of four holes ⊕ with two water molecules:

$$4 \oplus + 2\,H_2O \rightarrow O_2 + 4\,H^+ \qquad (4)$$

(3) the cooperative reaction of two electrons e with PQ and water:

$$2\,e + PQ + 2\,H_2O \rightarrow PQH_2 + 2\,OH^- \qquad (5)$$

This overall reaction scheme shows that photosynthetic water cleavage in PS II can be considered as a solar water electrolysis which comprises three functional elements: (i) a solar battery referred to as PS II photogenerator, (ii) a highly oxidizing microscopic reaction vessel which is designated as the water oxidizing complex (WOC), and (iii) a reducing site where $PQH_2$ formation takes place. This site in the PS II complex is referred to as the $Q_B$-site.

Based on these general considerations it is clear that an understanding of photosynthetic water oxidation requires a detailed analysis of the molecular mechanisms of reaction sequences (3)–(5) and the properties of the functional elements which catalyze these reactions.

## 3. Light absorption and transformation into electrochemical free energy in photosystem II

### *3.1. General principles*

The reaction sequence described in Eqn. 3 comprises at least two generally different types of processes, i.e. the generation of an excited state by light absorption and the photoredox reaction of an excited pigment molecule P* with a functionally coupled acceptor molecule A. The basic, underlying concept of light absorption and transformation into electrochemical free energy was developed in 1932 through the key experiments by Emerson and Arnold. They discovered that only one molecule of dioxygen is evolved per 2500 chlorophyll molecules and per flash when chlorella cells were illuminated with repetitive short flashes (10 $\mu$s) of saturating intensity. This finding and calculations by Gaffron and Wohl (1936) led to the conclusion that the vast majority of chlorophyll molecules are without photochemical activity. They form together with other pigments (designated as accessory pigments) an antenna system which funnels the absorbed light energy into a special pigment $P_T$ within the reaction center where the photochemistry takes place. The average size of the antenna could be defined as the stoichiometric ratio of the number of chlorophylls per photoactive reaction center. There was a long standing debate about whether this ratio only reflects a statistical value or a well defined structural entity referred to as photosynthetic unit (PSU). It was later found that the antenna system consists of an aggregate of pigment protein subunits, i.e. the photosynthetic unit can be considered as a meaningful entity although a more complex pattern arises if one considers the detailed arrangement (see section 3.3.3 and 3.3.4). Regardless of its structural organization and the problems associated with the determination of its exact size, the concept of the PSU implies that the early steps of light transformation summarized by Eqn. 3 comprise two separate reaction sequences: (a) generation of an electronically excited state by light absorption at a pigment in the antenna and transfer of the excitation energy to the photoactive pigment $P_T$ which is located within an operational unit referred to as reaction center, and (b) photoreaction of the electronically excited $P_T^*$ giving rise to an electric charge separation within the reaction center. The excited states in the ordered pigment arrays are designated as excitons. (The term "exciton" describes excited states in solids which are characterized by both definite energy and momentum. In photosynthesis the term is used in a looser sense as an excited electronic state that resides on more than one pigment of a group of molecules during its lifetime.)

Both of these processes take place in pigment-protein complexes. An understanding of the molecular mechanism of these reactions in photosynthetic organisms requires detailed knowledge of the pigment-pigment and pigment-protein interactions. This is illustrated in Fig. 1 by a very simple scheme which shows two different ways of reactive interactions between an electronically excited molecule $P_1^*$ and a second molecule $P_2$ in its electronic ground state. Depending on the mode of functional coupling, the interaction leads either to excitation energy transport ("antenna" function) or to electron transfer ("reaction center" function). In general, the mode of reaction depends: (i) on the nature of the pigments $P_1$ and $P_2$ and (ii) on the electronic and vibronic coupling between these molecules and of the chromophors with their surrounding protein matrix.

The rate constant of the corresponding processes can be calculated by using Fermi's Golden Rule:

$$k_{12} = \frac{2p}{\hbar} |V_{12}|^2 \times \text{FCWD} \qquad (6)$$

where $V_{12}$ is the matrix element of the electronic coupling between the states 1 ($P_1^*$ $P_2$) and 2 ($P_1$ $P_2^*$ or $P_1^+$ $P_2^-$ in Fig. 2), FCWD represents the Franck-Condon weighted density of states, $\hbar$ = h/2π (h = Planck's con-

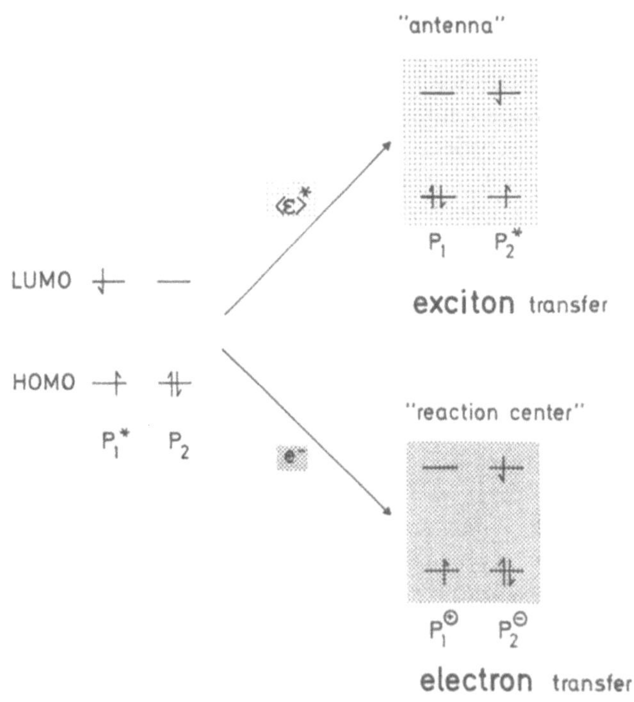


Fig. 1. Schematic representation of the electronic coupling between two pigment molecules in lowest excited state ($P_1^*$) and ground state ($P_2$), respectively, giving rise either to excited state (top) or electron (bottom) transfer from $P_1^*$ to $P_2$. For the sake of simplicity, only the highest occupied (HOMO) and lowest unoccupied (LUMO) molecular orbitals of the pigments are considered.

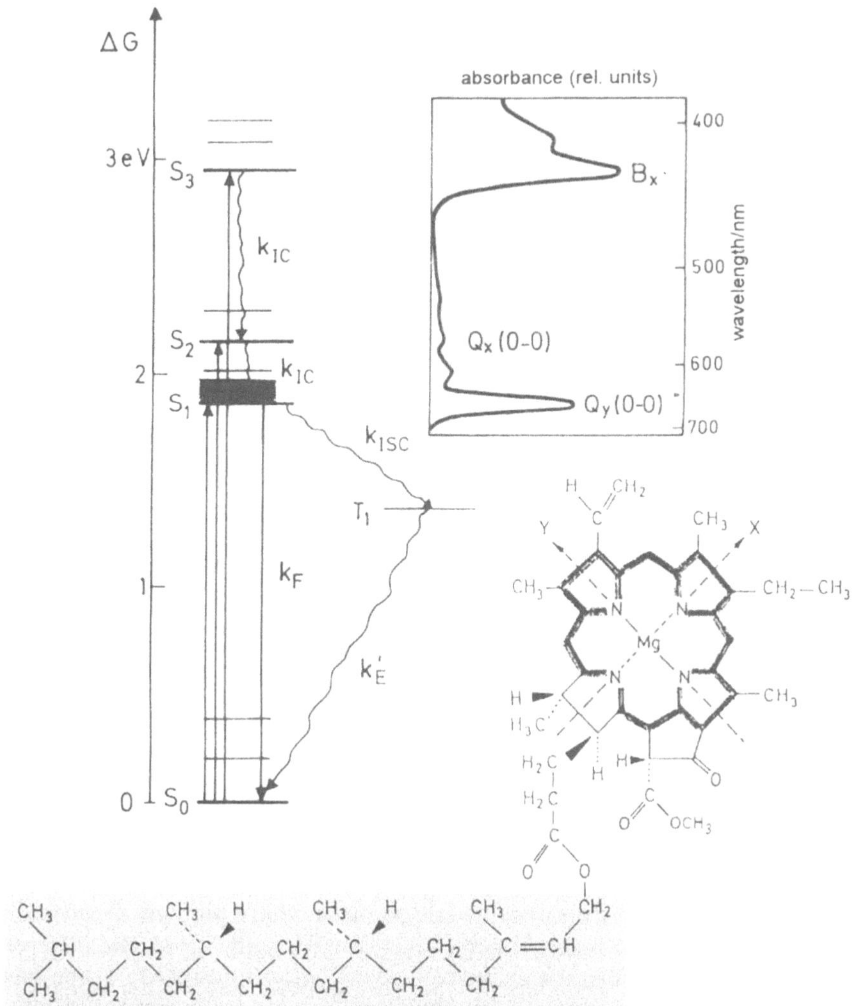


Fig. 2. Molecular structure, energy levels and absorption spectrum of chlorophyll-a in solution. The conjugated $\pi$-electron system is symbolized by dark gray areas, the phytyl chain by a light gray area (chlorophyll-b differs only in the substitution of the $CH_3$ group in pyrrole ring II by a $-C\genfrac{}{}{0pt}{}{\leq O}{\langle H}$ group). The energy levels of electronically excited states are characterized by thick bars, the vibrational states of the molecules by thin bars, whereas the phonons with their much smaller energy differences form a quasi continuum symbolized by a dark-gray area (only $S_1$ is considered). The absorption spectrum of chlorophyll-a in ethanol was taken from Petke et al. (1979)*. The optical transitions are symbolized by capital letters (B, Q) and their orientation within the frame of molecular axes by indices (x, y). The rate constants, $k_F$, $k_E'$, $k_{IC}$ and $k_{ISC}$ symbolize fluorescence, phosphorescence, internal conversion and intersystem crossing, respectively. For the sake of simplicity, only 0-0-transitions are shown, and radiationless transitions to the ground state are omitted.

*YD Petke, GM Maggiora, LL Shipman and RE Christoffersen, Photochem. Photobiol. 30 (1979) 203–223 stant). The value of $V_{12}$ is determined by the coupling of the donor and acceptor electronic wavefuncions and therefore strongly depends on the distance between the reactants. On the other hand, the term FCWD is given by the nuclear motions (of the reactant and their microenvironment) which accompany a particular reaction. Both parameters are determined by the structure of the protein matrix. Equation (6) can be applied to both processes, i.e. exciton migration and electron transfer. Explicit equations will be given in sections 3.3 and 3.4. Based on the above-mentioned considerations, it is reasonable to consider the polypeptides of the antenna and reaction centers as the apoenzymes of the pigments incorporated into these functional units, because the great variability of proteins in space and charge permits a perfect tuning of each reaction coordinate. Therefore, by using this "ideal" material practically any degree of specificity, efficiency and regulatory control required for a certain biological function could be achieved during the long evolutionary process.

With the above-mentioned ideas as guideline, first the basic properties of chlorophyll-a as the key chromophore of all oxygen evolving photosynthetic organisms are briefly summarized. Subsequently, excitation energy transfer from pigments to the photoactive chlorophyll $P_T$ and electron transfer within the reaction centers of photosystem II (PS II) will be discussed.

### *3.2. Properties of chlorophyll-a molecules in solution*

Chlorophyll-a (Chl-a) plays a unique role because, according to our current knowledge, this molecule is the only photoreactive pigment $P_T$ in both photosystems (PS I and PS II) of all oxygen-evolving photosynthesizing organisms. This situation markedly differs from that of the reaction centers in anoxygenic photosynthetic bacteria where pigment $P_T$ contains various bacterio-chlorophyll derivatives (BChl-a, -b, -g) in the different bacteria. In contrast to the exclusive use of Chl-a (and its $Mg^{2+}$-free base pheophytin a, see section 3.4) in the reaction centers, a variety of tetrapyrrol-pigment molecules are components of the different types of antenna systems (see section 3.3.3).

The structure and the basic spectral properties of Chl-a in dilute solution are summarized in Fig. 2. The molecule contains a conjugated $\pi$-electron system (symbolized by thick black bonds) and a long hydrophobic phytyl tail (gray box). The molecular orbital structure of the conjugated p-electron system determines the optical properties of Chl-a. The absorption spectrum of monomeric Chl-a in solution is characterized by strong bands in the red and blue (Soret band) due to spin- and symmetry-allowed $\pi\pi^*$-transitions. In bacteriochlorophyll the $\pi$-electron system is less quadratic because of the reduction of a double bond in ring II, giving rise to a shift of the $Q_y$ transition into the near infrared region and of the Soret-band into the UV. As chlorophyll-a contains heteroatoms (nitrogen, oxygen) with occupied non-bonding orbitals, in addition $n\pi^*$-transitions arise, but with markedly lower oscillator strength. The relative position of the energy levels of $\pi\pi^*$ and $n\pi^*$-transitions depend on the environmental polarity. Whether $n\pi^*$-transitions play a functional role in photosynthesis remains to be clarified.

The higher electronic singlet states of chlorophyll rapidly relax via a radiationless internal conversion process in the ps time domain down to the level of lowest excited $^1$Chl-a*. In ethanolic solution $^1$Chl-a* has a lifetime of about 5 ns. It decays with a probability of about 1/3 under emission of radiation (fluorescence) to the singlet ground state ($k_f = 6.5 \times 10^7$ $s^{-1}$) and with a probability of about 2/3 via intersystem crossing into the lowest triplet state $^3$Chl-a ($k_{ISC} = 1.3 \times 10^8$ $s^{-1}$). Radiationless dissipation of Chl-a* to the singlet ground state is almost negligible in protic solutions of monomeric Chl-a (in contrast, the radiationless decay is dominant in Chl-a aggegates). Because of the very fast internal conversion of higher singlet states, chlorophyll-a emits only in the red spectral region and also the photochemical reactions start from the lowest excited singlet or triplet state, regardless of the excitation wavelength. The triplet state plays an important role for *in vitro* reactions of Chl-a because of its much longer lifetime (order of 1 ms), whereas the redox reactions within the photosynthetic reaction centers directly start from the lowest excited singlet state of $^1P_T^*$ (see section 3.4).

Electronically excited chlorophyll-a and its $Mg^{2+}$ free derivative, pheophytin a, can either act as electron donor or acceptor depending on the type of reactants. These properties are of central relevance for the primary processes within the rection centers.

The central $Mg^{2+}$ of chlorophyll is coordinatively unsaturated. Therefore, chlorophylls form complexes with proper ligands and also with each other due to the donor properties of carboxylic groups and the acceptor function of the central $Mg^{2+}$ ion. The type of complex depends on the nature of the solvent. In general, apolar solvents favor the formation of chlorophyll aggregates, while in polar solvents complexation occurs either between monomeric chlorophyll-a and the solvent (nucleophilic polar) or oligomeric adducts are formed, comprising bifunctional polar liquids (e.g. $H_2O$) as a bridge. In photosynthetic organisms, the chlorophyll molecules are bound to proteins. This gives rise to a special ligation. Information about the type of coordination can be obtained by using modern spectroscopic techniques like Resonance Raman Spectroscopy and Fourier Transform Infrared (FTIR) Spectroscopy. Recent data showed that the $Mg^{2+}$ of chlorophyll-a in PS II reaction centers is penta-coordinated.

The optical and reactive properties of chlorophyll do not only depend on the mode of coordination by the protein matrix but also on the interaction with neighbor pigments. If the interaction becomes sufficiently strong, a band splitting arises referred to as Davydov-splitting. Of special interest are the properties of homodimers with specifically oriented pigment planes because such special pairs are often "used" as the photoactive pigments $P_T$ of the reaction centers (see ref. 3 and references cited therein).

### *3.3. Antenna systems in $O_2$-evolving photosynthetic organisms*

Optimal exploitation of solar radiation by photosynthetic organisms requires suitable adaptation mechanisms to different environmental conditions (light, temperature, minerals, etc.). Among these parameters the adaptation to the light conditions plays a key role. Photosynthetic organisms are exposed to widely varying photon flux densities (PFD). Leaves, in bright sunlight, are illuminated by a few milli-Einstein/$m^2$; whereas, plants growing at the bottom of tropical rain forests or deep sea algae have to satisfy their free energy demand with PFD values that are orders of magnitude lower. Moreover, the spectral distribution of the radiation can differ markedly, e.g., the lower parts of a tree canopy receive light not only diminished in its PFD but also affected by the preferential absorption of blue and red light in the upper part. The adaptation to the PFD conditions occurs at the excitonic level and, therefore, it is not surprising to find a great variety of antenna systems within the different oxygen-evolving photosynthetic organisms, i.e., cyanobacteria, algae and higher plants; whereas, at the electronic level, the water-cleavage machinery underwent only comparatively minor changes during evolution (see section 3.4). A detailed description of this interesting topic is beyond the scope of this chapter and, therefore, the following considerations can focus only on basic principles underlying the antenna function and discuss the structural organization for an arbitrarily selected example (for further reading, see refs. 2,3). The antenna system of green plants appears to be very suitable to illustrate the latter point.

*3.3.1. General problems of light adaptation:* The basic functional role of all antenna systems is the increase of the optical cross-section of the photoactive pigments in the reaction centers by about two orders of magnitude. As an additional regulatory function, in all oxygen evolving systems a proper distribution of excitation energy between the reaction centers of PS I and PS II, respectively, is required in order to optimize the overall electron transport under light-limiting conditions. In respect to higher plants, it should be kept in mind that almost all of them are practically fixed to their place. Except for leaf orientation, they cannot respond to varying light conditions by large-scale phototropic movements like some microorganisms. Accordingly, the antenna systems are adapted to the prevailing photon flux densities and markedly differ in their size for light and shadow plants.

Besides the adaptation of plants to prevailing PFD values, a proper exploitation of solar radiation also requires a chromatic adaptation. Chlorophyll-a predominantly absorbs in the blue and red, while the absorption in the green and yellow is comparatively weak (see Fig. 2). In order to fill this gap, the antenna systems contain different accessory pigments which transfer their excitation energy via Chl-a to the reaction centers. The nature of these pigments depend on the type of organism (see section 3.3.3).

The funneling of excitons to the reaction centers and their transformation into electrochemical Gibbs energy occurs exclusively via singlet states. The rate of the latter process is kinetically limited by the turnover rate of the whole machinery. Hence, at high PFD the antenna systems have an excess of excitons. This superfluous excitation is predominantly dissipated via fluorescence emission and radiationless decay. Part of it, however, leads to chlorophyll triplets via intersystem crossing (see Fig. 2), reacting with ground state triplet $O_2$ by formation of singlet oxygen, as described by

$$^3\text{Chl} + {}^3\Sigma_g\,O_2 \rightarrow \text{Chl} + {}^1\Delta_g\,O_2 \tag{7}$$

where $^3\Sigma_g$ and $^1\Delta_g$ are the term symbols of $O_2$ in its ground triplet and lowest excited singlet state, respectively. This singlet oxygen species can give rise to photodynamic destructions. Therefore, a protection mechanism is needed to eliminate this reaction. It has been found that carotenoids play a key role in this function, because carotenoid deficient mutants of photosynthesizing algae cannot survive under aerobic conditions. The protection occurs via triplet-triplet energy migration from $^3$Chl to carotenoids according to the reaction:

$$^3\text{Chl} + {}^1\text{Car} \rightarrow {}^3\text{Car} + {}^1\text{Chl} \tag{8}$$

The carotenoid triplets formed in this reaction dissipate the excess energy in the μs range via radiationless decay. In addition, carotenoids are also efficient quenchers of singlet oxygen according to the reaction:

$$\text{Car} + {}^1\Delta_g\,O_2 \rightarrow {}^3\text{Car} + {}^3\Sigma_g\,O_2 \tag{9}$$

Recently, the protein matrix of the antenna complexes was reported to provide an additional protection by partial screening of the pigments from access to $O_2$. In spite of these protective mechanisms, however, all oxygen evolving organisms are susceptible to degradative processes under strong illumination (light stress). This phenomenon designated as photoinhibition affects mainly PS II. There exist mechanisms permitting some dissipation of excess excitation energy by enhanced radiationless decay in the antenna. For the interesting physiological implications of light stress and strategies of protection and repair of PS II the interested reader should consult the literature (for a review, see ref. 4 and references cited therein).

*3.3.2. Principles of exciton migration in photosynthesis:* The antenna systems are aggregates of pigment-protein subunit each containing 3–20 pigment molecules. The nature of the pigment molecules and their number per photosynthetic unit greatly depend on the type of organism. In general, the antenna can be considered as a two- or three-dimensional array of pigment molecules containing the photoactive pigment, $P_T$, of the reaction center serving as a trap for the excitation energy. For the sake of simplicity the detailed structure of the pigment protein subunits and their association to form an antenna-reaction center complex will be neglected. In this case, a photosynthetic unit can be described by a lattice of pigment molecules $P_1$, ... $P_N$ which is connected with the photochemically active trap molecule $P_T$. The simple scheme of a 2-dimensional quadratic array with $P_T$ in its center is depicted in Fig. 3. Light absorption leads within about 1 femtosecond ($10^{-15}$ s) to the generation of an electronically ex-

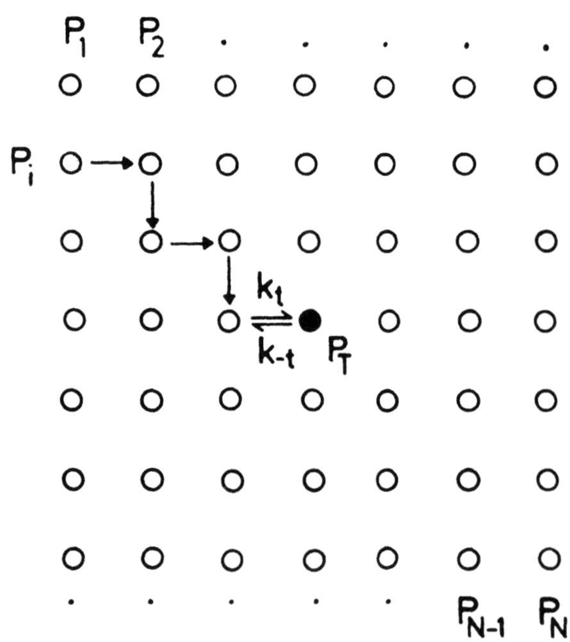


Fig. 3. Scheme of a photosynthetic unit consisting of N pigment molecules $P_1$ ... $P_N$. The central molecule is assumed to act as a photochemical trap, $P_T$. The rate constant $k_t$ and $k_{-t}$ describe the pairwise exciton transfer between a neighbor pigment molecule and the trap, $P_T$, in the forward and reverse direction, respectively. In the simplest simulation model, all pairwise rate constants for excitation transfer between two pigment molecules are assumed to be the same.

cited state at any pigment of this array, e.g. $P_i$. Depending on the strength of the electronic interaction between the pigments and the vibronic coupling, this excited state can become delocalized over the ensemble by two alternative mechanisms: (i) a wavelike coherent process, where the exciton cannot be localized at any time to a particular pigment of the ensemble or (ii) a rapid dephasing relaxation process takes place leading to loss of coherence. In the latter case an exciton can be considered to be localized for a very short time at a particular pigment $P_i$ and to migrate with a transfer rate constant $k_{ij}$ to a neighbor pigment $P_j$. In this way, the exciton arrives via an incoherent random walk process at the photoactive pigment $P_T$. In any random walk process the arrival of an exciton at $P_T$ depends on the value of the pairwise rate constant $k^*_{ij}$ and on the size and pigment topology of a particular lattice array. Therefore, first the pairwise rate constant will be described and then some problems of the random walk process from antenna pigments to the reaction center are discussed.

*3.3.2.1. Pairwise excitation energy transfer:* Taking into account the contributions of exchange and Coulomb interactions between pigments $P^*_i$ and $P_j$, application of Eqn. 6 to the problem of pairwise exciton transfer from an electronically excited pigment $P^*_i$ to a ground state pigments $P_j$ gives rise to the formulas

$$k^*_{ij} = \frac{2\pi}{\hbar} |U^{exch}_{ij}|^2 \int F_D(\nu)\, \varepsilon_A(\nu)\, d\nu \qquad (10)$$

where $U^{exch}_{ij}$ is the electronic matrix element for exchange interaction and $F_D(\nu)$ and $\varepsilon_A(\nu)$ are the spectral distributions of the normalized donor ($P^*_i$) emission and of acceptor ($P_j$) absorption, respectively, and:

$$k^*_{ij} = \frac{A\,\kappa^2}{\tau_o n^4 R^6} \int F_D(\nu)\, \varepsilon_A(\nu)\, \frac{d\nu}{\nu^4} \qquad (11)$$

where A is a constant, $\kappa^2$ the factor describing the mutual orientation of the transition dipole moments of the donor ($P^*_i$) and the acceptor ($P_j$), $\tau_o$ the intrinsic lifetime of $P^*_i$, n the refractive index, R the distance between the transition dipoles of $P_i$ and $P_j$ and $F_D(\nu)$ and $\varepsilon_A(\nu)$ have the same meaning as in Eqn. 10. It must be emphasized that derivation of Eqn. 11 additionally comprises the validity of the point dipole approximation.

Equation 10 derived by Dexter, describes the exchange mechanism, and Eqn. 11 results as dipole-dipole approximation of the Coulombic interaction referred to as Förster excitation energy transfer.

Both mechanisms (exchange and Coulomb) have a markedly different dependence on the distance between pigments $P^*_i$ and $P_j$ and also different spin selection rules. The exchange interaction depends on the overlap of the electronic wavefunctions. As a consequence, the efficiency of the Dexter mechanism is normally restricted to rather short distances (< 1 nm). In marked contrast, the Coulomb-type mechanism permits radiationless excitation energy transfer over significantly longer distances. On the other hand, regarding the spin selection rules, the Förster mechanism allows only excitation energy transfer between pigments of the same spin state in the ground and excited state of molecules $P_i$ and $P_j$, respectively, while the exchange interaction also permits transfer from triplet state to a pigment in singlet ground state. Therefore, the protective carotenoid reaction for quenching chlorophyll triplets (Eqn. 8) requires an exchange type mechanism (Dexter), i.e., a special arrangement and close contact of the pigments participating in the process. This restriction is not necessary for singlet-singlet excitation energy transfer, which can take place via the Coulomb-type Förster mechanism.

In practice, the Förster equation is often used in the form

$$k_{ij} = \frac{1}{\tau_0}\left(\frac{R_0}{R}\right)^6 \qquad (12)$$

where $R_0$ is a characteristic distance (Förster radius) for the specific system. Comparison with Eqn. 11 shows that

$$R_0^6 = \frac{A\kappa^2}{n^4}\int F_D(\nu)\,\varepsilon_A(\nu)\,\frac{d\nu}{\nu^4} \qquad (12a)$$

$R_0$ values of 4–10 nm were calculated for systems containing chlorophyll as pigment constituent. Using these equations, the random walk process can be evaluated in principle for any particular pigment topology of the lattice.

*3.3.2.2. Exciton migration from antenna pigments to the reaction center:* A complete description of the fate of an exciton created by light absorption within a photosynthetic unit requires detailed knowledge about the structural organization of the pigments within the different polypeptides of the antenna and reaction centers as well as of the electronic and vibronic coupling between the pigments themselves and the polypeptide matrix. A straight-forward solution of this extremely complex problem is far from being available. Numerous attempts have been reported in the literature to describe the exciton dynamics for regular lattice arrays of pigments (see Fig. 3) as models for antenna systems. In the extreme case of very strong pigment-pigment interaction, the excited state would be distributed over the whole ensemble, while in the case of weak or very weak coupling an incoherent hopping process takes place, where the excitation can be considered as a localized entity being transferred from pigment to pigment by a sequence of pairwise transfer steps with a rate constant described in the former section (3.3.2.1). There is a long standing dispute of localized versus delocalized excitons in photosynthesis. Theoretical studies led to the conclusion that immediately after its creation, the motion of an exciton is coherent (like in a strongly coupled homodimer) but after a short time (< 100 fs for bacteriochlorophyll) this coherence is lost. Based on these results, the Förster-type exciton transfer mechanism is inferred to provide a proper description of the excitation-energy migration within the photosynthetic units. It cannot be entirely excluded, however, that at least in some cases, the coherent exciton transfer could also be relevant, e.g. for the processes between the pigments located inside of the same subunit of the antenna system. In the following some general features are described which result if excitation energy migration is assumed to occur via an incoherent hopping random walk process.

Without discussing details, a general treatment within the framework of incoherent excitation energy transfer reveals that the average overall lifetime of an exciton within the photosynthetic unit, $<\tau>$, can be approximated by the expression (for furher details, see e.g. refs. 5,6):

$$<\tau>^{-1} = <\tau_{loss}>^{-1} + [<\tau_{migr}> + <\tau_T>]^{-1} \qquad (13)$$

where $<\tau_{loss}>$ results from exciton loss processes (fluorescence emission, radiationless decay, irreversible transformation into electrochemical free energy) in the system, $<\tau_{migr}>$ is designated in the literature as the first passage time and $<\tau_T>$ as the trapping time.

The first passage time describes the average time that an exciton created at any antenna pigment needs to reach for the first time the photoactive pigment $P_T$ of the reaction center. Its value depends on the structure of the lattice and on the magnitudes of the pairwise transfer-rate constants. The trapping time is the time required for an exciton to become photoconverted into a radical pair after its first arrival at $P_T$. $<\tau_T>$ is identical with the mean lifetime of an exciton that is directly created by light absorption at the photoactive pigment, $P_T$. This characteristic trapping time is given by

$$<\tau_T> = \frac{1}{q_T k_t}\left[1 + \frac{k_{-t}}{k_t}(N-1)\right] \qquad (14)$$

where $q_T$ is a quenching parameter which is determined by the photoconversion process at $P_T$, $k_t$ and $k_{-t}$ are the (Förster) rate constants of exciton transfer between the nearest neighbor pigments of $P_T$ in the forward and backward direction, respectively, and N the number of pigment molecules of the lattice (for the sake of simplicity, all pairwise transfer rate constants between antenna pigments i and j are assumed to be the same). In PS II, the reaction center is inferred to be a rather shallow trap; i.e. $k_t$ and $k_{-t}$ are of the same order of magnitude. As a consequence of $k_t \approx k_{-t}$, the trapping time $<\tau_T>$ is almost a linear function of the antenna size. The quenching parameter $q_T$ is determined by the function of the reaction center (see section 3.4).

In the description presented so far, the system was simplified by the assumption that the antenna consists of identical pigments. If, however, the existence of different pigment forms is taken into account, the rate constants of the forward and back exciton transfer between pigments $P_i$ and $P_j$, $k^*_{ij}$ and $k^*_{ji}$, respectively, are different. If the Stokes shift of both molecules is similar, $k^*_{ij}$ and $k^*_{ji}$ are related by the energy difference, $\Delta\varepsilon_{ij}$, of the excited states:

$$k^*_{ij} = k^*_{ji} = e^{-\Delta\varepsilon_{ij}/kT} \quad (15)$$

This effect introduces a "directionality" of exciton migration from the shorter to the longer wavelength pigments forms. This is of special relevance for the interaction between the accesory pigments and Chl-a in the antenna. Also, differently absorbing Chl-a (due to the interaction with the protein matrix) with longer wavelength forms in the neighborhood of the photoactive pigment, $P_T$, would improve the funneling of excitation towards the reaction centers. On the other hand, this directionality could hamper the escape of the excitons from the neighborhood of a functionally inactive (closed) and transfer to an open reaction center. The effect of antenna pigment heterogeneity on the diffusion rate of excitons and its consequences for the "communication" between different photosynthetic units of PS II and the regulation of exciton distribution between PS I and PS II have been thoroughly analyzed by model calculations.

*3.3.3. Structural and functional organization of PS II antenna systems in oxygen-evolving systems:* Essentially two types of antenna systems can be distinguished in PS II: (a) the reaction center complex itself contains pigments that form the inner core antenna (see also section 3.4), and (b) light harvesting complexes. During the evolutionary process leading from cyanobacteria to higher plants, the adaptation of the oxygen-evolving photosynthesizing organisms to different environmental light conditions led to the development of quite different light-harvesting complexes connected with an almost invariant inner core antenna. Two basically different types of outer and distal antenna complexes (see Fig. 4) can be distinguished: (i) phycobilisomes as the prototype of light-harvesting complexes in cyanobacteria and red algae, and (ii) integral light-harvesting complexes of green algae and higher plants.

Both types of light harvesting systems exhibit remarkable differences in their pigment composition and structural organization (see Fig 4). The chromophores of phycobilisomes are bile pigments (phycocyanobilins and phycoerythrobilins) which are covalently linked with cysteine residues of the protein via thioether bonds. Phycobilisomes which exhibit either hemidiscoidal (cyanobacteria and some red algae) or hemispherical (sev-

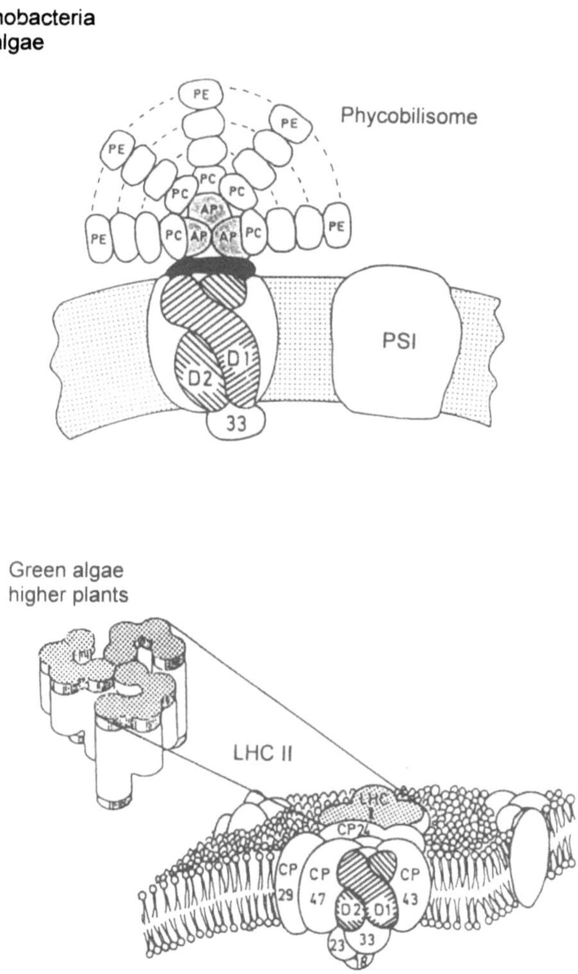


Fig. 4. Simplified scheme of the organization of the antenna systems in cyanobacteria and red algae (top) and in green plants (bottom). The top part shows a hemispherical phycobilisome associated with the PS II complex via the linker protein symbolized by a black area (modified after Gantt, 1986)*. AP = allophycocyanin, PC = phycocyanin, PE = phycoerythrin. In the bottom part only the LHC II complex as the most abundant pigment-protein of the antenna system is explicitly shown in its general structural organization (modified after Kühlbrandt, 1984).** For a more detailed structure of the pigment array, see Kühlbrandt and Wang (1991).***

*E Gantt in *Encyclopedia of Plant Physiology,* New Series, Vol. 19, LA Staehelin and CJ Arntzen (eds.), Springer, Berlin, 1986, pp. 260–268.
** W Kühlbrandt, Nature 307 (1984) 478–480.
*** W Kühlbrandt, DN Wang and Y Fujimori, Nature 367 (1994) 614–621.

eral red algae) shape are located in well defined arrays at the outer surface of the thylakoids (Fig. 4, top). In contrast, the light-harvesting complexes of green algae and higher plants are integral membrane proteins embedded into the thylakoid membrane (Fig. 4, bottom). Their subunits contain in addition to Chl-a other accessory pigments like Chl-b (in green algae and higher plants), Chl-c derivatives (in brown algae, dinoflagellates and cryptomonades) and carotenoides (see Chapter 5). Among the different chlorophyll (plus accessory pigments) proteins forming the PS II antenna, a complex referred to as LHC II (light harvesting complex) is the most abundant pigment protein of the thylakoid membrane. At present, LHC II which forms a trimer of triangular shape (see Fig. 4, bottom) is also the structurally best characterized part of the PS II antenna in higher plants (3.4 Å resolution by electron microscopy of 2-dimensional crystals). Each subunit contains 12 (or 13) chlorophylls (seven Chl-a, five Chl-b and several carotenoid molecules (xanthophylls)). The chlorophyll head groups are arranged on two levels which form an upper and lower leaflet with respect to the bilayer of the thylakoid membrane (seven Chls in the upper, five in the lower leaflet, a direct distinction of Chl-a and Chl-b could not yet be achieved). The porphyrin rings are oriented roughly perpendicularly to the membrane plane. As the LHC II polypeptide forming three transmembrane helices contains only three histidines, most of the Chls are bound to other residues or seem to be suspended between the polypeptides of the membrane surface and the $\alpha$-helices. This LHC II is connected with the reaction center complex (see section 3.4) via other chlorophyll proteins, referred to as CP24, CP26 and CP29. The architecture of the whole PS II antenna is not yet precisely known. It is possible that due to heterogeneities of the lateral distribution of pigment protein complexes within the thylakoid membrane PS II antenna of different average size are established.

As the formation of electronically excited chlorophylls inevitably comprises a probability term of radiative emmision (see section 3.2) the analysis of the fluorescence emission kinetics with very high time resolution provides invaluable information on the exciton dynamics within the photosynthetic unit. Complementary to time-resolved fluorometry, mechanism and kinetics of excitation energy transfer from the antenna to the PS II reaction center can be also analyzed by application of laser spectroscopy with high spectral resolution (e.g. hole burning spectroscopy). Significant progress was made during recent years by the development of lasers generating ultrashort light pulses and of suitable detection systems like the pump and probe pulse technique. As an example, the singlet exciton transfer from the accessory pigment carotenoid to Chl-a, was recently found to take place in 100–300 femtoseconds. Time-resolved measurements of the fluorescence decay kinetics which reflects the overall lifetime of excitons revealed that their equilibration within a PS II antenna of 200–300 chlorophyll molecules occurs in picoseconds. Accordingly, very fast exciton transfer processes are involved. The mechanistic details of exciton dynamics within the PS II antenna still remain to be clarified.

So far, we have considered the antenna only as a static pigment array which assures an effective capture of light and exciton funneling to the photoactive pigment $P_T$ within the reaction center. However, optimal adaptation to widely varying illumination conditions requires dynamic structural changes that allow to vary the antenna size and the light distribution between PS I and PS II.

*3.3.4. Lateral mobility of light-harvesting complexes as a regulatory mechanism:* The size of the PS II antenna depends on the illumination conditions, i.e. on the PFD (light versus shadow plants) and on chromatic effects like growing in blue or red light. Furthermore, the pigment protein subunits exhibit a heterogeneity in their lateral distribution depending on illumination. Enzyme-catalyzed phosphorylation/dephosphorylation processes induce movement of LHC II complexes in plants, thereby modulating the antenna properties. This event is assumed to be responsible for short-term regulation of excitation-energy distribution between PS I and PS II via a phenomenon referred to as state 1-state 2 transitions. In addition to the short-term adaptation to varying illumination conditions (order of hour(s)) by lateral movement of pigment protein complexes giving rise to state 1-state 2 transitions, long-term adaptations (order of day(s)) do exist at the level of the biosynthesis. In the latter case the number of pigment protein complexes, i.e. the total antenna size, is adapted via a genetically controlled feedback cascade to illumination conditions which persist for longer times. The topic of light adaptation is a matter of current research and is not fully understood in many of its facets. For further reading on the topics described in section 3.3, see refs. 2-7.

### *3.4. Reaction center of PS II*

In general, the trapping of excitons and their transformation into electrochemical Gibbs energy takes place in all photosynthetic reaction centers by rapid electron transfer from the excited singlet state of the photoactive pigment $P_T$ to the functionally coupled primary electron acceptor A and the subsequent stabilization of the charge separation through rapid reoxidation of $A^-$ by a secondary acceptor. The components $P_T$, A and the secondary acceptor of the PS II reaction center have been identified as a special chlorophyll-a complex referred to as P680, a pheophytin molecule (Pheo) and a specifically bound plastoquinone molecule $Q_A$ associated with a high spin nonheme iron $Fe^{2+}$. Accordingly, if one takes into account that P680 acts as a shallow exciton trap (see 3.3.2.2), the overall reaction described by Eqn. 5 can now be substantiated in the following way:

$$\begin{array}{ccccc} {}^1\mathrm{Chl}^*\mathrm{P680\ Pheo\ Q_A} & \underset{k_{-t}}{\overset{k_t}{\rightleftharpoons}} & \mathrm{Chl}\,{}^1\mathrm{P680}^*\mathrm{Pheo\ Q_A} & \underset{k^*}{\overset{k_{PC}}{\rightleftharpoons}} & \mathrm{Chl\ P680^+\ Pheo^-\ Q_A} \\ h\nu \uparrow\!\downarrow k_\Sigma + k_F & & h\nu \swarrow\!\nearrow k'_\Sigma + k'_F & & \downarrow k_{stab} \\ \mathrm{Chl\ P680\ Pheo\ Q_A} & & & & \mathrm{Chl\ P680^+\ Pheo\ Q_A^-} \end{array} \qquad (16)$$

where Chl symbolizes the antenna chlorophylls coupled via exciton transfer with P680, $k_F$ ($k'_F$) = rate constant for radiative emission ($k_F = \tau_0^{-1}$, with $\tau_0$ = reciprocal internal life time) $k_\Sigma$ ($k'_\Sigma$) = sum of all rate constants of radiationless decay, (subscript ' refers to P680), $k_t$ and $k_{-t}$ are the rate constant of forward and reverse exciton transfer between nearest antenna neighbors and P680 (see Eqn. 14), $k_{PC}$ is the rate constant of primary radical pair formation from $^1$P680* and k* the reverse reaction, $k_{stab}$ is the rate constant of the electron transfer form $Pheo^-$ to $Q_A$ which is indispensible in PS II for the stabilisation of the primary charge separation.

The rate constants of the pairwise excitation energy transfer were specified in the previous section (Eqns. 10 and 11). In an analoguos way the rate constants of electron transfer reactions can be derived by using again Fermi's Golden Rule (see Eqn. 6). Therefore, a few general comments on redox reactions in biological system will be presented before discussing the PS II reaction center in more detail.

*3.4.1. Electron transfer rates in biological systems:* Electron transfer reactions in biological systems take place via vibrationally coupled electron tunneling. In principle, the rate constant of the reaction DA (state 1) → $D^+A^-$ (state 2), $k_{12}^{ET}$, can be calculated by using Eqn. 6 in an analogous way as briefly outlined in section 3.3.2.1 for the transfer of excitation energy. The initial (1) and final state (2) represent a multidimensional system of vibrations including the redox groups and their environments. In a highly simplifying way the potential of states 1 and 2 can be characterized by one-dimensional harmonic oscillators as is shown in Fig. 5. According to the Franck-Condon principle, an electron transfer can only take place, if state 1 and state 2 attain the same special nuclear geometry (transition state). The mode of the reaction depends on the magnitude of the electronic coupling. In the case of strong coupling a marked energy splitting occurs at the crossing point of the potential curves 1 and 2 (see Fig. 5, right side). As a consequence of this effect, potential curve 1 rides over to curve 2 at the surface marked by "a" in Fig. 5, right side. This phenomenon implies that the electron transfer occurs within the same potential curve, i.e. the process is adiabatic. On the other hand, if the splitting is rather small, the electron transfer comprises a crossing of the potential curves and therefore the process is non-adiabatic (Fig. 5, left side).

The calculation of the FCWD factor (Eqn. 6) via a classical treatment of the vibrations in a harmonic oscillator approximation led for nonadiabatic thermally activated electron transfer to the widely used Marcus

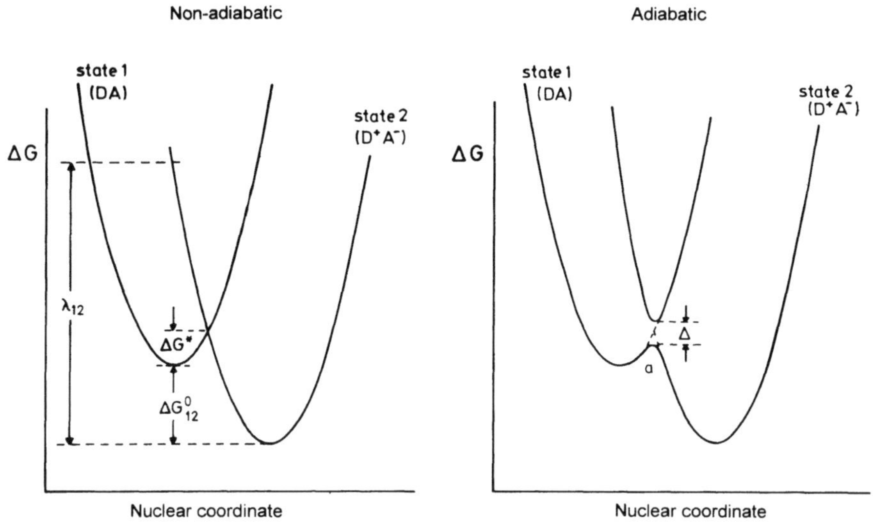


Fig. 5. Coupling of nuclear motion with electron transfer. The ordinate represents the Gibbs energy of the nuclei of the whole system, i.e., donor plus acceptor plus surrounding. The multidimensional potential energy surface is extremelly simplified by a one-dimensional harmonic potential. The abscissa describes the nuclear geometry for the initial (DA) and the final ($D^+A^-$) state of the whole system. Left side: non-adiabatic electron transfer, i.e. the process occurs in the vicinity of the crossover of the potential curves of the initial and the final state with a comparatively low transmission coefficient (see text). Right side: adiabatic electron transfer, i.e., strong coupling leads to split of the energy (its magnitude it symbolized by $\Delta$) near the crossover nuclear geometry (region a) so that the initial state (DA) overrides to the final state ($D^+A^-$) at the same potential surface.

equation (the fundamental work of Marcus was awarded the Nobel Prize for Chemistry in 1992):

$$k_{12}^{ET} = \frac{2\pi}{\hbar} |V_{12}(R)|^2 \sqrt{4\pi\,\lambda_{12} \times k_B T}\,\exp\left(-\frac{(\Delta G_{12}^0 + \lambda_{12})^2}{4\lambda_{12} \times k_B \times T}\right) \tag{17}$$

where $k_B$ = Boltzmann constant, $V_{12}(R)$ represents the quantum mechanical matrix element coupling reactant and product electronic wave functions at distance R of the interacting redox group, $\lambda_{12}$ is the reorganization energy parameter which is related to the changes of the vibrational modes in the redox group itself and in their microenvironment (e.g. solvent polarization) and $\Delta G_{12}^0$ is the standard free energy gap between initial state 1 (DA) and the final state 2 ($D^+A^-$) (see Fig. 5, left side).

An inspection of Eqn. 17 readily shows that the rate constant $k_{12}^{ET}$ in redox enzymes can be tuned by the protein matrix through the variation of three parameters: $|V_{12}(R)|^2$, $\Delta G_{12}^0$ and $\lambda_{12}$.

The matrix element, $V_{12}$, depends on the overlap of the electronic wave functions and, therefore, decreases exponentially with the distance R between the reacting groups (for further reading see, for example, refs. 8, 9). The dependence on R can be phenomenologically described by:

$$|V_{12}(R)|^2 = |V_{12}(0)|^2 \exp(-\beta R) \tag{17a}$$

where $V_{12}(0)$ is the matrix element for the case of redox groups in direct van der Waals contact (R = 0). In biological systems $|V_{12}|^2$ is sufficiently small so that practically all redox reactions take place via nonadiabatic electron transfer. A thorough analysis of data reported for electron transport processes in biological systems led to an interesting result. It was found that within a wide range of rate constants (more than 10 orders of magnitude) a linear relationship exists between ln $k_{12}$ and R, i.e., $\beta$ in Eqn. (17a) has virtually the same magnitude for redox groups bound to protein matrices. A value of $\beta = 14$ nm$^{-1}$ was derived, i.e. a variation of R by 1.7 Å causes a change of the rate constant $k_{12}$ by a factor of 10 (for details, see ref. 10). This finding shows that the distance between the reactants is of central relevance for biological adaptation of rate constants. A fine tuning is achieved by modulating the Franck-Condon factor. The magnitude of $\lambda_{12}$ depends on the nature of the redox center and its environment. Comparatively large values (~1.5 eV) arise if the charge(s) is (are) localized at small redox centers in a polar environment. Correspondingly, $\lambda_{12}$ is rather small (down to 0.25 eV) if the charge(s) is (are) highly delocalized in redox groups that are embedded in an environment of minimal polarizability. The reorganizational energy parameter, $\lambda_{12}$, provides information on the coupling of the electron tunneling with different vibrational modes of the prosthetic redox groups and of the protein matrix. In principle, the energy of the dominant vibrational modes for an electron transfer step can be obtained by measurements of the rate constants as a function of $\Delta G^0_{12}$ at different temperatures. In this case, however, the semiclassical approach of Eqn. (17) can be only used if the energy of the contributing modes is sufficiently low ($h\nu \ll k_BT$, where $\nu$ is the frequency of the mode). If this condition is not satisfied, the contribution of the vibrations has to be calculated by complete quantum mechanical treatment. In the simplest case of coupling with a single vibrational frequency ($\nu$), one obtains:

$$k_{12}^{ET} = \frac{2\pi}{\hbar^2} |V_{12}|^2 e^{-S(2n+1)} [(n+1)/n]^{P/2} I_{|P|} [2S\sqrt{n(n+1)} \tag{18}$$

where $S = \lambda_{12}/h\nu$, $n = (e^{h\nu/k_BT} - 1)^{-1}$, $P = \Delta G^0_{12}/h\nu$

and $I_{|p|}(x)$ the modified Bessel function.

One important implification of the quantum mechanical treatment of nuclear motion is the possibility of electron transfer coupled with nuclear tunneling as is schematically shown in Fig. 6. As the population of the vi-

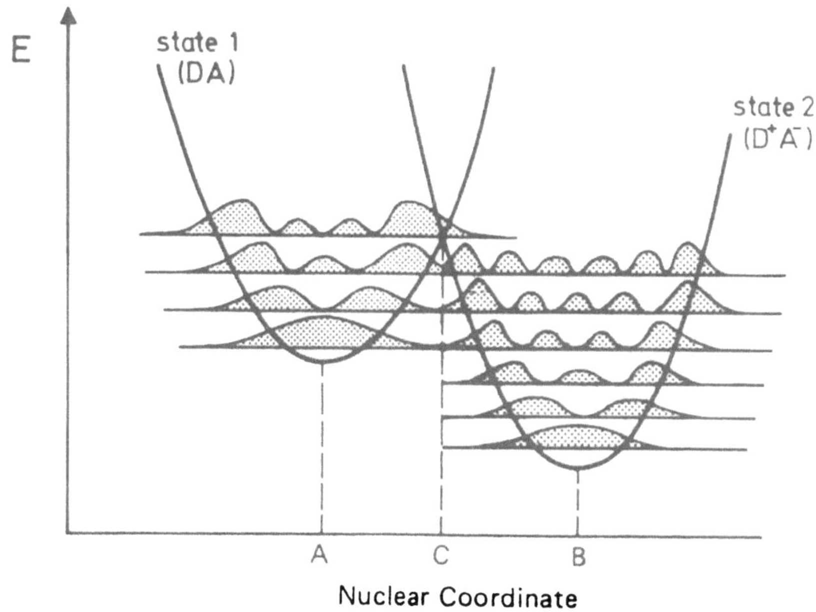


Fig. 6. Nuclear tunneling. The nuclear wave functions are presented for the lowest 4 (initial state) and 7 (final state) vibrational levels. It is seen that the crossover geometry is reached even when the total energy is below that of the potential energy at point C (modified according to ref. 8).

brational levels depends on temperature, nuclear tunneling becomes the dominant factor for electron transfer at very low temperatures. A recent detailed study on bacterial reaction centers revealed that the participation of frequency modes with middle and higher energies cannot be neglected, especially at low temperatures. In a final remark it should be mentioned that the use of potential energy curves in a two-dimensional description is not always appropriate to provide even qualitative explanations for a multistep system.

A key question which concerns the role of the protein matrix is the possibility of a specific regulation of the electron transport via the different electronic properties of the amino acid residues. Aromatic amino acid residues could be of functional significance in photosynthetic reaction centers, because the primary charge separation as well as its subsequent stabilization is a photochemical redox reaction sequence including only molecules with $\pi$-electron systems as prosthetic groups. Therefore, aromatic amino acid residues of the matrix intercalated between redox groups could significantly enhance the electronic coupling via a super-exchange mechanism (for more detailed information on the mechanism of electron reactions in biological system, see refs. 8–10).

After this brief summary of some general principles of electron transfer in biological systems we will now proceed to discuss first the kinetics and

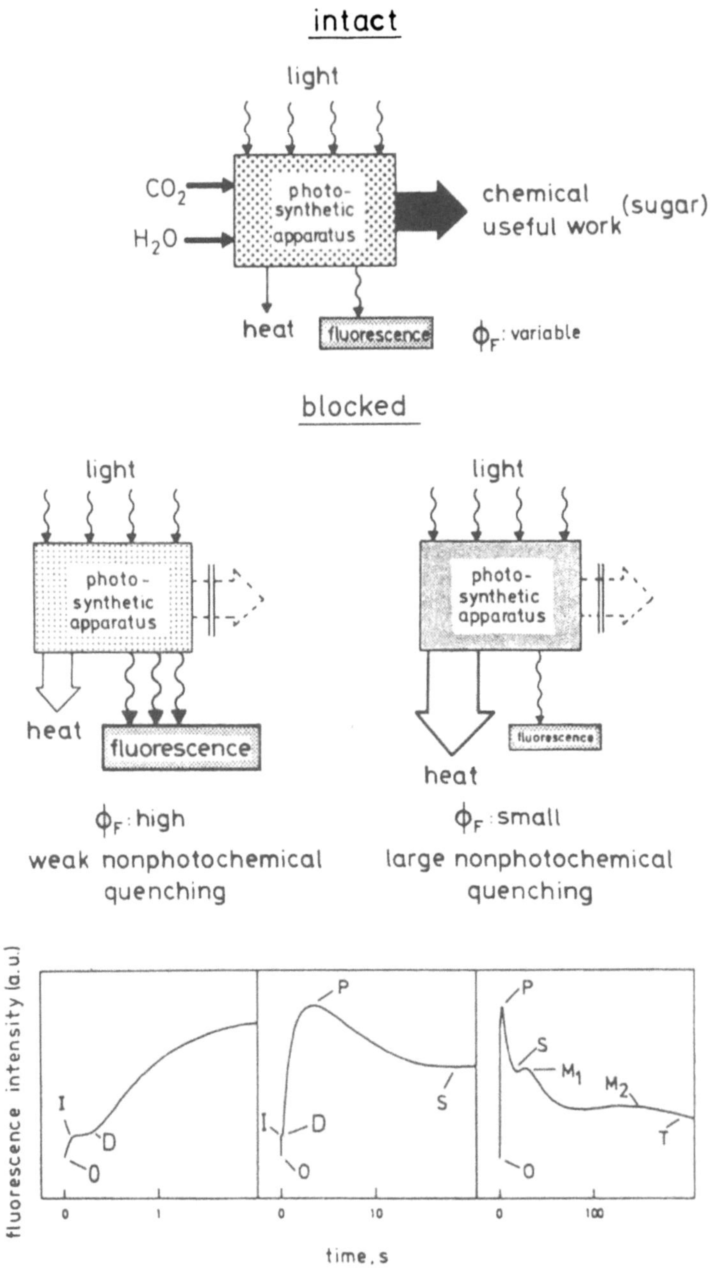


Fig. 7. Schematic representation of light transformation into chemical useful work, radiative emission (fluorescence) and dissipation into heat in intact (top) and blocked (middle) photosynthetic apparatus. The lower part shows the fluorescence emitted from the upper surface of a spinach leaf as a function of actinic CW-illumination at three different time scales. The letters denote characteristic points of the fluorescence induction curves (redrawn from ref. 11).

thermodynamics and subsequently the structural and functional organization of the PS II reaction centers.

*3.4.2. Dynamics of exciton transformation into electrochemical Gibbs energy in the PS II reaction center:* The application of time-resolved flash spectroscopy with high time resolution provided kinetic information on the electron transfer steps which are involved in exciton trapping and transformation into electrochemical Gibbs energy within the PS II reaction center. Measurements in PS II preparations deprived of their antenna complexes (distal, proximal and core antenna, see 3.3. and 3.4.3) permitted the direct determination of the rate constant $k_{PC}$ for the primary radical pair formation (see Eqn. 16). Values of the order of $(3\ ps)^{-1}$ were obtained. The rate constant $k_{PC}$ however, does not represent the actual exciton trapping time $<\tau_T>$ *in vivo* because the PS II reaction acts as a shallow trap (see 3.3.2.2). Therefore, according to Eqn. 14 the value of $<\tau_T>$ should be dependent on the number of Chl-molecules which participate in the rapid distribution of the excited singlet states between the antenna and P680. This idea is consistent with previous findings of overall trapping times in the order of 200 ps in the PS II-preparations which contain 60 Chl per reaction center. The primary radical pair $P680^+\ Pheo^-$ in its spin-coupled singlet state can recombine back to $^1P680^*$ unless the electron from $Pheo^-$ is not removed by a rapid transfer to $Q_A$. The rate constant $k_{stab}$, of this stabilization step of the primary charge separation was recently determined to be of the order of $(300\ ps)^{-1}$. It is very interesting to note that rate constants $k_{PC}$ and $k_{stab}$ closely resemble those of the corresponding reactions in anoxygenic purple bacteria. This finding reveals that the reaction centers of purple bacteria and PS II exhibit striking similarities in their functional and structural organization of exciton transformation into electrochemical Gibbs energy. The essential difference between both types of reaction centers is the thermodynamics of the cation radical $P^+$ formed in the primary charge separation step. The $P^+$ state of the special pair in purple bacteria exhibits only a moderate oxydizing power ($E_{m.7} \approx +0.5$ V), while $P680^+$ is a strongly oxidizing species ($E_{m.7} \approx +1.1$ V). The "invention" of P680 by nature was the thermodynamic prerequisite for the evolutionary development leading to oxygen photosynthesizing organisms. The high oxidation potential of $P680^+$ is not simply a consequence of incorporating Chl-a into the reaction centers but more intimately related to the structure of P680 and its microenvironment (see section 3.4.4).

If reoxidation of $Q_A^-$ by secondary quinones is blocked, e.g. by addition of certain herbicides (see section 5), a stable charge separation is prevented and therefore these reaction centers are "closed" for useful photochemical exciton transformation. In this case, the rate constant $k^*$ (see Eqn. 16) becomes increased by a factor of 5–10 due to the electrostatic interaction of $Q_A^-$ with the radical pair $P680^+\ Pheo^-$ (see Fig. 8). In systems with closed reaction centers photochemical exciton trapping is prevent-

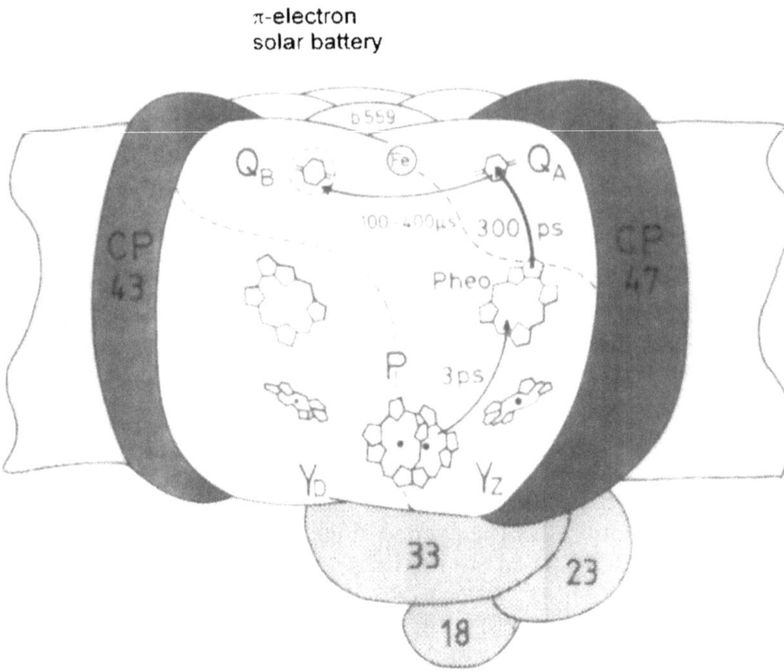


Fig. 8. Simplified scheme of PS II in higher plants. Based on the assumption of close structural homologies (see text), as a working hypothesis, the functional redox groups within the D1/D2 heterodimer are arranged in the same way as in the resolved crystal structure of anoxygenic purple bacteria. Except for CP 47 and CP 43 (dark gray areas), Cyt b559 and the extrinsic regulatory proteins of the water oxidizing complex (dotted areas symbolized by numbers 33, 23 and 18) the other polypeptides of the whole PS II complex are not explicitly shown. $Y_Z$ and $Y_D$ are redox active tyrosine residues of polypeptides D1 and D2, respectively. For further details see text.

ed, i.e. the excitons dissipate completely via radiationless decay and radiative emission. The latter effect gives rise to changes in steady state fluorescence emission (see Fig. 7). As this phenomenon is relevant for a number of useful analytical fluorometric techniques, the basic principles will be illustrated on the basis of the most simple model consisting of a single photosythetic unit which contains the antenna pigments and only one PS II reaction center. The fluorescence quantum yield of a photosynthetic unit with an open reaction center attains a minimum fluorescence level with a quantum yield $\Phi_F^0$ according to the relation:

$$\Phi_F^0 = \frac{k_F}{k_F + k_\Sigma + k_{PC}} \tag{19a}$$

with the rate constants $k_i$ as defined in Eqn. 16. In Eqn. 19a the formation of $P680^+$ by direct absorption and its decay via $k'_F$ and $k'_\Sigma$ has been neglected because *in vivo* the number of antenna chlorophylls exceeds that of P680 by about two orders of magnitude. The decrease of fluorescence yield due to exciton trapping via stable radical pair formation in PS II is referred to as photochemical fluorescence quenching.

Analogously, in the case of a system with a closed reaction center ($k_{PC} = 0$) the fluorescence quantum yield reaches a maximum level:

$$\Phi_F^{max} = \frac{k_F}{k_F + k_\Sigma} \tag{19b}$$

Based on a similar kinetic consideration of Eqn. 16, the quantum yield of photochemical trapping is given by:

$$\Phi_{PC} = \frac{k_{PC}}{k_F + k_\Sigma + k_{PC}} \tag{20}$$

A simple algebraic rearrangement of Eqns. 19a,b and insertion into Eqn. 20 leads to the expression:

$$\Phi_{PC} = \frac{\Phi_F^{max} - \Phi_F^0}{\Phi_F^{max}} \tag{21}$$

The relations (19)-(21) are valid for the microscopic states of a single photosynthetic unit with one reaction center in either the open or the closed state, respectively. The experimentally detectable parameter, however, is the fluorescence emission of the macroscopic ensemble which leads to an average fluorescence yield by the relation:

$$\Phi_F = \frac{F}{I_{abs}} \tag{22}$$

where F is the fluorescence and $I_{abs}$ the absorbed light intensity of the sample.

The value $\Phi_F$ is determined by the functional properties of all photosynthetic units, i.e. it depends on the population probabilities of open and closed reaction centers, respectively, and on the functional connection of different photosynthetic units via excitation energy transfer (see section 3.3). It is easily understandable that the latter effect can give rise to complicated relations. Regardless of the complexity of the detailed analytical expression describing the fluorescence quantum yield as a function of the population probability W(0) of open reaction centers, (the probability of closed reaction centers, W(cl), is simply given by 1–W(0)), fluorescence techniques provide useful information on the functional integrity of PS II because the fluorescence emission from PS I is comparatively small. For a detailed analysis, in addition to the photochemical quenching (described by $k_{PC}$ at a microscopic level) also effects have to be taken into account which give rise to nonphotochemical fluorescence quenching. A prominent example is the acidification of the lumen which gives rise to an increased probability of radiationless decay (other efficient nonphotochemical quenchers are radicals like P680$^+$ and Pheo$^-$). A simplified scheme of the energy balance in systems with a functionally competent (top) and blocked (bottom) photosynthetic apparatus is shown in Fig. 7. Illumination of dark adapted samples with actinic light gives rise to a change of W(0) so that a characteristic fluorescence induction curve emerges. In whole leaves the shape is dependent on many factors affecting photochemical and nonphotochemical quenching. It is important to note that in contrast to the sophisticated and expensive equipment with high time resolution required for fluorescence life time studies, the relative fluorescence quantum yield reflecting the functional state of PS II can be measured with a rather simple setup even in intact plants. Therefore, the latter fluorescence techniques provide a very useful tool to screen for the photosynthetic activity in field studies. However, it has to be emphasized that a careful analysis of the data is required in order to avoid the numerous pitfalls of interpretation (for further details, see ref. 11 and literature therein). If applied in correct manner, fluorescence can be used as an invaluable analytical tool of the photosynthetic apparatus *in situ* comparable with the electrocardiogram in the medicine for heart diagnosis.

As a summary of the discussion on the kinetics, it should be kept in mind that the photochemical exciton trapping in PS II leads in less than 1 ns to stable charge separation in the form of the radical pair P680$^+$ Pheo$Q_A^-$. The quantum yield of this process is high (it can approach values close to 1) for open reaction centers because $k_{PC} \gg k_F, k_\Sigma$ (see Eqn. 21). This high quantum yield originates from a proper selection of the rate constants of the individual steps in the overall scheme of Eqn. 16. This goal was achieved by the tuning of the reaction coordinates through the distances ($V_{12}(R)$) and thermodynamics ($\Delta G^0$) of the redox components participating in these reactions and their vibronic coupling with the protein matrix ($\lambda_{12}$). Accordingly, this matrix has to be considered as an apoprotein of functional relevance for efficient exciton transformation into electrochemical Gibbs energy.

After the brief discussion of the PS II reaction center dynamics it appears wortwhile to add some basic remarks on the thermodynamics of this system.

*3.4.3. Thermodynamics of exciton trapping and transformation into Gibbs energy in PS II and in the photosynthetic apparatus:* As a general remark it has to be emphasized that thermodynamic considerations of photosynthesis require a precise definition of the system. The energetics of an individual reaction center *per se* has to be analyzed in terms of the configurational energy rather than of the Gibbs energy. The configurational energy does not contain concentration dependent contributions. On the other hand, the thermodynamics of a macroscopic ensemble, e.g., a chloroplast suspension or a whole plant has to be described on the basis of the Gibbs energy transformation within this system. Both problems will be addressed in a brief description.

The energetics of the PS II reaction center is based on the configurational energy, $\Delta G^c$, of $^1P680^*$. If one assumes that the electron is ejected from the lowest excited singlet $^1P680^*$ in its vibrationally relaxed state, the energy of the 0–0 transition provides the proper basis. Its value is 1.83 eV. Of course, the Gibbs energy of $^1P680^*$, $\Delta G^0$, is lower due to the distribution probability of the singlet exciton between P680 and antenna chlorophylls (see Eqn. 16). Accordingly, if for the sake of simplicity spectral differences are neglected, i.e. $k_t = k_{-t}$, the Gibbs energy $\Delta G^0$ ($^1P680^*$/P680) is given by:

$$\Delta G^0 (^1P680^*/P680) = \Delta G^c (^1P680^*/P680) - RT \ln (N + 1) \quad (23)$$

where $\Delta G^c$ ($^1P680^*$/P680) is the configurational energy and N the number of chlorophylls participating in the rapid exciton distribution between the antenna and P680 (see Eqn. 16). The configurational energy of the primary radical pair is not precisely known but based on the rate constants $k_{PC}$ and $k^*$ reported in the literature, the relation $\Delta G^c (P680^+Pheo^-/^1P680^*Pheo) = -RT \ln (k_{PC}/k^*)$ would give rise to a value of $-0.17$ eV. With the relation $\Delta G^c (P680^+ Pheo^-/P680\ Pheo) = -[\Delta G^c (^1P680^*/P680) + \Delta G^c (P680^+ Pheo^-/^1P680^*\ Pheo)]$ the former value is estimated to be $-1.66$ eV. If one assumes that $\Delta G^c$ and $\Delta G^0$ are practically the same for the states $P680^+ Pheo^-$ versus P680 Pheo, Eqn. 23 would give rise to $\Delta G^c$ values of $-0.025$ to $-0.050$ eV for the primary charge separation in a PS II reaction center connected with an antenna consisting of 100–300 chlorophylls with very similar spectral properties. This value nicely fits with data reported in the literature. These calculations illustrate the necessity of a clear distinction between the terms configurational and Gibbs energy.

With respect to the structure of its microenvironment, the initially formed primary radical pair $P680^+ Pheo^-$ could decrease its energy by rapid transition into a relaxed state. A transition $[P^+A^+]_{initial} \rightarrow [P^+A^-]_{relaxed}$ is well known in reaction centers of purple bacteria with relaxation energies of the order of 0.04 – 0.08 eV. According to recent data derived from fluorescence lifetime and quantum yield measurements, relaxation processes with similar enegetics could also occur in PS II. In contrast to the primary charge separation, the subsequent stabilization by electron transfer from $Pheo^-$ to $Q_A$ is accompanied by a much more pronounced loss of configurational and Gibbs energy which is of the order of 0.5 to 0.6 eV.

Measurements down to 4 K revealed that the rate constant $k_{PC}$ is almost independent of temperature in the range of $4\ K < T < 300\ K$ ($k_{PC}^{-1} \approx 3$ ps at 300 K and 1.7 ps at 4 K). Likewise, in reaction centers of purple bacteria the rate constant $k_{stab}$ was also found to exhibit a similar temperature dependence. These findings suggest that these reactions are virtually activationless, i.e. $\Delta G^c$ compensates for the reorganization energies coupled with the electron transfer steps from $^1P680^*$ to Pheo and from $Pheo^-$ to $Q_A$ (see Eqn. 17).

A markedly different pattern, however, arises for the thermodynamics of the overall machinery in an algal cell or a plant leaf. Within the framework of irreversible thermodynamics, the overall system can be considered as a photoelectrochemical energy converter of sunlight which drives with the aid of an exergonic photon flux, $J_{h\nu}$, an endergonic electron flux $J_R$ against an electrochemical potential difference $\Delta\mu$ (for details, see Chapter 2).

Rapidly growing plants (e.g. sugar cane) under optimal conditions reach maximum efficiencies of about 5% for solar energy exploitation (for thermodynamic considerations of solar energy exploitation in photosynthesis, see e.g. ref. 12 and references therein).

*3.4.4. Structure of PS II reaction center:* A breakthrough in the understanding of photosynthetic reaction centers was achieved by the crystallization and x-ray structure analysis of isolated reaction centers from purple bacteria *(Rhodopseudomonas viridis)* by Michel and Deisenhofer (see Chapter 7). This fundamental work (Deisenhofer, Huber and Michel) was awarded the Nobel Prize in 1988. Unfortunately, so far direct structural information of comparable details could not be obtained for PS II but progress can be expected in the future, because recently successful crystallization of PS II complexes has been reported, which is a prerequisite for x-ray analysis. In this respect, it has to be mentioned that preparations of isolated PS II reaction centers are not yet available which exhibit an analogous functional integrity as those from purple bacteria *(vide infra).* Despite these shortcomings, indirect information on the structural organisation of PS II are obtainable. In addition to striking functional similarities (see section 3.4.2) it was found that polypeptides D1 and D2 of PS II exhibit characteristic sequence homologies with the L and M subunits, respectively, which form as a heterodimer the apoprotein of the purple bacteria reaction centers. It is therefore attractive to conclude that the redox active groups P680, Pheo-a and $Q_A$ ($Fe^{2+}$) of PS II (see Eqn. 16) are arranged within a heterodimer of polypeptides D1 and D2 in an analogous way as the corresponding reactants (special pair P, BPheo, QA × $Fe^{2+}$) in the L- and M-subunit of purple bacteria. This idea has gained strong support by the isolation of a complex consisting of polypeptides D1, D2, the subunits of cytochrome b559 and at least one small polypeptide. These D1/ D2/ b559 preparations resembling in their pigment content (4-6 Chl-a, 2 Pheo-a, 1-2 β Car) those of anoxygenic purple bacterial reaction center are able to form the light-induced primary radical pair $P680^+$ $Pheo^-$. However, they are deprived of $Q_A$. This shows that D1/ D2/ b559 preparations are certainly not a perfect analogue of PS II to isolated bacterial reaction centers. On the other hand, it appears to be clear that the functional redox groups of the PS II reaction center (P680, Pheo and $Q_A$ ×

$Fe^{2+}$) are incorporated into an apoprotein matrix that is essentially the D1/ D2 heterodimer. Accordingly, in an extension of the currently widely accepted paradigm of structural analogy, the well resolved structure of purple bacteria reaction centers is often used as a guideline for model simulations of the PS II structure. As a consequence of these studies, the spatial arrangement of P680, Pheo and $Q_A \times Fe^{2+}$ in PS II is inferred to correspond in many aspects to that of the functional analogues in purple bacteria (special pair P, BPheo, $Q_A \times Fe^{2+}$). In the most simplifying case of perfect structural homology, an overall geometry of these groups would emerge that is depicted in Fig. 8. However, it has to be emphasized that the similarity should not be overinterpreted and therefore Fig. 8 actually provides a tentative structural proposal which should be considered as nothing more than a useful working hypothesis. Certainly this picture has to be improved (and modified) by future studies. In this respect, it has to be emphasized that a number of key questions still require an unambiguous answer: (a) What is the structure of P680? (b) How many chlorophylls in addition to P680 do exist in D1/D2? (c) What is the role of the chlorophyll- containing polypeptides CP 47 and CP 43? (d) What is the role of cytochrome b559?

Among these problems, two are of central relevance: the structure of P680 and the role of CP 47 (CP 43). The nature of P680 is of paramount importance because the high oxidation potential of this photoactive pigment is the indispensable thermodynamic prerequisite for photosynthetic water oxidation to dioxygen. In anoxygenic photosynthesizing bacteria the oxidized special pair $P^+$ ($E_{m,7} \lesssim +0.5$ V) cannot energetically support this reaction. At first glance, one might assume that the substitution of bacteriochlorophyll in the special pair by chlorophyll-a in P680 is the key point of the functionally relevant shift towards markedly higher oxidation potentials. However, the Chl–a molecule *per se* is not the clue, as convincingly shown for the example of the PS I reaction centers. The photoactive pigment P700 which also consists of Chl–a (probably in a dimer form) exhibits a redox potential of $E_{m,7} \approx +0.5$ V (see Chapter 9) which is far below the energetic requirements for water oxidation. A comparison with P700 clearly illustrates that the rather high redox potential of P680 ($E_{m,7} \approx +1.1$ V) certainly requires a special microenvironment, established by the aggregation state (monomer versus dimer) of the photoactive Chl-a and the interaction with the protein matrix. The structural determinants of this central P680 function are not yet resolved. Latest studies in isolated reaction centers from purple bacteria mutants revealed that the redox potential of $P^+$ could be enhanced by genetically engineered introduction of amino acid residues which can form hydrogen bonds with the BChl moieties. It remains to be analyzed as to what extent this tool has been used by nature in order to achieve the high oxidizing power of $P680^+$. At present, neither the aggregation state nor the orientation of the porphyrin plane(s) with respect to the thylakoid membrane are unambiguously clarified.

Another related problem is the arrangement of pigment molecules near P680. In order to prevent their "undesired" oxidation by $P680^+$, the distance should be sufficiently long compared with that between P680 and $Y_Z$ (see section 3.4.1). On the other hand, however, a rapid excitation energy transfer from the antenna to P680 requires a sufficiently short distance (see section 3.3.2.1). How this problem has been solved by nature is an as yet unanswered question (for further reading see articles of Schelvis and van Gorkom and Renger in ref. 13).

A similar level of ignorance exists also for the role of polypeptides CP47 and CP43 which are symbolized in Fig. 8 by dark gray areas. It is widely assumed that these Chl-containing polypeptides form the core antenna of the PS II reaction centers. The "invention" of a core antenna could provide the soluton for achieving an efficient exciton transfer to P680. However, independent lines of evidence suggest that at least CP47 is of additional relevance for the structure and function of PS II. In all attempts reported so far, the separation of CP47 from the D1/ D2 heterodimer led to a loss of $Q_A$. Reconstitution of $Q_A$ could be only achieved in CP47/D1/D2/b559 preparations. In addition to these biochemical studies, recent mutagenesis experiments revealed that CP47 is indispensable for the assembly of a stable PS II core complex. Apart from these data another argument appears worth considering. During the evolutionary development from $O_2$-evolving cyanobacteria to higher plants the PS II reaction centers did not undergo significant changes of their CP47 (CP43)/D1/D2/Cyt b559 core structure while the antenna systems were modified dramatically (see section 3.3 and Fig. 4). It is therefore not easily comprehensible why at the polypeptide level of the reaction centers, the key evolutionary step from anoxygenic to oxygenic photosynthesis should consist entirely of the "construction" of a new core antenna system in the form of CP47 and CP43, rather than the development of new functional elements for water oxidation (see section 4).

## 4. Water oxidation

### *4.1. General organization scheme of photosynthetic water oxidation*

A brief inspection of Eqn. 4 readily shows that, regardless of the detailed mechanism, two indispensible prerequisites have to be satisfied in order to permit water oxidation to dioxygen: (i) the generation and (ii) the cooperation of four redox equivalents of sufficient oxidizing power. The first goal is achieved by the light-induced charge separation within the PS II reaction center which leads within 1 ns to the formation of a sufficiently stabilized and highly oxidizing $P680^+$ cation radical (see section 3.4). The basic principles of satisfying the second prerequisite of photosynthetic water oxidation i.e. the mode of functional cooperation of four holes gen-

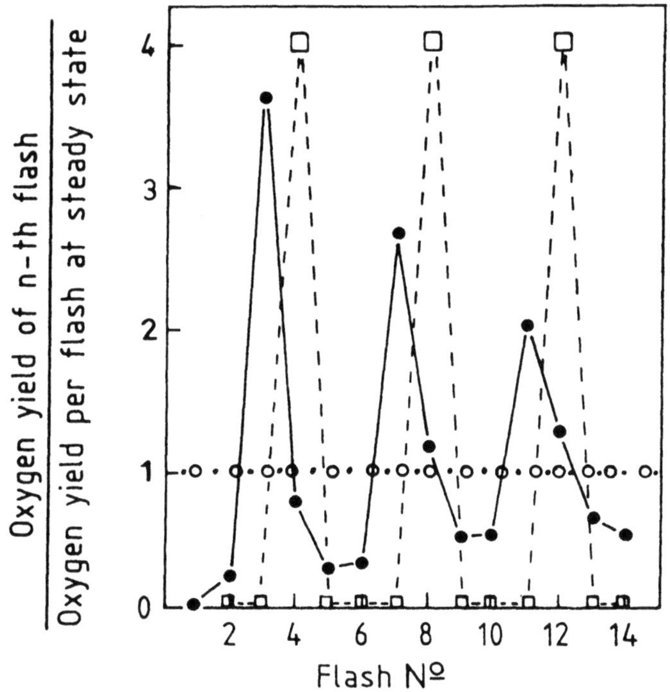


Fig. 9. Average oxygen yield per flash as a function of flash number in dark adapted thylakoids from spinach at pH = 7.0 and 7°C. For a comparison the theoretical pattern is given (open squares, dashed curve) that would arise in the case of the WOC relaxing completely to $S_0$ in the dark and in the absence of damping due to misses and double hits (see text). Closed and open circles represent experimental data in dark-adapted samples and under steady state conditions.

erated at $P680^+$ were unraveled about 25 years ago by the fundamental work of Kok, Joliot and coworkers (for a recent review, see ref. 13). The key result of these studies was the finding that upon excitation of dark adapted samples (algae, isolated chloroplasts) with a train of saturating short flashes, a pronounced oscillation pattern of the oxygen yield arises which exhibits a striking periodicity of four. A typical trace is shown in Fig. 9. The characteristic feature of a period four oscillation provides unambiguous evidence for a sequence of univalent redox steps energetically driven by $P680^+$ as oxidant. A stepwise accumulation of oxidizing equivalents requires a suitable operational unit which permits the intermediary storage of holes and catalyzes the overall process of photosynthetic water oxidation. Regardless of its detailed structural and functional organization, this unit will be referred to as the water oxidizing complex (WOC). A particular redox state of the WOC is symbolized by $S_i$, where i = 0, ... 4 describes the number of stored oxidizing redox equivalents. According to Eqn. 4 dioxygen can be only formed and released after attaining the formal redox state $S_4$ in WOC. An additional finding of significant relevance is the observation that the shape of the period four oscillation pattern is virtually invariant to the number of functionally competent WOC's within the thylakoid membrane. Consequently, each PS II reaction center is functionally coupled with only one WOC. This conclusion has also an important structural implication, i.e. both operational units form a multimeric complex.

Based on simple considerations, further interesting information can be gathered from an inspection of the characteristic oscillation pattern (see Fig. 9). At first, one might expect that the oxidizing redox equivalents stored in WOC have a finite lifetime. Therefore, after sufficiently long dark adaptation the WOC would be anticipated to relax into the lowest redox state $S_0$. Secondly, it seems reasonable to assume that a sufficiently short saturating flash (single turnover flash) would give rise to a univalent oxidation $S_i \rightarrow S_{i+1}$ in in each WOC. If both conditions were satisfied, oxygen release would be expected to arise only after the 4th, 8th, 12th etc. flash (open squares and dashed line in Fig. 9). This is obviously not the case. Two marked deviations have to be taken into account: (i) the experimental pattern (closed circles and solid line) exhibits a marked damping, i.e. oxygen formation is not exclusively formed after four flashes only, and (ii) the $O_2$ maxima appear after the 3rd, 7th etc. flash (see Fig. 9). The former phenomenon (i) can be explained by possibility of misses (no redox transition $S_i \rightarrow S_{i+1}$ even at saturating flash intensity) and double hits (double oxidation $S_i \rightarrow S_{i+2}$) occuring due to excitation with a short saturating flash. The latter effect (ii) suggests that after sufficiently long dark adaptation the water oxidizing complex attains the redox state $S_1$ rather than $S_0$. This feature implies that the states $S_2$ and $S_3$ decay into $S_1$ and that $S_0$ becomes oxidized to $S_1$ in the dark.

Table 1. Typical life times of PS II donor side reactions at room temperature and neutral pH

| Redox state $S_i$ | Type of reaction | | | |
|---|---|---|---|---|
| of WOC | $P680^+ Y_Z \rightarrow P680 Y_Z^{ox}$ | $Y_Z^{ox} S_i \rightarrow Y_Z S_{i+1}$ | $Y_D^{ox} S_0 \rightarrow Y_D S_1$ | $Y_D S_{i+1} \rightarrow Y_D^{ox} S_i$ |
| $S_0$ | 20 ns[a] | (30 μs)*[b] | 10 min[c] | –[d] |
| $S_1$ | 20 ns | 100 μs | – | – |
| $S_2$ | 50 ns, 250 ns | 250 μs | – | 10 s |
| $S_3$ | 50 ns, 250 ns | 1.3 ms | – | 7 s |

*The rate constant of this particular reaction is not yet clarified (data in the literature vary over two orders of magnitude).

Data taken from:

[a]K Brettel, E Schlodder and HT Witt, Biochim. Biophys. Acta 766 (1984) 403–415, H-J Eckert and G Renger, FEBS Lett. 236 (1988) 425–431.

[b]G Renger and B Hanssum, FEBS Lett. 299 (1992) 28–32, F Rappaport, M Blanchard Desce and J Lavernge, Biochim. Biophys. Acta 1184 (1994) 178–192.

[c]J Messinger and G Renger, Biochemistry 32 (1993) 9378–9386.

[d]J Messinger, WP Schröder and G Renger, Biochemistry 32 (1993) 7658–7668.

The next step in unraveling the functional organisation of photosynthetic water oxidation is the resolution of the kinetics. Application of different time resolved spectroscopic techniques revealed that the reduction kinetics of $P680^+$ is by two to three orders of magnitude faster than the univalent redox transitions $S_i \rightarrow S_{i+1}$ in WOC (see Table 1). Therefore, an intermediary redox component does exist which functionally connects the WOC with $P680^+$. This component often symbolized in the literature as $Y_z$ has recently been substantiated as a redox active tyrosine. In the oxygen evolving cyanobacterium (Synechocystis sp. PCC 6803) $Y_z$ was identified by site directed mutagenesis as tyrosine 161 of the polypeptide D1 (see Fig. 8). The involvement of $Y_z$ in the PS II reaction sequence is of general relevance. It reveals that the protein matrix does not only provide the proper ligation of prosthetic groups but can also actively participate in redox reactions. In addition to $Y_z$ as the unique direct oxidant to WOC, other endogenous redox groups were shown to interfere with the $S_i$-states. Analogously to $Y_z$ in D1, also the polypeptide D2 contains a redox active tyrosine residue referred to as $Y_D$ (identified as Tyr 160 in Synechocystis sp. PCC 6803). Although, according to recent structural modelling, $Y_z$ and $Y_D$ are almost symmetrically arranged with respect to P680 (see Fig. 8), $Y_D$ is not involved in the main electron transport pathway for thermodynamic and kinetic reasons: the markedly lower redox potential of this residue does not permit a $Y_D^{ox}$-mediated oxidation of the WOC in redox state $S_i$ with $i \geq 1$. In contrast, the reduced form $Y_D$ gives rise to a fast decay ($\tau$ = order of few seconds at room temperature) of $S_2$ and $S_3$, and (ii) the oxidation of $S_0$ into $S_1$ by $Y_D^{ox}$ is a rather slow reaction (order of minutes) (see Table 1). Interestingly, the redox active manganese cluster exhibits markedly different magnetic interactions with $Y_Z^{ox}$ and $Y_D^{ox}$, respectively. Based on this finding, the manganese cluster is assumed to be asymmetrically arranged with respect to the redox active Tyr-residues in D1 and D2 (see Fig. 8). Values of the order of 15 Å and 30–40 Å were estimated for the distances between the manganese cluster and $Y_z$ and $Y_D$, respectively. Latest findings suggest that $Y_Z$ is rather close (~4.5 Å) to the manganese.

Based on data reported so far, the general functional organization of photosynthetic water oxidation and its kinetic pattern can be summarized by the scheme depicted in Fig. 10 and Table 1, respectively. The scheme of Fig. 10, however, does not provide any information on the actual chemical mechanism of water oxidation because it leaves the redox states $S_i$ as unresolved symbols. It is clear that the resolution of the chemical mechanism of water oxidation requires the unraveling of the electronic configuration and the nuclear geometry of the water oxidase in the different redox states $S_i$.

In terms of the chemical nature it is important to note that manganese is an indispensable component of the water oxidizing complex. Detailed stoichiometric studies revealed that PS II with a functionally competent

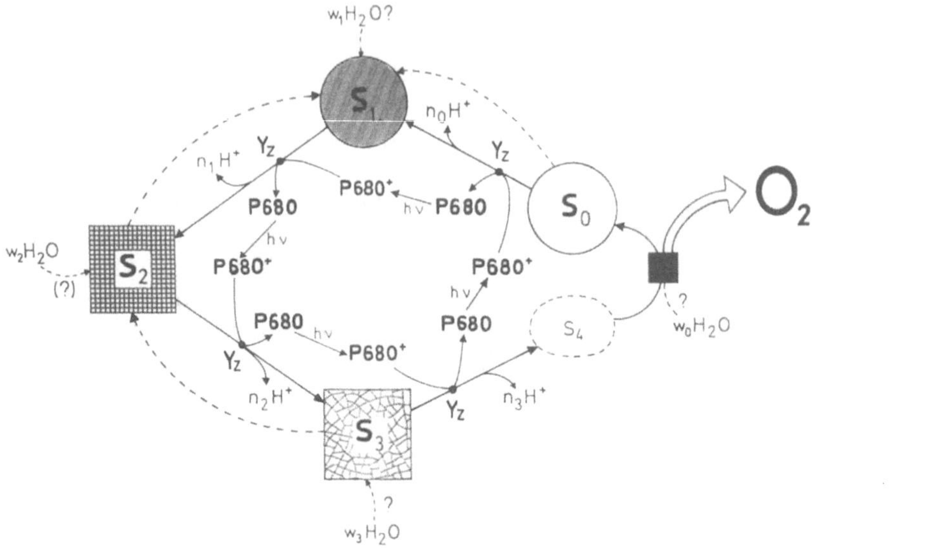


Fig. 10. Functional scheme of photosynthetic water oxidation via a four-step univalent redox sequence energetically driven by photooxidized $P680^+$. The stabilities of the different redox states of the hole storage unit HSU(Mn) are symbolized by a gray circle for the dark stable state $S_1$ and by a thin circle for the metastable $S_0$. The transient redox states $S_2$ and $S_3$ are symbolized by squares, while the kinetically undetectable $S_4$ state is characterized by a dashed oval. The possibility of a separate water-oxidizing site is symbolized by a black box. Likewise, the unresolved stage(s) of substrate water entry into the redox cycle is (are) symbolized by question marks (see text). The variable stoichiometry of proton release is symbolized by coefficients $n_i$ ($i = 0, \ldots 3$) (see text).

WOC contains four manganese. As manganese can attain different redox states, it appears reasonable to consider the WOC as a manganoprotein. With this idea as a starting point for our considerations, a number of more detailed questions arise:

(a) What is the nature of the catalytic site of water oxidation?
(b) What is the nuclearity of the functional manganese complex and its electronic configuration in the different $S_i$-states and why is $S_1$ the thermodynamically stable state in sufficiently dark adapted samples?
(c) What are the nature and the redox properties of the structural ligands to the functional manganese? Which polypeptide(s) provide(s) the ligands and how are the reaction coordinates regulated by the protein matrix?
(d) At which redox state $S_i$ does the substrate water bind and at which redox level of the substrate is the essential O–O bond formed?
(e) What types of protolytic reactions are coupled with the redox transitions in the WOC?
(f) What types of cofactors like $Ca^{2+}$ and $Cl^-$ are involved and what is their functional role?

As will be shown in the following description, answers to these questions are often very fragmentary and therefore the mechanistic resolution of photosynthetic water oxidation still remains an ambitious challenge to all experts working in this field.

### *4.2. The nature of the catalytic site*

In the preceeding description it was shown that the photosynthetic water oxidation comprises three different types of reactions: (i) the stepwise accumulation and intermediary storage of oxidizing redox equivalents, (ii) their eventual reaction with two water molecules to form dioxygen, and (iii) product ($O_2$)/substrate ($H_2O$) exchange. In general, two alternative modes have to be considered for the possible performance of the overall process: (a) a sequence of independent events, or (b) a sequence of intermingled reactions. At first glance, a sequential mechanism (type a) appears to be very attractive, because it permits separation of the hole storage process (i) and the chemical water oxidation (ii), thereby preventing undesired dissipation and/or deleterious side reactions. In this case, the following reaction pattern appeares to be most plausible: (i) sequential transfer and storage of oxidizing redox equivalents in a manganese-containing hole storage unit HSU(Mn) up to the redox level of four:

$$Y_z^{ox} + HSU(Mn)_i \rightarrow Y_z + HSU(Mn)_{i+1} + n_i H^+_{in} \qquad (24)$$

where i = 0, 1, ... 3 represents the number of oxidizing redox equivalents stored in the $HSU(Mn)_i$, $n_i$ = number of protons released into the lumen in each individual redox step as a consequence of partial or complete balancing the charge due to electron abstraction (see section 4.4).

(ii) concerted four electron abstraction from two substrate $H_2O$ molecules of a definite protonation state at a specific catalytic site refered to as water oxidizing site (WOS):

$$\begin{aligned} &WOS(2\,H_2^*O) + HSU(Mn)_{i=4} \rightarrow \\ &\rightarrow HSU(Mn)_{i=0} + WOS(O_2^*) + (4 - \sum_i n_i - m)\,H^+_{in} \end{aligned} \qquad (25)$$

m = number of protons released during the $O_2^*$/substrate water exchange at WOS (see Eqn. 26) and $O_2^*$ symbolizes coordinated dioxygen and $H_2^*O$ the bound substrate water of a definite protonation state which depends on the Lewis acid/base strength of WOS compared with that of water.

(iii) product ($O_2$)/ substrate (2 $H_2O$) exchange at WOS:

$$WOS(O_2^*) + 2\,H_2O \rightarrow WOS(2\,H_2^*O) + O_2 + mH^+_{in} \qquad (26)$$

At this stage it appears premature to discuss further details of this mechanism (e.g. the possibility that the functional entities HSU(Mn) and

WOS could be structurally distinguishable units and the interesting implications), because there exist kinetic problems. Based on different lines of experimental evidence, the rate of $S_3$ oxidation by $Y_Z^{ox}$ was found to be almost the same as the dioxygen release (1-2 ms). Therefore, the overall process including the concerted abstraction of four electrons from two water molecules under O–O bond formation (Eqn. 25) and the substrate/product exchange reaction (Eqn. 26) at the WOS needs to occur with a rate comparable with that of a single one-electron abstraction from the hole storage unit in redox state $HSU(Mn)_{i=3}$ by $Y_Z^{ox}$. This imposes serious mechanistic constraints. Accordingly, a sequence of intermingled reactions seems to be more likely. Two further interesting implications emerge from this idea: manganese itself (probably as a binuclear unit) provides the template (or at least part of it) for water oxidation (see 4.3.1); i.e. the HSU and WOS functions are exerted by the same unit within the water oxidizing complex (WOC), and secondly, it is possible that two different types of reaction occur at the same step, e.g. storage of oxidizing equivalents in form of intermediary redox states of the substrate (formation of bound peroxide at a level prior to $S_4$). Before addressing this mechanistically important question in more detail (see section 4.3.3), our current knowledge on the structure and redox chemistry will be briefly summarized.

### *4.3. Structure and redox chemistry of the manganese containing active site*

Each PS II functionally competent in water oxidation to $O_2$ is known to contain four manganese. Direct structural data based on x-ray or electron diffraction patterns are lacking (see section 3.4.4). However, indirect information is available. EXAFS (extended x-ray absorption fine structure) studies suggest that there does exist a distance of 2.7 Å between two manganese centers. In addition, another Mn ... Mn distance was found at 3.3 Å. These EXAFS data in combination with thorough analyses of EPR-spectra (see below) led to the conclusion that the four manganese form a spin-coupled tetranuclear cluster which can be best described by a dimer of two binuclear units. The manganese in each of these units is inferred to be connected via (one) two μ-oxo-bridge(s).

EXAFS studies further suggest that the vast majority of the other ligands (in addition to the μ-oxo-bridges) are also coordinated via O (N) atomes to the manganese centers. Therefore, glutamate and aspartate residues are the most likely candidates as ligand constituents of the protein matrix. Latest evidence gathered from pulsed EPR measurements in samples containing either normal ($^{14}N$) or $^{15}N$-labeled histidine revealed that also one or two of these residues contribute to the first coordination sphere of manganese (the possible assignment of amino acid residue ligands to specific polypeptides will be briefly outlined in section 4.5). At present, it is not yet clear whether in addition one chloride is attached as a terminal ligand (see also section 4.6).

With respect to the inherent potential to undergo redox reactions, a tetranuclear manganese cluster could be envisaged as a suitable storage device where each metal center undergoes a univalent oxidation. Models of this type have been discussed but this idea could not be confirmed experimentally. Before entering in a more detailed discussion on possible electronic configurations of the different $S_i$ states, some general considerations on the redox chemistry should be taken into account: (i) the four manganese can exhibit markedly different redox properties even in a highly symmetric array of the metal centers, if the surrounding protein matrix provides an asymmetric microenvironment, (ii) the oxidation steps can be metal- or ligand (including substrate water)-centered, and (iii) the protein matrix itself can contain redox active groups as potential candidates for hole storage.

*4.3.1. Functional heterogeneity of the manganese in WOC:* The key reaction in WOC is the formation of an O-O bond by oxidation of two substrate water molecules. It therefore appears attractive to postulate that a binuclear manganese group provides a suitable template for this process. This idea would imply that the four manganese are functionally not equivalent. Several lines of structural and functional evidence suggest that there exists a pronounced heterogeneity of the manganese. If one combines this piece of information with the dimer of dimer model derived from EXAFS studies *(vide supra),* the following model emerges for the WOC that is depicted in Fig. 11: (a) the four manganese are incorporated into a highly anisotropic protein matrix; (b) as a consequence of point (a), there exist two functionally entirely different binuclear centers referred to as $[Mn(n)Mn(m)]_{WOX}$ and $[Mn(p)Mn(r)]_C$; (c) the oxidative pathway from $S_0$ to $S_4$ leading to oxygen formation takes place at $[Mn(n)Mn(m)]_{WOX}$, where n and m denote the redox state of Mn in the different redox states $S_i$ (see 4.3.2) and WOX stands for water oxidase; (d) the group $[Mn(p)Mn(r)]_C$ is assumed to remain redox inactive during the oxidative pathway but can become reduced under special conditions, thus giving rise to "superreduced" states $S_{-1}$ and $S_{-2}$ of WOC (p and r represent the redox states of Mn).

The basic idea of redox tuning of the manganese center by an anisotropic protein matrix is supported by experimental data obtained with synthetic binuclear manganese complexes. It was shown that the redox potential can be shifted over a wide range ($\Delta E_m$ of more than 500 mV) not only by the nature of ligand substitution but also by protonation of the $\mu$-oxo-bridge(s). The inherent modulating potential of this protein microenvironment also implies the possibility of allosteric regulation of the reactivity.

*4.3.2. Electronic configuration of the redox states $S_i$ in WOC:* In principle, spectroscopic methods provide the most suitable tools to analyze the electronic configuration and nuclear geometry of functional redox groups containing transition metal centers. Several sophisticated techniques have been applied in order to unravel the properties of the WOC: magnetic resonance (cu-EPR, pulsed EPR; ENDOR, NMR), UV-VIS, FTIR (*F*ourier *T*ransform *I*nfra-*R*ed), and X-ray absorption (XANES, EXAFS). Despite numerous attempts and great efforts, however, only limited information is available. Therefore, before summarizing the rather fragmentary conclusions, it seems worth mentioning some inherent problems that severely complicate the investigations. This brief discussion not only describes the difficulties of current attempts in this field but also illustrates the challenges of future research. A general problem arises from the fact that the WOC operates in a rather narrow temperature range, while many spectroscopic measurements can be performed successfully only at unphysiologically low temperatures: the turnover of the oxidative pathway in WOC becomes blocked between 170 and 250 K, depending on the individual redox transition, but EPR signals characterizing some of the $S_i$ states are only detectable at very low (liquid helium) temperatures. Therefore, these data reflect the electronic configuration and nuclear geometry of a WOC that is functionally incompetent (a similar situation arises for EXAFS and XANES studies). In order to gather reliable information on the mechanism of the WOC from low temperature studies, it remains to be shown that the distribution of electron density between ligand and metal centers remains essentially invariant in the range from 4K up to physiological temperatures. This basic problem has not yet been solved. In addition, a few other difficulties should be mentioned: (a) some redox states seem to be EPR silent ($S_0$, normal $S_3$ state); (b) K-edge inflection point energies do not only depend on the valence state of manganese but are equally susceptible to redox and structural changes of the first coordination sphere; (c) regardless of difficulties in the deconvolution of the experimental data into difference spectra, $\Delta\varepsilon_{i+1,i}(\lambda)$ of the individual oxidation steps $S_i \rightarrow S_{i+1}$, the shape of these spectra in UV is rather structureless and does not permit any reliable conclusion on the nature of these processes.

If one takes these considerations into account for a critical survey of the wealth of reported spectroscopic data, it turns out that only limited sound information can be extracted. Without discussing all "pros" and "cons", the following conclusions can be drawn: (i) the reaction $S_1 \rightarrow S_2$ is a metal centered redox step which very likely comprises the oxidation of a Mn(III) center to Mn(IV) within a $\mu$-oxo-bridged binuclear entity; (ii) the redox transition $S_0 \rightarrow S_1$ is much less characterized but most of the information available is in line with the idea that this is also a metal-centered reaction. It seems most plausible to ascribe this process to an oxidation of Mn(II) into Mn(III) at the same binuclear manganese unit, and (iii) the

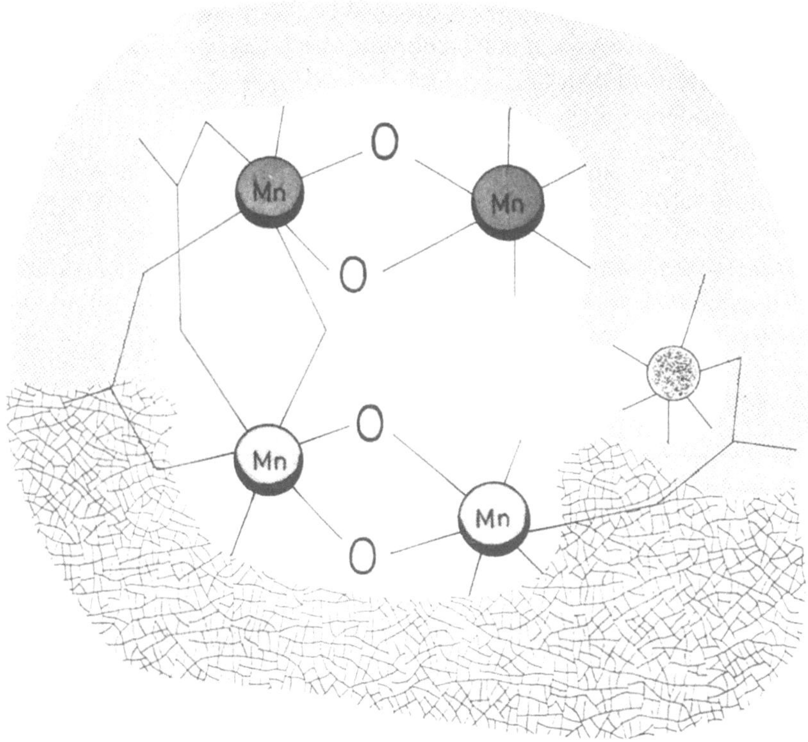


Fig. 11. Schematic representation of the arrangement of the tetranuclear manganese cluster within an asymetric protein matrix. The geometry of the manganese ions and their bridging ligands corresponds with a recent proposal (VK Yachandra, V De Rose, MJ Latimer, I Mukerjee, K Sauer and MP Klein, Science 260 (1993) 675–679). For further details, see text.

nature of the $S_2 \rightarrow S_3$ transition is a matter of controversy but the majority of the experimental findings favors a ligand-centered process comprising either a redox active ligand from the matrix (histidine?) or substrate water (see section 4.8). If one accepts the idea of functional heterogeneity of manganese (dimer of dimer model, see Fig. 11), the above-mentioned reactions are ascribed to the binuclear unit $[Mn(n)Mn(m)]_{WOX}$. Therefore, it remains to consider briefly the state of group $[Mn(p)Mn(r)]_C$. All information so far available is in line with $p = r \equiv III$ or $p = r \equiv IV$. A further distinction between these two alternatives lacks sound experimental evidence, although at present many groups prefer the latter, i.e., $p = r = IV$.

Regardless of this detail, it is clear that the WOC is a system that requires Gibbs energy for its formation. The assembly of a functionally competent WOC by incorporation and proper ligation of manganese is known to be a low quantum yield multi-step process. It comprises the light-driven oxidaton of $Mn^{2+}$ by $P680^+$ (via $Y_Z$) and a dark ligation step. This process, called photoactivation, is a clear indication for higher oxidation states of manganese in the intact WOC. Another reflection of this property is the observation that treatment with an excess of certain exogenous electron donors (e.g. $NH_2OH$, $NH_2NH_2$) leads to destruction of WOC by reductive loss of manganese. Apart from indirect information on the valence states of manganese, photoactivation experiments also revealed that $Ca^{2+}$ is an essential component probably acting as a cofactor (see section 4.6).

*4.3.3. Structural changes coupled with the oxidative turnover of the water oxidizing complex and substrate water entry:* Regardless of any detailed structure, the process of water oxidation necessarily comprises dynamic changes of the nuclear geometry. Several processes have to be considered: (a) the binding of two water molecules to a suitable template, (b) the possibility of hydrogen bond formation of substrate water with the protein microenvironment, (c) the approaching of the two O-atoms in order to form an O–O bond, (d) the movement of protons due to deprotonation reactions coupled with the oxidation steps, and (e) the product ($O_2$) release.

In principle, sophisticated spectroscopic techniques should provide the most powerful tools to analyze interactions between manganese and ligand nuclei and their changes due to redox transitions in the WOC. Time-resolved FTIR and RAMAN-spectroscopy appear to be most suitable for monitoring dynamic structural changes. The application of these techniques to processes in the WOC is just at the beginning due to the enormous complexity of the whole system and related technical problems. First results with conventional FTIR led to the conclusion that at 250 K structural changes take place upon the $S_1 \rightarrow S_2$ transition.

ENDOR spectroscopy and pulsed EPR permit the study of the interaction between electron spin states of manganese and the nuclear spin of associated ligands. In this case, a general problem arises for the WOC because the application to the most interesting O-containing ligands requires labeling with $^{17}O$ (the nuclear spin of the most abundant isotope $^{16}O$ and also of $^{18}O$ is zero). Another interesting nucleus is the proton, especially with respect to binding and properties of substrate water. So far, very few studies have been reported for the WOC. The data available are in favor with the binding of at least one substrate water molecule to manganese in redox state $S_2$.

In addition to these complicated and expensive direct methods, valuable information can be also gained from more indirect approaches. An analysis of the temperature dependence of the redox reactions $Y_Z^{ox}S_i \rightarrow Y_ZS_{i+1}$ within the framework of the classical Marcus theory (see section 3.4.1 and Eqn. 17) led to the conclusion that the reorganization energy of $S_2$ oxidation into $S_3$ exceeds that for $S_1 \rightarrow S_2$ by a factor of about 2.5 (see Table 2). This result can be considered as evidence for significant structural changes coupled with the formation of $S_3$. Independent support for this idea was obtained by using the reductants $NH_2OH$ and $NH_2NH_2$

Table 2. Energetics parameters of oxidation steps in the water-oxidizing complex

| Reaction | Driving force (meV) | Activation Gibbs energy (meV) | Reorganizaton energy (meV) |
|---|---|---|---|
| $Y_Z^{ox}S_0 \rightarrow Y_ZS_1$ | 250[a] | 65 (n.d.)[b] | 650 |
| $Y_Z^{ox}S_1 \rightarrow Y_ZS_2$ | 55 | 135 (115) | 630 |
| $Y_Z^{ox}S_2 \rightarrow Y_ZS_3$ | 40 | 385 (300) | 1600 |
| $Y_Z^{ox}S_3 \rightarrow (Y_ZS_4) \rightarrow Y_ZS_0 + O_2$ | 105 | 225* (185)* | 1100* |

*Values above a critical temperature of about 5°C (spinach) or about 15°C (thermophilic cyanobacteria, values in brackets).

Data taken from:
[a]MH Vos, HJ van Gorkom and PJ van Leeuwen, Biochim. Biophys. Acta 1056 (1991) 27–39, I Vass and S Styring, Biochemistry 30 (1991) 830–839.
[b]H Koike, B Hanssum, Y Inoune and G Renger, Biochim. Biophys. Acta 893 (1987) 524–533, G Renger and B Hanssum, FEBS Letters 299 (1992) 28–32.

(these compounds are isoelectronic with $H_2O_2$) as exogenous redox probes for the different $S_i$ states of the WOC. It was found that $S_3$ exhibits an unusual slow reactivity compared to $S_2$. The most simple explanation for this effect is a marked conformational change due to $S_3$ formation from $S_2$. It therefore appears attractive to postulate that in $S_3$ a re-rearrangement takes place leading to a peroxidic type electronic configuration and nuclear geometry (see section 4.8).

With respect to substrate water binding, it is not yet clear whether the product ($O_2$)/substrate ($H_2O$) exchange occurs at the same redox level or whether after $O_2$ release the catalytic site remains empty until the substrate enters into the catalytic site at a higher redox state of the WOC.

In Fig. 10 question marks were introduced in order to remind the reader that this is a problem that still remains to be clarified.

### *4.4. Proton release pattern coupled with the oxidative pathway in the water-oxidizing complex*

Latest data favor the idea that at least one substrate water molecule in an unresolved protonation state becomes ligated in redox state $S_0$ and/or $S_1$ of the WOC. In $S_2$ and $S_3$ one substrate molecule is slowly exchangeable (order of 0.5 s), the other one exhibits a fast replacement ($\leq$30 ms). The oxidation of two water molecules to dioxygen is indispensably coupled with the release of four protons. At first glance, one might therefore naively assume that the pattern of proton release coupled with the sequence of oxidation steps in the WOC could provide direct information on the mechanism of photosynthetic water oxidation. A closer inspection, however, reveals that this is not the case. If one takes into account that the functional redox groups are incorporated into a protein matrix containing protonizable amino acid residues, it is clear that electron transfer reactions will simultaneously change the electrostatic interaction with these groups. This gives rise to a pK shift that can cause either proton uptake or release depending on the direction of ΔpK. The extent of the measured effect is a function of the number of protonizable groups, their pK values, the magnitude of ΔpK and of the pH in the surroundings. Therefore, it was predicted that noninteger stoichiometries of proton release can be coupled with the oxidative $S_i$-state transitions and that these patterns should exhibit a dependence on the external pH. Likewise, a suitable modification of the protein matrix was expected to cause changes of this stoichiometry (14). Extensive experimental studies during the last half decade have fully confirmed these ideas. It is therefore concluded that the experimentally observed proton release is predominantly owing to the electrostatic compensation of positive charges at the PS II donor side. In PS II core complexes with high $O_2$ evolution capacity each univalent oxidation step leads to the release of one proton irrespective of a particular redox state $S_i$ in the WOC. In the light of this strong influence of the protein matrix, the actual proton release stoichiometry is meaningless in terms of the central redox chemistry of substrate water (for further reading, see, for example, the article of Lavergne and Junge in ref. 13).

### *4.5. Protein matrix and regulatory subunits*

The protein(s) surrounding the manganese cluster exert at least two functions: (i) they provide ligands for the coordination of the manganese and can possibly form hydrogen bonds with the substrate water, and (ii) they act as regulatory entities by modulating the redox properties, controlling the substrate ($H_2O$) accessibility, and by shielding the manganese cluster against undesired side reactions with reductants in the lumen.

The polypeptide(s) providing ligands for the manganese cluster is (are) not yet unambiguously identified. Different lines of evidence suggest that the reaction center polypeptide D1 (see section 3.4.4) is of central relevance. Latest studies with genetically engineered mutants of the cyanobacterium *Synechocystis* PCC 6803 revealed that Asp 170 of D1 is essential for photoligation of manganese (see section 4.3.2). Its possible role as direct ligand to the mature tetranuclear manganese cluster remains to be confirmed experimentally. Other potential candidates among the amino acid residues of D1 are Glu 189, His 332, Glu 333, His 337, and Asp 342. It has to be emphasized that other polypeptides of the PS II complex cannot be excluded to contribute constituents of the ligand sphere. With respect to the regulatory subunits, the extrinsic 33 kDa polypeptide encoded by the psbO gene (therefore often referred to as PS II-O protein) plays a key role. It is present in all oxygen-evolving organisms and contains a conserved cystine bridge. Reductive cleavage of this bridge causes a loss of the oxygen- evolution activity without affecting the manganese binding. Oxidative reformation leads to restoration of the function. Therefore, this structural element is inferred to be essential. The underlying mechanism of its regulatory effect remains to be clarified. In addition, plants contain further polypeptides (18 and 23 kDa polypeptides, see Fig. 8) which affect the rate of oxygen evolution by modulating the demand of $Ca^{2+}$ and $Cl^-$.

The function of other smaller subunits is also at present not known. Another unresolved mystery is the role of cytochrome b559 which tightly binds to the D1/D2-heterodimer (see section 3.4).

In summary, the polypeptide pattern of PS II is rather complex. Extensive work is required to answer questions about the unit(s) forming the apoprotein of the WOC and the structural and functional role of a number of regulatory polypeptides (for further reading, see ref. 15 and article of Vermaas et al. in ref. 13 and literature cited therein).

The tuning of reaction coordinates in the WOC by proteins via direct and indirect effects is a stimulating topic of future research.

### *4.6. The role of cofactors ($Ca^{2+}$, $Cl^-$)*

It is now well established that $Ca^{2+}$ and $Cl^-$ play an essential role for photosynthetic water oxidation. However, despite a wealth of experimental data reported in the literature, the underlying mechanism for the functional role of these two ions is still not clarified. One problem, which so far hampered a real breakthrough, is the possibility that these ions can exert a multitude of effects including the modulation of protein structures. Therefore, many procedures used to generate a $Ca^{2+}$ and/or $Cl^-$ demand in oxygen evolution activity may introduce secondary, artificial protein modifications which are partially or fully compensated by the addition of these ions. In spite of these problems, it is clear that $Ca^{2+}$ is inherently required to establish a functionally competent WOC, as deduced from the need of $Ca^{2+}$ in the activation of $O_2$ evolution after photoligation of manganese (see 4.3.2). Based on these consideration $Ca^{2+}$ may be assumed to be required as a structure determinant for the proper arrangement of the catalytic site. In a more specific way, $Ca^{2+}$ could also be involved in substrate water binding. Furthermore, it is possible that in connection with $Cl^-$ binding these cofactors participate in protolytic reactions due to electrostatic modulation of pK-values (see section 4.4). In line with this idea, $Cl^-$ and/or $Ca^{2+}$ could also act as counterion for balancing electric charges and/or competitively controlling $HO^-$ binding to the manganese cluster in order to prevent undesired side reactions. However, a more specific role of $Cl^-$ (e.g. as a manganese ligand that affects the redox potential gap between $S_i$ redox states) cannot be entirely ruled out (for further reading see ref. 15 and articles in ref. 13).

*4.7. Energetics of photosynthetic water oxidation*

The energetics of water oxidation to dioxygen is pH dependent. At pH = 7.0 the process requires a minimum redox potential of +0.815 V. The redox potential of P680/P680$^+$ under working conditions has not been determined experimentally, but estimations of the order of +1.1 V provide a realistic measure. If one takes into account a redox gap between P680/P680$^+$ and $Y_Z/Y_Z^{ox}$ of the order of 0.1 V, the driving force at the level of $Y_Z^{ox}$ is sufficient to permit water oxidation at pH = 7.0 but reaches the thermodynamic limit at pH values of 4.5 to 5.0. The average redox potential of the WOC itself is even lower. Although precise data on the energetics of the individual redox transitions $S_i \rightarrow S_{i+1}$ are still lacking, indirect lines of evidence provide reliable estimations. At the current stage of knowledge, the reaction coordinates of the oxidation pathway in the WOC can be described by a scheme presented in Fig. 12 and the data summarized in Table 2. However, it has to be emphasized that a detailed description requires information on the configurational energies of the individual redox steps. These values are not known but thermodynamic considerations seem to favor the idea that the key step of the overall reaction, i.e. the O–O bond formation, takes place via a concerted two-electron abstraction which gives rise to a peroxidic state. (For thermodynamic aspects of photosynthetic water oxidation, see ref. 16 and literature presented

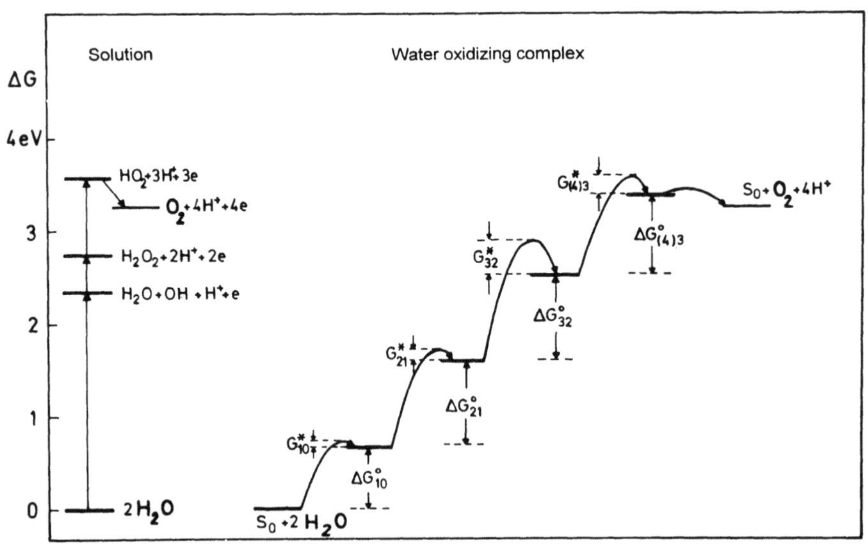


Fig. 12. Energetics of water oxidation to dioxygen in aqueous solutions (left side) and of the oxidation pathway in the water-oxidizing complex (WOC). For further details, see text and Table 2.

there.) It is a matter of current debate whether this reaction takes place only in $S_4$ or at a lower redox state of the WOC (see section 4.8).

Another unresolved problem is the comparatively low redox potential of the $S_0/S_1$ couple. As the water oxidation to dioxygen driven by $P680^+$ (via $Y_Z$) requires a delicate energetic balance, it is not easy to understand that a redox gap of 250 mV in the $S_0$ oxidation to $S_1$ should be dissipative. It is, therefore, reasonable to assume that this part of the Gibbs energy is used for the overall process. The underlying mechanism of this phenomenon remains to be clarified.

### *4.8. Mechanism of photosynthetic water oxidation*

Although it is certainly premature to propose a molecular mechanism of photosynthetic water oxidation, it might be intriguing to summarize the facets described in the former sections (4.2–4.7) in the form of a hypothetical model. An attemt is presented in Fig. 13. This scheme tacitly as-

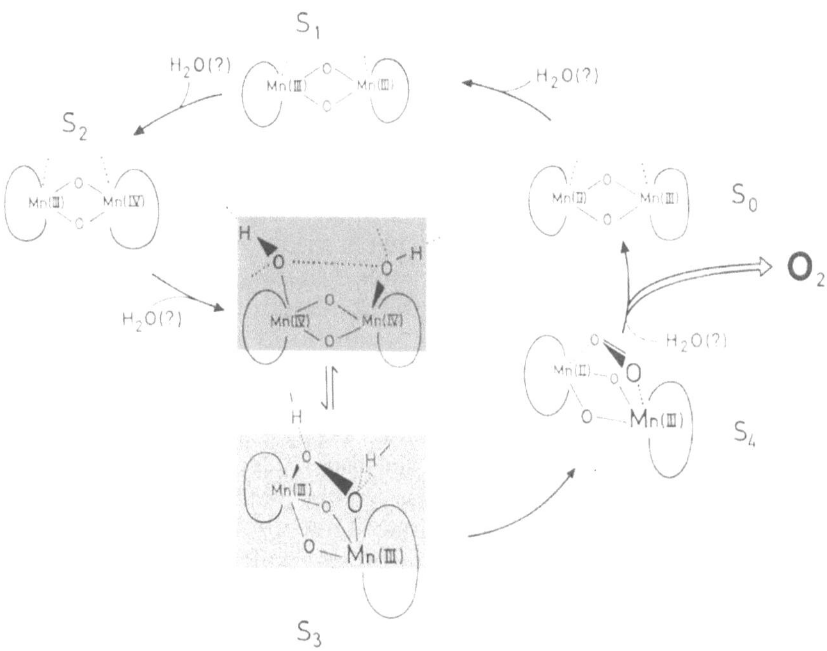


Fig. 13. Model of photosynthetic water oxidation at a binuclear manganese template $[Mn(n)Mn(m)]_{WOX}$. For the sake of simplicity, the other ("inert") manganese are omitted. The ensemble of ligands other than the two presumed μ-oxo-bridges and the substrate are symbolized by ovals. A substrate ligation is explicitly shown only for $S_3$ (in the other redox states a possible water ligation is symbolized by dotted lines). The proposed hydrogen bonding between the protein matrix and ligand water molecules is also symbolized by dotted lines. For further details, see text.

sumes that water oxidation takes place at a binuclear manganese template containing probably two μ-oxo-bridges. For the sake of simplicity, the other two manganese of the tetranuclear cluster are not shown. They are assumed to be redox inert during the oxidative pathway from $S_0$ to $S_4$ (see section 4.3.1). As discussed in 4.3.2, the redox transitions $S_0 \rightarrow S_1$ and $S_1 \rightarrow S_2$ are considered to be metal-centered electron abstraction steps. The essential point of this mechanism is the postulate that redox isomerism equilibrium exists in $S_3$ that comprises a peroxidic type configuration and two hydroxyls complexed at the binuclear manganese group (see Fig. 13). The two hydroxyls exhibit quite different exchange rates with bulk water molecules as shown by labeling experiments with $H_2{}^{18}O$. $Y_Z^{ox}$ is inferred to induce an oxidant-induced reduction of the manganese in the peroxidic type state, leading to an $S_0$ configuration of the binuclear manganese group complexed with $O_2$ (this state corresponds to $S_4$). For further reading, see review articles in ref. 13 and literature quoted therein.

## 5. Plastoquinone reduction

In the preceding sections the formation of reducing and sufficiently oxidizing redox equivalents within the reaction center (see section 3.4) and of water oxidation (see section 4) have been described. Now it remains to briefly summarize the reductive pathway of PS II leading to plastohydroquinone formation. The reductive formation of plastohydroquinone ($PQH_2$) from plastoquinone (PQ) requires the cooperation of two electrons (see Eqn. 5). Analoguously to water oxidation, the reduction of PQ to $PQH_2$ takes place via a sequence of one-electron redox steps. In this process, one PQ molecule of the pool is statistically and noncovalently associated with a special binding site referred to as $Q_B$-site. After accepting one electron from $Q_A^-$ of the reaction center (see section 3.4), the newly formed semiquinone, which is designated as $Q_B^-$, is highly stabilized at the binding site. A second electron transfer from $Q_A^-$ leads to a redox state at $Q_B$ which corresponds to that of plastohydroquinone. The protonation pathway leading to plastohydroquinone is not resolved in detail but it is clear that amino acid residues from the protein matrix play a key role. The eventually formed $PQH_2$ is relatively weakly bound and becomes replaced by another oxidized PQ molecule of the pool thereby accomplishing the redox cycle at the acceptor side. The binding site of $Q_B$ is located within the D1 polypeptide at the stroma exposed side of the reaction center complex (see Fig. 8). A high-spin $Fe^{2+}$, which is assumed to be coordinately bound to four histidine residues of the polypeptides D1 and D2 (in analogy to the coordination of the corresponding iron center in purple bacteria) is located between $Q_A$ and $Q_B$ (Fig. 8). At first sight, this $Fe^{2+}$ could participate in the reaction sequence by undergoing a transient valence change to $Fe^{3+}$ via a so-called "push-pull" mechanism, thereby directly supporting the electron transport from $Q_A^-$ to $Q_B$ (or $Q_B^-$). Recent investigations in isolated reaction centers of photosynthetic purple bacteria have shown, however, that the kinetics of electron transfer are practically unaffected by depletion of $Fe^{2+}$. Reconstitution experiments revealed that neither $Fe^{2+}$ itself nor another bivalent cation is required for the fast electron transfer from $Q_A^-$ to $Q_B$ (or $Q_B^-$) in anoxygenic purple bacteria. Therefore, it appears very likely to assume that also in PS II the non-heme $Fe^{2+}$ center does not directly participate as redox active group in the reoxidation of $Q_A^-$ by $Q_B(Q_B^-)$.

On the other hand, it was shown that the properties of the PQ-reducing enzyme are changed after detergent treatment. This finding suggests that the structure of the microenvironment is susceptible to structural changes of functional relevance. In this respect, the interaction with specific lipids could be of significance for modulation of the $Q_B$-site.

A large number of chemicals are known to block the reoxidation of $Q_A^-$ by $Q_B$ ($Q_B^-$). Different classes of these substances are of practical relevance because they are used as PS II herbicides. It was found that these agents bind to the $Q_B$-site thereby preventing the reductive turnover of the endogenous plastoquinone by electron transfer from $Q_A$. Mutants generated and selected under herbicide stress as well as obtained by site directed mutagenesis revealed that the exchange of only one amino acid residue (e.g., Ser 264 of polypeptide D1) can drastically reduce the susceptibility to PS II herbicides (for further reading on the topics in section 5, see, for example, refs. 17, 18).

## 6. Concluding remarks

In this chapter an attempt was made to provide a snapshot of our current knowledge on photosynthetic water cleavage. It certainly cannot provide an encyclopedic description of all detailed information. Therefore, our efforts were concentrated on an illustration of the basic principles and their exploitation in photosynthesizing organisms. A number of problems appears to have been solved, other important questions remain to be answered, especially with respect to the structural organization and the detailed mechanism of the water oxidizing complex. With the possibility of modifying the apparatus selectively by genetic engineering together with improvements in membrane biochemistry and further development and refinement of spectroscopic techniques new research roads are being opened in the fascinating field of photosynthetic water oxidation.

## Suggestions for further reading


1. DL Gilbert (ed.), *Oxygen and Living Processes,* Springer, New York, 1981.
2. AR Holzwarth, Quart. Rev. Biophys. 22 (1986) 239–326.
3. G Renger in *The Photosystems: Structure, Function and Molecular Biology,* J Barber (ed.). Elsevier, Amsterdam, 1992, pp. 45–99.
4. EM Aro, I Virgin and B Andersson, Biochim. Biophys. Acta 1143 (1993) 113–134.
5. RS Knox in *Primary Processes of Photosynthesis,* J Barber (ed.), Elsevier, Amsterdam, 1977, pp. 55–97.
6. RM Pearlstein in *Photosynthesis,* Govindjee (ed.), Academic Press, New York, 1982, Vol. 1, pp. 292–330.
7. JF Allen, Biochim. Biophys. Acta 1098 (1992) 278–335.
8. D De Vault, *Quantum-Mechanical Tunneling in Biological Systems,* 2nd edn., Cambridge University Press, New York, 1984.
9. RA Marcus and N. Sutin, Biochim. Biophys. Acta 811 (1985) 265–322.
10. CC Moser, JM Keske, K Warncke, RC Farid and L Dutton, Nature 355 (1992) 796–802.
11. G Renger and U Schreiber in *Light Emission by Plants and Bacteria,* Govindjee, J Amesz and DC Fork (eds.), Academic Press, New York, 1986, pp. 587–619.
12. JB Bell in *The Photosystems: Structure, Function and Molecular Biology,* J Barber (ed.), Elsevier, Amsterdam, 1992, pp. 17–43.
13. G Renger and Govindjee (eds.) *How Plants and Cyanobacteria Make Oxygen: 25 Years of Period Four Oscillations,* Special Issue Photosynth. Res. Vol 38 (1993).
14. G Renger, Photosynthetica 21 (1987) 203–224.
15. RA Debus, Biochim. Biophys. Acta 1102 (1992) 269–352.
16. LI Krishtalik, Bioelectrochem. Bioenerg. 23 (1990) 249–263.
17. AR Crofts and CA Wraight, Biochim. Biophys. Acta 726 (1983) 149–185.
18. W Oettmeier in The *Photosystems: Structure, Function and Molecular Biology,* J Barber (ed.), Elsevier, Amsterdam, 1992, pp. 349–408

Bioenergetics
ed. by P. Gräber and G. Milazzo†


# *CHAPTER 8*
# The photosystem I reaction center in oxygenic photosynthesis

Paul Mathis

*DBCM/SBE, CEA-Saclay, 91191 Gif-sur-Yvette Cedex, France*

## 1. Introduction

In the photosynthetic apparatus of green plants, two types of reaction centers named Photosystem-I and Photosystem-II (PS-I, PS-II) work in series to transfer electrons from water to $NADP^+$ (see Chapter 5). This endergonic reaction is activated by electronic excitation energy resulting from light absorption. In the process, the PS-I reaction center constitutes the low-potential, i.e. reducing pole; it drives electrons from small metallo-proteins, usually plastocyanin but also sometimes cytochrome *c,* at $E_m \approx +370$ mV, to $NADP^+$ ($E_{m,7} = -320$ mV). Plastocyanin and/or cytochrome *c* are reduced by PS-II via plastoquinone and the cytochrome b/f complex. Like all reaction centers, PS-I is a complex membrane protein made of several polypeptides, of pigments and of redox-active centers. It is found in the thylakoid membrane of all photosynthetic oxygenic organisms, with properties nearly identical in cyanobacteria and in higher plants. It is rather analogous to the reaction center of some non-oxygenic bacteria, green sulfur bacteria and heliobacteria, as detailed below. Many properties of PS-I have been described in more extensive or more specialized reviews (see for instance refs. 1–7). A few of these properties are now well established. They will be summarized in this chapter, starting with functional aspects (electron and energy transfer) and continuing with structural aspects, and especially with the role of the polypeptides.

It is now clear that all reaction centers have a common mode of functioning: after absorption of light, excitation energy is transferred to a species named P (P-700 in PS-I) which is both a trap for energy and the primary electron donor. An electron leaves excited P and travels in an array of several acceptors with progressively less negative redox potentials (Fig. 1). The oxidized donor ($P^+$) is re-reduced by an electron donor. In addition to this common scheme, it is more and more obvious that all reaction centers have an overall analogous structure, so that the three-dimensional structure of the reaction center of the purple bacterium *Rhodopseudomonas viridis* can serve as a model for PS-I, for which no detailed three-dimensional (3-D) structure has been obtained to date.

## 2. Energy transfer in PS-I

The PS-I reaction center always contains a large number of pigment molecules; the smallest unit which can be isolated includes about 110 chlorophyll *a* and 15 β-carotene in its native state (many of these pigments can be artificially removed, for instance with organic solvents). Most of these pigment molecules perform a so-called antenna function: they absorb light and they very quickly convey excitation energy to P-700. This aspect will be only briefly described in this review.

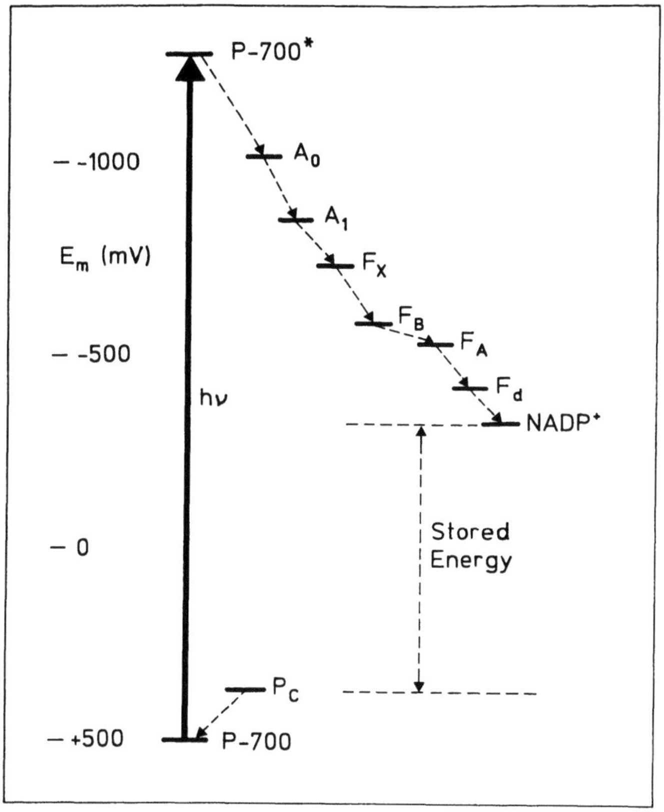


Fig. 1. Presentation of redox centers in PS-I arranged according to their redox potentials (versus NHE; vertical scale). The arrows indicate the effect of electronic excitation of P-700 and the successive electron transfer steps. The final stored energy corresponds to a couple with oxidized plastocyanin (Pc) and reduced NADP.

## *2.1. Structure of the PS-I antenna*

The PS-I antenna has a different structure in chlorophyll *b* containing organisms (plants, algae) and in cyanobacteria (which have no chlorophyll *b,* but which have additional pigmented complexes, the phycobilisomes). All organisms, however, have a similar core structure.

*2.1.1. CP1: the core structure:* The core complex of PS-I [3, 8], named CP1, consists of two large polypeptides which hold the primary reactants for electron transfer and antenna pigments: chlorophyll *a* and β-carotene (about one per 6–10 chlorophyll *a*). It also includes several small polypeptides (see below). In cyanobacteria this complex includes all the PS-I antenna. It comprises about 110 chlorophyll *a* per P-700, i.e. per reaction center. These pigments are held by the two large polypeptides of ≈ 83 kDa each, resulting in a very high density of chlorophylls in the protein: about one per 1.5 kDa on average. It is usually thought that histidine residues are involved in chlorophyll binding. Each of the 83 kDa polypeptides contributes around 40 histidine residues, but other residues might be implicated as well. In plants and green algae, the core complex can be prepared only by a rather harsh treatment with detergents, in order to remove the other antenna complexes. Perhaps for that reason, the chlorophyll content varies from preparation to preparation: between 40 and 110 chlorophyll *a* per P-700. It is reasonable to think that the core complex includes about 110 chlorophyll *a*, like in cyanobacteria, but that a more or less important part of them can be lost during the detergent treatment [9, 10]. During controlled greening, PS-I assembles with 95 chlorophyll *a* per P-700 in barley chloroplasts [11]. The situation in various classes of algae is described in detail by Hiller et al. [12].

*2.1.2. Other PS-I antenna complexes:* It is now recognized that the antenna of all photosynthetic organisms is highly complex. This is also the case for PS-I. In higher plants, the core complex is surrounded by several chlorophyll *a*/*b* binding proteins, constituting a "native" PS-I complex named PS-I-200, with about 200 chlorophylls [8–10]. After its isolation, this complex can be dissociated into the PS-I core and a pure antenna complex named LHC1, the analysis of which reveals the presence of four different polypeptides of about 20–24 kDa (reviewed in refs. 13, 14). These polypeptides bear chlorophylls *a* and *b* in a ratio around 4. Corresponding genes or cDNA have been sequenced, leading to interesting speculations on the mode of folding of these polypeptides and on their evolutionary relationships [14]. It seems that PS-I antenna complexes bear significant homology with the major PS-II light-harvesting complex, which is much better known and which may thus offer a structural model: mode of chlorophyll binding, 3-D structure, and organization of the several complexes with the reaction center.

### *2.2. Energy transfer in the PS-I antenna*

With photons being absorbed by any one of the pigment molecules in the antenna, energy transfer first occurs within one single antenna complex. Excitation then migrates from complex to complex, finally to the reaction center, and ultimately to the trap (P-700 in our case). The major problem with respect to energy transfer is to determine the path followed by excitation, and whether there are structural and spectroscopic properties which define a directionality toward the trap. Several factors are relatively favorable for such studies in PS-I: (i) it is possible to obtain rather native complexes with different antenna sizes: core complex (especially in cyanobacteria) and PS-I-200 from plants; (ii) P-700 is a rather deep trap; excitation energy arrives and induces charge separation, without any significant back energy transfer like in PS-II where P-680 is a much more shallow trap; (iii) P-700 and some of its close chlorophyll molecules have spectroscopic properties different from the major antenna: a long wavelength of absorption ($\approx$ 700 nm instead of 670–680 nm) and in correlation a long wavelength of fluorescence at low temperature (730 nm instead of $\approx$ 695 nm).

The dynamics of energy transfer have been studied essentially by four methods: fluorescence decay, picosecond flash absorption, exciton annihilation and time-course of flash-induced photovoltage. I will just summarize a few key features (see ref. 7 for a recent detailed review):

- individual energy transfer steps are very rapid (much faster than 1 ps) and excitation energy may be distributed within the core antenna within a few tens of picoseconds.
- a fast phase of fluoroscence decay ($\approx$ 50 ps) is attributed to the time for energy transfer in the core antenna and excitation quenching by P-700.
- excitation quenching by P-700 does not change significantly with its redox state and the important fluorescence induction well-known in PS-II has no well-defined equivalent in PS-I.

## 3. Electron transfer in PS-I

The sequence of electron transfers within the reaction center can be schematized as follows: P-700 $\Rightarrow$ $A_0$ (chlorophyll *a*) $\Rightarrow$ $A_1$ (phylloquinone) $\Rightarrow$ $F_X$ (iron-sulfur center) $\Rightarrow$ $F_B$, $F_A$ (iron-sulfur centers) (see Fig. 2). Before we detail the redox centers and the reactions, a few general remarks seem to be worthwhile:

(a) the sequence above is widely accepted, but many of its aspects are not safely established, in particular the precise chemical nature and the respective roles of $A_1$ and $F_X$. There is evidence for parallel or short-cut paths, as for instance from $A_0$ to $F_X$.

(b) the forward rates of electron transfer are hardly known, but the back reactions, from any given reduced acceptor to P-700$^+$, have been studied in detail and largely used in mapping and identification. Unfortunately, there is no good inhibitor for PS-I, nothing like the herbicides which act as specific inhibitors at the $Q_B$ site in PS-II.

(c) a comparison with purple bacteria may incite to search for a structural symmetry in PS-I and for the occurrence of an "inactive branch" of potential electron carriers. There are several indications in these respects: P-700 is most probably a dimer of chlorophyll; there are two molecules of phylloquinones per center, one which is extractable without impairing the function; the iron-sulfur centers $F_A$ and $F_B$ make up a nearly symmetric couple, and, finally, the core of the reaction center comprises two homologuous large polypeptides which carry P-700, $A_0$, $A_1$ and $F_X$. The recently obtained three-dimensional structure of the PS-I reaction center provides a strong support for these speculations [15].

(d) the contours of PS-I cannot be uniquely defined since the reaction center core is surrounded by peripheral polypeptides and antenna complexes in a manner which varies *in vivo* and still much more *in vitro* according to the preparation procedure. Following Reilly and Nelson [5] we could define the PS-I reaction center as the "minimal structural unit which catalyzes the photoreduction of ferredoxin with plastocyanin or cytochrome *c* as the electron donor". We should add, however, that some part of the acceptor side may be involved in the

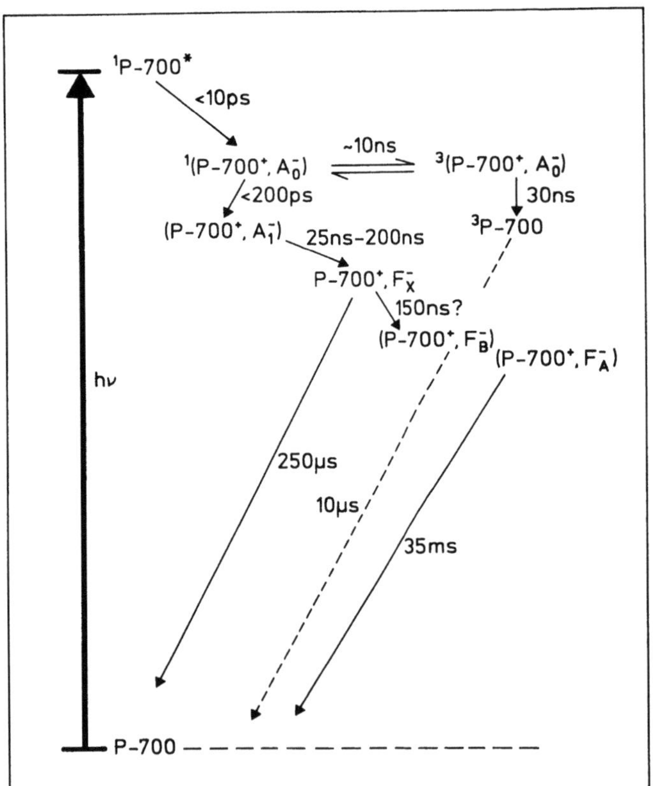


Fig. 2. Simplified kinetic scheme for electron transfer in the PS-I reaction center, at room temperature.

ferredoxin-NADP$^+$-reductase activity, and also that the minimal structure for electron transfer may be surrounded, in a native state, by specific antenna complexes which can be considered as part of the reaction center.

### *3.1. P-700: the primary electron donor*

Like in all reaction centers, the primary donor P is named according to the peak wavelength of its red absorption band: oxidation of P-700 induces a maximum bleaching around 700 nm. The primary donor is a key species because it traps excitation energy and starts the electron transfer process. Its properties are discussed in detail in refs. 1 and 7. They include:

- chemical nature. P-700 is very probably a dimer of chlorophyll *a*. Several suggestions of a different nature are either weakly supported or contradicted by experiments: monomeric chlorophyll, enol form of chlorophyll, epimer named chlorophyll *a'*, or hydroxychlorinated chlorophyll. Chlorophyll *a'* seems to be a bona fide component of PS-I but its function is not certain. The detailed structure of the dimer has not been established. It can be speculated, however, that the chlorophylls have no direct binding interaction and that, like in purple bacteria, they are bound to the protein which maintains them at a rather close distance.
- absorption spectrum. The absorption spectrum of P-700 is not known. We only know the difference spectrum induced when P-700 gets from the neutral ground state into an oxidized state (radical cation P-700$^+$) or into a triplet state ($^3$P-700), a process which will be discussed below. In both cases the difference spectra are similar to the analogous difference spectra obtained for chlorophyll *a in vitro,* with a few significant differences, however, in particular the long wavelength of the red peak which lies at 700 nm instead of 670-678 nm. Like all chlorophyll cations, P-700$^+$ has a weak absorption band in the near infra-red, peaking around 810 nm, which is very useful for flash kinetic studies. Among the new developments in optical methods, vibrational methods appear very promising, especially for determining the binding of chlorophylls to the protein and the difference in structure between the neutral and radical cation states. Differential Fourier-transform infra-red (FT-IR) and resonance Raman spectroscopy (RRS) start providing useful information, although the interpretations of the spectra are still largely in progress [16, 17].
- oxidation-reduction potential. The $E_m$ of the P-700/P-700$^+$ couple is about +490 mV. This is much lower than monomeric chlorophyll *a in vitro*, the $E_m$ of which is around +800 mV. This lowering is probably

associated with the function of P-700 since the P-700*/P-700$^+$ couple must be highly reducing in order to start the electron transfer in the sequence of low-potential electron carriers. The molecular reason for the relatively low $E_m$ of P-700 is not known.

- EPR spectrum of P-700$^+$. This spectrum has a great historical importance since it was among the first EPR spectra to be discovered in biological systems, under the name "Signal I". It is also at the origin of the fruitful "special pair" theory which states that primary donors are dimers of pigments [18]. The basis of the proposal is that the EPR linewidth of P-700$^+$ is narrower than that of oxidized chlorophyll *a* by a factor $2^{1/2}$, as expected for a dimer with perfect delocalization of the unpaired electron over the two halves. A refined analysis showed that the spin sharing is not equal, indicating that the two chlorophylls in the pair are not equivalent (as discussed in ref. 7).
- Excited states of P-700. The lowest singlet excited state of P-700 is the one which induces electron transfer. This excited state is thus efficiently quenched, with the consequences that its lifetime is shorter than 20 ps and that fluorescence due to its radiative deexcitation has not been observed with certainty. Stark effect has recently been reported for P-700, leading to the conclusion that the excited state has a rather large dipole moment attributed to internal charge transfer. A significant charge-transfer character was also concluded from hole burning experiments.

The triplet state of P-700 has been the object of many studies. It is populated by the Radical-Pair Mechanism (see below) and detected by optical and magnetic spectroscopies (reviewed in ref. 20). These studies are relatively easy because the triplet state is long-lived, around 1 ms at low temperature, and is populated with a rather high yield. A particularly clear information derived from triplet studies is the orientation of P-700, which is perpendicular to the thylakoid membrane plane; this conclusion was reached from EPR measurements with PS-1 particles oriented in the magnetic field of the spectrometer [21]. The shape of the EPR spectrum of the triplet should also be informative on the chemical nature of P-700, i.e. monomer or dimer of chlorophyll. The zero-field splitting parameters of $^3$P-700 resemble those of the triplet state of monomeric chlorophyll. This behavior seems to contradict the conclusions from other methods which lead to the proposal that P-700 is a dimer. It can be argued, however, that in P-700 the triplet character is localized on only of the two chlorophylls of the dimer, or that the two chlorophylls in the dimer have their three magnetic axes parallel [7]. In a recent work [22], it has been possible to study $^3$P-700 at room temperature and to deduce structural properties of P-700 under those conditions.

### *3.2. $A_0$: the primary electron acceptor*

In the normal mode of functioning, $A_0$ is rapidly reduced following light excitation, and then rapidly reoxidized by electron transfer to the next electron acceptor: $P^* A_0A_1 \Rightarrow P^+ A_0^-A_1 \Rightarrow P^+ A_0A_1^-$. These processes should be observable by picosecond flash absorption spectroscopy. As a matter of fact, these observations are not easy in PS-I. A main experimental difficulty arises from the fact that PS-I has always a rather large number of chlorophyll molecules (at least 40) which perform an antenna function. Picosecond flash experiments are perturbed in two manners: (i) the first step is artifactually lengthened by the time needed for excitation to arrive at P-700; (ii) the chlorophyll singlet excited state has absorption properties which are not obvious to distinguish from P-700$^+$ or $A_0^-$. There is nevertheless some agreement to say that $A_0$ is reduced in about 10 ps and reoxidized in 30 to 200 ps. Reduction of $A_0$ induces absorption changes consistent with the reduction of a chlorophyll *a* molecule having a red absorption band around 690 nm [23].

The (P-700$^+$, $A_0^-$) radical pair has a much longer lifetime when electron transfer is blocked away from $A_0$. The blocking is most often realized by removal of the secondary acceptor, the phylloquinone $A_1$, by means of controlled treatment with organic solvents or with detergents. It is also rather easy to fully reduce $A_1$ to the hydroquinone state, in which case also electron transfer away form $A_0^-$ is blocked. Under these conditions, the (P-700$^+$, $A_0^-$) state lasts for 20–100 ns. From the nanosecond flash-induced difference spectrum it is possible to substract the difference spectrum due to P-700 $\Rightarrow$ P-700$^+$, leaving the spectrum due to $A_0 \Rightarrow A_0^-$. The features of that spectrum are consistent with $A_0$ being a chlorophyll *a* molecule absorbing around 694 nm, in agreement with the picosecond data.

Historically, the experiments aiming at observing $A_0$ were conducted differently, by the so-called photoaccumulation technique. This technique, which was successful for the stable photoreduction of the primary acceptor (pheophytin or bacteriopheophytin) in PS-II or in purple bacteria, consists in illuminating a sample under reducing conditions, in the presence of an electron donor. Electron acceptors which are not chemically reduced before illumination become progressively reduced under light, although with a low yield, because electron donation to P-700$^+$ can partly compete with back-reactions. In PS-I, a species accumulated, corresponding to a bleaching around 670 nm and to the appearance of an EPR signal of a radical with a g-value around 2.0025. Both properties are consistent with the reduction of a chlorophyll *a* molecule. The reason for the difference in absorption properties between the pico- or nanosecond flash techniques and photoaccumulation is not clear. We suggested [24] that the species which accumulates under light at low redox potentials is not $A_0^-$ but rather a neighbor chlorophyll molecule absorbing at 670 nm which would be progressively reduced by $A_0^-$. In any functional scheme of PS-I, $A_0^-$ has to be highly reducing and might be able to reduce other pigments in the neighborhood. The photoaccumulation technique can work well in PS-II or purple bacteria because the equivalent (primary) electron acceptor is a pheophytin and thus is much less reducing. The redox potential of the Chl *a*/Chl $a^-$ couple has been measured *in vitro* [25]. It is about −0.9 V, versus the NHE. The mid-point potential of the $A_0/A_0^-$ couple has been estimated to be −1.01 V by an indirect method based on the effect of an electric field on luminescence [26].

The properties of the (P-700$^+$, $A_0^-$) radical pair are worth being discussed in some detail for the case where forward electron transfer is blocked away from $A_0$. The initial reactive species is the singlet excited state of P-700, and thus the radical pair is born out in a singlet state, i.e. the unpaired spins on both partners are correlated and anti-parallel. The fate of the radical-pair is governed by two properties (see ref. 20):

(i) nuclear hyperfine interactions induce the radical-pair to evolve to a triplet state, and then back and forth at a frequency of the order of $10^8$ Hz. In terms of energy levels, the singlet state and the three triplet states are nearly identical. Degeneracy among the triplet sub-levels can be removed by application of an external magnetic field, in which case only one state remains nearly isoenergetic with the singlet state.

(ii) since electron transfer is impossible, the radical-pair eventually decays by charge recombination. This leads to three possible states: two singlets (the ground state and the singlet excited state of P-700) and one triplet ($^3$P-700). A final singlet or triplet state is reached when recombination takes place from the respective state (singlet or triplet) of the radical-pair.

Two external factors influence the radical-pair behavior: temperature and magnetic fields. Temperature has an effect on the lifetime ($\approx$ 30 ns at room temperature; $\approx$ 120 ns at low temperature) and somewhat on the triplet yield. Magnetic fields influence the triplet yield in a complex manner: Increasing the magnetic field, the triplet yield first increases, up to a field of about 50 G, and then decreases with a half-field value of about 200 G [27]. The low-field increase is a consequence of the zero-field splitting between singlet and triplet states while the high field decrease reflects the Zeeman effect on triplet sub-levels in the radical-pair.

The properties of the radical-pair are of basic interest because they are intimately related to physical properties of the partners (P-700 and $A_0$) which govern the primary step of electron transfer. The properties of $^3$P-700 are also of great interest. This state has a very distinctive EPR spectrum, because of the mechanism of its formation: the $T_0$ level is much more populated that the $T_{+1}$ or $T_{-1}$ levels. The triplet state $^3$P-700 is rather short-lived at room temperature ($t_{1/2} \approx 10$ µs), but it lasts much longer ($t_{1/2} \approx 0.8$ ms) below 100 K (the effect of temperature on the triplet lifetime has been studied in detail in ref. 28 and erroneously attributed to a charge recombination). Relaxation between the triplet sub-levels also becomes very slow at low temperature, so that light-induced EPR signals of $^{3}$P-700 are quite large and easy to measure. They have often been used as a tool to study PS-I: properties of P-700, as mentioned above, and properties of the electron acceptors in various biochemical preparations. Formation of the triplet indeed indicates that the primary photochemistry is still functioning, but also that electron transfer to $A_1$ is not working, so that $A_1$ should be either missing or pre-reduced.

### *3.3. $A_I$: a phylloquinone molecule*

Is there a need for an intermediary electron carrier to permit electron flow from $A_0$ to the first iron-sulfur center $F_X$? In terms of thermodynamics, the $E_m$ of $A_0/A_0^-$ and $F_X/F_X^-$ being about −1.01 and −0.73 V, respectively, an intermediary acceptor has a quite limited range of redox potential available. The first experimental evidence for $A_1$ perhaps came from our own flash absorption studies with several PS-I preparations at low temperature: P-700 gets oxidized, and then is re-reduced with a half-time of about 120 μs [29] which fits neither with the back-reaction with $A_0^-$ (≈ 100 ns) or $F_X^-$ (hundreds of ms) nor with the triplet state. An acceptor $A_1$ was postulated, although at that time the properties of $A_0$ were not known. These results also showed that, at low temperature, electrons go to $A_1$ and then have the highest probability to return to P-700$^+$ and not to more forward along the normal route to the iron-sulfur centers.

The chemical nature of $A_1$ has been the object of much controversy (see refs. 1 and 6 for a more complete discussion). There is now a rather large agreement to identify $A_1$ with one phylloquinone (vitamin $K_1$) molecule. PS-I reaction centers contain two phylloquinone molecules and no other quinone. Treatments with organic solvents which extract one phylloquinone have no major effect on electron transfer, but harsher treatments which extract both phylloquinones largely block electron transfer away from $A_0$. Nearly normal electron transfer can be reconstituted with a small addition of pure phylloquinone. Reconstitution experiments with different quinones permit to appreciate the molecular properties which are required for a correct binding and functioning: appropriate side chains, presumably for the positioning, and naphthoquinone ring, provide a low redox potential. The site can apparently be occupied by competitive inhibitors, like herbicides do at the $Q_B$ site in PS-II, but also by functional molecules which have a structure very different from naphthoquinones. Flash-induced absorption changes measured at low temperature for the 120-μs phase are also consistent with $A_1$ being a phylloquinone.

A species named $A_1^-$ has been produced by the photoaccumulation method. Its EPR spectrum has some features of a quinone radical-anion (see below, and ref. 1), but this approach may suffer from the same criticism as for the photoaccumulated species considered as $A_0^-$, as discussed above. The proposal that $A_1$ is a phylloquinone raises some diffuculties with regards to some experiments such as UV irradiation, which seems to destroy phylloquinone without impairing electron transfer, and such as photoinduced electron transfer still taking place to the iron-sulfur centers at low temperature, although with a reduced efficiency, when phylloquinone has been extracted. A recent work by Sétif and Bottin strongly supports the view that $A_1$ is a molecule which, like quinones ($Q + 2e^- + 2H^+ \Rightarrow QH_2$), can be doubly reduced to a neutral state [30]. It was shown indeed that electron transfer is blocked at the level of $A_0^-$ under reducing conditions where no radical-ion like $A_1^-$ is detected by EPR, and even where $F_X$ is not reduced. It can be proposed that the mid-point potential is much less negative for the double reduction to a neutral state than for the single reduction to the charged semi-quinone which will be highly unfavored in a hydrophobic protein. The proposal of a double reduction of $A_1$, taking place slowly at not very low redox potentials, may explain a large number of earlier inconsistencies in redox titrations and in photoreactions induced by continuous light.

As mentioned before, electron transfer from $A_0^-$ to $A_1$ takes place in the sub-nanosecond time scale. This reaction, however, has never been the object of a detailed and convincing study. Electron transfer from $A_1^-$ to the iron-sulfur centers takes place in the nanosecond time scale. It has been studied by flash absorption kinetics, and half-times of 15 ns or 200 ns have been reported. Kinetics in the 200-ns time range have been reported by time-resolved EPR at room temperature. A recent work [31] compares the electron transfer kinetics in PS-I particles from cyanobacteria and plants, and shows that electrons go from $A_1^-$ to an Fe-S center in a multiphasic manner.

The radical-pair (P-700$^+$, $A_1^-$) displays a few striking properties. It normally evolves by forward electron transfer to $F_X$, in the nanosecond time range. Forward electron transfer, however, is blocked under several circumstances. At low temperature the (P-700$^+$, $A_1^-$) charge recombination has a half-time of about 120 μs when $F_X$ is oxidized, which accelerates to 25 μs when $F_X$ is reduced. The acceleration is probably attributable to electrostatic repulsion. At room temperature, when $F_X$ is reduced, the radical-pair decays much faster, with a half-time around 250 ns. Surprisingly, the decay is nearly entirely via the triplet route: P-700$^+$, $A_1^- \Rightarrow {}^3$P-700, $A_1$. The triplet is not formed, however, when the equivalent reaction takes place at low temperature. These observations probably indicate that the (P-700$^+$, $A_1^-$) radical-pair state has an energy somewhat below the triplet state, the population of which requires an activation energy. These data have been interpreted in energetics terms [30]; with an $E_m$ of +0.49 V for P-700/P-700$^+$ and assuming that $^3$P-700 lies 1.27 eV above the ground state, it can be concluded that the $A_1/A_1^-$ couple has an $E_m$ below –0.79 V, in agreement with its functional properties.

In all cases the radical-pair (P-700$^+$, $A_1^-$) lives longer than 10 ns. This rather long life is at the origin of another important property of PS-I, i.e. electron spin polarization (ESP) (older literature is cited in refs. 1, 2, 32). A well-studied case of electron spin polarization is that of triplet states (like $^3$P-700) formed by the RPM, as discussed above. The phenomenon also takes place in radicals, during the lifetime of a radical pair or even later on in one of the radicals, which keeps for some time the souvenir of the interaction which it was experiencing in the radical-pair. For the PS-I reaction center, an ESP has been discovered as a large but transient perturbation of the EPR spectrum of P-700$^+$ and of its counterpart radical-anions: the polarization of P-700$^+$ lasts for a few microseconds at room temperature, when the involved radical-anion has already disappeared by forward electron transfer. With our present knowledge, it appears that ESP is essentially due to the (P-700$^+$, $A_1^-$) radical-pair: the primary radical-pair (P-700$^+$, $A_0^-$) is too short-lived for ESP to develop, and the interaction may be too weak in the (P-700$^+$, $F_X^-$) radical-pair. Recent experiments have shown that a normal ESP is strictly correlated to the presence of phylloquinone at the $A_1$ site: it disappears after phylloquinone reduction, it reappears after quinone addition, but only phylloquinone can reconstitute the normal spectrum [33]. The shape of the ESP spectrum can provide chemical and structural information on $A_1$ and its position with regard to P-700$^+$. This type of approach is in progress; it requires difficult and still uncertain theoretical treatments, but there is no doubt that it will be very informative in the future.

### *3.4. $F_X$: a low-potential iron-sulfur center*

Iron-sulfur centers (Fig. 3) play important roles in PS-I. The terminal acceptor is a soluble ferredoxin, with one 2Fe-2S center, which resembles its numerous counterparts in biological electron transfer. Within the reaction center itself, the terminal acceptor is either of a couple of iron-sulfur centers named $F_A$ and $F_B$, with rather low redox potentials: –0.55 to –0.59 V. These electron carriers have been first evidenced by their low-temperature EPR spectra: all iron-sulfur proteins with 2Fe-2S or 4Fe-4S clusters indeed display characteristic EPR spectra when they are reduced [34]. Reduced PS-I reaction centers display these spectral properties, which are attributed to the centers $F_A$ and $F_B$, as detailed below. Under conditions of low ambient redox potential (e.g. –0.6 V or less), it was found that illumination induces additional and rather unusual EPR features which were attributed to the light-induced reduction of an electron acceptor (later on named $F_X$) more primary than $F_A$ or $F_B$ [35]. The detailed properties of $F_X$ are described in refs. 1 and 3. A few of them will be briefly presented.

*3.4.1. EPR properties:* The EPR signal attributed to reduced $F_X$ was observed after low-temperature illumination, slowly decaying simultaneously with the charge separated pair (P-700$^+$, $F_X^-$). Reduced $F_X$ can also be trapped as a stable species, by illumination and cooling. Its EPR spectrum displays g-values of 2.04, 1.88 and 1.78. These values differ from those of other Fe-S centers. Also the relaxation properties are quite unusual, giving rise to a very high power saturation. These unusual properties were the origin of many controversies on the chemical nature of the redox center. It is now clear that $F_X$ is a 4Fe-4S center, the unusual properties of which are essentially due to its binding by two polypeptides, and not by a single one as for most Fe-S centers.

*3.4.2. Redox potential:* $F_X$ has a redox potential much lower than other Fe-S centers. Measurements gave values of –0.70 V (EPR) and –0.67 or –0.73 V (kinetic absorption spectroscopy). These values are at the limit of what can be safely titrated with a protein.

*3.4.3. Room temperature functional studies:* Flash-absorption revealed that, when $F_A$ and $F_B$ are reduced, P-700 is oxidized by a flash and then re-reduced with $t_{1/2} \approx 250$ µs [36]. This was taken as evidence for the existence of a more primary electron acceptor than $F_A$ and $F_B$, which should most probably be equated with $F_X$. It is important to note that this reaction takes place in nearly all reaction centers after a single flash, showing that $F_X$ can be a very efficient acceptor. By contrast, in low temperature experiments, the photoreduction of $F_X$ has a low efficiency, presumably because electron transfer from $A_1$ from $F_X$ is slowed down.

*3.4.4. $F_X$ in the PS-I core:* It has been recently possible to study $F_X$ in greater detail with the availability of a PS-I "core complex", i.e. a reaction center from which peripheral polypeptides have been removed, including the psaC gene product which carries $F_A$ and $F_B$ (see ref. 1 for a detailed discussion). EPR properties are much more simple than in intact PS-I, and the charge recombination between P-700$^+$ and $F_X^-$ can be observed without the low-potential redox poising required for the previous reduction of $F_A$ and $F_B$. The core complex contains 4.6 acid-labile sulfide per P-700, and no other Fe-S center was detected apart from $F_X$. One interesting property of this system is that it is possible to denature $F_X$ by treatment with urea and potassium ferricyanide. It is then possible to reconstitute $F_X$ by addition of $FeCl_3$, $Na_2S$ and mercaptoethanol to the denatured core complex. This ensemble of results provide good evidence that $F_X$ is a 4Fe-4S center bound by the two large subunits of PS-I.

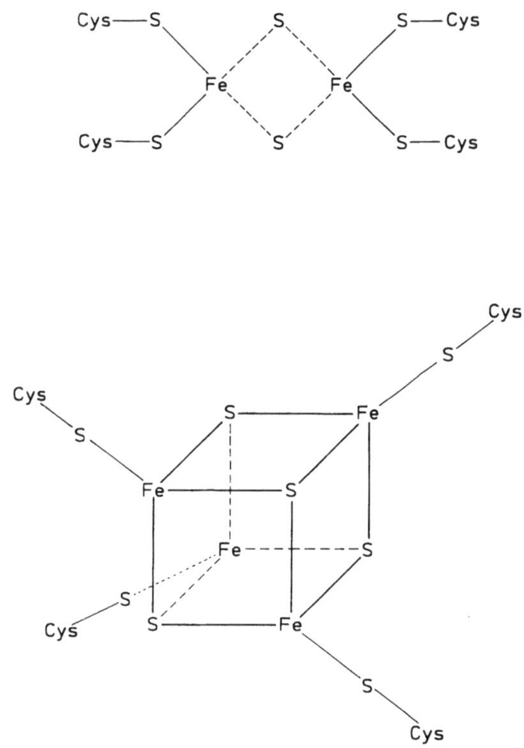


Fig. 3. Chemical structure of Fe-S clusters: a 2Fe-2S cluster and a 4Fe-4S cluster.

*3.4.5. Studies by other spectroscopic techniques:* Uncertainties on the chemical nature of $F_X$ have stimulated its study by various spectroscopic techniques [1]. EPR was extensively used; one of the last pieces of information gained was the stoichiometry, establishing that there is only one $F_X$ per reaction center. The availability of a core complex with $F_X$ present but devoid of $F_A$ and $F_B$ also made possible fully significant studies by X-ray absorption and by Mössbauer spectroscopy. The data are in agreement with the hypothesis of one 4Fe-4S cluster.

*3.4.6. Functional properties of $F_X$:* It has not yet been possible to prove that $F_X$ is directly involved in electron transfer from P-700 to centers $F_A$ and $F_B$. This will certainly be hard to do because with flash absorption–which is the best suited technique–it is nearly impossible to distinguish $F_X$ from $F_A$ or $F_B$. But why should we doubt the functional role of $F_X$ when there are good arguments, such as: (i) an appropriate mid-point potential to reduce $F_A$ or $F_B$; (ii) a full extent of charge separation at room temperature when $F_A$ and $F_B$ are pre-reduced and also when they are removed, in the core complex; (iii) the occurrence of $F_X$ as an intrinsic constituant of PS-I in all studied organisms? The suspicion perhaps may be tracked to the older experiments done at low temperature. When $F_A$ and $F_B$ are pre-reduced, a single saturating flash creates the (P-700$^+$, $F_X^-$) pair in only about 20% of the reaction centers. At that time it was thought that full charge separation could be elicited by a single flash when $F_A$ and $F_B$ were pre-oxidized, thus the idea that $F_X$ is not functioning very efficiently. As a matter of fact, it has now been clearly shown that when $F_A$ and $F_B$ are pre-oxidized, a single saturating flash induces charge separation between P-700 and $F_A$ or $F_B$ in only about 30% of the reaction centers. Thus it seems that there is no serious reason to suspect a *bona fide* role of $F_X$ in electron transfer, although this remains to be directly proved.

### *3.5. $F_A$, $F_B$: the terminal bound electron acceptors*

The low-temperature photoreduction of membrane-bound iron-sulfur centers in PS-I was first reported by Malkin and Bearden [37]. The reaction is practically irreversible below 77 K. It was later on found by EPR that two different Fe-S centers can be photoreduced, according to the temperature, redox potential, the species under study, and some experimental factors such as glycerol added to the medium. The two centers were named $F_A$ (g-values: 2.05, 1.94, 1.86) and $F_B$ (g-values: 2.065, 1.93, 1.88). These spectral lines are observed when either $F_A$ or $F_B$ is reduced in each reaction center. When both Fe-S centers are reduced, they interact magnetically and a new set of lines appear, at g-lines: 2.05, 1.94, 1.92 and 1.89. Redox titrations of $F_A$ and $F_B$ have been performed by several groups, giving $E_m$ around −530 and −560 mV, respectively, with some variation around these values (see refs. 1–4 and 38 for a more complete presentation).

Also, in the early 1970s, flash absorption studies revealed that, in membranes enriched in PS-I, charge separation takes place between P-700 and a terminal acceptor named P-430 [39]. Difference spectra showed that P-430 may well be an Fe-S center, and further analysis established that P-430 is indeed $F_A$ or $F_B$. In the absence of other electron donors or acceptors, the charge separation is reversible with a half-time around 35 ms. This back-reaction slows down when temperature is decreased; it extrapolates to a nearly irreversible reaction below 77 K, in agreement with the EPR data.

The spectroscopic properties of $F_A$ and $F_B$ indicate that both are 4Fe-4S clusters. It has been shown conclusively that both clusters are carried by one single polypeptide of about 9 kDa, which can be reversibly dissociated from the reaction center. Golbeck and his coworkers were also able to denature the clusters and to reconstitute them with iron and sulfide under reducing conditions (see ref. 1). The polypeptide is coded by the psaC gene. It contains eight cysteine residues appropriately located to make two Fe-S clusters.

It is still not understood why there are two Fe-S clusters, and what their respective roles are. The most frequent ordering of redox potentials is with $F_B$ more negative: $F_B$ and $F_A$ would thus be linearly arranged, with electrons flowing from $F_X$ to $F_B$, then to $F_A$ and then away from PS-I. However, the ordering can be inverted, and it can also be envisioned that $F_A$ and $F_B$ play equivalent roles, or that they can be on parallel electron paths leading to different external acceptors. This question has been addressed by selective denaturation: $F_B$ can indeed be denaturated rather selectively by several chemical treatments and, after inhibition by mercurials, it was shown that $F_A$ remains fully photo-reducible. This shows that $F_B$ in not a mandatory intermediate in the reduction of $F_A$. Structural studies, however, showed that the line joining $F_A$ and $F_B$ is nearly perpendicular to the membrane plane, supporting the idea of a sequential electron transfer between these two redox centers [41]. The X-ray 3-D structure also supports this proposal [15].

### *3.6. Soluble electron donors and acceptors*

In thylakoids of plant chloroplasts, electrons coming from PS-II via the plastoquinone pool and the cytochrome b/f complex reach PS-I by means of a small copper protein, plastocyanin. This protein of 10 kDa possesses one copper atom of class I (oxidized: $Cu^{2+}$, reduced: $Cu^{+}$, $E_m$ = +370 mV). Its properties have been studied in great detail: 3-D structure from X-ray crystallography, NMR, electron transfer reactions *in vitro,* etc. [42–45]. Its interaction with the PS-I reaction center, however, is still poorly known. From a kinetic point of view, it appears that electron transfer takes place by two mechanisms [46, 47]: first-order rapid donation from a bound state ($t_{1/2}$ = 12 μs), and second order diffusion-controlled reaction ($k = 1 \times 10^8\ M^{-1}\ s^{-1}$ under optimum conditions). Plastocyanin thus appears to be present under two forms, bound or unbound to the PS-I reaction center, in equilibrium with a dissociation constant which depends very much upon experimental conditions, such as the ionic composition, temperature or the method of preparation of the PS-I.

Plastocyanin is undoubtedly the immediate electron donor to P-700 in plant chloroplasts. In many organisms, specially green algae and cyanobacteria, it can be replaced by a c-type cytochrome, cytochrome $c_{552}$ (or $c_6$) [48]. In organisms where this cytochrome is normally present in much lower concentration than plastocyanin, its concentration can be greatly increased (and plastocyanin concomitantly decreased) by using a growth medium devoid of copper. Two problems are presently studied in relation with this alternative function: (i) how does copper regulate the biosynthesis of plastocyanin and cytochrome $c_{553}$? (ii) are there common structural motifs in these two proteins, which would permit their correct docking to the b/f complex and to the PS-I reaction center? It may also be pointed out that both proteins are localized in the thylakoid lumen, where is also oriented the donor side of the PS-I reaction center. So they are localized in very restricted volumes, at a high concentration, but presumably with barriers to their diffusion. In addition to functional aspects, this localization raises interesting questions, now well studied, concerning the post-synthesis processing and addressing of these proteins.

Electron acceptors from PS-I are also small soluble proteins. Ferredoxin is the most common acceptor. This is a small protein of about 10 kDa, with one 2Fe-2S cluster, and a redox potential about −420 mV. Ferredoxins are very widespread in biological systems where they play the role of low-potential one-electron redox mediators (see ref. 49 for a review). Ferredoxin is present in the chloroplast stroma. It is normally largely oxidized, but it becomes reduced by PS-I upon illumination, giving rise to a distinctive EPR spectrum at low temperature. Electron transfer from PS-I to ferredoxin has not yet been studied in detail. It is assumed that ferredoxin binds to the reaction center via one or two of the subunits (see below) and gets an electron from $F_A$ or $F_B$. The study of that electron transfer reaction is rendered difficult by the similar absorption spectra of all iron-sulfur centers. One of the major reactions of photoreduced ferredoxin is to reduce $NADP^+$ via the ferredoxin-$NADP^+$-oxidoreductase, an enzyme which seems to be associated to the PS-I reaction center.

Like in the case of plastocyanin and cytochrome $c_{553}$, ferredoxin is provided with an alternate protein of identical function, flavodoxin. This is a small protein (14–23 kDa according to the species) which contains FMN as redox-active cofactor. It has two reduced states, the semiquinone and the fully reduced hydroquinone, with the following mid-point potentials: −443 mV for the first reduction, −238 mV for the second reduction (pH 7.8, protein from the cyanobacterium Synechocystis PCC 6803) [50]. Flavodoxin has been found essentially in cyanobacteria where its concentration is greatly increased when iron is deficient in the growth medium, under conditions where ferredoxin synthesis is greatly reduced. The functional properties of flavodoxin in PS-I have not been thoroughly studied.

Small molecules can accept electrons from the PS-I reaction center (see ref. 39 for an old valuable review). Viologen dyes (especially methyl- and benzyl-viologen) are very efficient in that respect. Thus, PS-I activity is often assayed by measuring the oxygen uptake in the presence of viologen, according to the reactions: $V + 2e^- + 2H^+ \Rightarrow VH_2$ and $VH_2 + O_2 \Rightarrow V + H_2O_2$. The eventual occurrence of $O_2$ reduction by PS-I *in vivo* is now actively studied in relation with the possible photodamages associated with PS-I function.

## 4. Polypeptides of the PS-I reaction center

Recent years have seen great advances in the knowledge of the polypeptides of PS-I, like of all other photosynthetic membrane complexes. Gene sequencing and peptide sequencing with micro-quantities have provided the amino-acid sequences of all polypeptides, but progress is much slower to identify their structure and their function. In this brief description, polypeptides will be named according to their genes, for which a stricter terminology has been defined [51]. A solid understanding of PS-I polypeptides arose from the pioneering work of Bengis and Nelson [52, 53]. These authors found six polypeptides for which they proposed a roman numbering which will not be used here. For detailed references see [1].

### *4.1. psaA and psaB gene products*

These gene products are two large polypeptides of 83 kDa which migrate anomalously around 60 kDa in SDS-polyacrylamide gel electrophoresis and which make up the heart of the reaction center. They are very hydrophobic, and their folding pattern may include as many as 11 transmembrane $\alpha$-helices [54]. They display a large extent of mutual sequence homology, as also do the $D_1$ and $D_2$ polypeptides of PS-II or the L and M polypeptides of purple bacteria. These two polypeptides perform the essential functions of holding in the right geometry all the primary reactants

Table 1. List and properties of polypeptides of the PS-I reaction center

| psa | MW (kDa) | Gene location | Extrinsic/ integral | Orientation | Function |
|---|---|---|---|---|---|
| A | 83 | Chl. | Integral | Transmembrane | Primary electron transfer and core antenna |
| B | 83 | Chl. | Integral | Transmembrane | Idem |
| C | 9 | Chl. | Extrinsic | Stroma | Fe-S centers $F_A$ and $F_B$ |
| D | 16–18 | Nucl. | Extrinsic | Stroma | Docking of ferredoxin |
| E | 8–10 | Nucl. | Extrinsic | Stroma | Docking of psa D gene product? |
| F | 17 | Nucl. | Extrinsic | Lumen | Docking of plastocyanin |

The leftmost column is the gene name (psa X). The gene location refers to the chloroplast genome in eucaryots (Chl.: chloroplast; Nucl.: nucleus).

(P-700, $A_0$, $A_1$, $F_X$) and the core antenna of about 110 chlorophyll *a*. A few of their structural properties can be inferred:

- both polypeptides are closely associated in a very stable structure which is resistant to proteases and to low SDS concentration. It has been hypothesized that leucine residues may be involved in maintaining the structure by formation of a "leucine zipper" but this was not confirmed by site-directed mutagenesis.
- two redox centers are located at the interface between the two polypeptides: P-700, the two chlorophylls of which being liganded by histidine residues; and the iron-sulfur center $F_X$ which requires four cysteine residues, two of them being provided by each of the sub-units. Each polypeptide contains two fully conserved cysteine residues which are part of a conserved sequence of 12 amino-acids typical of iron-sulfur clusters. It is thus now certain that $F_X$ is a 4Fe-4S center located at the interface between the two large sub-units.
- a large number of conserved histidine residues is probably involved in binding the ≈ 110 chlorophyll molecules which fulfill an antenna function.
- in the native PS-I complex, the two large polypeptides also serve to hold all the other polypeptides of PS-I, presumably by means of specific sites of interaction. The identification of these interactions will be a major objective of future research.

### *4.2. psaC gene product*

This polypeptide of about 10 kDa contains the two iron-sulfur clusters $F_A$ and $F_B$. The gene (which is located in the chloroplast genome in eucaryots)

Table 2. List of small polypeptides in the PS-I reaction center (see Table 1)

| psa | MW (kDa) | Gene location | Extrinsic/ Integral | Orientation | Function |
|---|---|---|---|---|---|
| G | 10 | Nucl. | Extrinsic (cyano) plants? | Stroma | Unknown |
| H | 10 | Nucl. | Extrinsic | Stroma? | Docking of LHC-1? |
| I | 4 | Chl. | Integral | Transmembrane | Unknown |
| J | 5 | Chl. | Integral | Transmembrane | Unknown |
| K | 8 | Nucl. | Integral | Transmembrane | Unknown |
| L | 14 | Nucl. | Integral | ? | Unknown |
| M? | 17.5 (app) | ? | ? | ? | Unknown |
| ? | 5 (app) | ? | Integral | Transmembrane | Docking of LHC-1? |
| ? | 9 (app) | Nucl.? | ? | ? | Unknown |

has been sequenced in many organisms, and the polypeptide also has been sequenced. Two major features emerge from the sequence: (i) the polypeptide has no potential $\alpha$-helix and it is overall quite hydrophilic, as can be expected for a stroma-oriented peripheral polypeptide; (ii) a large part of the sequence is fully conserved from cyanobacteria to higher plants; it includes nine cysteine residues. It was easy to recognize in these features a duplicate of the sequences which are involved in the formation of 4Fe-4S iron-sulfur centers in bacterial ferredoxins which have a known 3-D structure.

The psaC gene product is present in most types of isolated PS-I preparations. It can be dissociated, however, and in this process the two Fe-S clusters are usually denaturated. However, by careful dissociation under reducing conditions, it has been possible to keep the centers and to study their properties in the isolated protein. They are highly comparable to those attributed to $F_A$ and $F_B$ in native PS-I and the assignment can thus be considered as definitive. Denaturated Fe-S clusters in the isolated protein can also be renatured upon binding to the PS-I core complex, under appropriate conditions. It thus became possible to make cross-species reconstitutions, and also to re-incorporate the polypeptide synthesized in *E. coli*. These achievements open the way for several kinds of future studies of the structure-function relationships for this polypeptide: attribution of centers $F_A$ or $F_B$ in the protein, mode of interaction with the PS-I large sub-units and with other peripheral proteins (psaD and E gene products, ferredoxin, flavodoxin), etc.

### *4.3. psaD and psaE gene products*

Two polypeptides of the PS-I reaction center are involved in the binding of the psaC polypeptide and in its interaction with soluble ferredoxin. They are encoded by nuclear genes in eucaryots: psaD and psaE. These genes and their polypeptides products have been sequenced in several higher plants and cyanobacteria. They correspond to subunits II and IV, respectively, according to Bengis and Nelson's nomenclature [52, 53]. The sequence indicate a rather high hydrophilicity and give no sign of the occurrence of a transmembrane $\alpha$-helix. Several arguments lead to conclude that they are membrane-extrinsic, stroma-oriented, and involved in the binding of the stroma-located ferredoxin:

- the gene sequences indicate that the precursor proteins in eucaryots include one N-terminal transit sequence, which is involved in crossing the chloroplast envelope. They are however devoid, in eucaryots and in cyanobacteria, of the transit peptides which might be involved in their addressing to the thylakoid lumen.
- selective extraction by detergents, by salts at high pH or by chaotropic agents is fully consistent with both polypeptides being extrinsic and oriented toward the stroma.

- cross-linking experiments of PS-I ferredoxin have clearly shown that the psaD polypeptide is the nearest neighbor to bound ferredoxin [55, 56]. The psaE polypeptide was not reported to be cross-linked with ferredoxin, but it is also involved in electron transfer from the bound iron-sulfur centers to ferredoxin [57].
- starting from a complete PS-I reaction center, it has been possible to dissociate the psaC, D and E polypeptides in a controlled manner, so that reconstitution was possible [1]. It appeared that reconstitution goes together for all three polypeptides. PsaD however, is more important than psaE in the process since it is necessary for psaC rebinding although the psaE subunit is not.
- both polypeptides are positively charged, a feature which may be involved in a function of facilitating their interaction with the negatively charged ferredoxin and with the membrane.
- gene inactivation of psaD and E has been performed in cyanobacteria (Synechocystis 6803). Insertional inactivation of the psaD gene produces a mutant which is capable of photoautotrophic growth, but at a greatly reduced rate, although the PS-I reaction center is assembled. In the case of psaE, a deletion mutant grows nearly like the wild type. This may indicate that the psaE gene product is not essential under normal growth conditions, although there are many examples where a deletion mutant grows well thanks to the compensatory expression of a normally weakly expressed gene.

The functions of the psaD and E polypeptides are not easy to precisely determine. This objective may require a careful examination of their structure. The psaD polypeptide has a MW of about 18 kDa in plants and 16 kDa in cyanobacteria. A few regions in the sequences are absolutely conserved: they may be implicated in the function, but more probably via structural requirements. Of special importance may be a few highly positively charged domains. The psaE gene product has a MW of about 10 kDa in eucaryots and 8 kDa in cyanobacteria, but it migrates on SDS-polyacrylamide gels with a higher apparent MW of about 15 kDa. Although the sequences have a rather high extent of homology, putative, functional or structural, features have not yet been deciphered (see a discussion in ref. 1). In a recent work a cross-linked complex of PS-I with ferredoxin was found to have electron transfer maintained. Sequencing of the peptides obtained after hydrolysis of the complex showed that Glu 93 of ferredoxin was linked to Lys 106 of the psaD subunit [56].

### *4.4. psaF gene product*

This polypeptide corresponds to subunit III [52, 53]. It is encoded by a nuclear gene in eucaryots and has a MW around 17 kDa. The gene and the

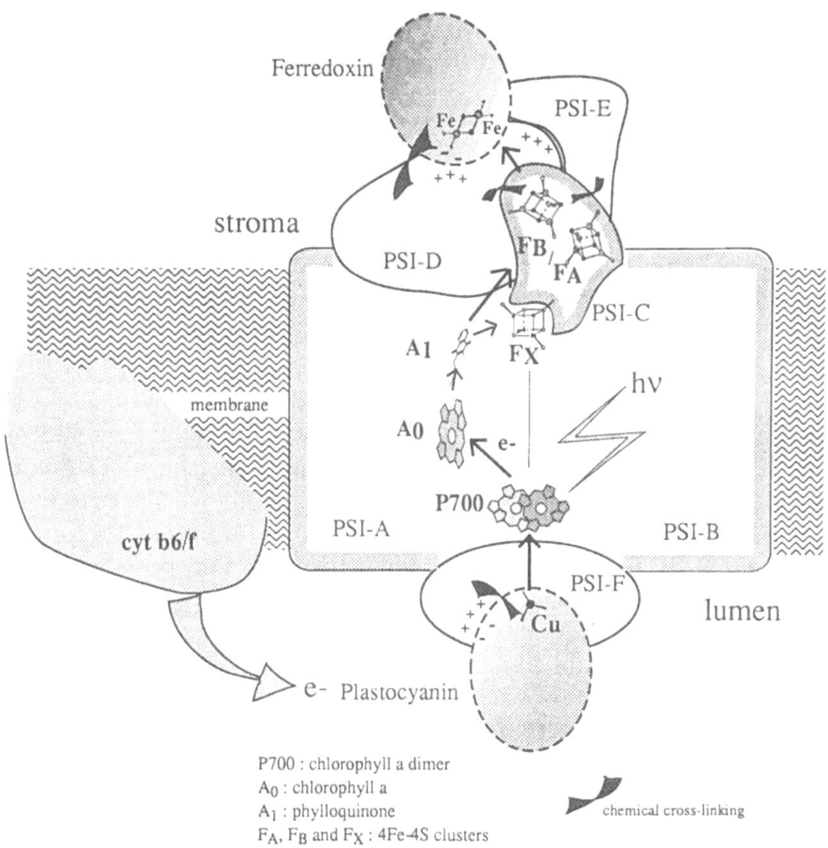


Fig. 4. Model of structure of the PS-I reaction center showing the two large subunits and the four peripherical subunits which are involved in the binding of ferredoxin (electron acceptor) and plastocyanin (electron donor). Also shown are the thylakoid membrane and the cytochrome b6/f complex which re-reduces plastocyanin. Most of the small subunits are missing, as well as the antenna complexes. (Model kindly provided by B. Lagoutte and P. Sétif.)

polypeptide have been sequenced in several organisms (reviewed in ref. 1). The properties of that subunit are still rather controversial (1), and in this simple review I will simply adopt the most generally accepted point of view.

Polypeptide psaF appears to be located at the lumenal side of the PS-I reaction center, with properties indicating a rather extrinsic location. In support of these proposals are (i) the structure of the transit peptide, which is analogous to the transit peptide from other lumenal polypeptides; (ii) the sequence does not have the hydrophobic properties of membrane intrinsic proteins, and the polypeptide is removed by treatments known to remove membrane-extrinsic subunit; (iii) the polypeptide becomes cross-linked to plastocyanin (and to its counterpart in cyanobacteria (cytochrome $c_{553}$).

The psaF polypeptide is positively charged, and many of its recently determined properties support the function initially proposed for subunit III by Bengis and Nelson [52, 53], i.e. to participate in the docking of plastocyanin (or cytochrome $c_{553}$) to the large subunits of PS-I. It appears, however, that the subunit is not indispensable in Synechocystis 6803, since a deletion mutant is still capable of photoautotrophic growth.

### *4.5. Other low molecular weight polypeptides: psaG–M gene products*

Analysis of PS-I preparations by polyacrylamide gel electrophoresis, especially adapted for the resolution of low MW polypeptides, have shown the presence of a surprisingly large number of subunits. Polypeptide and gene sequences have been obtained, at least partially, for all of them [1, 90–94], but in many instances the data are still controversial as to the occurrence of the subunits, and still more as to their function. The situation is presented in Table 2, in a simplified manner. The possible roles of polypeptides psaG–M have been guessed on the basis of several properties: (i) are they integral or extrinsic? Integral subunits are usually considered to be more probably related to primary processes, and extrinsic ones related to the interaction of PS-I with other polypeptides (often soluble ones), (ii) are they oriented toward the stroma (i.e. toward electron acceptors) or toward the lumen (i.e. electron donors)? (iii) are they present in the smaller PS-I-100 particles or only in the bigger PS-I-200 particles which also include the peripheral antenna? According to the presently accepted concepts, the subunits found in PS-I-100 of eucaryots should also be present in the PS-I particles from cyanobacteria, which have no chlorophyll-containing peripheral antenna.

The psaG gene has been sequenced in several organisms, and its product only partly sequenced. It is not membrane-intrinsic, but it is easily dissociated from plant membranes. It is rather acidic (pI = 4,5), by contrast with other extrinsic subunits of PS-I. The psaH gene is a more interesting case because it has been found in eucaryots (nuclear gene), but not in cyanobacteria. It is suggested that the polypeptide, which is extrinsic and stroma-oriented, has a function related to the interaction of the LHC-1 antenna with the reaction center core, although this is highly hypothetical. The products of genes psaI, J and K are small hydrophobic polypeptides which are assumed to have a transmembrane location. The formation of one transmembrane-helix has been predicted for the psaL and J products and it is not excluded for psaK. These polypeptides are part of the PS-I reaction center core and they have been implicated in pigment or quinone binding, although the evidence is rather weak (see a discussion in ref. 1). The psaL gene product is a newly found polypeptide of 14 kDa (apparent) MW. A partial amino-acid sequence has been obtained, and the gene may have been sequenced and confused with a second psaE gene (as discussed in ref. 58). The polypeptide belongs to the PS-I core. It has a blocked N-terminus and was reported to be in a less than stoichiometric amount.

It thus appears that the PS-I reaction center includes a large number of polypeptides. The small ones have perhaps not all been discovered. The same complexity has been found in other large membrane complexes, such as PS-II, the antenna complex, cytochrome oxidase, ATP-ases, etc. The function of many of these polypeptides, especially the small ones, is difficult to investigate. Some answers will certainly be obtained soon thanks to mutagenesis, either by gene deletion or by point mutations.

## 5. Structure of the PS-I reaction center

Our current view of the PS-I reaction is largely based on the known structure of the reaction center of purple bacteria. These two classes of reaction centers, however, are largely different, so that the structure shown in Fig. 4 is largely hypothetical. It is based on the knowledge of the subunits, of their stoichiometry, of their orientation with respect to the membrane, and on limited knowledge on their proximity, based for example on cross-linking experiments, as discussed above. Several properties lead to propose a rather symmetric structure for the core, like in purple bacteria. The two large polypeptides psaA and B are of similar size and they display an important homology which leads to propose that their genes derive from duplication of an initial unique gene. These subunits are rather symmetric at the level of $F_X$ and they presumably fit together to make the dimeric structure of P-700. As mentioned above, symmetry properties can be found in phylloquinones (two molecules) and in the psaC gene product with centers $F_A$ and $F_B$. A putative axis of pseudosymmetry would thus be perpendicular to the membrane and would run from P-700 to $F_X$ and then through the psaC subunit. The distance between redox centers might be obtained, along the normal to the membrane plane, by photoelectric measurements such as photovoltage or electroluminescence; these methods have not yet given their full potential.

The overall shape of PS-I has been examined by electron microscopy, using image analysis of isolated particles [59, 60]. In a recent report [61], images were obtained from small ordered arrays formed directly in thylakoid membranes. The images give some indications of two large bodies (presumably psaA and B subunits) with complex features around, which can be more or less successfully related to the small subunits (see refs. 60–62). Still more information is to be expected even with the present (low) resolution. Structural studies revealed that the PS-I reaction center is either monomeric or trimeric, and perhaps sometimes dimeric. Is this a preparation artefact or does PS-I adopt a functionally significant multimeric structure *in vivo?* Zero-angle neutron scattering measurements showed that the solubilized PS-I reaction center of a cyanobacterium is monomeric, with a molecular weight of 217 kDa [62]. A monomeric state was also reported in the electron microscopy of thylakoid membranes [61]. Electron microscopy of isolated PS-I and of 2-D arrays reveals monomers, dimers or trimers, whereas the organization is monomeric or trimeric in 3-D crystals. Formation of oligomers could be induced by the preparative method (see a good analysis in ref. 60), but Almong et al. [63] reported that an additional subunit is present when PS-I organizes into trimers. There is still no final answer as to the state(s) *in vivo.*

Our knowledge on the structure of the PS-I reaction center has been transformed by the resolution of the 3-D structure by X-ray crystallography [15]. The present structure provides a 6Å resolution, but the quality of the crystals should provide a 4Å resolution. At the present stage, the structure essentially supports the aspects described above, which were obtained by more uncertain methods. In the analyzed crystals, the reaction centers were organized into trimers. Among the strong information provided by the structure is the position of the iron-sulfur centers. These centers are organized roughly as shown in Fig. 4 with the innermost center ($F_X$) located at 15 and 22Å of the two other centers ($F_A$ and $F_B$, which cannot be distinguished). Although a few things can be said about the primary partners (P-700, $A_0$, $A_1$), much more will be gained from a more refined analysis.

## 6. Comparison with green sulfur bacteria and heliobacteria

Photosynthetic reaction centers can be grossly classified into those which have two quinones $Q_A$ and $Q_B$ as secondary electron acceptors, and those which have iron-sulfur centers. The second category includes PS-I of oxygenic organisms, and two classes of non-oxygenic bacteria: the green sulfur bacteria (chlorobacteriaceae, such as Chlorobium sp., ref. 64) and heliobacteria (heliobacteriaceae, ref. 65). Without going into a detailed comparison (see e.g. ref. 64–66), let us mention a few common properties.

(1) In *Chlorobium limicola,* the reaction center core contains two large subunits with an apparent MW of 65 kDa, like in PS-I. Büttner et al. sequenced an entire operon which supposedly should have included the genes for the two large subunits [67]. However, they found only one gene for a large subunit (732 amino acids), in addition to a gene for a small subunit. The large subunit contains a domain homologous to that making up the Fe-S center $F_X$ in PS-I, and the small subunit is homologuous to psaC carrying two 4Fe-4S centers. It is thus highly probable that the reaction center core is made of a homodimer of two large subunits in *Chlorobium,* instead of a heterodimer in PS-I. It can-

not be totally excluded that the second gene is located outside of the operon but protein sequencing leaves little probability for that hypothesis. The gene coding for the large subunits has been cloned and sequenced in the heliobacteria *Heliobacillus mobilis* [68]. Here also one single gene was found to be coding for large subunits, but no evidence was found for an equivalent to psaC.

(2) In the chain of electron acceptors, three bound iron-sulfur centers were found in *Chlorobium* and heliobacteria, with properties very similar to $F_A$, $F_B$ and $F_X$ of PS-I: EPR spectra, orientation, magnetic interaction between two of them (equivalent to $F_A$ and $F_B$), kinetic behavior (in a first approximation, and as far as they are known). Redox properties are only slightly different, but the Fe-S centers have more negative potentials in these bacteria than in PS-I. The equivalents to $F_A$ and $F_B$ are carried by cysteine-rich small polypeptides, like the psaC subunit of PS-I. An important point in the comparison resides in the polypeptidic sequences which are involved in forming the $F_X$ center: like in PS-I, they have been found in the large subunits (on the basis of the gene sequences) of *Chlorobium*.

(3) like in PS-I, low-potential quinones have been implicated as secondary acceptors in these bacteria. Electron transfer is blocked under conditions [69, 70] analogous to those which were interpreted as bringing a double reduction of A1 in PS-I [30].

(4) The situation is less clear for the primary partners [66]: the primary acceptor may be very similar, but the primary donors display somewhat different properties. The primary donor, however, is at the interface between the acceptor side (very similar in PS-I, *Chlorobium* and heliobacteria) and the donor side where more variability has been found.

(5) the three classes of reaction centers have a core which possesses a large number of chlorophyll-type molecules with an antenna function. By contrast, PS-II and purple bacteria have only a few pigment molecules in their reaction center core.

These comparative properties are quite important for providing a basis for speculations on the evolution of photosynthetic organisms [66, 71], and clues for experiments. The reaction centers of *Chlorobium* and heliobacteria revealed to be very difficult to study, and it was (and it will certainly be in the future) very useful to use the knowledge on PS-I as a source of research ideas for their understanding.

## 7. Conclusions

This review does not cover all the properties of the PS-I reaction center. A few important aspects have not been detailed:

(i) the biosynthesis of the reaction center which is especially interesting since it involves (in eucaryots) the coordinated expression of nuclear and chloroplastic genes, the processing of several polypeptides, and the incorporation of many pigments and other cofactors;
(ii) the degradation of the reaction center, with some indications being gained on processes of photoinhibition;
(iii) the so-called cyclic electron transfer which may involve soluble cofactors, but perhaps also species closely associated with the PS-I reaction center;
(iiii) the thermodynamic properties of the system, since several methods have been used to evaluate the fraction of the photon energy which is stored in various states of the system.

Many questions are open, from the overall structure to many specific aspects of the local structure, and their relations to the functional properties. It is no surprise that the PS-I reaction center is becoming the object of more and more intense research efforts.

The literature on all aspects of the PS-I reaction center has developed considerably during recent years. Many references can be found in the chapter on photosystem I in: P. Mathis (ed.) "Photosynthesis: from Light to Biosphere", vol. II, 1995, Kluwer Academic Publishers, Dordrecht.

## References

1. JH Golbeck and DA Bryant, Curr. Top. Bioenerg. 16 (1991) 83.
2. P Mathis and AW Rutherford, in New Comprehensive Biochemistry, J Amesz (ed.) vol. 15 "Photosynthesis", Elsevier, Amsterdam, 1987, p. 63.
3. JH Golbeck, Biochim. Biophys. Acta 895 (1987) 167.
4. R Malkin, in The Light Reactions, J Barber (ed.), Elsevier, Amsterdam, 1987, p. 495.
5. P Reilly and N Nelson, Photosynth. Res. 19 (1988) 73.
6. B Lagoutte and P Mathis, Photochem. Photobiol. 49 (1989) 833.
7. P Sétif, in The Photosystems: Structure, Function and Molecular Biology, J Barber (ed.) Elsevier, Amsterdam, 1992, p. 471.
8. JP Thornber, Ann. Rev. Plant Physiol. 26 (1975) 127.
9. JE Mullet, JJ Burke and CJ Arntzen, Plant Physiol. 65 (1980) 814.
10. RM Wynn and R Malkin, Biochemistry 27 (1988) 5863.
11. RE Glick and A Melis, Biochim. Biophys. Acta 934 (1988) 151.
12. RG Hiller, JM Anderson and AWD Larkum, in Chlorophylls, H Scheer (ed.), CRC Press, Boca Raton, 1991, p. 529.
13. BR Green, Photosynth. Res. 15 (1988) 3.
14. BR Green, E Pichersky and K Kloppstech, Trends Biochem. Sci. 16 (1991) 181.
15. N Krauss, W Hinrichs, I Witt, P Fromme, W Pritzkow, Z Dauter, C Bentzel, K Wilson, HT Witt and W Saenger, Nature 361 (1993) 326.
16. E Nabedryk, M Leonhard, W Mäntele and J Breton, Biochemistry 29 (1990) 3242.
17. P Moënne-Loccoz, B Robert, I Ikegami and M Lutz, Biochemistry 29 (1990) 4740.
18. JR Norris, RA Uphaus, HL Crespi and JJ Katz, Proc. Natl. Acad. Sci. USA 68 (1971) 625.
19. S Krawczyk and W Maksymiec, FEBS Lett. 286 (1991) 110.

20. DE Budil and MC Thurnauer, Biochim. Biophys. Acta 1057 (1991) 1.
21. AW Rutherford and P Sétif, Biochim. Biophys. Acta 1019 (1990) 128.
22. I Sieckmann, K Brettel, C Bock, A van der Est and D Stehlik, Biochemistry 32 (1993) 4842.
23. VA Shuvalov, AM Nuijs, HJ van Gorkom, HWJ Smith and LNM Duysens, Biochim. Biophys. Acta 850 (1986) 319.
24. P Mathis, I Ikegami and P Sétif, Photosynth. Res. 16 (1988) 203.
25. I Fujita, MS Davis and J Fajer, J. Amer. Chem. Soc. 100 (1978) 6280.
26. MH Vos and HJ Van Gorkom, Biochim. Biophys. Acta 934 (1988) 293.
27. K Brettel and P Sétif, Biochim. Biophys. Acta 893 (1987) 109.
28. P Mathis, K Sauer and R Rémy, FEBS Lett. 88 (1978) 275.
29. P Mathis and H Conjeaud, Photochem. Photobiol. 29 (1979) 833.
30. P Sétif and H Bottin, Biochemistry 28 (1989) 2689.
31. P Sétif and K Brettel, Biochemistry 32 (1993) 7846.
32. AJ Hoff, Quart Rev. Biophys. 7 (1984) 153.
33. RR Rustandi, SW Snyder, LL Feezel, TJ Michalski, JR Norris, MC Thurnauer and J Biggins, Biochemistry 29 (1990) 8030.
34. TG Spiro (ed.) *Iron-sulfur proteins,* Wiley, New York, 1982.
35. MCW Evans, CK Sihra, JR Bolton and R Cammack, Nature 256 (1975) 688.
36. K Sauer, P Mathis, S Acker and J van Best, Biochim. Biophys. Acta 545 (1979) 466.
37. R Malkin and AJ Bearden, Proc. Nat. Acad. Sci. USA 68 (1971) 16.
38. R Cammack, Meth. Enzym. 167 (1988) 427.
39. B Ke, Biochim. Biophys. Acta 301 (1973) 1.
40. JH Golbeck, J Mehari, K Parrett and I Ikegami, FEBS Lett. 240 (1988) 9.
41. B Guigliarelli, J Guillaussier, C More, P Sétif, H Bottin and P Bertrand, J. Biol. Chem. 268 (1993) 900.
42. AG Sykes, Chem. Soc. Rev. 14 (1985) 283.
43. JM Moore, DA Case, WJ Chazin, GP Gippert, TF Havel, R Powls and PE Wright, Science 240 (1988) 314.
44. MR Redinbo, D Cascio, MK Choukair, D Rice, S Merchant and TO Yeates, Biochemistry 32 (1993) 10560.
45. EI Solomon, MJ Baldwin and MD Lowery, Chem. Rev. 92 (1992) 521.
46. W Haehnel, A Pröpper and H Krause, Biochim. Biophys. Acta 593 (1980) 384.
47. H Bottin and P Mathis, Biochemistry 24 (1985) 6453.
48. PM Wood, Eur. J. Biochem. 87 (1978) 9.
49. DB Knaff and M Hirasawa, Biochim. Biophys. Acta 1056 (1991) 93.
50. H Bottin and B Lagoutte, Biochim. Biophys. Acta 1101 (1992) 48.
51. RB Hallick, Plant. Mol. Biol. Rep. 7 (1989) 266.
52. C Bengis and N Nelson, J. Biol. Chem. 250 (1975) 2783.
53. C Bengis and N Nelson, J. Biol. Chem. 252 (1977) 4564.
54. LE Fish and L Bogorad, J. Biol. Chem. 261 (1986) 8134.
55. G Zanetti and G Merati, Eur. J. Biochem. 169 (1987) 143.
56. C Lelong, P Sétif, B Lagoutte and H Bottin, J. Biol. Chem. 269 (1994) 10034.
57. F Rousseau, P Sétif and H Bottin, EMBO J. 12 (1993) 1755.
58. M Ikeuchi and Y Inoue, FEBS Lett. 280 (1991) 332.
59. RC Ford and A Holzenburg, EMBO J. 7 (1988) 2287.
60. J Kruip, EJ Boekema, D Bald, AF Boonstra and M Rögner, J. Biol. Chem. 268 (1993) 23353.
61. A Hefti, RC Ford, M Miller, R Cox and A Engel, FEBS Lett. 296 (1992) 29.
62. ME Schafheutle, E Setlikova, PA Timmins, H Johner, P Gutgesell, I Setlik and W Welte, Biochemistry 29 (1990) 1216.
63. O Almog, G Shoham, D Michaeli and R Nechushtai, Proc. Nat. Acad. Sci. USA 88 (1991) 5312.

64. U Feiler and G Hauska, in *Anoxygenic Photosynthetic Bacteria,* R.E. Blankenship, M.T. Madigan and C.E. Bauer (eds.) Kluwer Academic Publishers, 1995, pp. 665–685.
65. J Amesz, in *Anoxygenic Photosynthetic Bacteria,* RE Blankenship, MT Madigan and CE Bauer (eds.). Kluwer Academic Publishers, 1995, pp. 687–697.
66. W Nitschke and AW Rutherford, TIBS 16 (1991) 241.
67. M Büttner, DL Xie, H Nelson, W Pinther, G Hauska and N Nelson, Proc. Nat. Acad. Sci. USA 89 (1992) 8135.
68. U Liebl, M Mockensturm-Wilson, JT Trost, DC Brune, RE Blankenship and W Vermaas, Proc. Nat. Acad. Sci. USA 90 (1993) 7124.
69. W Nitschke, U Feiler and AW Rutherford, Biochemistry 29 (1990) 3834.
70. W Nitschke, P Sétif, U Liebl, U Feiler and AW Rutherford, Biochemistry 29 (1990) 1079.
71. P Mathis, Biochim. Biophys. Acta 1018 (1990) 163–167.

Bioenergetics
ed. by P. Gräber and G. Milazzo†


# *CHAPTER 9*
# The cytochrome *b*6*f*/*bc*1-complexes

Günter Hauska und Michael Büttner

*Department of Cell Biology and Plant Physiology, University of Regensburg, Regensburg, Germany*



## 1. Introduction

Oxidation of isoprenoid quinols by cytochrome complexes constitutes the central reaction in biological electron transport chains. This reaction represents another example of nature's fascinating universality. It takes place not only in respiration, but also in photosynthesis, although these processes counteract each other in the major cycle of living matter (Fig. 1a). In the eucaryotic respiratory chain of mitochondria the cytochrome *bc*1-complex functions as ubiquinol-cytochrome *c* oxidoreductase, in photosynthetic electron transport of plant chloroplasts the cy-

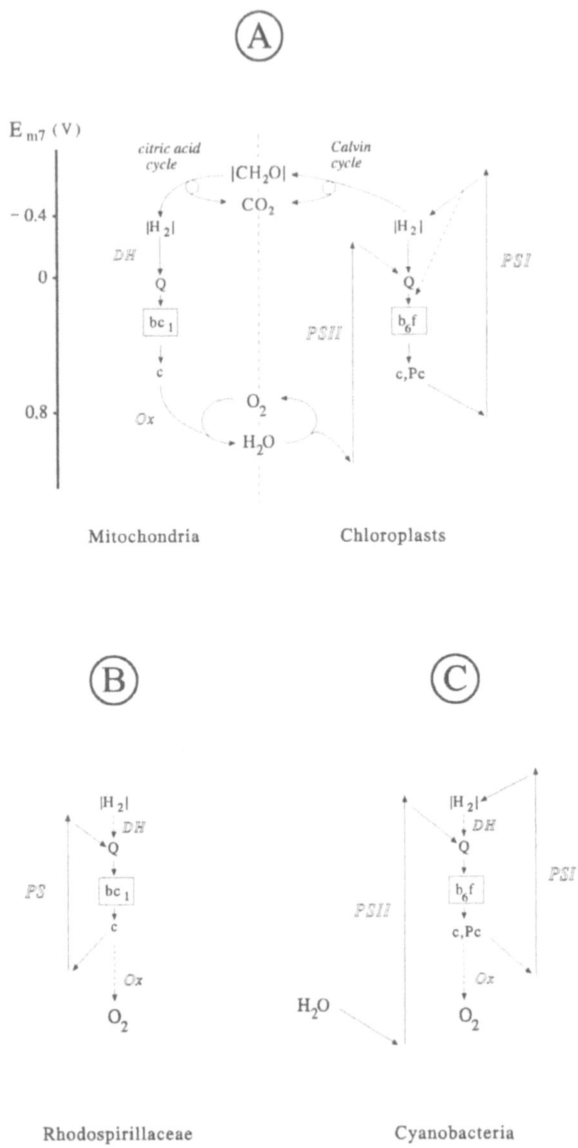


Fig. 1. Cytochrome *bc*1/*b*6*f*-complexes in photosynthesis and respiration. (A) Cycle of C, H and O between mitochondrial respiration and chloroplast photosynthesis. The scheme is drawn approximately to the redox potential scale on the left, except for the upward arrows in photosynthesis. The symbols represent: DH - dehydrogenase, OX - cytochrome oxidase, PSI and PSII - photosystem I and II, [H2] - bound forms of hydrogen, c - cytochrome c, Pc - plastocyanin. The dashed arrow represents cyclic electron transport around PSI. (B) and (C) Dual electron transport in purple photosynthetic bacteria *(Rhodospirillaceae)* and in bluegreen algae *(Cyanobacteria),* respectively. The dashed arrows represent respiration.

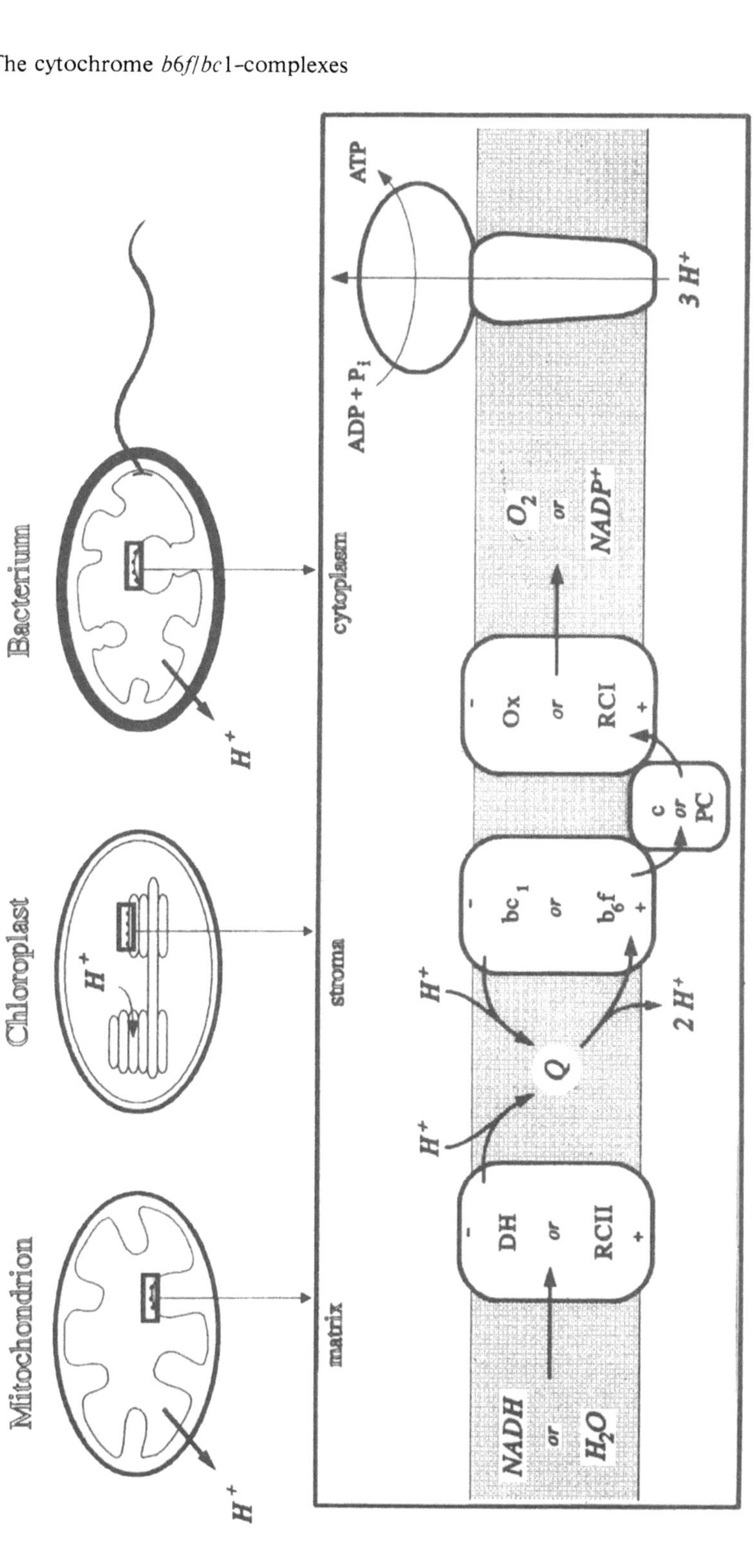


Fig. 2. Universal topography of electron transport in mitochondria, chloroplasts and bacteria. RCI and II stand for the reaction centers of photosystem I and II, respectively. Other abbreviations resemble the ones for Fig. 1. Only the protons translocated by the "Q-cycle" of the cytochrome *bc*-complexes are shown.

tochrome *b6f*-complex acts as plastoquinol-plastocyanin oxidoreductase. In photosynthetic procaryots the quinol oxidoreductases may function in both processes in one and the same membrane, respiration in the dark and photosynthesis in the light. Thus, as shown in Fig. 1bc, the cytochrome *bc*1-complex of purple photosynthetic bacteria *(Rhodospirillaceae)* and the cytochrome *b6f*-complex of bluegreen algae *(Cyanobacteria)* represent metabolic zwitters. This dual function, together with the relatedness of cytochrome *bc*1- and *b6f*-complexes, including the primary structures of their subunits, supports the current notion that respiration evolved from photosynthesis; [see 1 and note 1 added on proof]. The still controversial "chlororespiration" of chloroplasts possibly represents a relic of the cyanobacterial duality.

The membrane topography of chloroplasts, mitochondria and bacteria is organized according to universal principles (Fig. 2). Protons are translocated by electron transport out of the mitochondrial matrix, and out of the bacterial cytoplasma, or into the thylakoid lumen in chloroplasts. The resulting electrochemical proton gradient drives ATP-synthesis on the coupling factor complex with a stoichiometry of 3, or even 4 $H^+$/ATP (see Chapter 12). Both the cytochrome *bc*1 –as well as the cytochrome *b6f*-complexes– operate very efficiently in energy conservation, oxidizing quinol in a coordinated mechanism, the so-called Q-cycle, which translocates two protons per transferred electron.

This chapter is intended to summarize our current notion on structure and function of the cytochrome *bc*-complexes, complementing the overviews on chloroplasts (Chapter 5) and on mitochondria (Chapter 6). The key differences between the *bc*1- and the *b6f*-types will be elaborated. For more details the reader is referred to the numerous reviews [2–8].

## 2. Occurrence

### *2.1. Organisms*

Cytochromes are prominent constituents of electron transport chains. They are heme-proteins and are classified according to the long wavelength absorption (α-peak) of the reduced forms into three types. Reduced cytochrome *a* is green absorbing around 600 nm, while reduced cytochrome *b* and *c* are red and absorb at about 560 and 550 nm, respectively. Cytochromes *c* contain a covalently bound heme, while cytochrome *a* and *b* bind the heme group non-covalently (see Chapter 6).

In eucaryotic cells the major portion of the cytochrome complement is functioning in mitochondria, in plants also in chloroplasts. Both obligatorily comprise a cytochrome *bc*-complex, either the respiratory cytochrome *bc*1- complex, or the photosynthetic cytochrome *b6f*-complex.

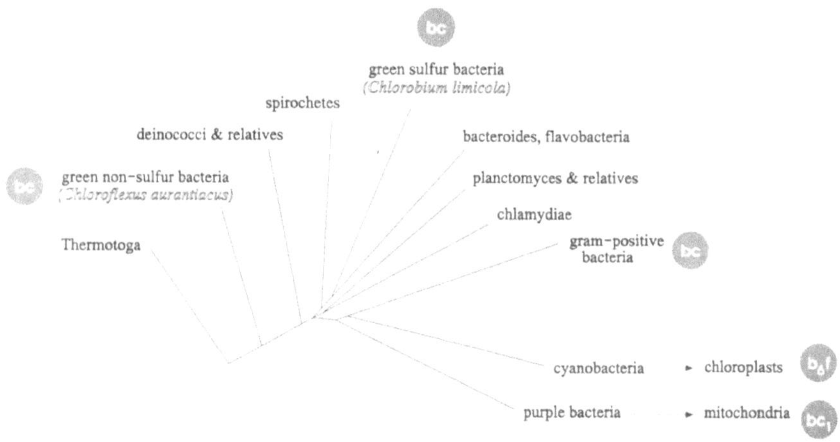


Fig. 3. Occurence of cytochrome *bc*-complexes among Eubacteria. The phylogenetic map was modified from Woese [10]. The distances are inversely proportional to a similarity coefficient of ribosomal RNA sequences. Ribosomal RNA is among the most highly conserved structures in biology, and allows to estimate the degree of phylogenetic relation between very distant organisms.

Almost since their rediscovery by John Keilin in the 1920s cytochromes are known to occur not only in eucaryots, but also in bacteria [9], but procaryotic respiration is less bound to the function of a cytochrome *bc*-complex. Already *E. coli* is lacking it. If a *bc*-complex is present, it may function in only one of several electron transport branches from quinol to oxygen, probably in the one most efficient in energy conservation. Its presence may depend on environmental conditions, reflecting the adaptive flexibility of bacteria.

A strong indication for the presence of a cytochrome *bc*-complex is the joint occurrence of cytochromes *b* and *c* together with the Rieske FeS-center, plus sensitivity of electron transport to antimycin A, or other characteristic inhibitors. Even more compelling is the observation of "oxidant-induced reduction of cytochrome *b*", which is a reflection of the Q-cycle mechanism (see section 4. 2. below). Figure 3 shows the presence of cytochrome *bc*-complexes within the major eubacterial phyla in a distance map, which is based on the analysis of ribosomal RNA-sequences [10]. Cytochrome *bc*-complexes have been isolated from mitochondria and chloroplasts, and from several gram-negative bacteria [2-7].

In view of the endosymbiotic theory it is not surprising that the *bc*1-type of mitochondria is also found in the gram-negative, "purple" bacteria, and that the *b6f*-type of chloroplasts also occurs in cyanobacteria [1, 12]. More intriguing is the indication that gram-positive bacilli may contain a *b6f*-type complex (see note 2 added on proof). However, the division into *bc*1- and *b6f*-types has its limitation, the *bc*-complex of *Chlorobium* limicola, a green sulfur bacterium, combining features of both [11].

The occurrence of a cytochrome *bc*-complex in archaebacteria is uncertain, although a Q-cycle may be operative, and a Rieske FeS-center has been reported for *Sulfolobus*; for further details, see [12 and note 1 added on proof].

### *2.2. Cellular location*

Within the eucaryotic cells the cytochrome *bc*-complexes are confined to the endosymbiotic organelles. The cytochrome *bc*1-complex is found in the inner membrane of mitochondria. The cytochrome *b6f*-complex is localized in the membranes of plastids (chloroplasts and etioplasts). In mature chloroplasts it is distributed over the grana and the stroma lamellae [4], in contrast to photosystems I and II, which are concentrated in the stroma lamellae, or in the grana, respectively. More precisely the *b6f*-complex may be concentrated in the margins between grana and stroma lamellae, where it could connect photosystems I and II.

In bacteria the *bc*-complexes are also membrane bound, but their lateral distribution is largely not investigated. Interestingly, however, for certain cyanobacteria it has been claimed that a cytochrome *b6f*-complex is located in the thylakoid membrane, whereas a cytochrome *bc*1-complex is concentrated in the cytoplasmic membrane (see note 3 added on proof).

## 3. Components and structure

The functionally intact, isolated cytochrome *bc*-complexes all contain four redox centers, one heme *c*, one 2Fe2S-center and two hemes *b*, which are bound to proteins. In addition to protein the complexes contain lipids (and detergent) and some residual quinone, a part of which, especially in the case of *bc*1-complexes, may be firmly bound. Even the purest cytochrome *b6f*-complex retains about one molecule residual chlorophyll *a* (see note 4 added on proof).

### *3.1. Polypeptides*

The four metal redox centers are bound to three proteins: Cytochrome *c*1 or *f* (both are c-type cytochromes, *f* standing for "folium" after its discovery in leaves), the Rieske FeS-protein, and cytochrome *b* or *b*6, which are double heme proteins. These three proteins constitute the minimum structure for a cytochrome *bc*1-complex as it is indeed found in bacteria [5, 7]. The cytochrome *b6f*-complexes, in addition contain subunit IV of about 17 kDa, which together with cytochrome *b*6 of 23 kDa adds up to the cytochrome *b* of cytochrome *bc*1-complexes. Subunit IV corresponds to the C-terminal part of cytochrome *b* [2]. Thus also the *b6f*-complexes can be regarded to represent the minimum structure. Some of the isolated bacterial cytochrome *bc*1-complexes also contain another additional subunit of about 14 kDa, but this is an extra one of unknown function [5, 7].

All of the complexes have small additional, hydrophobic proteins (see note 4 added on proof). However, the mitochondrial *bc*1-complexes contain a high number of further proteins, in addition to cytochromes *b* and *c*, and the Rieske FeS-protein – eight for mammals and seven for fungi (see Chapter 6). These supernumerary subunits are characteristic for mitochondrial membrane complexes, and probably reflect the control of protein assembly in the complicated interplay between the nucleus and the organelle genome, which during evolution has yielded a major part of its genes to the nucleus. The two larger of these subunits, around 50 kDa, are called core protein I and II (plant mitochondria contain three core proteins). These are not participating in the transfer of electrons, but surprisingly at least one of them is involved in proteolytic processing of mitochondrial precursor proteins (PEP = processing enhancing protein [13]).

### *3.2. Genetic organization*

In gram-negative bacteria the genes for the three redox center carrying polypeptides are organized in the *fbc*-operons, also called *pet*ABC [5, 7]. The gene for the Rieske FeS-protein (*pet*A) is followed by the genes for cytochrome *b* (*pet*B) and for cytochrome *c*1 (*pet*C). The additional subunit, occasionally present in bacterial *bc*1-complexes is encoded in a separated gene.

For the cytochrome *b6f*-complex in cyanobacteria two transcription units are known, one coding for the Rieske FeS-protein and cytochrome *f*, and one for cytochrome *b*6 and subunit IV [3]. Recently a transcription unit in the green S-bacterium *Chlorobium limicola* has been discovered, which has lost the third gene for cytochrome *c*1 or *f*, but retains the gene for a Rieske FeS-protein followed by the gene for a complete cytochrome *b*. Although the latter gene is continuous, its sequence exhibits characteristics which are closer to cytochrome *b*6 plus subunit IV [10]. Thus, according to the phylogenetic distances indicated in Fig. 3, the splitting of the gene for a cytochrome *b* into two, one for cytochrome *b*6 and one for subunit IV, must have occured after the separation into complexes with *bc*1- and *b6f*-type characteristics. It is still uncertain whether *Chlorobium* contains an equivalent for cytochrome *c*1 or *f*, but it is possible that yet another type of cytochrome takes over in green S-bacteria, as has been found recently for *Bacillus* (see note 2 added on proof).

Chloroplasts have retained the cyanobacterial transcription unit for cytochrome *b*6 and subunit IV on their genome, but from the other one the gene for the Rieske FeS-protein has been transferred to the nucleus. There it attained a leader sequence for transport of the protein into the chloroplast. The gene for cytochrome *f* remained on the plastome, which also harbors one of the small, hydrophobic subunits in a separate location. Mitochondria only kept the gene for cytochrome *b* on their genome, all the other proteins are synthesized from nuclear genes as precursors and have to be taken up into the organelle, where they are processed to the mature forms (see Chapter 6).

### *3.3. Cytochrome* c1 *and cytochrome* f

The heme of cytochrome *c*1 or *f* transfers electrons from the Rieske FeS-center to cytochromes *c* or *c*2, or to plastocyanin (see Figs. 9 and 10). The redox difference spectra of the cytochrome *b*6*f*-complex isolated from spinach chloroplasts for the α-band region of the cytochromes are shown in Fig. 4. Cytochrome *f* can selectively be reduced by ascorbate (Fig. 4a), while the reduction of both the hemes in cytochrome *b*6 require dithionite (Fig. 4a and b). The α-band of cytochrome *f*, as well as of cytochromes *c*1, is asymmetric. The shoulder on the low wavelength side forms a separate peak at low temperature.

Cytochromes *c*1 or *f* are both c-type cytochromes around 30 kDa, with the heme group covalently bound to the protein by addition of the vinyl groups of the tetrapyrol ring to the SH-groups of cysteines. The conserved heme-*c* binding motif CXXCH is found close to the N-terminus. At the C-terminus a pronounced hydrophobic stretch of 20 residues, flanked by charges, is conserved, which fixes the protein in the membrane (see Chapter 6 and Fig. 7). Cytochromes *c*1 and *f* have very few residues in common. Besides VCXXCH for heme binding there are only four more around the putative C-terminal transmembrane helix. This is in contrast to the identities and similarities among cytochromes of either type alone [2]. Cytochromes *f* are shorter at the N-terminus, but are substantially longer in the middle than cytochromes *c*1. A most significant difference between cytochromes *c*1 and *f* is that the six ligand to the heme-Fe is the S of a methionine in the former, but the N form the N-terminus in the latter, as documented by x-ray crystallography of cytochrome *f* [32]. Taken together, all this indicates that cytochromes *c*1 and *f* evolved from different ancestors. The possibility that the cytochrome complex in *Chlorobium* neither contains cytochrome *c*1 nor *f*, but another type of cytochrome instead, is in line with this conclusion (see note 2 added on proof).

Mitochondrial cytochrome *c*1 of eucaryots is synthesized on cytoplasmic ribosomes as a precursor with an N-terminal signal peptide, which is cleaved off in two steps during transport to its functional location [13].

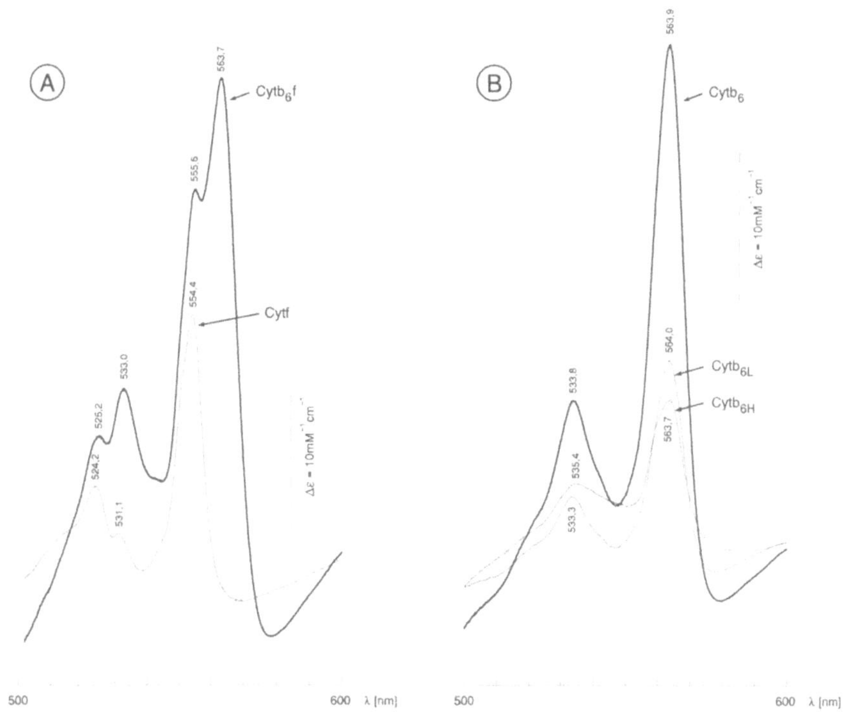


Fig. 4. Redox difference spectra of the cytochrome *b6f*-complex. The cytochrome *b6f*-complex purified from spinach chloroplasts is shown. For gross estimation of concentrations a differential extinction coefficient of 20 $mM^{-1}cm^{-1}$ for the α-band of reduced and oxidized heme suffices. (A) Fat spectrum (cyt b6f): Cytochrome *b*6 plus cytochrome *f* - fully reduced by dithionite minus oxidized by ferricyanide. Thin spectrum (cyt f): Cytochrome *f* - reduced by ascorbate minus oxidized. (B) Fat spectrum: Total cytochrome *b*6 (cyt b6) - reduced by dithionite minus reduced by ascorbate. Thin spectrum (cyt $b6_H$): High potential heme of cytochrome *b*6 - first 30% of cytochrome *b*6 reduced by dithionite minus reduced by ascorbate. Thin spectrum (cyt $b6_L$): Low potential heme of cytochrome *b*6 - fully reduced minus 70% reduced by dithionite.

Also procaryotic cytochromes *c*1, as well as cytochromes *f* are first built as precursors with N-terminal extensions, although they are not synthesized from external genes. However, they have to be translocated to the periplasmic space, or into the thylakoid, whereby the extensions are cleaved off. The final, processing protease specifically cleaves after alanine (see note 5 added on proof).

### *3.4. Rieske FeS-protein*

The Rieske FeS-protein was discovered in mitochondria by J. S. Rieske [14], and functions as the primary oxidant of quinol in cytochrome *bc*-

complexes. It contains a 2Fe2S-center with an unusually positive redox potential, so that it can conveniently be monitored by EPR after reduction with ascorbate. The EPR-spectrum for the spinach protein is shown in Fig. 5. It is typical for an anisotropic, rhombic g-tensor for the unpaired electron in the reduced FeS-cluster which is characterized by the values $g_z$, $g_y$ and $g_x$. The spectrum can be used to very sensitively monitor interactions of quinones and inhibitors with the FeS-cluster. The spectra for the mitochondrial and bacterial proteins are very similar, with a slight shift to lower field, especially obvious for the $g_x$-value.

The high redox potential of the 2Fe2S-cluster is caused by N-ligation of the iron, in addition to S-ligation. Indeed, each of two universally conserved peptides, CTHLGC and CPHCG, in the C-terminal half of the protein contains a histidine (see Chapter 6 and [2–8]). Interestingly, also two cysteines are present in each of these peptides, however, and it has been demonstrated by directed mutations that all four of them are required for a functional Rieske FeS-protein [7, 8], although only two of them are involved in S-ligation of the cluster (see note 6 added on proof).

Over 20 sequences of Rieske FeS-proteins are known to date. Recently, the one for *Chlorobium limicola* was added, which represents a menaquinol-oxidizing, low-potential form [see 10 and note 2 added on proof]. The proteins of the *b6f*-complex have a shorter N-terminus, accomodating one putative transmembrane helix only, while the proteins from the *bc*1-complexes have been considered to form an N-terminal hairpin

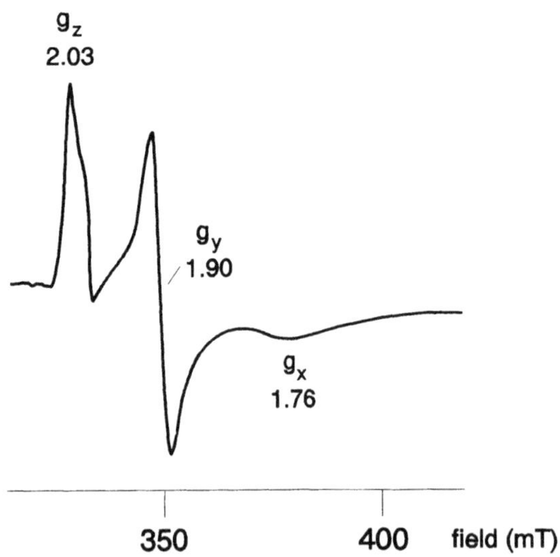


Fig. 5. EPR-spectrum of the Rieske FeS-center in the cytochrome *b6f*-complex from spinach. The three characteristic values of the rhombic g-tensor and the scale of the magnetic field strength are given. The spectrum was measured at 25°K, microwave frequency 9.27 GHz, modulation amplitude 1.6 mT, and microwave power 6.3 mW.

structure through the membrane (see Chapter 6). The relative ease of dissociation from the bc-complexes argues for a more peripheral binding, however (Fig. 7). Overall identity is significant, and similarity is high in the C-terminal half only, which carries the FeS-cluster.

The Rieske FeS-proteins of chloroplasts and mitochondria are encoded in nuclear genes as precursor forms with N-terminal signal peptides, which are processed to the mature proteins on the way to the functional sites in the organelles. The bacterial Rieske FeS-proteins, in contrast to bacterial cytochromes *c*1, are not processed beyond cleavage of the N-terminal methionine [7, 13].

*3.1.4. Cytochrome* b, *cytochrome* b6 + *subunit IV:* Cytochrome *b* is the most hydrophobic part of the *bc*1-complexes. It is a double-heme protein and corresponds to cytochrome *b*6 plus subunit IV of *b6f*-complexes [2–8]. The two hemes of cytochrome *b* are non-covalently bound and have redox potentials around +90 and –50 mV. They are designated high- and low-potential cytochrome *b* (cytochrome $b_H$ and $b_L$), and can be spectrally distinguished. For fungal and mammalian mitochondria, cytochrome $b_H$ has a single α-peak at 562 nm (cytochrome b-562), while cytochrome $b_L$ has a split α-peak at 566 with a satellite at 558 nm (cytochrome *b*-566). These absorptions are somewhat changed in plant mitochondria. The corresponding two hemes of cytochrome *b*6 are about 100 mV more negative, with potentials around –50 and –150 mV, and with a less pronounced difference in the α-peak absorptions (Fig. 4b). Both, cytochrome $b6_H$ and $b6_L$ absorb at 564 nm, but the peak for $b6_H$ is asymmetric.

From sequence comparisons of cytochromes *b* [2, 7, 15] a transmembrane organization of the cytochrome *b*-hemes could be concluded. Folding models based on these comparisons are depicted for cytochrome *b* of *Rhodobacter capsulatus* in Fig. 6a, and for cytochrome *b*6 plus subunit IV from spinach chloroplasts in Fig. 6b. Decisive for the designation of the four heme-binding histidine residues in the second and fourth membrane spanning helix was the sequence of the shorter cytochrome *b*6 (Fig. 6), which contains the two hemes, but only four conserved histidines. Originally, nine transmembrane helices have been suggested for cytochrome *b,* however the location of inhibitor resistance sites and expression experiments with fusions of cytochrome *b* truncates and alkaline phosphatase favor an arrangement with eight transmembrane helices "A-H" and one parallel helix "cd". For further details see [7] and note 6 added on proof.

A host of cytochrome *b* sequences is known by now, which include a wide scope of phyla in which cytochrome *bc*-complexes occur. They add up to several hundred, if all the mitochondrial sequences are counted [15]. The sequence for *Rhodobacter* with 436 amino acids (47 kDa) is 51 residues longer than the one from yeast (385 residues, 42 kDa), which

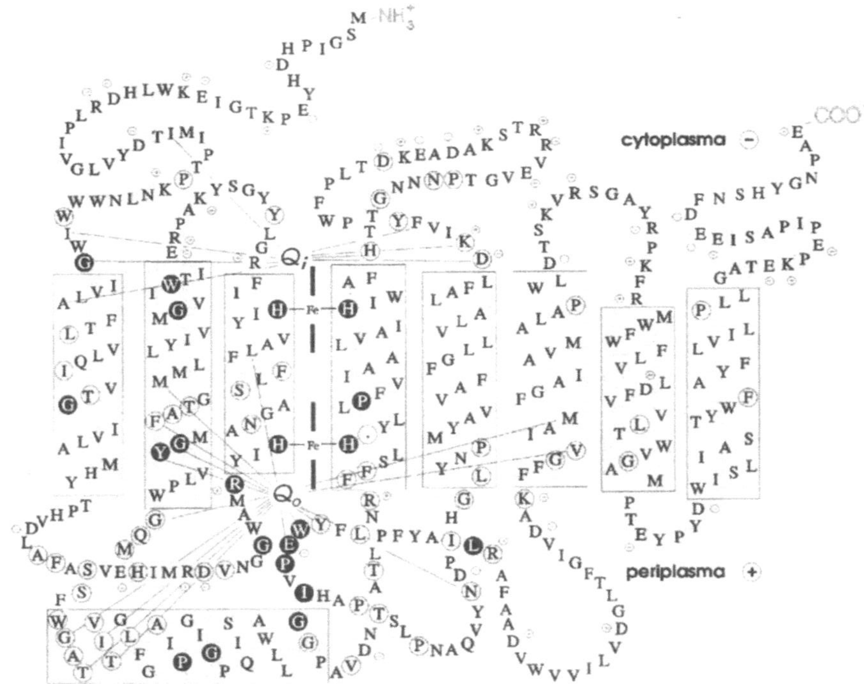


Fig. 6. Folding models of cytochrome *b* and of cytochrome *b6*/subunit IV. (A) Cytochrome *b* from *Rhodobacter capsulatus,* (B) Cytochrome *b6* plus subunit IV from spinach. The parts of the sequences which correspond to cytochrome *b* from yeast mitochondria are underlaid in gray. The eight putative transmembrane helices and one putative parallel helix are boxed. The second and fourth transmembrane helices are arranged to hold the two heme groups by four histidines. Fully conserved residues are given in white letters on black circles, residues typical for either cytochrome *b* or cytochrome *b6*/subunits IV (fully con-

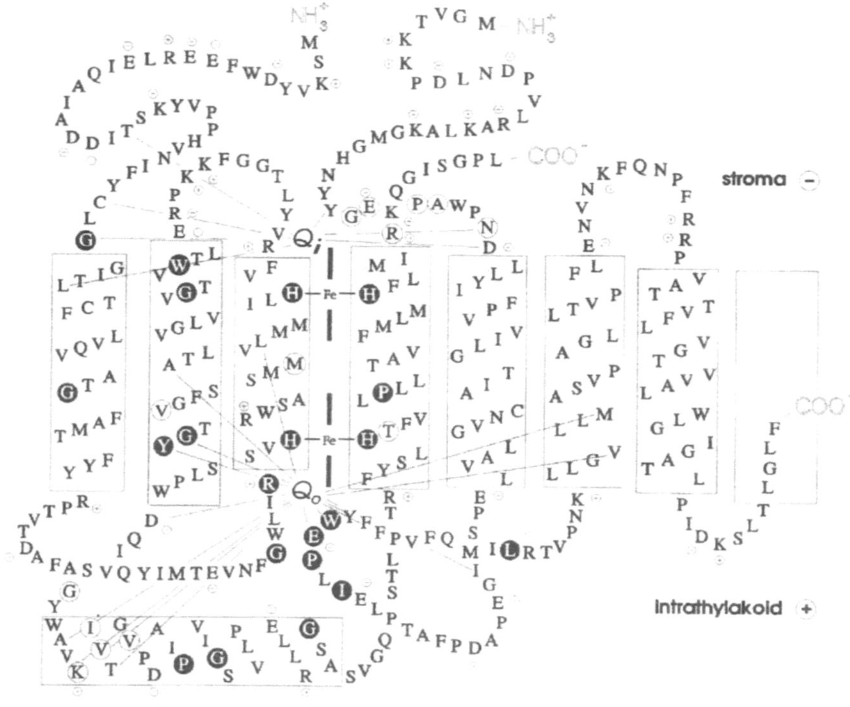


served in each group, but different) are given on white circles. Residues additionally conserved in cytochromes *b* are encircled in (A). This is not shown for cytochromes *b*6/subunits IV in (B), because they are about 90% identical, from cyanobacteria to higher plants [2]. $Q_o$ and $Q_i$ stand for the quinol oxidation- and the quinone-reduction site, respectively. Contributions of amino acid residues to these sites, as indicated by spontaneous inhibitor resistant mutants and by site-directed mutations for cytochrome *b* of the *bc*1-complexes of yeast, mouse and *Rhodobacter* [7, 8], are shown by lines in (A), and for comparison of the residues in equivalent positions also in (B).

corresponds to the portions underlaid in gray in Fig. 6a. The additional peptides in *R. capsulatus,* N- and C-terminal extensions plus two insertions (one in the loop between fourth and fifth, and one between fifth and sixth transmembrane helix), are not underlaid. Only 21 amino acid residues out of some 400 are conserved in all the sequences obtained by now [11]. These are indicated in white letters on black circles in Fig. 6. Next to the four crucial histidines, seven glycines and three prolines are of structural significance. The conserved aromatic residues may participate in quinone binding and/or electron transfer.

The information about amino acid residues contributing to the sites of quinol/quinone interaction ($Q_o$ and $Q_i$), either from spontaneous inhibitor resistant mutants, or from site-directed mutants, is growing fast [7, 8; see note 7 added on proof], and is indicated by lines to the $Q_o$- and $Q_i$-sites in Fig. 6. The two-dimensional models of Fig. 6 have to be imagined to fold back in the third dimension, so that these amino acid residues come close at these sites on the two opposite membrane surfaces. Noteworthy is that a few of the residues interacting with either $Q_o$ or $Q_i$ are found rather deep in the membrane, like a leucine in helix B and a methionine in helix C, both interacting with $Q_o$. If the folding in Fig. 6 is valid, these residues must contribute from a distance. It has been further concluded that the first four putative transmembrane helices form a bundle, the four conserved glycines, two each in helices A and C, providing space for the two hemes [7].

Cytochrome *b*6 (215 residues, 23 kDa) and subunit IV (160 residues, 17 kDa) add up to almost a full cytochrome *b* with 40 kDa. Cytochrome *b*6 ends after the 4th transmembrane helix. The two most characteristic differences between cytochromes *b* of *bc*1- and cytochrome *b*6/subunit IV of *b*6*f*-complexes, next to the split of the protein, are the occurrence of a threonine as an additional 14th amino acid between the two histidines in helix D of cytochrome *b*6, and the lack of helix H in subunit IV (compare a and b in Fig. 6). The additional amino acid between the histidines in helix D must influence the geometry of heme ligation, and may be responsible for the lower redox potentials of hemes in cytochrome *b*6. Another significant difference comprises several residues involved in inhibitor binding [7, 8], which are in line with the insensitivity of the cytochrome *b*6*f*-complexes to antimycin and myxothiazol and its sensitivity to stigmatellin [11].

Cytochromes *b* are not processed from precursor forms during insertion into the membranes, only the N-terminal methionines are cleaved off.

### *3.6. Overall structure*

A scheme for a transmembrane, monomeric arrangement of the three redox center carrying polypeptides and subunit IV of the cytochrome *b*6*f*-

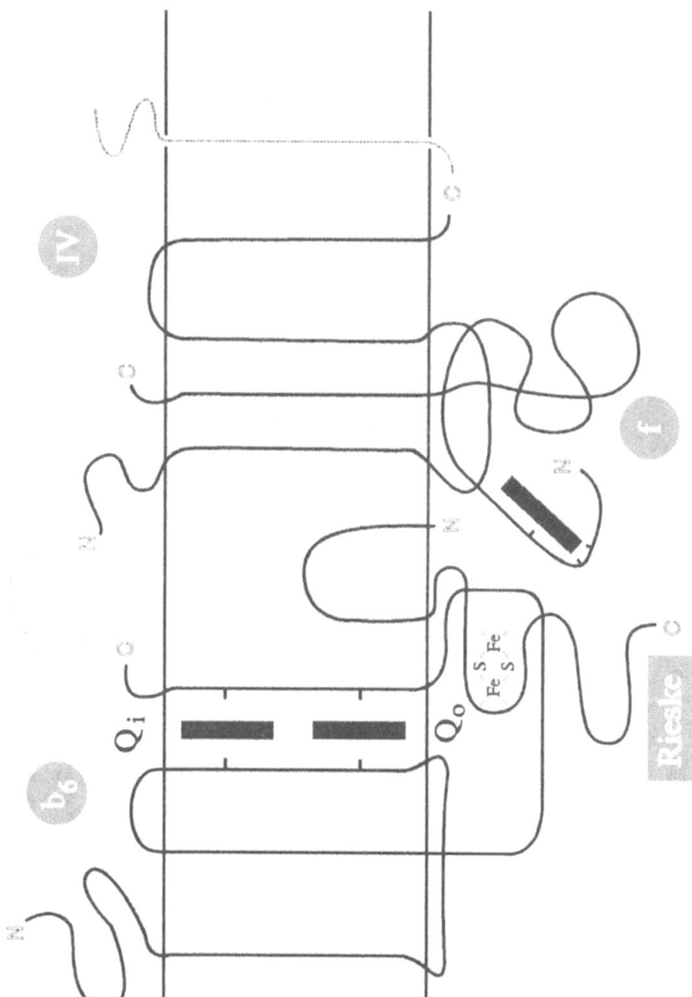


Fig. 7. Overall folding model of the cytochrome *b6f*-complex in the membrane. N- and C-termini are indicated by N and C, $Q_i$ and $Q_o$ stand for quinone reduction- and quinol-oxidation sites, respectively. The three heme groups are represented by shaded bars. The upper part of the scheme represents the stromal surface of the membrane. The connection of cytochrome *b6* and subunit IV to form a cytochrome *b*, and the missing C-terminal part of cytochrome *b* are also indicated.

complex is drawn in Fig. 7. A 3D-structure of a cytochrome *bc*1-complex is available now [30] and the structure of cytochrome *f* has been obtained [32; see note 6 added on proof].

In the isolated state the cytochrome *bc*1-, as well as the cytochrome *b*6*f*-complexes often are dimers, but it is under debate still, whether this is of functional significance [6].

## 4. Function

The cytochrome *bc*1-complex functions as ubiquinol-cytochrome *c* oxidoreductase in the respiration of mitochondria, the cytochrome *b*6*f*-complex functions as plastoquinol-plastocyanin oxidoreductase in photosynthesis of chloroplasts. In some algae the acceptor protein can be cytochrome *c* instead of plastocyanin, either alternatively or exclusively. In procaryots these quinol oxidoreductases may function in photosynthesis as well as in respiration, as detailed above (Fig. 1). The question of a respiratory function in chloroplasts (chlororespiration), possibly involving the *b*6*f*-complex is not settled. The activities of a NAD(P)H-plastoquinone oxidoreductase and of a plastoquinol oxidase are negligible in mature chloroplasts from higher plants. However, a triggering by light, or a transient function of such activities during plastid differentiation cannot be excluded.

The immediate reductant of the acceptor proteins, cytochrome *c* or plastocyanin, is cytochrome *c*1 or *f*. The interaction is largely governed by electrostatic forces.

Cytochrome *c*1 is rereduced by electrons from the Rieske FeS-center. This has been demonstrated conclusively in flash experiments with *Rhodobacter sphaeroides,* in which the effect of specific inhibitors on the extents of photooxidation of cytochromes *c*2 plus *c*1 was studied (see Fig. 11). These cytochromes were fully oxidized by flashing light only when electron transfer from the Rieske FeS-center was blocked. Up to now a corresponding demonstration of electron transfer from the Rieske FeS-center to cytochrome *f* in chloroplasts is lacking, in spite of repeated efforts [6]. Cytochrome *f* is always fully photooxidized by plastocyanin/P700, with no observable rereduction. How does the FeS-center escape detection in flashing light, or is it feasible that it does not participate at all in electron transfer through the *b*6*f*-complex? What else may be its function?

The turnover numbers of carefully isolated, pure cytochrome *bc*1-complexes, either from mitochondria, or from bacteria, are about 10 times higher (more than 1000 $s^{-1}$) than the ones reported for cytochrome *b*6*f*-preparations. It is not settled whether this is an inborn difference, or reflecting inactivation of the latter during isolation. The rate limit of electron transport in chloroplasts is around 5 ms, which is in support of a limitation by the intrinsic turnover of the cytochrome *b6f*-complex.

In addition to electrons cytochrome *bc*-complexes transfer protons through the membrane, and this is intimately linked to the redox chemistry of the quinone system.

### *4.1. Redox potentials*

The four metal centers in cytochrome *bc*1/*b6f*-complexes operate in two pairs, two of them at relatively high, and two at relatively low redox potential. The FeS-center has an exceptionally high redox potential (about +300 mV). It is the immediate oxidant of quinol, and transfers the electron to the heme *c* of cytochrome *c*1 or *f* (+240/+340 mV). The two hemes *b* of cytochrome *b* or *b*6 have comparatively lower redox potentials (about +90 and –50 mV for *bc*1 - and about –50 and –150 mV for *b6f*-complexes), and accept the electron from the resulting semiquinone. This relation of the high potential redox pair in the *b6f*-complex to the first half reaction of the plastoquinol/-quinone system, and of the low potential pair to the second half reaction is elaborated in Fig. 8. The redox potentials for the two single electron transfers, and for the overall two-electron transfer of the quinone system is shown over the whole pH-range, considering the protonation of quinol and semiquinone. The diagram was adopted from P. Rich [16], and can be easily modified for the ubiquinone or menaquinone systems. As a reflection of the first law of thermodynamics the overall redox potential always is half the sum of the two potentials of the single electron transfer events. The diagram also demonstrates the instability of the semiquinone through most of the pH-range: Because it tends to lose, as well as to take up an electron, the redox potential of quinol/semiquinone is more positive than the one of semiquinone/quinone. Only at very high pH is this relation turned around, and the semiquinone becomes a stable species. In the physiological pH-range the redox potentials of the two half reactions of the quinone system come close to the FeS-center plus cytochrome *f* on one hand, and to the low-potential cytochrome *b*6 on the other, and this provides for a key observation with all the cytochrome *bc*-complexes, which was enigmatic at first. When an electron is pulled out from cytochrome *c*1 or *f* by oxygen via the oxidase, by light through photosystem I, or even by ferricyanide cytochrome *b* or *b*6 goes reduced. This so-called oxidant-induced reduction of cytochrome *b* has been explained by Wikström to reflect the concerted oxidation of quinol via the semiquinone intermediate [17]. The high potential duo oxidizes quinol to semiquinone, which reduces the low potential duo for additional work. Indeed, two protons were found to be translocated per electron transfer through the quinol/cytochrome *bc*-system, twice of what is expected for quinol alone.

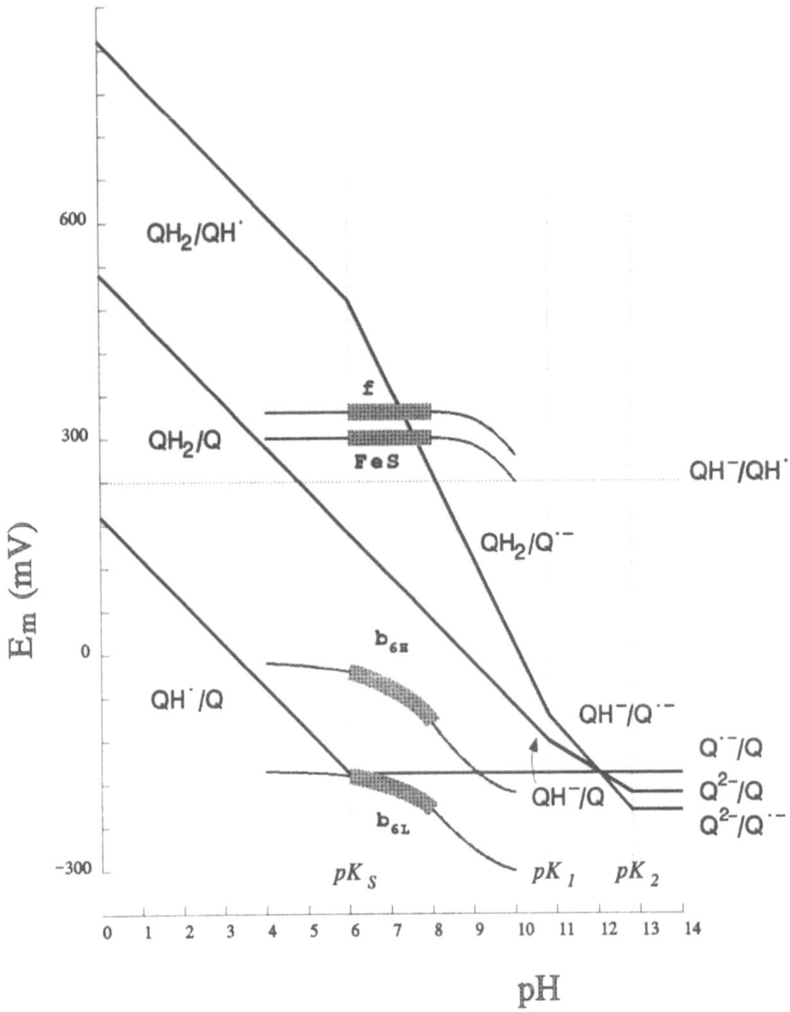


Fig. 8. Redox potentials of the plastoquinone/cytochrome *b6f*-system. The pH-dependence of the redox potentials of plastoquinone, and of the four redox centers of the cytochrome *b6f*-complex are shown. Q stands for plastoquinone, f for cytochrome *f*, FeS for the Rieske FeS-center, and b6L and b6H for the low- and the high-potential heme of cytochrome *b*6, respectively. The three pK-values correspond to the dissociation from the semiquinone ($pK_s$), and to the first and second dissociation from quinol.

## *4.2. The Q-cycle*

The original explanation of oxidant-induced reduction by Wikström was carried further by Mitchell's formulation of the Q-cycle [18], which also provides for the observed $H^+/e^-$-ratio of two during one turnover of cytochrome *bc*-complexes. He conceived two sites for quinone interaction, one for the concerted quinol oxidation, called $Q_o$ ("o" for *o*xidaton or *o*utside) or $Q_p$ ("p" for *p*ositive), with a low stability of the semiquinone intermediate, and one for quinone reduction which stabilizes the intermediate semiquinone, called $Q_i$ ("i" for *i*nside), $Q_r$ ("r" for *r*eduction), or $Q_n$ ("n" for *n*egative). For the interplay of these two quinone reaction sites with different semiquinone stability the term "redox seesaw" was coined. Figure 9a shows this for the plastoquinone/*b6f*-system, Fig. 9b for the ubiquinone/*bc*1-system. The redox potential difference between the first and second electron transfer to the quinone reflects the stability of the intermediate semiquinone. The relation is described by the equation (E1–E2) = RT/nF × ln $K_s$. If E2 is more positive than E1, ln $K_s$ becomes negative, $K_s$ is smaller than 1 and the intermediate semiquinone is an unstable species. For Fig. 9 it is assumed that the semiquinone in not stabilized at the site of quinol oxidation ($Q_o$), the stability constant being the same as in solution.

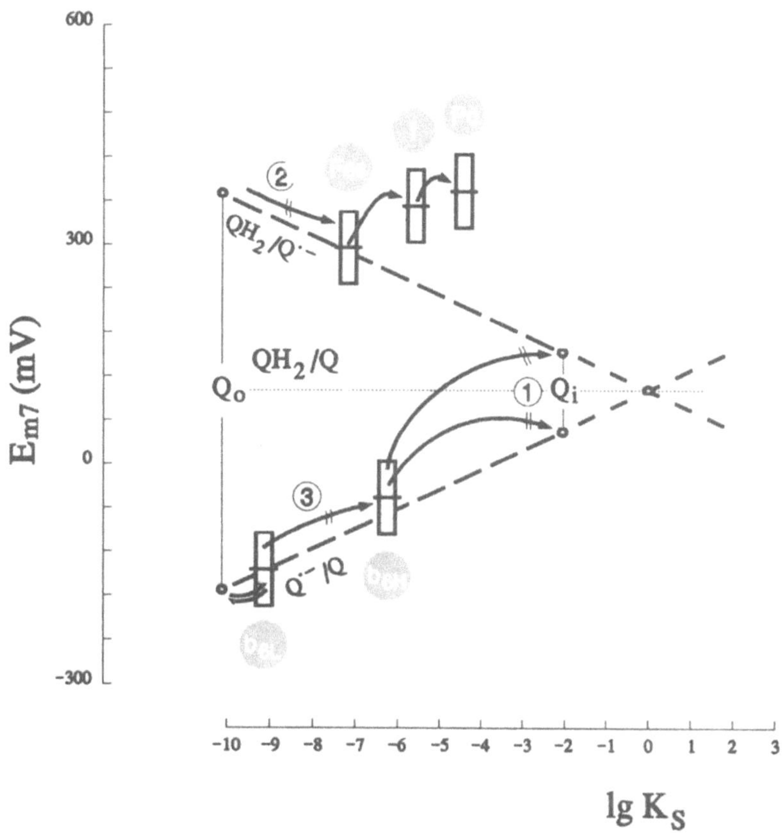


Fig. 9A. Stability constant of the semiquinone and redox potentials in cytochrome *bc*-complexes. The stability constant ($K_S$) is given on a log-scale. Other abbreviations are as for the previous figures. Fig. 9A represents the plastoquinol/*b6f*-system.

With the *b*6*f*-complex and plastoquinol the concerted reaction at $Q_o$ is close to equilibrium, the redox potentials of the FeS-center and of $b6_L$ both matching one of the electron transfers from the quinol-system (Fig. 9a). Consequently, the semiquinone cannot be much stabilized at the $Q_o$-site in the *b*6*f*-complex, because it could not reduce cytochrome $b6_L$. This is not so stringent for the oxidation of ubiquinol by the *bc*1-complex, where cytochrome $b_L$ is clearly above $Q^{\cdot -}/Q$ in the scheme (Fig. 9b). Rather, cytochrome $b_H$ and the $Q_i$-site are in equilibrium. Indeed, in this case the semiquinone intermediates at both sites, can be detected by EPR. The one at the $Q_i$-site is very stable and sensitive to antimycin (inhibitor 1 in Fig. 9), the one at the $Q_o$-site is less stable, but insensitive. For the *b*6*f*-system the documentation of any semiquinone is much more difficult [6]. This difference in the redox potential relations of the *b*6*f*- and the *bc*1-system cannot be rationalized up to now.

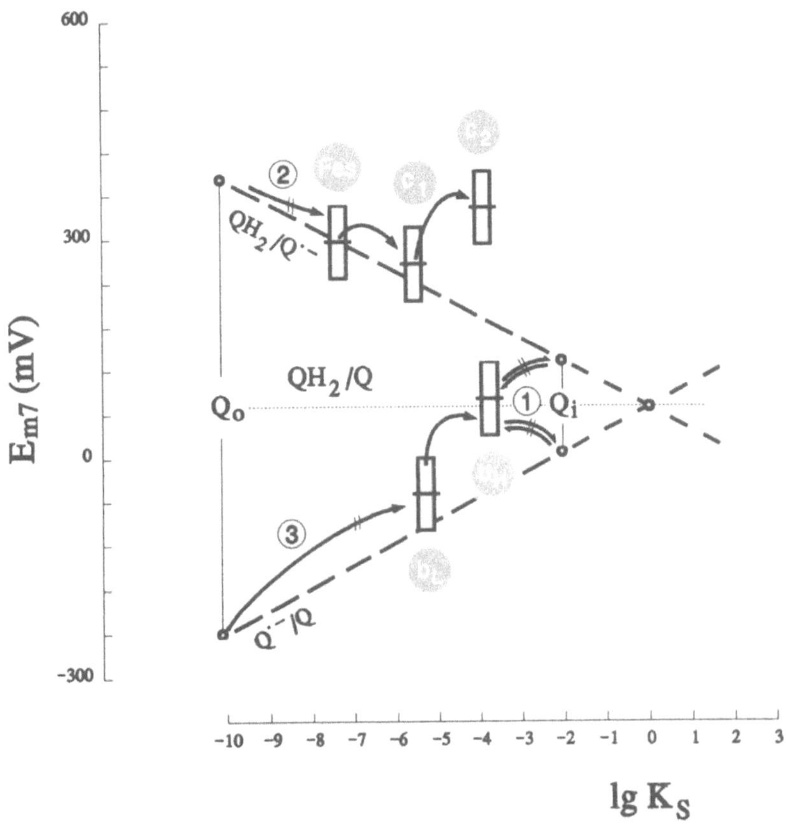


Fig. 9B. The ubiquinol/*bc*1-system. For the four redox centers of *bc*-complexes the redox range from 10-90% oxidation is indicated by bars. The $K_s$ at the $Q_o$-site equals the value of the unbound form, for $Q_i$ a substantially higher value was assumed, reflecting stabilization of the semiquinone by binding at this site. Three modes of inhibition are shown by numbers.

The original formulation of the Q-cycle also viewed the reaction at the $Q_i$-site as concerted, with two different electron donors to quinone. The variant elaborated in Fig. 10 (and already presented in Fig. 9) holds that only the oxidation of quinol at the $Q_o$-site occurs in a concerted transfer of two electrons, while reduction of the quinone at the $Q_i$-site occurs in two consecutive steps [7, 19]. A full turnover can better be comprehended if it is divided into two halves, as in Fig. 10 a and b. Upon receiving a second electron $Q_B^{\cdot -}$ on the reaction center binds two protons and leaves the site as $QH_2$. At the same time a $QH_2$ from the quinone pool is oxidized at the $Q_o$-site, first by the Rieske FeS-center to the semiquinone anion plus two protons. The semiquinone anion at this site then reduces cytochrome $b_L$ which transfer the electron via cytochrome $b_H$ to the quinone reduction site $Q_i$, which stabilizes the resulting semiquinone anion. From the Rieske FeS-center the electrons go through cytochromes *c*1 and *c*2 to rereduce the photooxidized dimer of bacteriochlorophyll in the reaction center (Fig. 10a). In the second half reaction (Fig. 10b) a second $QH_2$ from the pool is oxidized at $Q_o$ as before, but now the semiquinone anion at $Q_i$ is reduced and binds two protons to form $QH_2$, which leaves the site. The cycle is completed by another light-dependent electron reansfer $(BChl)_2$ to $Q_B$, followed by rereduction of $(BChl)_2^{\cdot +}$ as before. The net effect is the translocation of four protons through the membrane per two electrons running through the cycle. The overall Q-cycle in mitochondrial respiration (see Chapter 6) and in chloroplast electron flow can be delineated accordingly. It represents a fascinating example of how a leading reaction triggers the formation of a reactive intermediate which is channelled to work by the molecular machinery. The interactions of Q at $Q_B$ and $Q_i$, and of $QH_2$ at $Q_o$ from the pool follow second order kinetics. The contributions of transmembrane electron and proton transfer steps to membrane potential formation are indicated on the side of the schemes in Fig. 10. They are fast in the reaction center and slow in the *bc*-complexes [20]. More complex versions of quinol oxidation are possible (see note 8 added on proof).

The two sites of quinone interaction in the cytochrome *bc*-complexes operate independently from each other, as shown conclusively for *bc*1-complexes by specific inhibitors and by site-directed mutants [7]. This excludes alternative mechanisms as discussed still for *b6f*-complexes in a recent review [6], with only one site of quinone interaction, which assumes different states during turnover, or with two mutually dependent sites. The Q-cycle mechanism was also formulated for a dimer of the complexes, and their structure could be dimeric indeed (see above). Careful titrations with tightly binding inhibitors are contradictory. One part of the results is in favor of one inhibitor molecule per dimer, the other is in favor of one molecule per monomer to be required for inhibition. The latter can be interpreted that each momomeric half *per se* is active, and that the dimeric state is of structural rather than functional importance. The issue remains open still [6].

10A

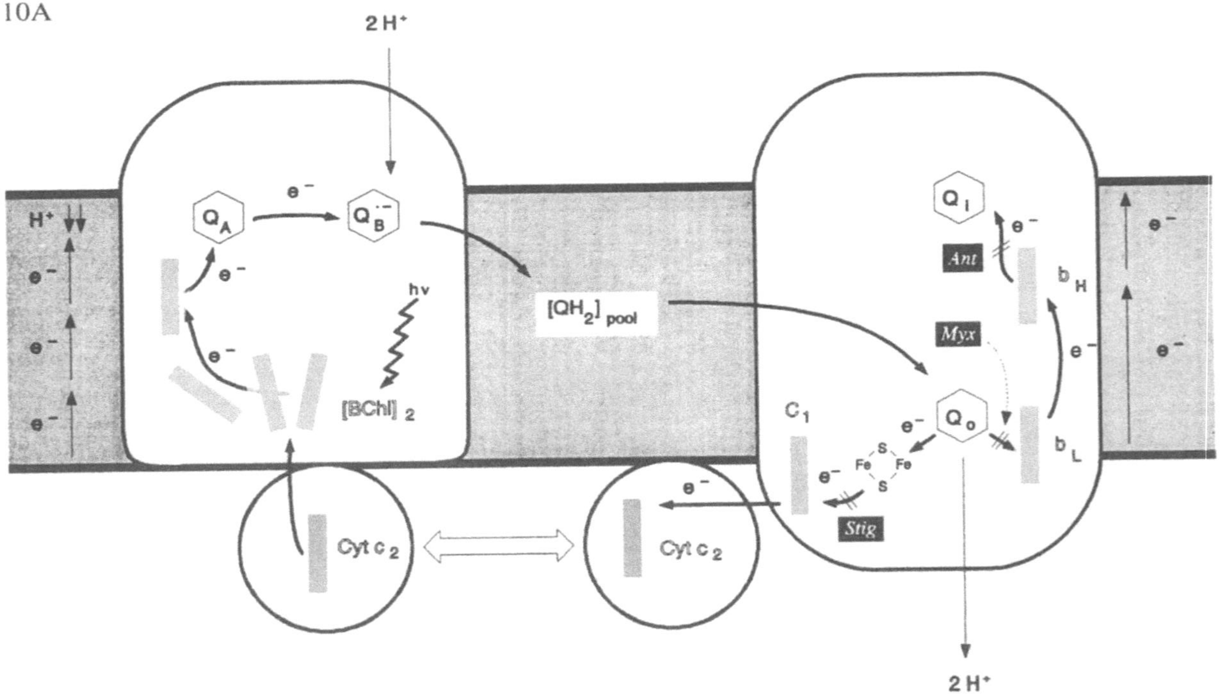


Fig. 10. The Q-cycle of photosynthetic bacteria. The scheme was modified from [19], and shows the photosynthetic reaction center of *Rhodobacter capsulatus* on the left and the cytochrome *bc*1-complex on the right, connected by the hydroquinone pool and cytochrome *c*2 in a cyclic electron transport system. A and B represent the first and second half of a complete turnover of two electrons. The quinone binding sites are indicated by hexagons, and by the denotations $Q_A$ and $Q_B$ for the primary and secondary quinone acceptor in the reaction center, and by $Q_o$ and $Q_i$ for the quinone oxidizing- and the quinol-reducing site in

10B

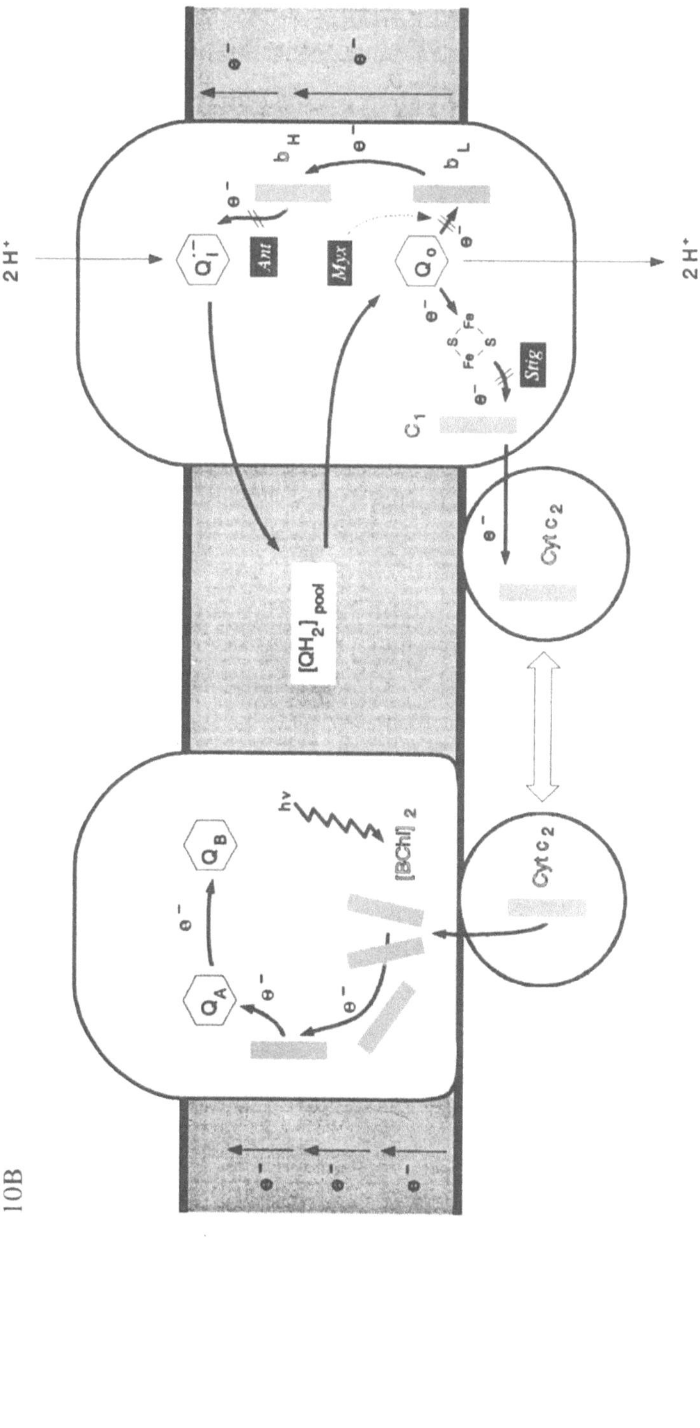


the *bc*1-complex, respectively. Tetraporphyrins of hemes, bacteriochlorophylls and pheophytin are indicated by bars. $[BChl]_2$ stands for the special pair of bacteriochlorophylls in the reaction center, cyt $c_2$, $c_1$, $b_L$ and $b_H$ stand for cytochromes *c*2 and *c*1, and for the low- and high-potential hemes of cytochrome *b*, respectively. The Rieske FeS-center is indicated by the 2Fe2S-cluster. Inhibition by antimycin A, myxothiazol and stigmatellin are symbolized by Ant, Myx and Stig, respectively. The steps contributing to the membrane potential formation [20] are represented by arrows, on the left for the reaction center, on the right for the *bc*1-complex.

The Q-cycle is permanently operative in the cytochrome *bc*1-complexes, but this is not agreed on for the cytochrome *b*6*f*-complexes. Evidence and arguments for and against its operation at high light intensity (= high proton potential across the membrane) have been put forward [6]. The Q-cycle and its possible down regulation have profound influence on the estimations of how photosynthetic electron flow meets the ATP requirement in chloroplasts. With the Q-cycle operative 6, but without it only $4H^+/e_2$ are translocated during photoreduction of NADPH by water (see Chapter 5; the latter value still dominates the pertinent literature). At least three or rather four $H^+$/ATP are required formed by the ATP-synthase (see Chapter 12). This yields a $P/e_2$-ratio of 2 or 1.5, and 1.33 or 1, respectively, depending on whether the Q-cycle operates or not. For fixation of 1 $CO_2$ in the Calvin cycle 1.5 ATP/NADPH are required. Consequently, the $P/e_2$ in photoreduction of NADPH would suffice only if the Q-cycle contributes. If it does not, the shortfall in ATP must be complemented either by cyclic electron flow, or by reuptake of oxygen via photosystem I (Mehler reaction). The need for these complementing reactions would actually even be higher, because the Calvin cycle is not the only ATP-consuming reaction in chloroplasts. It is certainly higher in C4-plants, which require an extra ATP per fixed $CO_2$ (see Chapter 5).

*4.2.1. Inhibitors:* Three modes of inhibition are discerned for cytochrome *bc* complexes, which were very valuable in elucidating the electron transfer path [5, 7] (see Chapter 6). They are indicated by numbers in Fig. 9, and by the abbreviations "ant", "stig" and "myx" in Fig. 10. Mode 1 blocks the reduction of the quinone at $Q_i$. As a consequence the oxidant-induced reduction of cytochrome *b* is stimulated. Mode 2 blocks electron transfer to the Rieske FeS-center, while mode 3 inhibits reduction of cytochrome $b_L$. Mode 1 is exerted very efficiently by antimycin A, mode 2 by stigmatellin, and mode 3 by myxothiazol in the ubiquinol/cytochrome *bc*1-system (see Chapter 6 and [5, 7]). Unfortunately antimycin and myxothiazol do not inhibit the plastoquinol/*b*6*f*-system, a fact that hampered the study of this system substantially. Only stigmatellin is universally effective. Interestingly, MOA-stilbene, a compound related to myxothiazol, inhibits the cytochrome *b*6*f*-complex, but rather by mode 1, and not by mode 3 [21]. Most of these inhibitory compounds in part resemble the structure of the isoprenoid quinones, and it is therefore not surprising that they are not completely specific for only one quinone interaction site, and that the relative specificity for $Q_o$ and $Q_i$ may be changed in *b*6*f* compared to *bc*1. In accord with this, many of these inhibitors also block the $Q_B$-site in the reaction centers of bacteria (Fig. 10), or of photosystem II. MOA-stilbene may prove itself as very valuable in the study of *b*6*f*-complexes, for which only alkyl-quinone N-oxides were known to act as mode 1-inhibitors so far.

## 5. Regulation

Since quinol oxidation is a relatively slow reaction and is often rate determining in electron transport chains, the cytochrome *bc*-complexes may be involved in metabolic regulation. The possibility that they generally function as redox sensors in bacteria has been discussed lately [22]. Moreover, the core proteins of the mitochondrial cytochrome *bc*1-complexes are involved in the processing of proteins targeted for import [13], as mentioned above. A connection between this function and the electron transfer activity is not yet obvious.

### *5.1. LHCII-kinase*

Good evidence exists for the cytochrome *b6f*-complex of chloroplasts to act as a redox sensor, in addition to its function in energy conserving electron transport I and II [22]. A specific kinase for the light harvesting chlorophyll *a/b*-protein of photosystem II (LHC-II) is activated when the plastoquinone pool is reduced (Fig. 11). The phosphorylated form of

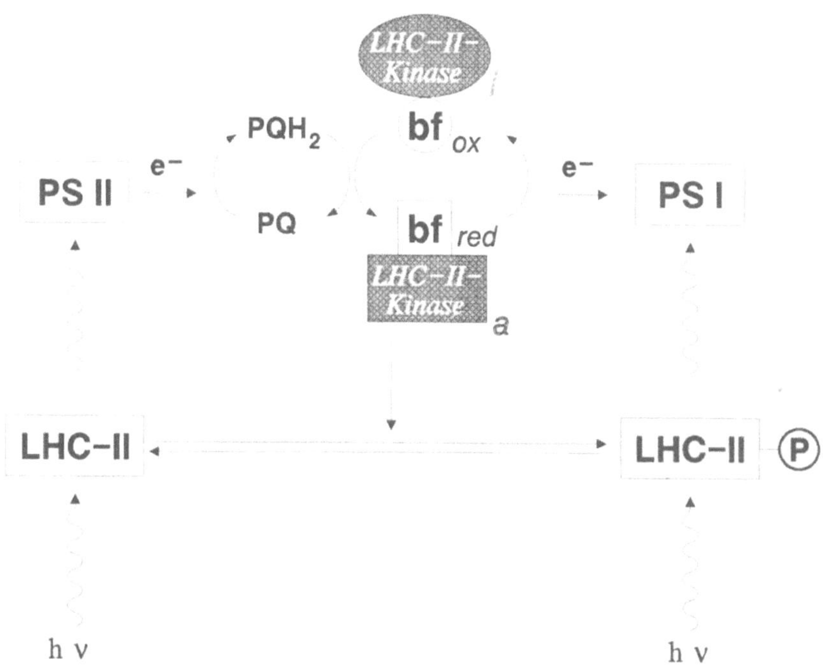


Fig. 11. Activation of the LHCII-kinase by the cytochrome *b6f*-complex. PSI, PSII, LHC-II, bf and PQ stand for photosystem I and II, for the light harvesting complex II (chlorophyll a, b-protein complex), for cytochrome *b6f*-complex and plastoquinone, respectively. The suffixes *i* and *a* stand for "inactive" and "active".

LHC-II is closer associated with photosystem I, which then receives more light and reoxidizes the quinone pool. Studies with inhibitors and with mutants pointed out that plastoquinol is not directly activating the kinase. The degree of plastoquinone reduction is sensed by the *b6f*-complex, and the resulting redox- or conformational change is transferred to the kinase (see note 9 added on proof).

## 6. Conclusions and open questions

At the present state the following conclusions can be drawn:

(1) The various cytochrome *bc*-complexes of respiratory and photosynthetic electron transport chains contain four metal redox centers on three membrane proteins - cytochrome *c*1 or *f*, the Rieske FeS-protein, and the double-heme protein cytochrome *b* or *b*6.

(2) For quinol oxidation the four redox centers cooperate in a universal mechanism called "Q-cycle". It involves a site for quinol oxidation ($Q_o$) and a site for quinone reduction ($Q_i$), which operate independently.

(3) This mechanism translocates two protons through the membrane per electron transferred, double the amount expected for quinol oxidation alone.

(4) The extra proton translocation is achieved by the second electron from quinol oxidation at the $Q_o$-site, which traverses the membrane via the hemes on cytochrome *b*/*b*6 to reduce quinone at the $Q_i$-site. Thus electron transport is branched and loops back into the quinone pool in the cytochrome *bc*-complexes. In this respect the term "electron transport chain" is a misnomer.

(5) The minimum structure for a functional cytochrome *bc*-complex requires the three redox proteins only, as realized in bacteria. For these proteins numerous primary structures are known by now. Their folding and interactions in the membrane show universal features. It was a rewarding experience that the sequence comparison of cytochromes *b*/*b*6 independently predicted a transmembrane orientation of the two hemes, as was postulated by the Q-cycle. Cytochrome *b6f*-complexes correspond to this minimal structure, which their cytochrome *b*-complement split into cytochrome *b*6 and subunit IV.

(6) The mitochondrial cytochrome *bc*1-complex contains supernumerary subunits which do not participate in electron transfer directly.

(7) The classification into *bc*1- and *b6f*-types, based on differential inhibitor sensitivities and characteristic features in the sequences, like the split into cytochrome *b*6 and subunit IV, has become less precise, since in *Chlorobium* the gene for cytochrome *b* is continuous but otherwise reveals *b6f*-type characteristics. During evolution the split must have occurred after the development of these characteristics.

The following central questions have yet to be solved:

(1) Is the tendency of the isolated complexes to form dimers of functional, or only of structural significance, or even an isolation artefact?

(2) All the isolated complexes additionally contain small, hydrophobic subunits. What is their role?

(3) Is the electron transfer mechanism universal indeed? Why then can the interaction between the Rieske FeS-center and cytochrome *f* not be observed in flash experiments with cytochrome *b6f*-complexes?

(4) Is the Q-cycle always operating in chloroplasts, or is it regulated down at high light intensity?

(5) How do the menaquinol oxidizing cytochrome *bc*-complexes of certain bacteria cope with the lower redox potential range? To what type do they belong?

(6) Are the predictions for the three redox proteins indeed valid for the 3D-structures in the membrane?

Undoubtedly the further elucidation of the cytochrome *bc*-complexes will remain in the center of bioenergetic interest for some time still.

## Notes added on proof

Since this article has been submitted pertinent progress has been made:

(1) The occurence of Rieske FeS-proteins [21A], and of double-heme cytochromes [22A] resembling cytochrome *b* of *bc*-complexes has been substantiated in *Archaebacteria*. Remarkably, this does not only further document a common origin of photosynthesis and respiration, but leads to the conclusion that respiration evolved before chlorophyll-based photosynthesis [23].

(2) An *fbc*-transcription unit has been sequenced for the gram-positive *Bacilli* [24, 25], which starts with the gene for a low-potential Rieske FeS-protein, followed by a gene for a cytochrome *b*6 plus one for a subunit IV-complement C-terminally fused to a cytochrome *c*. With this it is proven that yet another type of cytochrome *c*, in addition to *c*1 or *f*, can occur in *bc*-complexes.

(3) The evidences for the occurrence of both, a *b6f*-and a *bc*1-complex in cyanobacteria are accumulating [26].

(4) A highly active *b6f*-complex has been purified from *Chlamydomonas reinhardtii*, a preparation which leads the progress at present [27]. It is dimeric and contains three small, hydrophobic miniproteins, in addition the Rieske FeS-protein, cytochromes *f* and *b*6, and subunit IV. Moreover, it contains 1 molecule chlorophyll *a* per heme *f*, which does not exchange with radioactive chlorophyll added.

(5) An interesing feed-back inhibition on the synthesis of cytochrome *f* is exerted by its C-terminal domain, in contact with chloroplast ribosomes in the stroma [28].

(6) Crystal structures are available now for the water-soluble part of the Rieske FeS-protein at 1.5 [29], and for the whole *bc*1-complex at 3.4 Angströms resolution [30], both from beaf heart mitochondria. This impressing progress largely confirms the current structural model of *bc*-complexes, provides clarification of some open issues as well a surprise: a) The ligation of one of the two irons in the 2Fe2S-cluster in the Rieske FeS-protein by two histidines has become doubtless now; b) The *bc*1-complex has a dimeric structure; c) Surprisingly, the distance between the hemes of cytochrome *b* is 21, between the low potential heme of cytochrome *b* and the Rieske center is 27, and between the Rieske center and the heme of cytochrome *c*1 is 31 Angströms. The later is much too large for "through space" electron transfer fast enough.

Furthermore the interaction of the Rieske FeS-protein and cytochrome *f* in the *b*6*f*-complex has been modelled [31] from sequence comparisons and the available crystal structures [29, 32].

(7) An updated compilation of mutations in cytochrome *b* has been published recently [33].

(8) More complicated versions of quinol oxidation at the $Q_o$-site involving two quinol molecules are considered [34, 35].

(9) The trigger to activate the LHCII-kinase seems to be linked to the binding of plastoquinol to the $Q_o$-site, close to the Rieske FeS-center of the *b*6*f*-complex [36].

## References


1. E Broda, *The Evolution of the Bioenergetic Processes,* Pergamon Press, London, 1975.
2. G Hauska, W Nitschke and RG Hermann, J. Bioenerg. Biomemb. 20 (1988) 211.
3. R Malkin, Photosynth. Res. 33 (1992) 121.
4. JM Anderson, Photosynth. Res. 32 (1992) 341.
5. DB Knaff, Photosynth. Res. 35 (1993) 117.
6. AB Hope, Biochim. Biophys. Acta 1143 (1993) 1.
7. RB Gennis, B Barquera, B Hacker, SR van Doren, S Arnaud and AR Crofts, Biomemb. 25 (1993) 195.
8. KA Gray and F Daldal in *Advances in Photosynthesis, Anoxygenic Photosynthetic Bacteria,* RE Blankenship, MT Madigan and CE Bauer (eds.) Kluwer Acad. Publ., Amsterdam, 1995, p. 747.
9. EC Slater, in *Cytochrome Systems, Molecular Biology and Bioenergetics,* S Papa, B Chance and L Ernster (eds.) Plenum Press, London, 1987, p. 3.
10. CR Woese, Microbiol. Rev. 51 (1987) 221.
11. M Schütz, S Zirngibl, J le Coutre, M Büttner, D-L Xie, N Nelson, R Deutzmann and G Hauska, Photosynth. Res. 39 (1994) 163.
12. S Anemüller, CL Schmidt, G Schäfer and M Teixeira, FEBS Lett. 318 (1993) 61.
13. B Segui-Real, RA Stuart and W Neupert, FEBS Lett. 313 (1992) 2.
14. JS Rieske, Biochim. Biophys. Acta 456 (1976) 195.
15. M Degli Esposti, S De Vries, M Crimi, A Ghelli, T Patarnello and A Meyer, Biochim. Biophys. Acta 1143 (1993) 243.

16. PR Rich, Biochim. Biophys. Acta 768 (1984) 53.
17. MKF Wikström, Biochim. Biophys. Acta 301 (1973) 155.
18. P Mitchell, J. Theor. Biol. 62 (1976) 327.
19. DE Robertson and PL Dutton, Biochim. Biophys. Acta 935 (1988) 273.
20. AY Semenov, FEBS Lett. 321 (1993) 1.
21. PR Rich, SA Madgwick, S Brown, G von Jagow and U Brandt, Photosynth. Res. 34 (1992) 465.
21A. G Schäfer, WG Purschke, M Gleissner and CL Schmidt, Biochim. Biophys. Acta 1275 (1996) 16.
22. JF Allen, Biochim. Biophys. Acta 1098 (1992) 275.
22A. J Castresana, M Lübben and M Saraste, J. Mol. Biol. 250 (1995) 202.
23. J Castresana and M Saraste, Trends Biochem. Sci. (1995) 305.
24. J Yu, L Hederstedt and PJ Piggot, J. Bacteriol. 177 (1995) 6751.
25. N Sone, N Tsuchiya, M Inoue and S Noguchi, J. Biol. Chem. 271 (1996) 12457.
26. A Dworsky, B Mayer, G Regelsberger, S Fromwald and GA Peschek, Bioelectrochem. Bioenerg. 38 (1995) 35.
27. Y Pierre, C Breyton, D Kramer and JL Popot, J. Biol. Chem. 270 (1995) 29342.
28. F-A Wollmann, R Kuras and Y Choquet in *Photosynthesis: from Light to Biosphere* (P Mathis, ed.) Vol III, p. 737, Kluwer Academic Publishers, Dordrecht, The Netherlands, 1995.
29. S Iwata, M Saynovits, TA Link and H Michel, Structure 4 (1996) 567.
30. C-A Yu, J-Z Xia, AM Kachurin, L Yu, D Xia, H Kim and J Deisenhofer, Biochim. Biophys. Acta 1275 (1996) 47.
31. TA Link and S Iwata, Biochim. Biophys. Acta 1275 (1996) 54.
32. SE Martinez, D Huang, A Szepaniak, WA Cramer and JL Smith, Structure 2 (1994) 95.
33. G Brasseur, AS Saribas and F Daldal, Biochim. Biophys. Acta 1275 (1996) 61.
34. H Ding, F Daldal and PL Dutton, Biochemistry 34 (1995) 15997.
35. U Brandt, Biochim. Biophys. Acta 1275 (1996) 41.
36. AV Vener, PJM van Kan, A Gal, B Andersson and I Ohad, J. Biol. Chem. 270 (1995) 25225.

Bioenergetics
ed. by P. Gräber and G. Milazzo†


# *CHAPTER 10*
# Cytochrome *c* oxidase

Peter R. Rich and A. John Moody

*Glynn Research Foundation, Bodmin, Cornwall, United Kingdom*



The following convention is used for haem groups: when referring to a chemical class of haem compound, upper case letters are used e.g. haem A or haem C; when referring to a haem group in a specific protein location, lower case letters are used e.g. haems *a* and $a_3$ are A-type haems which are located in different positions in the subunit I polypeptide of cytochrome *c* oxidase.

## 1. Introduction

Cytochrome *c* oxidase is the terminal electron transfer complex of the mitochondrial electron transfer chain. It catalyses the transfer of electrons from cytochrome *c* to molecular oxygen. Water is produced and protons are consumed according to the following equation:

$$4 \text{ cyt } c^{2+} + O_2 + 4 \text{ H}^+ \rightarrow 4 \text{ cyt } c^{3+} + 2 \text{ H}_2\text{O} \qquad (1)$$

This chemical reaction would normally be highly exergonic, resulting in the liberation of excess energy as heat. However, the biochemical mechanism used by cytochrome oxidase to catalyse the reaction results in most of the excess energy being conserved in the form of a protonmotive force which can be used to drive other energy-requiring process such as ATP synthesis. The energy of the chemical reaction is conserved as a protonmotive force in two ways. Firstly, the spatial organisation of the enzyme is such that charges pass across the membrane as catalysis occurs and protons are taken up from one side of the membrane only. This action of the enzyme was originally envisaged by Mitchell [1]. Secondly, the chemical reaction cycle results in protons, other than those required for water production, being forced from one side of the membrane to the other, an action first detected by Wikström [2]. These processes result in transmembrane differences in charge and in proton concentrations - the two elements of the protonmotive force [1]. Hence the enzyme is said to be *protonmotive.*

The enzyme, particularly the mammalian mitochondrial form, has been intensively studied with a range of biochemical, biophysical and genetic techniques. Recently, progress has been enhanced by the study of simpler bacterial oxidases which are structurally related to cytochrome oxidase and which form a "superfamily" of homologous terminal oxidases (Fig. 1). Particularly surprising has been the finding that another major type of bacterial terminal oxidase, cytochrome *bo,* is also a member of this superfamily. It has become clear that the structure and mechanism of its oxygen-reducing and proton-translocating centres are remarkably simi-

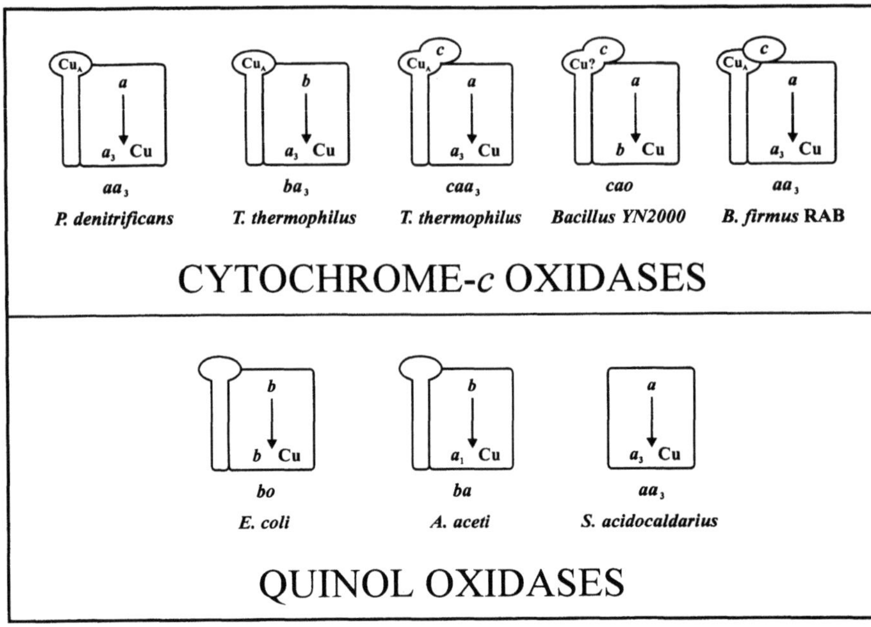


Fig. 1. The oxidase superfamily. A schematic diagram showing the metal centres present in bacterial terminal oxidases. Subunit I and II, where present, are shown, as is the co-purifying cytochrome *c* in the case of *B. firmus* RAB, together with the appropriate haem and copper centres. Note that the relative positions of the low spin haem and the binuclear centre are not intended to indicate their likely topology (see Fig. 6). After Saraste et al. [4].

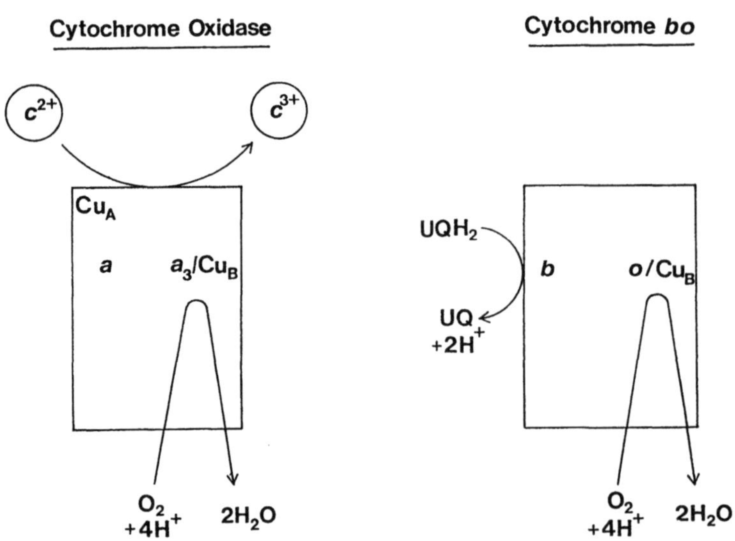


Fig. 2. Comparison of the overall reactions of cytochrome *c* oxidase and cytochrome *bo*.

lar, even though it is a ubiquinol oxidase rather than a cytochrome *c* oxidase (Fig. 2).

In this review our discussions will be centred around the mammalian mitochondrial form of cytochrome *c* oxidase, but important complementary data from other systems is included throughout. Unless stated otherwise, references to specific spectroscopic and other features, and the numbering system for amino acid residues, will refer to the bovine mitochondrial enzyme. The aim is to provide a basic overview of the major areas of research into the structure, function and electrochemistry of this central class of energy-transducing enzymes. For those readers who wish to enquire further, reference [3] is an invaluable guide to the extensive literature on cytochrome *c* oxidase up to about 1980, and many aspects of recent progress in the elucidation of the structure [4–12] and mechanism [13–21] of the enzyme have been reviewed in detail elsewhere.

## 2. Structure

### *2.1. Polypeptides and topology*

Mammalian mitochondrial cytochrome *c* oxidase spans the inner mitochondrial membrane and is a protein complex composed of 13 polypeptides. The three largest (Subunits I, II and III; $M_r$ 57, 26 and 30 kDa, respectively) are encoded in the mitochondrial genome. The rest, which range in size from 5 kDa for subunit VIII to 17 kDa for subunit IV, are nuclear-encoded. Subunits I, II and III are common to all eukaryotic cytochrome oxidases, and to many from prokaryotes, and form the "catalytic core" of the enzyme. The number of additional subunits is variable between organisms. They have no known catalytic functions. They might have a structural role, perhaps being required for correct assembly or for anchoring the enzyme in the membrane in some way, or a regulatory role, by modulating the activity of the enzyme in response to the binding of allosteric effectors. The notion of subtle mechanistic influences is partly supported by the occurrence of enzyme isoforms. These are often tissue-specific, e.g. "liver-type" and "heart-type" bovine oxidase have different subunits $VI_a$ and VIII, but there is also evidence that two forms may occur within a single tissue, e.g. in rat heart and in human skeletal muscle. The presence of isoforms is not confined to mammals: the slime mould, *D. discoideum,* for example, has two forms of oxidase which differ in the smallest subunit and which are expressed differently under different physiological conditions.

The enzyme is likely to be dimeric *in situ,* and is often dimeric when isolated. The possibility has been considered that this might have functional significance. However, it appears that the monomeric reaction cycle is rapid and coupled to proton translocation, a deduction supported by the finding that the protonmotive cytochrome *bo* from *E. coli* appears to be monomeric in the natural membrane. Perhaps the most likely reason for the dimeric structure of many respiratory and photosynthetic complexes, including the terminal oxidases, is the enhanced opportunity to shield the biochemical reactions from unwanted side-reactions by the binding together of the faces of the protein which are closest to the redox centres.

A general model of the juxtaposition of subunits of the mitochondrial enzyme has been proposed, based mainly on chemical cross-linking and antibody binding studies. Furthermore, a gross topological structure has been deduced from electron microscopy of two-dimensional membrane crystals. It is highly asymmetric, having a large domain projecting up to 5 nm into the cytoplasmic aqueous phase but with almost no projection into the matrix phase. A more refined topology at 20 Å resolution has been derived from electron diffraction studies (Fig. 3). Unfortunately, there are some inconsistencies in these topological models and none throw light on the location of the metal centres. Additionally, since only 50–80% of the known enzyme mass can be accommodated within the observed electron densities, the possibility of further as yet undetected features remains.

### 2.2. Metal content

Mammalian cytochrome *c* oxidase contains copper, iron, zinc and magnesium with the stoichiometry of 2.5–3Cu/2Fe/1Zn/1Mg per monomer, i.e. per 13 polypeptide unit. The irons are present as two haems A (Fig. 4), haem *a,* which is low spin, and haem $a_3$, which is mostly high spin. One copper ion, $Cu_B$, is within 5 Å of haem $a_3$, and together with haem $a_3$ forms a strongly interacting binuclear centre which is the site of oxygen reduction. Haem *a* and the other copper centre, $Cu_A$, are involved in transferring electrons from cytochrome *c* to the binuclear centre. The functions of the remaining metals are not certain. It has been suggested that the additional copper, sometimes termed $Cu_X$, is associated with subunit III and is involved in stabilising the dimeric form of the enzyme. However, this was at a time when the $Cu_A$ centre was thought to consist of a single copper atom. It now seems most likely that it is a binuclear $Cu^{2+}Cu^{1+}$ pair, which would itself account for the additional copper. Evidence for the binuclear $Cu_A$ site arises from the striking similarities of its EPR spectrum with the known bimetallic copper centre of nitrous oxide reductase; from the sequence similarities of the copper-binding motifs; from the inability to simulate the fine structure of the $Cu_A$ EPR spectrum on the basis of a mononuclear copper species; from the isolation and me-

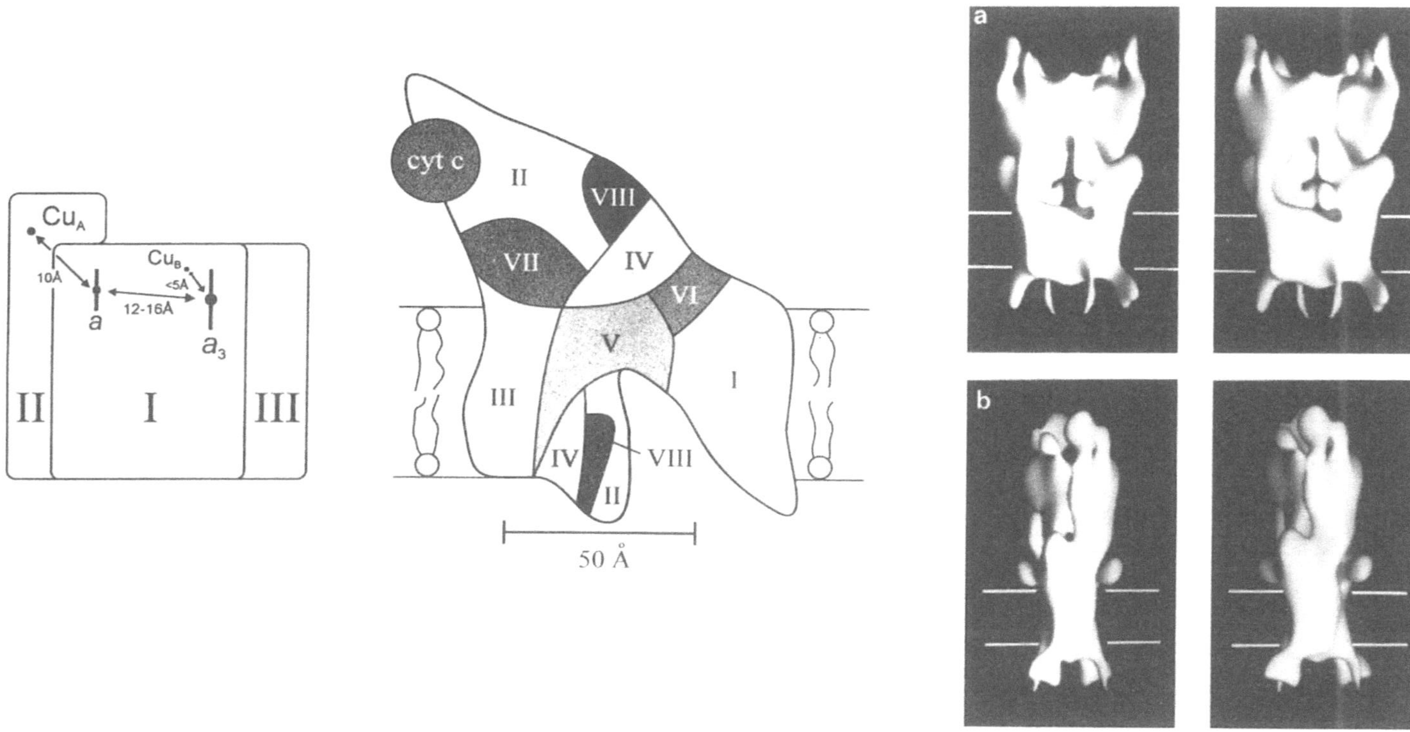


Fig. 3. Summary of topological aspects of mammalian cytochrome *c* oxidase. Several aspects of topology are illustrated. The relative positions of subunits and the overall topology are reproduced with permission from [9] and [35], respectively. A summary of the relative positions of the metal centres, deduced from a variety of sources, is also included.

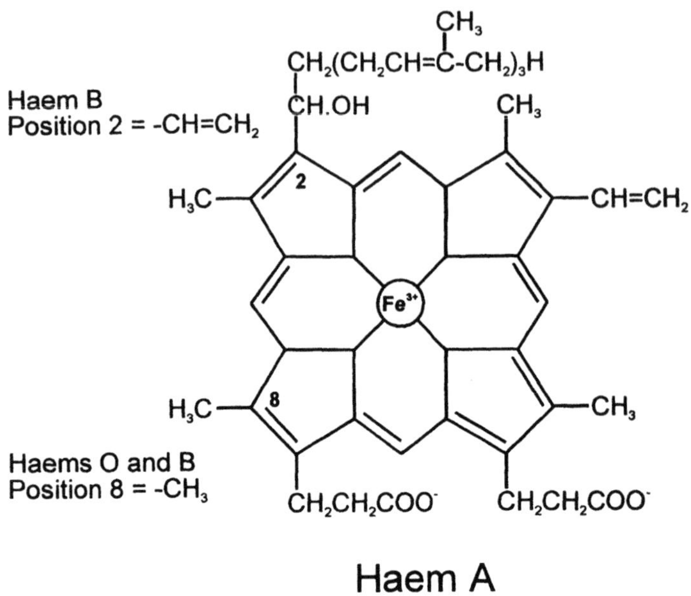


Fig. 4. Structures of haem types A, B and O [36].

tal quantitation of the soluble copper-binding domain of cytochrome oxidase from *Paracoccus denitrificans;* and from the demonstration that two copper atoms bind to a modified form of subunit II of cytochrome *bo* when the putative copper-binding motif is inserted by genetic manipulation.

The zinc, by bridging between a pair of cysteine residues, one on subunit $VI_a$ and the other on subunit $VI_b$, may also have a structural role, and a role for the magnesium in an ATP binding site has been suggested.

### *2.3. Relationship to bacterial terminal oxidases*

Many bacteria, e.g. *Paracoccus denitrificans,* possess a cytochrome *c* oxidase which is similar to the mammalian one, with three subunits which show homology with subunit I, II and III of the mammalian enzyme. However, other bacterial terminal oxidases are found in which there are variations on the common structural theme of a low spin haem and a Cu/high spin haem binuclear centre (Fig. 1). Overall, bacterial oxidases of this type can be subdivided on a functional basis into those that accept electrons from cytochrome *c* and possess a copper centre analogous to $Cu_A$, and those that accept electrons from quinols and do not have a $Cu_A$ analogue. Of the latter type, cytochrome *bo* from *Escherichia coli* has studied most extensively and, like several bacterial cytochrome *c* oxidases, has been shown to be proton translocating (Fig. 2). Cytochrome *bo* contains haem O, which is a modified form of haem B with a hydroxyethylfarnesyl side-chain at position 8. Haem A is structurally close to haem O, but has, in addition, a formyl group at position 2 (Fig. 4). Both haem A and haem B/O may occur within the same enzyme, either as the low spin and high spin haem components, respectively, as in a quinol oxidase from *Acetobacter aceti,* or *vice versa* as in a cytochrome *c* oxidase from *Bacillus YN-2000.*

Some bacterial Cu/Fe oxidases, in addition, contain stoichiometric amounts of haem C. An enzyme from *Thermus thermophilus* has two subunits. One subunit consists of a fusion of a cytochrome *c* domain onto the carboxyl terminal of a subunit II that is otherwise like subunit II of mammalian cytochrome *c* oxidase. In contrast, in a three-subunit oxidase from *Bacillus firmus RAB,* the third subunit appears to be a conventional low molecular weight ($M_r$ = 14 kDa) cytochrome *c* that is tightly bound and co-purifies with the enzyme.

### *2.4. Three-dimensional structure*

Only two of the core subunits have redox-active catalytic sites – subunit I, which contains haem *a* and the binuclear copper-iron centre which is the site of oxygen reduction, and subunit II. Subunit II appears to be essential for catalytic activity, although the sequence comparisons of this subunit from different species show much less conservation than is observed with subunit I. It is likely to have two membrane-spanning α-helices (Fig. 5) and a globular hydrophilic part located on the positively charged side of the membrane. In those oxidases of the superfamily which oxidise cytochrome *c,* this globular part contains the EPR- and optically-visible $Cu_A$ centre, together with at least a part of the docking site for cytochrome *c.* However, in members of the oxidase superfamily which oxidise a quinol rather than cytochrome *c,* the binding sites for cytochrome *c* and $Cu_A$ are absent, and the role of subunit II is not known.

Subunit III does not contain redox active centres and appears not to be essential for catalytic activity although, like the nuclear-encoded subunits, it may be important for the assembly and/or stability of the enzyme, or may have a role in modulating the function of the oxidase. It is a very hydrophobic polypeptide with 5–7 predicted membrane-spanning regions (Fig. 5).

Subunit I is the most highly conserved polypeptide of the oxidase superfamily and is the binding site for three metal redox centres: the low spin, six-coordinate haem (haem *a* in the mammalian enzyme) and the haem-copper binuclear centre (haem $a_3$ and $Cu_B$ in the mammalian enzyme). Subunit I from bovine mitochondria has 12 predicted transmembrane helices (Fig. 5) whereas the *E. coli bo*-type oxidase may have 15.

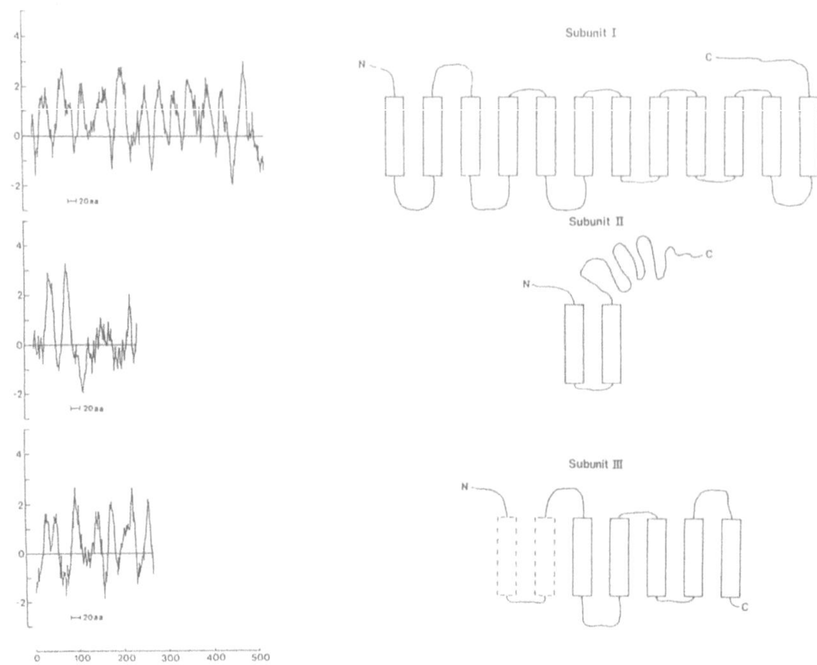


Fig. 5. Predicted polypeptide folding of subunits I, II, III of cytochrome *c* oxidase. Hydropathy plots of the sequences of subunits I, II and III of the bovine enzyme are shown. The bar represents 20 amino acids, the length required for a membrane-spanning α-helix. Possible folding models of the proteins derived from such plots are also shown.

From the comparison of a wide range of primary sequences, it was noted that only seven histidine residues were highly conserved throughout the superfamily. This constrained possible structural models, since six of these are probably required as ligands to the metal centres – two for haem *a*, one for haem $a_3$ and three for $Cu_B$. A structural model analogous to the probable haem arrangement in cytochrome *b* of *bc* complexes was originally proposed in which the two haem groups are held between the same pair of α-helices and span the membrane.

More recently, six of the conserved histidines have been assigned to specific metal centres and the seventh has been ruled out as a metal ligand, mainly from results obtained by site-directed mutagenesis of bacterial enzymes. A model of subunit I structure in which the haem groups are held at roughly the same level in the protein structure between different pairs of α-helices has been developed. This relatively compact grouping of the metal centres in subunit I is situated towards the cytoplasmic surfa-

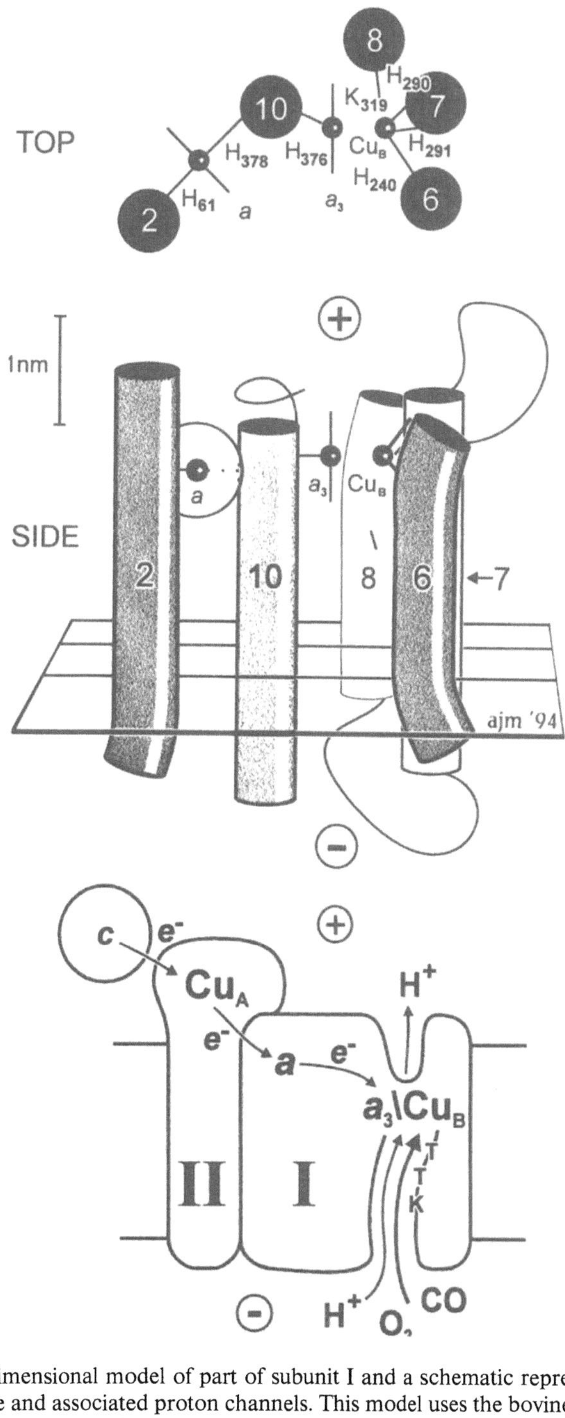


Fig. 6. Three-dimensional model of part of subunit I and a schematic representation of the binuclear centre and associated proton channels. This model uses the bovine sequence numbering, and is based on the figure in [37].

ce, with a proton channel connecting it to the negative aqueous phase so that appropriate protonation reactions can occur (see section 6) in concert with the electron transfer processes (Fig. 6).

At present we have little idea of how subunits II and III interact with subunit I, or of how the other helices of subunit I might be arranged.

## 3. Spectroscopy

The illustrative data used in this review come mainly from optical and EPR spectroscopy, so a brief description of near-UV, visible, near-IR and EPR spectra of cytochrome oxidase is warranted. However, a wide range of other spectroscopic techniques have been successfully applied, including Resonance Raman (RR), magnetic circular dichroism (MCD), electron nuclear double resonance (ENDOR), Mössbauer, Fourier transform infra red (FTIR) and extended X-ray absorption fine structure (EXAFS). Information on the contributions made by these techniques to our understanding of the enzyme can be found in the reviews cited in section 1. The additional reviews in [22–27] may also be of use.

### *3.1. Near-UV/visible/near-IR spectroscopy*

The near-UV/visible spectrum of cytochrome oxidase (Fig. 7, A) is dominated by typical absorption bands of the two haems. The principal bands of the oxidised form of the enzyme have maxima at 425 nm (the Soret band) and 598 nm (the alpha or visible band), although these positions can vary in different forms of the isolated enzyme (see section 4). On reduction, these bands change in their position and intensity in a manner typical of other haem compounds, giving rise to the characteristic reduced *minus* oxidised difference spectrum shown in Fig. 7, B. The contributions of haems $a$ and $a_3$ to this difference spectrum differ considerably, mainly because haem $a$ is low spin whereas haem $a_3$ is mostly high spin. Both contribute about equally to the peak at 444 nm but most (> 80%) of the band at 606 nm is caused by haem $a$.

A weak absorption band around 655 nm is also seen in the spectrum of the oxidized enzyme. It arises from a charge transfer transition between the porphyrin ring of haem $a_3$ and its high spin ferric iron. The band is lost on reduction or when haem $a_3$ becomes low spin (see section 4), and is also sensitive to the ligation state of ferric haem $a_3$.

The near-IR spectrum of oxidized cytochrome oxidase (Fig. 7, C) is dominated by a broad absorption band attributed to $Cu_A$. This band is lost when the $Cu_A$ is reduced and other, weaker bands, at 710 and 784 nm, appear. The 784 nm band is attributed to a charge transfer transition in high spin ferrous haem $a_3$ and the 710 nm band may be associated with ferrous haem $a$.

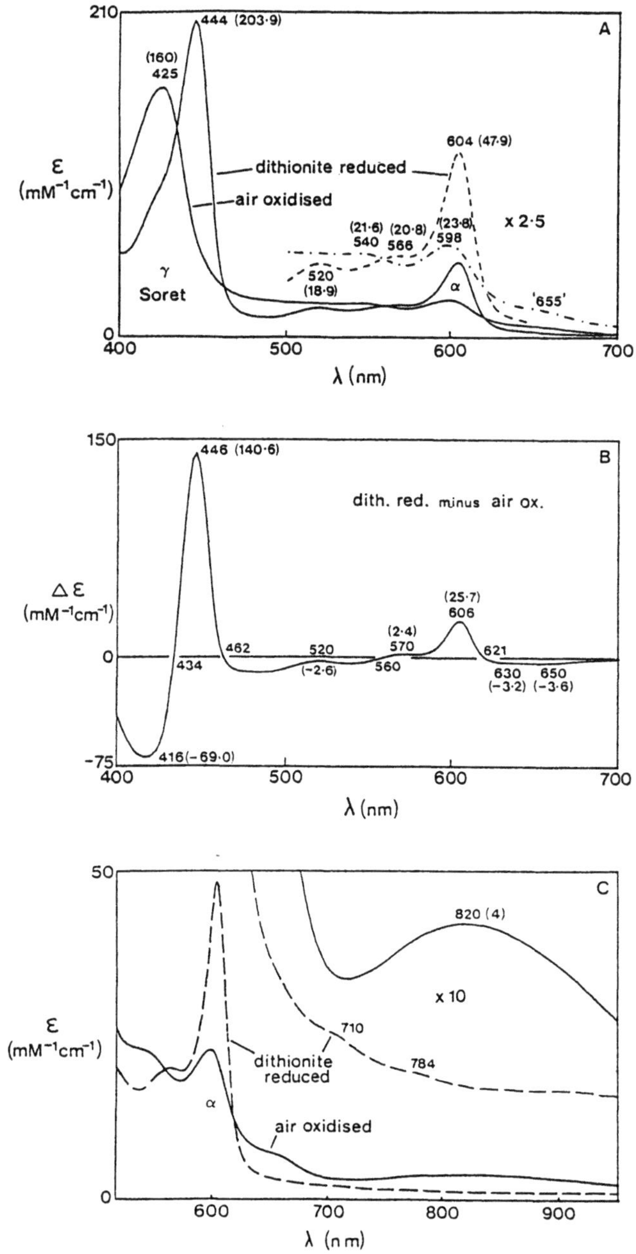


Fig. 7. Optical spectra of cytochrome *c* oxidase. A. Near-UV/visible spectra of the enzyme from bovine heart as prepared, and after anaerobic reduction with excess sodium dithionite. Peak positions are labelled, with extinction coefficients (relative to a value of 160 $mM^{-1}s^{-1}$ for the Soret maximum in fully-oxidized enzyme) shown in brackets. B. The dithionite-reduced minus air-oxidized difference spectrum. C. Visible/near-IR spectra of the enzyme as prepared, and after anaerobic reduction with excess sodium dithionite (redrawn from [38]).

### *3.2. Electron Paramagnetic Resonance (EPR) spectroscopy*

All the metal centres in fully oxidized cytochrome oxidase might be expected to be paramagnetic. However, the low temperature EPR spectrum of the oxidized enzyme consists mainly of resonances attributed only to low spin ferric haem *a*, at $g$ = 3.0, 2.2 and 1.4, and to $Cu_A^{2+}$, in the $g$ = 2 region (Fig. 8). Haem $a_3^{3+}$ and $Cu_B^{2+}$ are EPR silent because their spins are coupled. However, a typical spin haem signal at $g$ = 6 which arises from

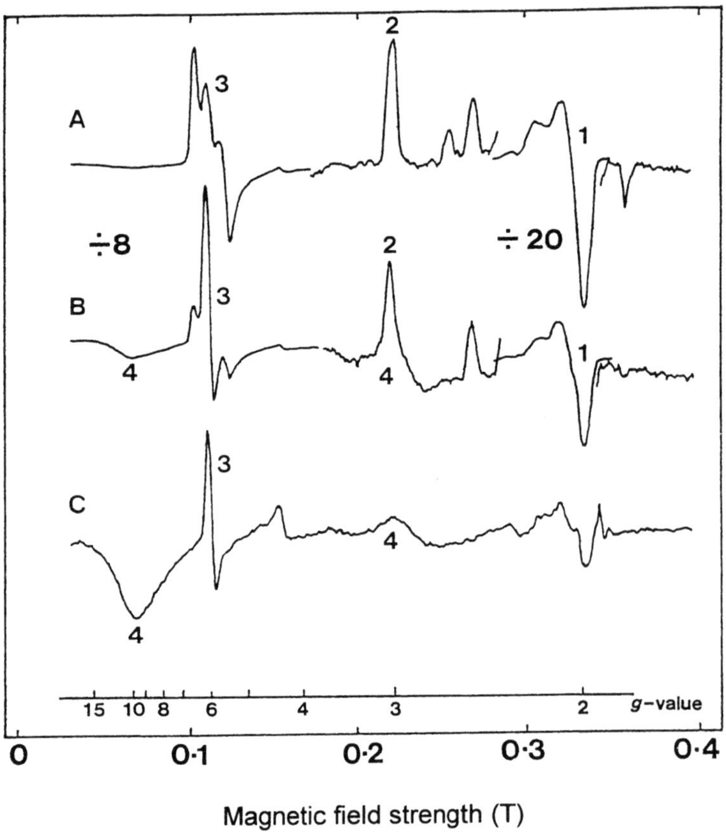


Fig. 8. Electron paramagnetic resonance (EPR) spectra of cytochrome *c* oxidase at pH 7.4. A. Enzyme (300 μM) partially reduced (2 $e^-/aa_3$) with NADH and phenazine methosulphate as mediator under anaerobic conditions. B. After anaerobic addition of 1 M sodium formate to A. C. Enzyme (180 μM) reduced with ascorbate and cytochrome *c* in the presence of formate. In A. signals from $Cu_A^{2+}$ (1), low-spin haem $a^{3+}$ (2), and high-spin ferric haem (3) can be seen. In B. and C. the $g'$ = 12 (and $g'$ = 2.95) signal (4) associated with "slow" oxidase can be seen (but here induced by formate). The *g* value of this signal is dependent on microwave frequency, the true value being 8. It was first described at $g'$ = 12 because it crosses the baseline at this value when observed at 9 GHz microwave frequency (X-band), as in this case. Spectra redrawn from [39].

haem $a_3^{3+}$ can be observed if this coupling is broken by reduction of the $Cu_B$. EPR signals from $Cu_B^{2+}$ are not normally observed, even when the enzyme is partially reduced, and can only be observed in unusual conditions.

Some preparations of oxidized cytochrome oxidase also show an unusual broad EPR signal in the $g = 12$ region, together with a feature at $g = 2.95$, that underlies the $g = 3$ signal from low spin haem $a^{3+}$. Although there is uncertainty as to the metal/ligand chemistry that produces such a signal, it is certain that it arises from the binuclear centre of one form of the enzyme, termed the "slow" form, which can accumulate during purification. The properties and origin of different forms of the enzyme are discussed in the following section.

## 4. Ligand binding reactions

The binuclear centre of cytochrome oxidase, as well as reacting with dioxygen, can react with a variety of other ligands, many of which are inhibitory. The strength of binding depends on the redox state of the binuclear centre. For example, cyanide ($CN^-$) binds tightly when haem $a_3$ is oxidised, but more weakly when haem $a_3$ is reduced ($K_d = 10^{-7}$–$10^{-8}$ M and $10^{-3}$–$10^{-4}$ M, respectively), whereas formate ($HCOO^-$) binds only when both haem $a_3$ and $Cu_B$ are oxidised ($K_d = 10^{-3}$ M), and carbon monoxide (CO) binds only when both haem $a_3$ and $Cu_B$ are reduced ($K_d = 10^{-6}$–$10^{-7}$ M). These different affinities for different redox states of the enzyme are reflected in the very large effects that ligands can have on the equilibrium redox potentials of the metal centres (section 7).

### *4.1. Spectroscopic effects of ligand binding*

Most ligands induce changes in the spectroscopic features of cytochrome oxidase, often as a result of spin state changes in haem $a_3$. Cyanide, carbon monoxide and nitric oxide (NO), all of which can bind to the fully reduced enzyme, induce a high spin to low spin transition in haem $a_3^{2+}$, resulting in a shift of the Soret band to a higher wavelength (red shift) and the appearance of a strong alpha band from low spin haem $a_3^{2+}$ in the 585–600 nm region, depending on the ligand (e.g. 590 nm for cyanide, Fig. 9, A). Ligands which bind to the fully oxidized enzyme can induce a high spin or a low spin shift of haem $a_3^{3+}$. For example, cyanide binding promotes the low spin state and results in a red shift of the Soret band and an increased size of the alpha band, whereas formate favours the high spin state and results in a blue shift of the Soret band and a decreased size of the alpha band (Fig. 9, B).

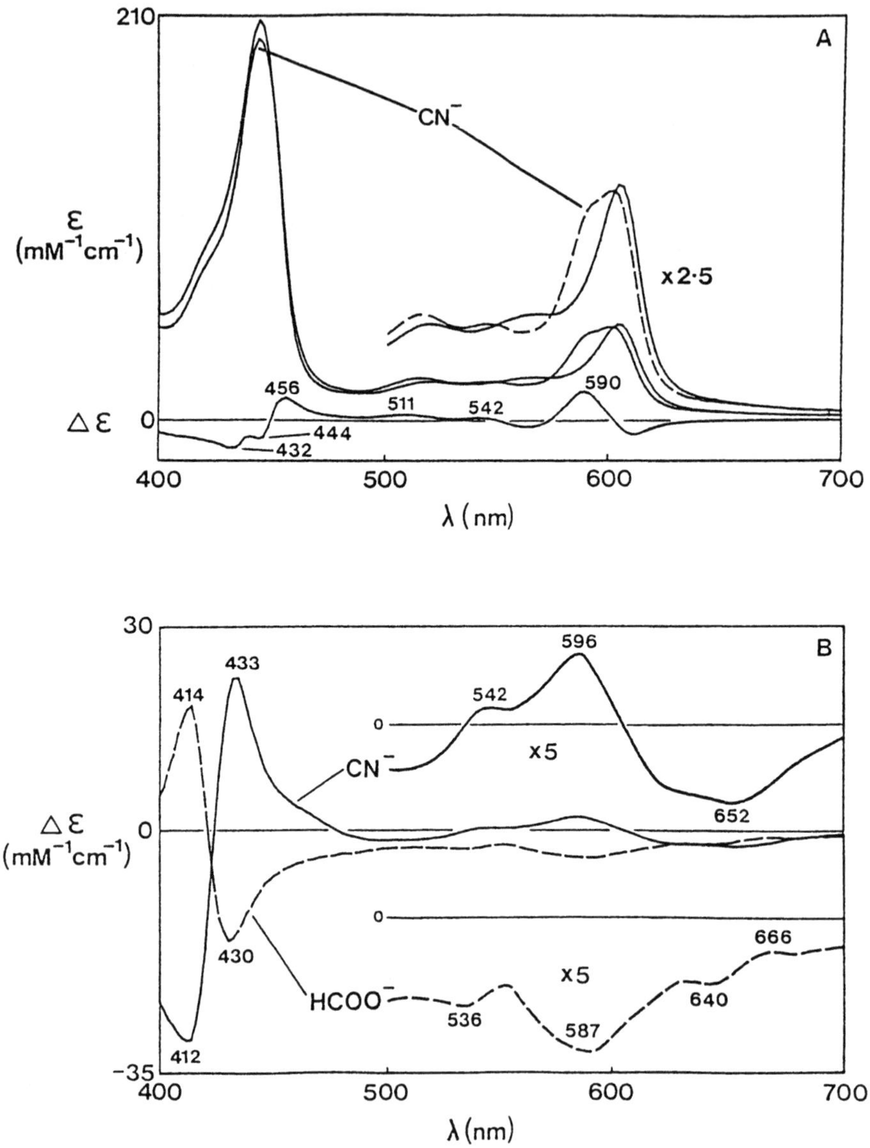


Fig. 9. Effects of ligands on the optical spectrum of cytochrome *c* oxidase. A. Near-UV/visible spectra before and after addition of 40 mM potassium cyanide to enzyme at pH 6.5 that had been reduced and made anaerobic with excess dithionite. The lower spectrum is the difference between these two, i.e. the cyanide-binding spectrum. The peak at 456 nm and trough at 432 nm in the Soret region are consistent with a shift in $\lambda_{max}$ of haem $a_3^{2+}$ from 444 to 446 nm. The trough at 444 nm is caused by a slight oxidation (< 5%) of haem $a_3$. B. Difference (binding) spectra obtained on incubation of oxidized enzyme with 20 mM cyanide or formate. The peaks and troughs in the Soret region are consistent with a shift in haem $a_3^{3+}$ to pure low spin, with $\lambda_{max}$ at about 428 nm, with cyanide, and to pure high spin, with $\lambda_{max}$ at about 415 nm, with formate.

Some ligands have more complicated effects. For example, low concentrations of azide ($N_3^-$) induce spectral changes similar to those induced by formate, but at high concentrations an additional change occurs, and the Soret band becomes red shifted.

### *4.2. Sites of ligand binding*

The "free" axial position of pentacoordinate high spin haem $a_3$ is the likely ligand-binding site in cytochrome oxidase and, certainly, the effects of ligands on the spectroscopic features of haem $a_3$, by comparison with similar ligand effects on simpler haemoproteins, such as metmyoglobin, are broadly consistent with this view: cyanide, a strong-field ligand, would be expected to form a low spin complex with $Fe^{3+}$, whereas a weak-field ligand, such as formate, would be expected to form a high spin complex.

However, $Cu_B$ may also play a part in ligand binding. The possibility of ligand-binding to both haem $a_3$ and $Cu_B$ leads to some complications, for example:

(i) Haem $a_3$ and $Cu_B$ may be occupied simultaneously by the same type of ligand, e.g. sulphide ($S^{2-}$ or $SH^-$) is thought to ligate both haem $a_3^{3+}$ and $Cu_B^+$ at the same time, with the low spin haem $a_3^+$-$S^{2-}$ exhibiting characteristic EPR resonances at $g$ = 2.6, 2.2 and 1.9.

(ii) Haem $a_3$ and $Cu_B$ may be bridged by a single ligand. For example, there is evidence from MCD and FTIR spectroscopies that cyanide may be bridged between $Cu_B$ and haem $a_3$ when cytochrome oxidase is fully oxidized. The cyanide is ligated to the iron by its carbon atom and the haem $a_3$ becomes low spin. The structure is EPR silent, but a band at 1946 nm arising from the cyanide adduct of low spin ferric haem $a_3$ can be observed by MCD which has magnetic properties consistent with a ferromagnetic interaction with $Cu_B$.

(iii) Haem $a_3$ and $Cu_B$ may be occupied simultaneously by different types of ligand, e.g. cyanide and nitric oxide are thought to occupy haem $a_3^{3+}$ and $Cu_B^{2+}$, respectively, at the same time. The NO ligand on $Cu_B$ breaks the antiferromagnetic coupling, and the low spin haem $a_3^{2+}$-$CN^-$ exhibits a characteristic EPR resonance at $g$ = 3.5.

(iv) A single ligand may bind either to haem $a_3$ or to $Cu_B$ depending on conditions, e.g. after photodissociation from haem $a_3^{2+}$, carbon monoxide has been shown by fast time-resolved spectroscopies and by FTIR at low temperatures to move first to $Cu_B^+$, before equilibrating with the bulk solution.

It should be noted that the ligand binding sites of both haem $a_3$ and $Cu_B$ are likely to have $H_2O$ or $OH^-$ bound normally and that these are displaced when the other ligands bind.

### *4.3. Electron transfer reactions with ligands*

Several ligands, in addition to binding, may undergo electron transfer reactions with cytochrome oxidase. For example, hydrogen peroxide binds initially to form a "peroxy" compound. A second hydrogen peroxide then acts as electron donor to generate further species. Carbon monoxide can donate two electrons to fully oxidized cytochrome oxidase, yielding carbon dioxide and an enzyme with two electrons which can proceed to bind either an oxygen or a second carbon monoxide molecule.

### *4.4. Kinetics of ligand binding: Enzyme heterogeneity*

Preparations of cytochrome oxidase often show a marked heterogeneity with respect to ligand-binding reactions. For example, up to three kinetic phases of cyanide binding to oxidized enzyme can be discerned, with second order rate constants at pH 6.5 of about 2, 0.2, and 0.02 $M^{-1}s^{-1}$, respectively. These three phases appear to arise from three distinct subpopulations of the enzyme:

(a) "Fast" oxidase, which is responsible for the fast binding phase, seems to be the fully-active, native form of the enzyme. In this form, haem $a_3$ appears to be in a mixture of high spin and intermediate spin states. The Soret maximum shifts from 424 to 417 nm on binding formate (high spin complex), and from 424 to 428 nm on binding cyanide (low spin complex) (Fig. 9, B).

(b) "Slow" oxidase, which is responsible for the slow binding phase, may appear during the isolation procedure if the pH is too low. Indeed, "fast" oxidase spontaneously converts to "slow" oxidase on incubation at pH values below about 8 (Fig. 10, A). In the "slow" form, haem $a_3$ seems to be entirely high spin, and the Soret maximum is about 418 nm. Other characteristics of "slow" oxidase include: (i) unreactivity with hydrogen peroxide and carbon monoxide; (ii) slow reduction of the binuclear centre; and (iii) the $g = 12$ EPR signal noted before. These properties are similar to those of "fast" oxidase after treatment with formate. Hence, it has been suggested that an endogenous ligand similar to formate, e.g. a glutamate side chain of the protein, might be affecting the binuclear centre of the "slow" form. "Slow" oxidase can be fully reactivated; once the binuclear centre is reduced, the endogenous ligand is displaced and the enzyme displays properties typical of the "fast" form of the enzyme.

(c) The chloride-ligated form of oxidase appears to be responsible for the intermediate phase of cyanide binding in at least some prepa-

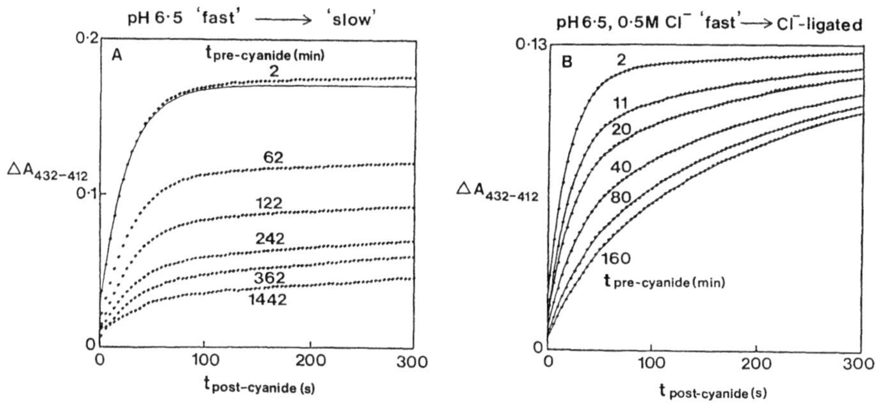


Fig. 10. The conversion of the "fast" form of cytochrome oxidase into the "slow" and chloride-ligated forms as monitored using the kinetics of cyanide binding. A. "Fast" oxidase was incubated at pH 6.5 and 20°C and samples taken for measurement of the cyanide binding kinetics at the times indicated. Note the loss of the fast phase of cyanide binding and the appearance of a much slower phase. B. as in A but with 0.5 M KCl present. Note in this that the loss of the fast phase is accompanied by the appearance of intermediate phase of binding. Data taken from [40].

rations where chloride is present during the isolation procedure. "Fast" oxidase is converted to chloride-ligated form by incubation with chloride, again at pH values below about 8. The spin state of haem $a_3$ in the chloride-ligated form appears to be the same as haem $a_3$ in the "fast" form, in that the Soret maximum is relatively unaffected by chloride, but the chloride-ligated form, as well as showing modified cyanide-binding kinetics (Fig. 10, B), is unreactive towards formate and carbon monoxide.

The ability to prepare such forms of the enzyme was originally widely interpreted in terms of forms which had different global conformational states of the protein. However, it now seems that all of these properties are most reasonably accounted for simply in terms of the ligand and the redox state of the binuclear centre itself.

## 5. Electron transfer reactions

### *5.1. Reaction with cytochrome* c

The electron source for cytochrome *c* oxidase is cytochrome *c*. This small, water soluble, protein is located in the aqueous space between the outer and inner mitochondrial membranes. It contains one covalently-linked haem C group in a protein structure which has been well-resolved and intensively studied as a model haem-containing protein. Cytochrome $c$ is itself reduced by cytochrome $c_1$ of the cytochrome $bc_1$ complex and diffuses in its reduced form to cytochrome oxidase where it is reoxidised. It therefore acts as an electron shuttle between the last two protein complexes of the mitochondrial respiratory chain.

Chemical cross-linking studies gave the first indications that the binding site for cytochrome $c$ is located on subunit II. A motif typical of a copper binding site can be identified in this region (Fig. 11) and this is likely to be the site of ligation of $Cu_A$. This deduction is supported by the fact that the motif is absent from the sequences of subunit II of the quinol oxidases, which do not have a $Cu_A$ centre. A cytochrome $c$ binding site close to $Cu_A$ seems reasonable since there is growing evidence that $Cu_A$ is the primary acceptor of electrons from cytochrome $c$, based upon rapid kinetic measurements of both forward and reverse electron transfer rates between components, and based on the fact that $Cu_A$ is located close to the aqueous phase in which cytochrome $c$ is dissolved. The view is strengthened by some bacterial $caa_3$ oxidases, where the cytochrome $c$ is an integral part of the enzyme, being present as a C-terminal extension of subunit II.

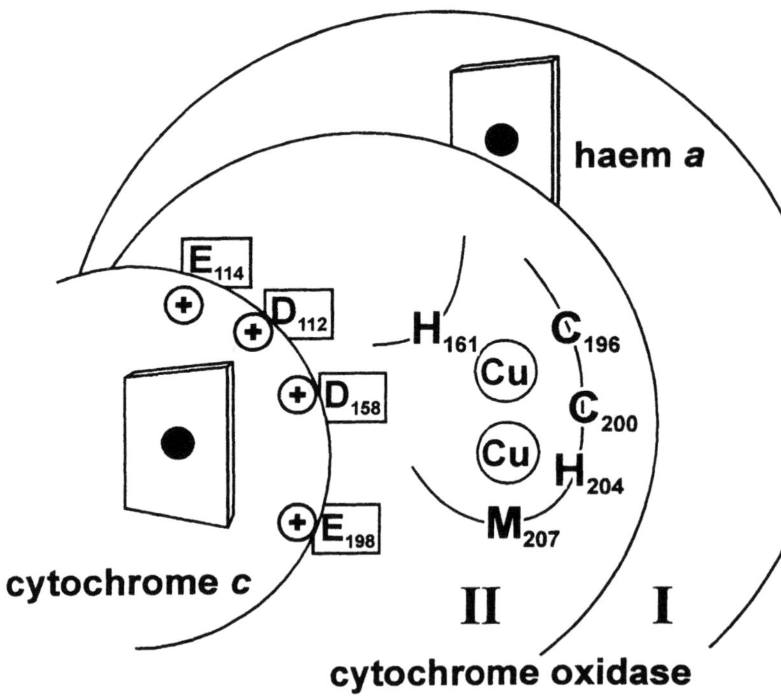


Fig. 11. Predicted structural model of subunit II and the binding site for cytochrome $c$. Redrawn from [12], with permission, and updated with information from [41].

The reaction of cytochrome *c* with mammalian cytochrome *c* oxidase is ionic strength dependent, indicative of important electrostatic effects in the mechanism of binding, although some hydrophobic interactions also appear to be involved. The interaction has been shown to involve a ring of lysine residues around the haem edge of cytochrome *c*, and these must form ion pairs with negatively charged residues on the oxidase. From the shielding effect of cytochrome *c* on the labelling of oxidase carboxylate groups by a water-soluble carbodiimide, three possible candidates have been identified in subunit II viz. Glu 129, Asp 131 and Glu 198. One of these is close to the putative $Cu_A$ binding site and the other two are in another highly conserved region of subunit II.

There remain a number of controversies and complications in the reaction between cytochrome *c* and $Cu_A$ which are not yet resolved. Firstly, some groups still consider that direct electron transfer from cytochrome *c* to haem *a* may occur, either as the only route or as a second route of electron entry. In our opinion, these possibilities are unlikely, but cannot yet be definitively ruled out. A second complication which may relate to this is the phenomenon that an Eadie-Hofstee plot of oxidation rates versus cytochrome *c* concentrations indicates that there may be two binding sites for cytochrome *c* of low and high affinities. The kinetics of oxidation of cytochrome *c* exhibit other phenomena which indicate complications. These details and their interpretation are beyond the scope of the present article and have been reviewed elsewhere [18–20]. Nevertheless, it seems reasonable to accept provisionally that the major site for binding of cytochrome *c* is a negatively-charged surface of subunit II close to the binding site, and that electrons are initially donated to the $Cu_A$ centre before they are passed to haem *a* and hence onto the binuclear centre.

### *5.2. Internal electron transfer to the binuclear centre*

From the above considerations on $Cu_A$, it seems likely that haem *a* transfers electrons from $Cu_A$ to the binuclear haem $a_3/Cu_B$ centre, a result consistent with measurements of the dielectric location of the redox centres. Because $Cu_B$ appears to be in redox contact with cytochrome *c* even when cyanide is present to prevent haem $a_3$ turnover, and because of the known redox interaction between haem *a* and $Cu_B$ (section 7), we have generally favoured a simple linear electron transfer model with the electron transfer sequence:

$$Cu_A \rightarrow \text{haem } a \rightarrow Cu_B \rightarrow \text{haem } a_3$$

However, the data which show redox contact between $Cu_B$ and cytochrome *c* when haem $a_3$ turnover is prevented by cyanide were all taken on slow (seconds) timescales and so may not be applicable to very rapid internal electron transfer processes. Indeed, some groups favour a more complex, branched, electron transfer sequence in which the electron transfer pathway may change during the four-electron cycle. In the recent consensus structural model of subunit I (Fig. 6), both haem groups are ligated to the same transmembrane helix X, with a close enough proximity of the haem groups to allow rapid (> $10^6\ s^{-1}$), direct interhaem electron transfer. In this model, $Cu_B$ is necessarily on the far side of haem $a_3$, so that the electron transfer sequence above may have to be modified. However, neither the detailed internal electron transfer sequence nor the major structural features of oxidase have been settled at present.

Direct measurements have been made of the rate of haem/haem electron transfer by exploiting the photolability of the carbon monoxide compound of reduced haem $a_3$. When carbon monoxide is bound, the midpoint potentials of haem $a_3$ and $Cu_B$ are substantially raised (section 7). This allows preparation of a "mixed-valence" enzyme in which haem $a_3$ and $Cu_B$ are reduced and haem *a* and $Cu_A$ are oxidised. Laser-induced photodissociation of carbon monoxide results in a lowering of the midpoint potential of the haem $a_3$ to its unliganded value and concomitant electron redistribution between the metal centres, a process which can be monitored optically. In this way it has been established that electrons can be transferred from the binuclear centre to haem *a* with a rate constant of around $10^5\ s^{-1}$ and that they then equilibrate with the $Cu_A$ centre with a rate constant of around $10^4\ s^{-1}$. However, discrimination between a direct pathway of haem/haem electron transfer and one which involves $Cu_B$ has not been possible.

### *5.3. The catalytic intermediates of oxygen reduction*

Electron transfer between haem *a* and the binuclear centre is complicated by the fact that oxygen reduction chemistry takes place in four chemically distinct steps. In each of these, the thermodynamics and chemistry of the intermediates is quite different. There has been considerable progress in identification by spectroscopic techniques of the stable catalytic intermediates of the binuclear centre which are formed during the reaction cycle. These intermediates are shown in Fig. 12 and their principal optical features are shown in Fig. 13. Evidence for the occurrence of these intermediates in the natural catalytic cycle and some of their other properties have been reviewed recently [14].

The first intermediate is the one-electron-reduced or E state. In "fast" oxidase preparations the electron is shared between the haem *a* and the binuclear centre, in a manner consistent with relative redox potentials of the four metal centres (section 7). At this stage oxygen probably does not bind and the electron transfer from cytochrome *c* is reversible.

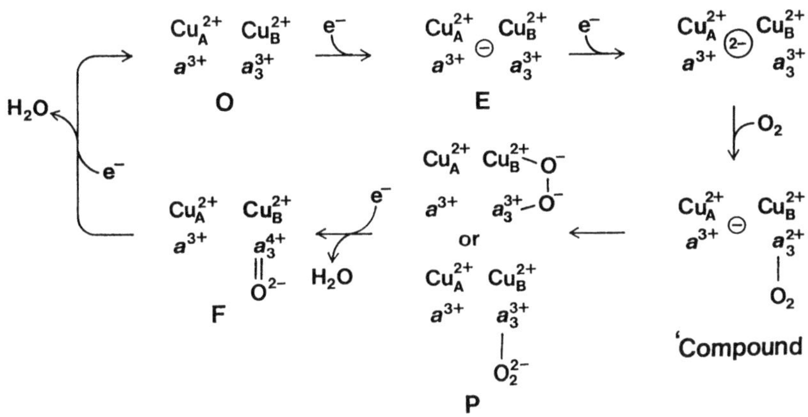


Fig. 12. The catalytic reaction cycle of oxygen reduction. Details of intermediates are given in the text. Protonation changes are not included because of uncertainties of the steps at which they bind (see section 6).

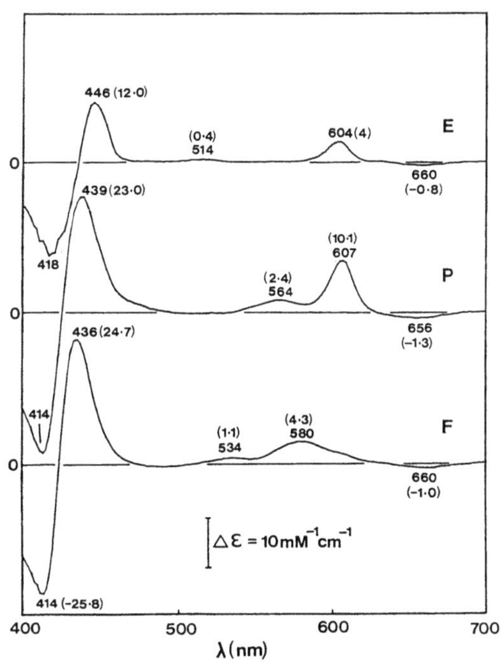


Fig. 13. Optical difference spectra of the 1 $e^-$ (E), 2 $e^-$ (P), and 3 $e^-$ (F), reduced intermediates of cytochrome *c* oxidase relative to the fully-oxidized (O) enzyme. E was generated at pH 8.5 by limited photochemical reduction (< 0.2 $e^-/aa_3$, using phenazine methosulphate/EDTA) of O. P was generated at pH 7.5 by addition of carbon monoxide to O in the presence of oxygen (see section 4). F was generated at pH 8.5 by adding hydrogen peroxide (2 mM) to O (see section 4). Note that there are small contaminations (< 5%) in each case: E by P, P by E, and F by P.

A subsequent electron transfer produces the two-electron-reduced enzyme. In this state, chemical reaction with oxygen occurs. The bound oxygen is reduced to the peroxide level and a stable intermediate, termed the "peroxy" or P state, is formed. This structure may contain a bridging peroxide between the iron and copper metals of the binuclear centre or may be an end-on compound of oxygen and haem (Fig. 12). A third electron transfer from cytochrome *c* causes a further chemical alteration of the binuclear centre and its bound oxygen. It is generally thought that breakage of the oxygen-oxygen bond occurs at this stage yielding a stable oxyferryl or F state. A fourth electron transfer completes the oxygen reduction process with the generation of water and the formation of the starting O state of the enzyme (Fig. 12).

## 6. Proton transfer reactions

The catalytic cycle of Fig. 12 involves four sequential electron transfers. A number of protonation reactions are also associated with this cycle: firstly, the chemical intermediates of the catalytic cycle will themselves be most stable in appropriate protonated forms and, in any case, four protons must eventually be taken up in the formation of two water molecules from the oxygen. Secondly, certain steps in the catalytic cycle are linked to the movement of protons from one side of the membrane to the other. These protons must at some point be associated at least transiently with chemical species within the enzyme.

All protonation sites could be directly associated with the metals and their oxygen compounds, or might be associated with specific amino acid side chains whose properties are modulated by the chemical state of the binuclear centre. It is likely that such amino acids will be very close to the binuclear centre, although distant amino acids which are affected by long range allosteric effects cannot be ruled out. Proton channels may facilitate proton movements from the aqueous phase to protonation sites close to the buried binuclear centre.

### *6.1. Proton changes associated with the catalytic intermediates of oxygen reduction*

Detection of protonation changes of the intermediates can be achieved in one of three ways: a pH-dependency of a spectroscopic feature of an intermediate, a pH-dependency of the midpoint potential of a redox change, or the net change in proton concentration in the medium on transformation of one intermediate into another. All of these methods have contributed some information, although the picture is still incomplete.

Studies of pH-dependencies of midpoint potentials of oxidase components during anaerobic redox titrations have shown that the midpoint potentials of haem *a*, haem $a_3$, and $Cu_B$ all exhibit some pH-dependency, indicative of protonation associated with reduction, whereas redox changes of $Cu_A$ are pH-independent (section 7).

However, this does not help in establishing the protonation changes associated with the catalytic intermediates which occur during turnover (Fig. 12). Such estimations are difficult since equilibrium redox potentiometry of each redox couple cannot be performed. It has generally been thought that no net proton changes occur in the O → P transition, but that two protons are consumed in each of the P → F and F → O steps. These estimates have arisen from measurements of the pH-dependencies of the P/F and F/O ratios in intact, energised mitochondria. More recently, the net proton changes of the individual steps have been measured by direct assay of proton uptake or release in soluble, purified enzyme. In this case it has been concluded that the net protonation changes for O → P, P → F and F → O are 2, 1 and 1, respectively, with no indication of a change in these values between pH 6 and 8.5. These results appear inconsistent with those reviewed in [14], but they may not be completely incompatible if one considers that the measurements in mitochondria were made under conditions where a transmembrane electric field was present. Indeed, if a proton channel of low dielectric strength connects the internal aqueous phase to the sites of protonation, then an electric field along the channel could make the pH at the bottom of the channel far more alkaline than the aqueous medium connected to it. This phenomenon of a "proton well" was first noted, and its properties outlined, by Mitchell [28]. With such a "well" connecting the negative phase medium to the binuclear centre protonation sites, the pH might be close to 10 at these sites in the intact, energised system. Such a high pH could prevent protonation of those sites which are observed in the O → P transition of the soluble system between pH values of 6 and 8.5, provided that they have p*K* values in the 8.5–10 range. More experiments are required to resolve these questions.

It has been claimed that both the O and the P states of cytochrome *c* oxidase have pH-dependent spectroscopic features. However, it is not clear that these represent reversible, rapidly accessible, sites rather than secondary processes, and we have not incorporated these data in formulating our current scheme of protonations.

### 6.2. *Proton uptake and release associated with proton translocation*

It is generally agreed that under most conditions an *average* of one $H^+$ is translocated across the membrane for each electron transferred to oxygen. Although there is some evidence that the value may decrease, perhaps even to zero, under conditions in which the mitochondrial membrane is energised, there is no evidence to suggest that it ever exceeds a value of one.

An upper value of one means that four protons can be translocated per complete catalytic cycle of four electron transfers. However, there is no reason to suppose that one proton translocation should be associated with each of the four catalytic steps. Indeed, the energetics of the steps indicate that this cannot be the case since the first step of O → E in particular is not sufficiently exergonic to be associated with a proton translocation. Wikström [29] was the first to provide direct experimental support for an unequal proton translocation stoichiometry of the four catalytic steps. By measurement of the effects of protonmotive force on the P/F and F/O ratios in intact mitochondria, and armed with knowledge of the net protonation changes of the oxygen intermediates, he deduced that the P → F and the F → O transitions were each associated with the translocation of two protons. At present, this still remains the best estimate, although reservations must be held until the differences between the measurements of net protonation changes of intermediates in the soluble, and energised-membrane conditions have been properly reconciled.

Accurate definition of the number of protons translocated at each step of the catalytic cycle is essential if we are to understand the details of the relation between the protonmotive chemistry and the electron transfer events.

### *6.3. Models for the mechanism of proton translocation*

With the discovery of the protonmotive action of cytochrome *c* oxidase, it was noted that there were no components which could fulfil the role of a physically mobile hydrogen-carrying species. This is a requirement for the "redox-loop" type of coupling mechanism seen in the *bc* complexes where the protons are physically carried across the membrane by the diffusion of the freely mobile ubihydroquinone. Any direct mechanism of coupling of proton and electron transfer in cytochrome oxidase must instead involve relatively localised changes in a compact reaction core, with associated channels for proton movement through the protein structure between this core and the aqueous phases, as shown in Fig. 6.

Alternatively, much more indirect mechanisms were considered in which proton movements occurred at some distance from the electron transfer chemistry. In such mechanisms, long range protein conformational changes must link the exergonic electron transfer events to proton movements across the membrane dielectric barrier. Several groups have favoured the more indirect mechanisms of energy coupling and this view was strengthened by the data which indicated that the oxidase might indeed exhibit a variety of global conformational states (see e.g. section 4). However, these data have since been explained in terms of rather localized properties of different ligand and redox states of the binuclear centre itself, and it is now unnecessary to invoke global protein conformational changes. Because of this we do not consider such models further here.

The first proposal for a mechanism of proton translocation which emphasized very local ligand changes around a metal centre was that of Chan [17]. It was postulated that protonatable ligands of $Cu_A$ might form a conduction pathway for protons. The protons could be driven in the required direction because of pK shifts induced in the ligands in response to redox changes of the copper atom. However, because the energy for proton translocation must originate in the binuclear centre chemistry, and because proton translocation occurs even in the quinol oxidases which lack $Cu_A$, most subsequent models have focussed attention on the binuclear centre itself.

Mitchell was the first to propose such a mechanism, termed the "O-cycle". This model was based on the notion of two distinct oxygen reactive sites, each in contact with opposite sides of the membrane. Protonated oxygen intermediates such as $H_2O_2$ pass between them to provide a mobile hydrogen-carrying element, somewhat reminiscent of the hydrogen-carrying role of reduced ubiquinone in the Q-cycle mechanism of the *bc* complexes. However, there was a general lack of evidence for such a model, especially for two oxygen-reactive sites. Because of this, Mitchell later proposed an alternative model for proton translocation in which a protonatable ligand (oxide or hydroxide) rotates around $Cu_B$, in response to redox changes of the metal [30]. The model in Fig. 14 depicts this basic idea. Starting at point A, a hydroxide ligand of oxidised $Cu_B$ faces the negative aqueous phase in a pyramidal ligand co-ordination geometry. However, it can rotate *via* a square planar intermediate (B) into a pyramidal orientation (C) in which it faces the negative aqueous phase. Only in this orientation can reduction of $Cu_B$ by donor occur, and this reduction induces a pK shift in the hydroxide so that it becomes protonated with a proton from the negative phase to form a water ligand. Reoxidation of the $Cu_B^+$ requires rotation of the water through square planar intermediate E so that the water faces the positive aqueous phase (intermediate F). Only in this orientation can oxidation by the terminal acceptor occur, and this induces the loss of the proton to the positive aqueous phase and the regeneration of A.

Other groups have suggested variants of binuclear centre ligand chemistry which might promote redox-linked proton translocation, involving ligand exchanges on $Cu_B$ [15, 31], on haem $a_3$ [32], or even between both metals [33]. All of these proposals share a common theme that a relatively small movement of ligands, linked to the oxygen reduction chemistry, can result in the movement of protons between proton channels connected to opposite aqueous surfaces.

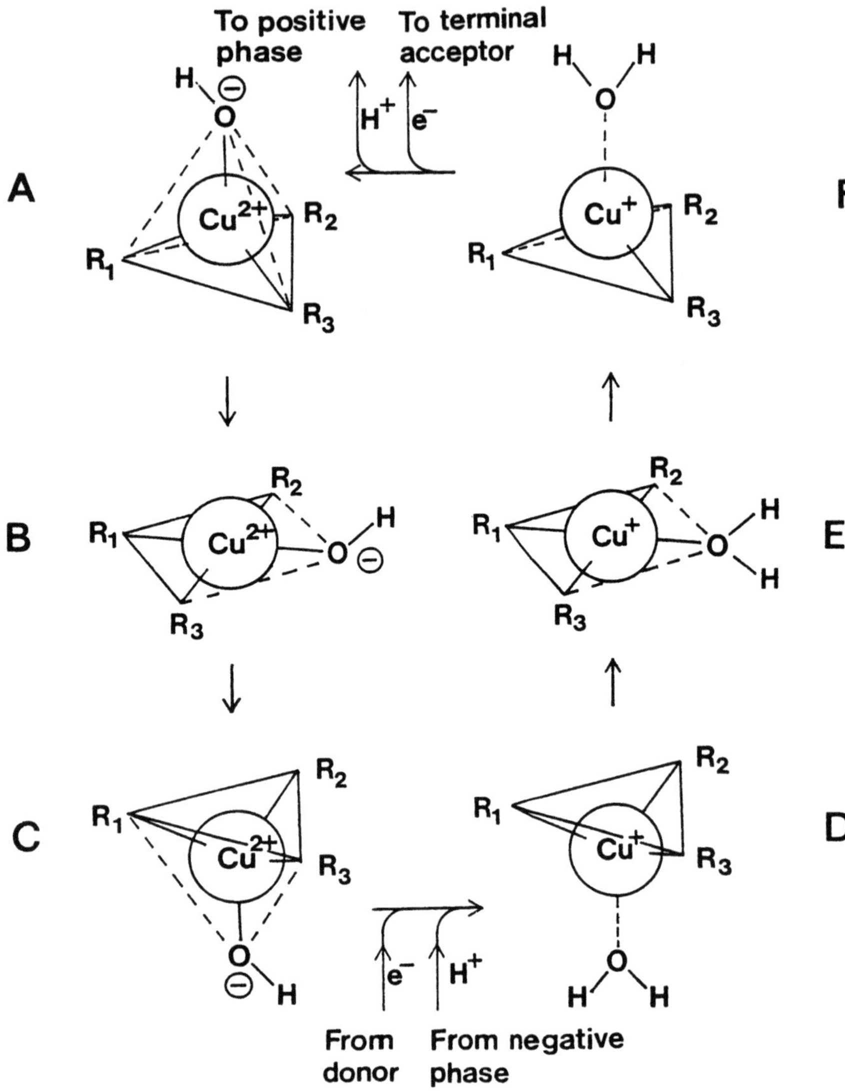


Fig. 14. Possible mechanism of protonmotive ligand rotation around $Cu_B$ in the protonmotive terminal oxidases. In this model, a hydroxide ligand on $Cu_B^{2+}$ can rotate to an orientation in which electron transfer from the substrate donor pathway can occur (A → B → C). A proton is bound to the hydroxide ligand as $Cu_B$ is reduced by donor (step C → D). The electron is supplied from the substrate, cytochrome *c* or ubiquinol, but the proton is delivered through the channel from the negative aqueous phase (see Fig. 6). Rotation of the resulting water ligand (D → E → F) is required before the electron can be transferred to the haem $a_3$/oxygen system. Reoxidation of $Cu_B$ (F → A) causes deprotonation into the channel leading to the positive aqueous phase. Hydroxide group rotation around $Cu_B^{2+}$ to the original orientation (A → B → C) is required before a further electron transfer from substrate is possible. Redrawn and simplified from [30].

The simple scheme in Fig. 14 describes only those protons which are translocated from one side of the membrane to the other, without consideration of how they are coupled with the individual steps of oxygen reduction which themselves require protons to be taken up to form water. The details of how these processes might be integrated are yet to be tested experimentally and are beyond the scope of this review.

### *6.4. Thermodynamics of oxygen reduction and proton translocation*

In the reaction catalyzed by cytochrome oxidase, there is a large redox potential difference between the source of electrons, cytochrome *c,* and the sink, the oxygen/water couple. The midpoint potential of cytochrome *c* is around +255 mV (pH-independent below pH 9) whereas the operative potential of the oxygen/water couple at atmospheric oxygen concentration and at pH 7 is +810 mV. As already noted, the first redox couple of the oxidase catalytic cycle, O/E, is not associated with oxygen reduction chemistry and is reversible. Its midpoint potential is reasonably close to that of cytochrome *c,* probably around +350 mV since this is roughly the average midpoint potential of the four metal centres amongst which the single electron is distributed.

By contrast, formation of the peroxy state is essentially irreversible. The reaction may be broken down into several component steps: the transfer of a second electron to the oxidase; the binding of oxygen; and the chemical reaction of oxygen reduction yielding a stable, partially reduced oxygen intermediate (Fig. 12). Transfer of the electron will occur at a potential a little lower than of the first electron (the potential is lower due to electrostatic interaction with the electron already present). Oxygen binding itself also occurs fairly reversibly to produce a compound rather equivalent to the oxygen compound of ferrous haemoglobin. In oxidase, this species was originally detected at low temperature by Chance with a low temperature "triple-trapping" technique [34]. It rapidly undergoes internal electron transfer reactions to produce the P state in which the majority of the electron density resides on the oxygen atoms. It is this step which is exergonic and makes the formation of P irreversible. An overall $E_{m7}$ value for the O/P couple of about +500 mV may be deduced.

Transfer of the third and fourth electrons and chemical transformation to the F and O states are also both highly exergonic. It has been possible to estimate the midpoint potentials of these couples by measuring the effects of variation of a ratio of ferricyanide/ferrocyanide on the ratios of P, F and O in energized mitochondria. It should be emphasized that the calculations rely on an accurate knowledge of both the protonation states of the intermediates and of the number of protons translocated with each step, factors which are not yet fully resolved. However, using values of

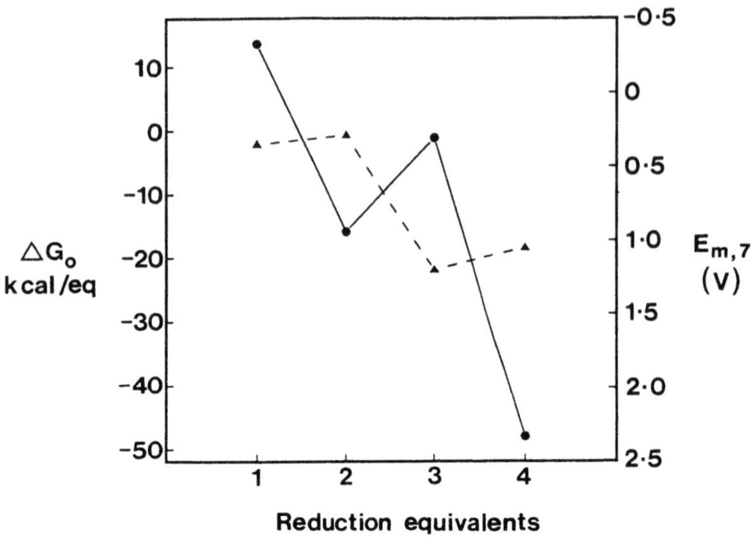


Fig. 15. The standard free energies for the four steps in the reduction of dioxygen to water: •, in aqueous solution, and ▲, in cytochrome *c* oxidase with ferrocytochrome *c* ($E_{m7}$ = +260 mV) as the reductant. Redrawn from Babcock and Wikström [14].

$2H^+$ bound and $2H^+$ translocated in both the P → F and the F → O transitions, Babcock and Wikström [14] have recently estimated $E_{m7}$ values of +1.2 and +1.05 V for the P/F and F/O couples respectively (Fig. 15). If the assignment of steps coupled to proton translocations is indeed correct, then the majority of the energy available in these two steps is conserved in the intact respiratory system as proton and charge translocations across the membrane. This would mean that only the step of bound oxygen reduction to the peroxy state is relatively irreversible in the complete reaction cycle.

## 7. Electrochemistry of the metal centres

### *7.1. Basic experimental methods*

These have been of two types. The first, stoichiometric titration, usually with reductant, has been used (i) to determine the number of redox components and their relative potentials; (ii) to determine the number of electrons that each component can accept; and (iii) to aid in the assignment of spectroscopic signals to individual components. The second, potentiometric titration, has been used to determine midpoint electrochemical potentials ($E_m$ values) of the centres and the redox interactions between them. Such experiments are possible only under strictly anaerobic conditions or in the presence of inhibitory binuclear centre ligands such as cyanide, carbon monoxide or azide, since otherwise catalytic turnover would occur. This is the reason that such techniques cannot be used to determine equilibrium midpoint potentials of the natural catalytic intermediates (section 5 and 6).

For both types of experiment the system should come to redox equilibrium. Equilibration is achieved by use of (bio)chemical mediators and for this purpose cytochrome *c,* the natural substrate of the enzyme, has proved most useful. In the case of potentiometric titrations, equilibration with the system used to monitor the electrochemical potential is essential, and again this is achieved by using an appropriate range of mediators.

### *7.2. Redox titration of cytochrome oxidase under anaerobic conditions: Anticooperativity*

Changes in the alpha band of cytochrome *c* oxidase at 606 nm arise mostly from changes in the redox poise of haem *a,* and so might be expected to titrate as a simple Nernstian n = 1 species. However, the band actually titrates as if there were two n = 1 electron carriers, with widely differing $E_m$ values, contributing about equally to the absorption changes (Fig. 16). This distortion of the haem *a* redox behaviour is caused by anticooperative redox interactions between the metal centres. These interactions appear to be primarily electrostatic, i.e. those electrons already on the enzyme repel entry of others. Consequently, haem *a* has eight possible $E_{m7}$ values ranging from about +340 mV, when all the other components are oxidized, to about +240 mV, when all the other components are reduced. The pattern of interactions is roughly consistent with the supposed topography of the metal centres (Figs. 3 and 6); there are large interactions between haem *a,* haem $a_3$ and $Cu_B$, but $Cu_A$ interacts primarily only with haem *a,* and so has only two $E_m$ values, one when *a* is oxidized, and the other when haem *a* is reduced.

A complete quantitative description of the redox properties of all the metal centres in cytochrome oxidase, which would require the assignment of values for the midpoint potentials of 14 redox couples, is not possible with the data currently available. However, it appears that haem $a_3$ is roughly isopotential with, and behaves in roughly the same manner as, haem *a*. The behaviour of haem $a_3$ has been determined most directly by potentiometric titrations using magnetic circular dichroism spectroscopy (MCD), a technique which allows the redox poise of the two haems to be monitored independently. $Cu_A$, monitored using the 820 nm absorption band or by EPR, titrates with a midpoint potential of +240 mV and is close to an n = 1 component.

The behaviour of $Cu_B$ is more difficult to assess, because of the absence of spectroscopic means of monitoring the redox poise of this centre.

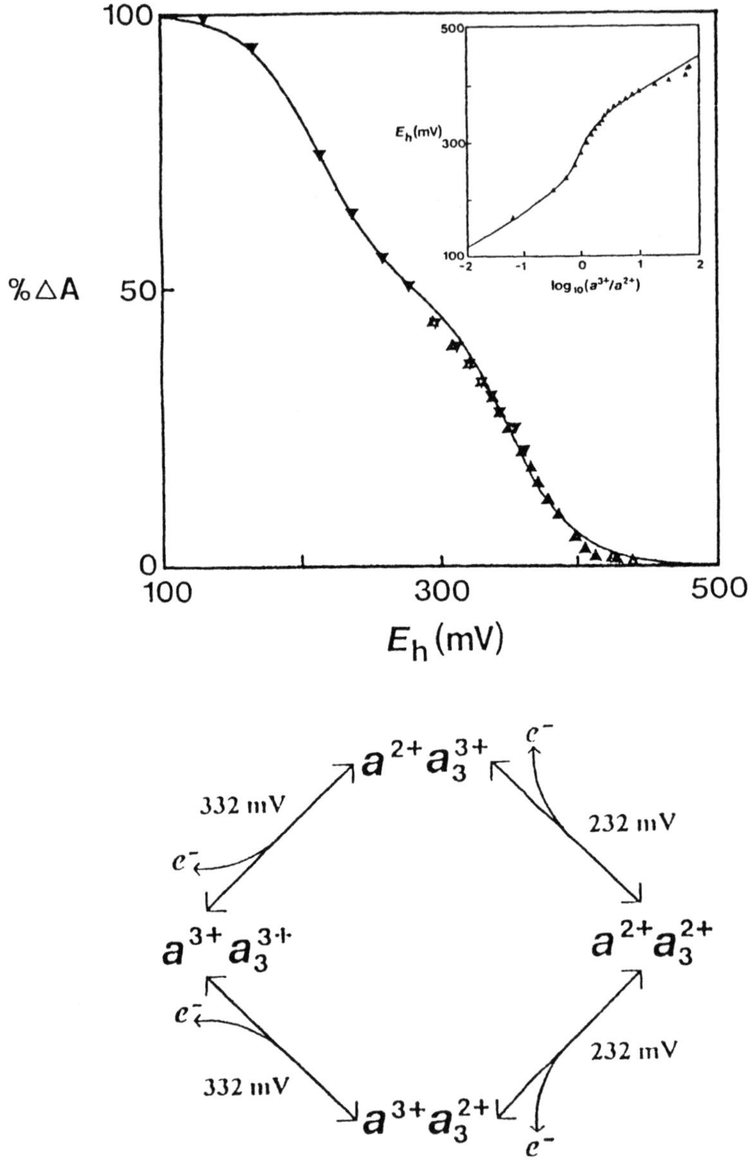


Fig. 16. Reductive titration of the alpha band of cytochrome *c* oxidase under anaerobic conditions at pH 7.5. Data are replotted from Mitchell et al. [42]. Enzyme was reduced photochemically (using riboflavin/EDTA). For potentials below about +290 mV cytochrome *c* ($E_m$ = +255 mV) was used as redox indicator (▼); for potentials above about +360 mV cytochrome *f* ($E_m$ = +365 mV) was used (▲); and for potentials between +290 and +360 mV both cytochromes *c* and *f* were used. The insert shows a Nernst plot of the same data. The continuous lines are best fit simulations using a simple model where (a) haem *a* interacts anticooperatively only with haem $a_3$, and (b) haem *a* is isopotential with haem $a_3$. The strength of the interaction is –100 mV.

Nevertheless, it can be determined indirectly, either by the appearance of EPR signals from ferric haem $a_3$ resulting from breakage of the spin coupling between ferric haem $a_3$ and reduced $Cu_B$, or by the distortion of the titration curve of haem *a* in cyanide-ligated oxidase caused by the anticooperative redox interaction between haem *a* and $Cu_B$ (Fig. 17). These are discussed below.

### *7.3. Redox titration of cyanide-ligated cytochrome oxidase*

As noted earlier (section 4), cyanide binds much more tightly to the enzyme when haem $a_3$ is oxidized than when it is reduced. As a result, the equilibrium midpoint potential of haem $a_3$ is lowered sufficiently in the presence of cyanide so that haem *a*, $Cu_A$ and $Cu_B$ can be redox titrated, while haem $a_3$ remains oxidized. Under these conditions, haem *a* again titrates over a much wider range of potential than exposed for a simple n = 1 species (Fig. 18). When the titration curve is analyzed in terms of redox interactions with $Cu_A$ and $Cu_B$, the four possible $E_{m7}$ values for haem *a*, ranging from about +330 mV, when both Cu centres are oxidized, to about +250 mV, when both centres are reduced, can be found. Moreover, the two possible $E_{m7}$ values for $Cu_B$ can also be determined, and are about +340 mV when haem *a* is oxidized and +285 mV when haem *a* is reduced. Hence, it appears that $Cu_B$ is similar in potential to both haems.

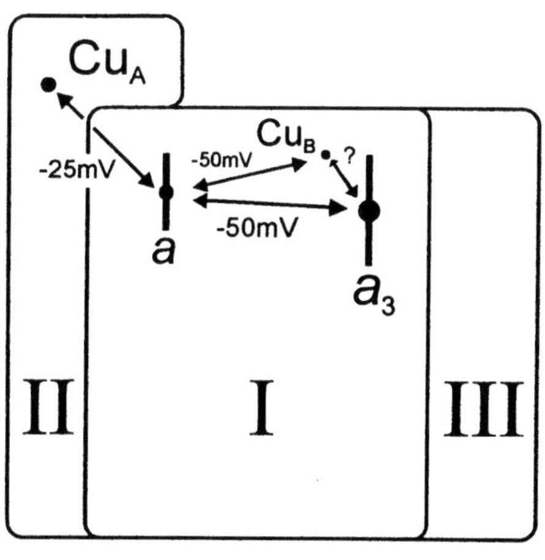


Fig. 17. Schematic diagram showing the anticooperative redox interactions between the metal centres in cytochrome oxidase. After Blair et al. [43]. The effect of the binuclear centre on haem *a* amounts to about −100 mV, roughly divided between haem $a_3$ and $Cu_B$. Blair et al. suggest an interaction of −35 mV between haem $a_3$ and $Cu_B$, but this is less certain.

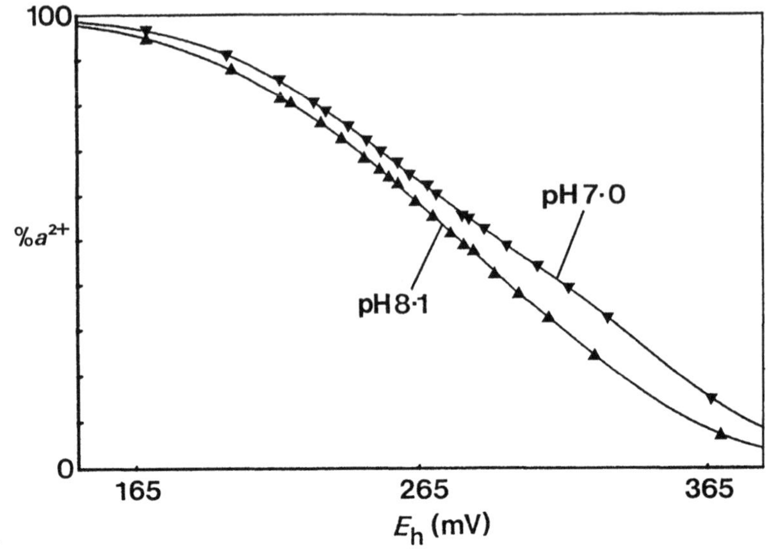


Fig. 18. Redox titrations of haem *a* in cyanide-ligated cytochrome oxidase. Data taken from Moody and Rich [44]. Enzyme was reduced photochemically (using riboflavin/EDTA). Cytochrome *c* was used as redox indicator. The solid lines are best fits assuming haem *a* interacts with two other components. One, $Cu_B$, is close in potential to haem *a*, which has an $E_{m7}$ value of +330 mV when the other components are oxidized. The other, $Cu_A$, has a much lower potential ($E_{m7}$ = +250 mV when haem *a* is oxidized). The strengths of the interactions between haem *a* and $Cu_B$, and haem *a* and $Cu_A$, are −55 mV and −25 mV, respectively.

## *7.4. Redox titration of CO-ligated cytochrome oxidase: Anticooperativity between haem* a *and* $Cu_A$

When cyanide is present, most of the distortion of the haem *a* titration curve is caused by the redox interaction with $Cu_B$. A weaker interaction with $Cu_A$ has less effect, especially since the $Cu_A$ itself titrates only at one end of the haem *a* redox curve. However, the interaction between haem *a* and $Cu_A$ can be seen in isolation when haem *a* is titrated in the presence of carbon monoxide. Since carbon monoxide binds only when the binuclear centre is fully reduced (see section 4), it raises the equilibrium midpoint potential of both $Cu_B$ and haem $a_3$ sufficiently for the other two redox centres to be titrated while both haem $a_3$ and $Cu_B$ remain reduced. Under these conditions the equilibrium midpoint potential of haem *a* is lower because of its interaction with the reduced binuclear centre. It moves closer to that of $Cu_A$, and it becomes possible to observe a slight distortion in the titration curves of haem *a* and $Cu_A$, caused by the redox interaction between these centres (Fig. 17).

### *7.5. Redox titration of CO-ligated cytochrome oxidase: Cooperativity between haem* $a_3$ *and* $Cu_B$

Because of the close proximity of haem $a_3$ and $Cu_B$, they are expected to exhibit an anticooperative redox interaction and, indeed, there is evidence that this is the case in the unligated enzyme. However, there is at least one case where haem $a_3$ and $Cu_B$ act cooperatively, namely, in the presence of carbon monoxide, a ligand which binds strongly only when both binuclear centre metals are reduced. Hence, the midpoint potential of the binuclear centre is raised by an amount dependent on the concentration of carbon monoxide. Since it does not bind if only one electron is present, this first electron reduction occurs at a potential identical to that in the unligated enzyme. The system then acts like an $n = 2$ redox centre, since the potential of the second electron transfer is raised above that of the first and haem $a_3$ and $Cu_B$ titrate in unison at a potential higher than that of the unligated enzyme.

### *7.6. pH-dependency of midpoint potentials*

Except for $Cu_A$, the midpoint potentials of the redox centres in cytochrome oxidase depend to some extent on pH in the range 6–9 indicating that protonation reactions occur in concert with the reduction process. However, this dependence is rather weak, between –10 and –30 mV change in the observed midpoint potentials for each unit change of pH. Since the pH-dependencies do not exhibit the approximate –60 mV/pH unit slope expected of simple systems which bind one proton per electron as reduction occurs, a simulation of the behaviour requires the involvement of at least two acid/base groups with different p$K$ values which are modulated by redox states of the centres.

From redox titrations of unligated enzyme, it is also clear that the total strength of the redox interactions of haem *a* with the other metal centres varies with pH (in the range –90 to –120 mV). The simplest interpretation of this variation is that the $pK_a$ of at least one acid/base group is affected by the redox state of haem *a* and at least one other redox component. A specific example is the pH-dependence of the estimated strength of the redox interaction between haem *a* and $Cu_B$ from redox titrations carried out in the presence of cyanide (Fig. 18). It appears that haem *a* and $Cu_B$, at least, are linked to the same acid/base group, which would have a $pK_a$ in the fully-oxidized enzyme at about 7.5, but > 8 when either haem *a* and/or $Cu_B$ are reduced.

*7.7. Redox titrations by EPR*

The results of redox titrations obtained using low temperature EPR to monitor the redox poise of the components are generally consistent with those found optically at higher temperatures (273–303 K). For example, the signal from $Cu_A^{2+}$ around $g = 2$ was found to titrate as a simple n = 1 Nernstian component with a midpoint at about +245 mV, independent of pH. (The quality of data is insufficient to detect any small redox interaction between $Cu_A$ and haem *a*). Other signals appear during redox titration at around $g = 6$ and arise from at least two high spin haem components. They can be attributed at least partly to high spin ferric haem $a_3$, which becomes paramagnetic when the spin coupling with $Cu_B$ is broken on reduction of $Cu_B$ (Fig. 19). This behaviour has been used to deduce the redox state, and therefore the midpoint potential, of the otherwise EPR-silent $Cu_B$.

The low spin haem signal at $g = 3.0$ can be confidently ascribed to low spin ferric haem *a*. Although it titrates at a potential roughly as expected from the optical titrations, the details of its behaviour leave some open questions. Disappearance of the signal could be caused either by reduction of haem *a,* or by a spin state change in haem *a* induced by the reduction of another centre. It appears that the former cannot be the only cause of signal loss, because the $g = 3$ signal is found to titrate as a simple n = 1 Nernstian component with $E_{m7}$ of approximately +390 mV. In fact, the disappearance of the $g = 3$ signal mirrors the appearance of high spin haem signals at $g = 6$. Hence, it has been suggested that one of $g = 6$ signals might be attributed to the appearance of a high spin haem *a* species at intermediate redox states of the system.

*7.8. Electrochemistry of related bacterial oxidases*

As already discussed above, mammalian cytochrome *c* oxidase is a member of a superfamily of proteins which, particularly in bacteria, are extremely diverse. However, the core structure of subunit I, folded around a low spin haem and a haem-copper binuclear centre (Fig 6), is likely to be well conserved and common to all of them. It therefore seems likely that the general pattern of redox interactions between these three metal centers which has been described for mammalian cytochrome *c* oxidase will also be a common feature. The work done so far on the electrochemistry of related bacterial oxidases tends to confirm this notion in general, although some details differ. The best studied bacterial enzyme to date is cytochrome *bo* from *E. coli.* The two haem groups of this quinol oxidase are rather lower in midpoint than their counterparts in cytochrome *c* oxidase, as might be expected since the quinol substrate ($E_{m7}$ = +90 mV) has

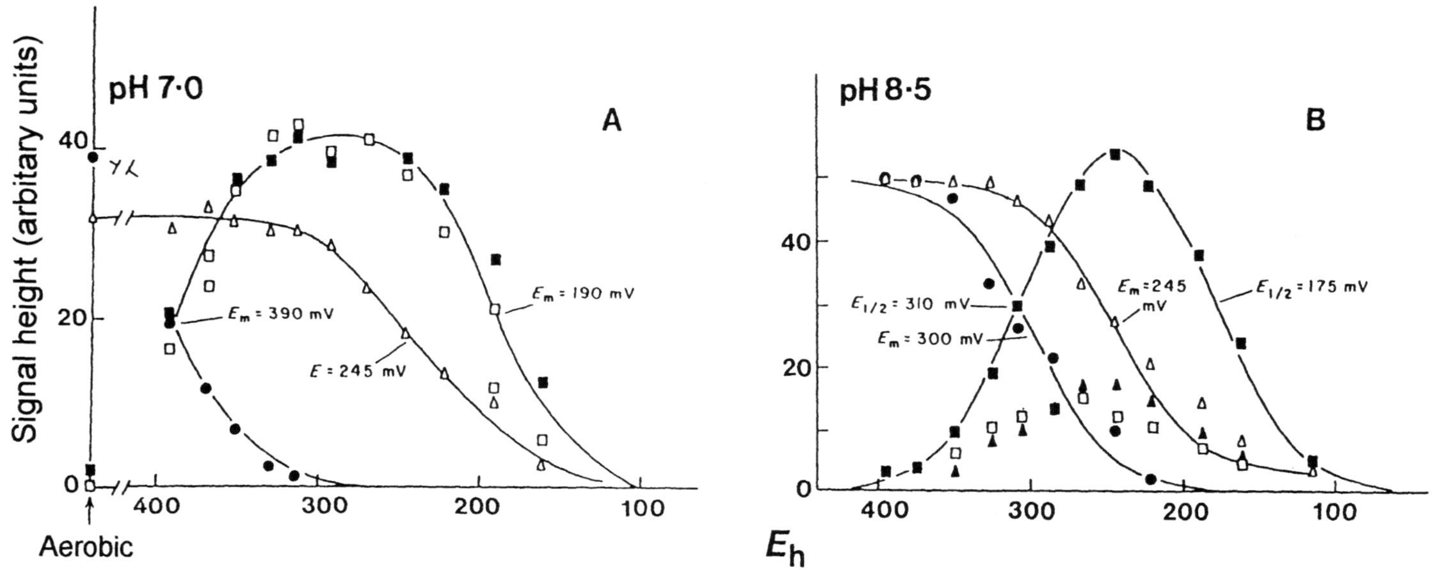


Fig. 19. Redox titrations of cytochrome oxidase in submitochondrial particles as monitored by EPR. Data taken from Wilson et al. [45] with permission. A. □, $g = 6.4$ maximum; ■, $g = 6.0$ maximum; ●, $g = 3$ maximum; and △, $g = 2$ region, $Cu_A^{2+}$ signal. B. ■, $g = 6$ maximum; ●, $g = 3$ maximum, △, $g = 2$ region, $Cu_A^{2+}$ signal; □, low-spin ferric haem signal at $g = 2.63$; and ▲, low-spin haem signal at $g = 2.57$. $E_{1/2}$ indicates where peak height has dropped to half its maximal observed value.

a lower potential than that of cytochrome *c* ($E_{m7}$ = +255 mV). Nevertheless, a generally similar pattern of magnetic and redox interactions between metal centres is found. For example, redox titrations of membrane-bound cytochrome *bo* have been performed in conjunction with EPR measurements. The g = 6 signal from the high spin haem (in this case a haem O) again shows a bell-shaped titration curve and, as with cytochrome *c* oxidase, it is thought that the appearance of this signal (with $E_{m7} \approx$ +370 mV) is caused by the prevention of the spin coupling between ferric haem *o* and oxidized $Cu_B$ by reduction of the copper. The g = 3 signal, from the low spin haem (in this case usually a haem B), titrates in a biphasic manner (with $E_{m7}$ approximately +180 and +280 mV), essentially coincident with the disappearance of the g = 6 signal, and is evidence for a redox interaction between the two haems. Recent redox titrations using optical spectroscopy have largely confirmed and extended this picture of redox interactions between the metals, although the picture remains incomplete and complicated by the tendency of the low spin haem site of the bacterial enzyme to be occupied by a mixture of haems O and B (section 2.3), each of which differ in their spectral and redox properties. Further work is necessary before the description of the bacterial system is as complete as that of the mammalian cytochrome *c* oxidase.

## 8. Conclusions

The cytochrome and quinol oxidases described here represent a key class of homologous metalloproteins which are widespread in nature and diverse in detail. They are extremely well suited for investigations by a large number of biophysical techniques and such studies have highlighted a rich range of phenomena relating to redox and other interactions between metals, structure and chemistry of metal-oxygen compounds, and fundamental mechanisms of coupling of electron transfer to proton translocation in a membrane-spanning protein structure. As these phenomena become more clearly defined, and especially if they can eventually be analyzed in the context of a definitive high resolution structural model of the protein, it seems likely that these oxidases may become the model systems for understanding and testing of principles of electron and proton transfer in multi-metallic membrane proteins in the same way that the globins and cytochrome *c* have become classic models for haemoprotein structure and function.

## Acknowledgements

The Glynn Research Foundation Ltd. provided the facilities and finances that made this review possible. Our experimental activities in this area are currently also founded by the Wellcome Trust (Grant 034505/2/91/2), the BBSRC (Grant GR/J28148) and the European Community (CEC Stimulation Contract CT90-0476). We thank Drs. Simon Brown and Roy Mitchell for useful comments and Mr. Robert Harper for artwork.

## References

1. P Mitchell, *Chemiosmotic Coupling in Oxidative and Photosynthetic Phosphorylation,* Glynn Research Ltd, Bodmin, 1966.
2. M Wikström, Nature 266 (1977) 271–273.
3. M Wikström, K Krab and M Saraste, *Cytochrome Oxidase A Synthesis,* Academic Press, London, 1981.
4. M Saraste, L Holm, L Lemieux, M Lübben and J van der Oost, Biochem. Soc. Trans. 19 (1991) 608–612.
5. S Ferguson-Miller, J. Bioenerg. Biomemb. 25 (1993) 69–79.
6. JP Hosler, S Ferguson-Miller, MW Calhoun, JW Thomas, J Hill, L Lemieux, J Ma, C Georgiou, J Fetter, J Shapleigh, MMJ Tecklenburg, GT Babcock and RB Gennis, J. Bioenerg. Biomemb. 25 (1993) 121–136.
7. BG Malmström and R Aasa, FEBS Lett. 325 (1993) 49–52.
8. RB Gennis, Biochim. Biophys. Acta 1101 (1992) 184–187.
9. CE Cooper, P Nicholls and JA Freedman, Biochem. Cell Bio. 69 (1991) 586–607.
10. B Kadenbach, A Stroh, FJ Hüther, A Reimann and D Steverding, J. Bioenerg. Biomemb. 23 (1991) 321–334.
11. A Azzi and M Müller, Arch. Biochem. Biophys. 280 (1990) 242–251.
12. RA Capaldi, Ann. Rev. Biochem. 59 (1990) 569–596.
13. GT Babcock and C Varotsis, J. Bioenerg. Biomemb. 25 (1993) 71–80.
14. GT Babckok and M Wikström, Nature 356 (1992) 301–309.
15. PR Rich, Biosci. Rep. 11 (1991) 539–571.
16. MT Wilson and D Bickar, J. Bioenerg. Biomemb. 23 (1991) 755–771.
17. SI Chan and PM Li, Biochemistry 29 (1990) 1–12.
18. CE Cooper, Biochim. Biophys. Acta 1017 (1990) 187–203.
19. BG Malmström, Arch. Biochem. Biophys. 280 (1990) 233–241.
20. BG Malmström, Chem. Rev. 90 (1990) 1247–1260.
21. K Krab and M Wikström, Biochim. Biophys. Acta 895 (1987) 25–39.
22. WS Caughey, A Dong, V Sampath, S Yoshikawa and XJ Zhao, J. Bioenerg. Biomemb. 25 (1993) 81–91.
23. DD Lemon, MW Calhoun, RB Gennis and WH Woodruff, Biochemistry 32 (1993) 11953–11956.
24. G Palmer, J. Bioenerg. Biomemb. 25 (1993) 145–153.
25. L Powers and BM Kincaid, Biochemistry 28 (1989) 4461–4468.
26. RA Scott, Annu. Rev. Biophys. Biophys. Chem. 18 (1989) 137–158.
27. WH Woodruff, RJ Kessler, NS Ferris, RF Dallinger, KR Carter, TM Antalis and G Palmer in *Electrochemical and Spectrochemical Studies of Biological Redox Components,* KM Kadish (ed.), American Chemical Society, Washington D.C., 1982, pp. 625–659.

28. P Mitchell, *Chemiosmotic Coupling and Energy Transduction,* Glynnn Research Ltd, Bodmin, 1968.
29. M Wikström, Nature 338 (1989) 776–778.
30. P Mitchell, FEBS Lett. 222 (1987) 235–245.
31. P Mitchell, Biosci. Rep. 11 (1991) 297–346.
32. DL Rousseau, YC Ching and J Wang, J. Bioenerg. Biomemb. 25 (1993) 165–177.
33. WH Woodruff, O Einarsdóttir, RB Dyer, KA Bagley, G Palmer, SJ Atherton, RA Goldbeck, TD Dawes and DS Kliger, Proc. Natl. Acad. Sci. USA 88 (1991) 2588–2592.
34. B Chance, C Saronio and JS Leigh, J. Biol. Chem. 250 (1975) 9226–9237.
35. R Henderson, JM Baldwin, TA Ceska, F Zemlin, E Beckmann and KH Downing, J. Mol. Biol. 213 (1990) 899–929.
36. W Wu, CK Chang, C Varotsis, GT Babcock, A Puustinen and M Wikström, J. Am. Chem. Soc. 114 (1992) 1182–1187.
37. S Brown, AJ Moody, R Mitchell and PR Rich, FEBS Lett. 316 (1993) 216–223.
38. PR Rich, AJ Moody and WJ Ingledew, FEBS Lett. 305 (1992) 171–173.
39. R Boelens and R Wever, Biochim. Biophys. Acta 547 (1979) 296–310.
40. AJ Moody, CD Cooper and PR Rich, Biochim. Biophys. Acta 1059 (1991) 189–207.
41. GCM Steffens, T Soulimane, G Wolff and G Buse, Eur. J. Biochem. 213 (1993) 1149–1157.
42. R Mitchell, P Mitchell and PR Rich, FEBS Lett. 280 (1991) 321–324.
43. DF Blair, WR Ellis, H Wang, HB Gray and SI Chan, J. Biol. Chem. 261 (1986) 11524–11537.
44. AJ Moody and PR Rich, Biochim. Biophys. Acta 1015 (1990) 205–215.
45. DF Wilson, M Erecinska and CS Owen, Arch. Biochem. Biophys. 175 (1976) 160–172.

Bioenergetics
ed. by P. Gräber and G. Milazzo†


# *CHAPTER 11*
# Bacteriorhodopsin

Lajos Keszthelyi

*Institute of Biophysics, Biological Research Center, Szeged, Hungary*



## 1. Introduction

In the early 1970s, a light-energy driven proton translocating system was discovered in *Halobacterium salinarium* (formerly *H. halobium*) [1]. This system was found to be used by the bacteria to produce a transmembrane electrochemical potential difference of protons. It was also found that the proton translocation is due to a single molecule: the bacteriorhodopsin (bR), a protein located in the plasma membrane of the bacteria. It is a very important molecule in understanding the mechanism of the light driven proton pump. The protein contains a retinal chromophore, which undergoes a cyclic reaction pathway after absorbing a photon, and during this photocycle a proton is translocated from the inside of the cell to the outside. It has been shown that the generated electrochemical potential difference of protons is used by the cell to synthesize ATP.

The study of bR as a model proton pump has many advantages: it is relatively easy to separate bR from the cell, it is stable and performs the proton pumping function for a long time *in vitro* if maintained properly. It is therefore not surprising that, since its discovery, a great number of studies have been performed to elucidate the detailed molecular mechanism of the bR proton pump. It is generally expected that this simple model system offered by nature will help towards a better understanding of the general principles of proton pumps.

## 2. The bacterium

*Halobacterium salinarium* belongs to the extreme halophilic branch of the archaebacteriae because they live and grow only in high concentrations of NaCl and Mg salt, and disintegrate when the salt concentration is lowered to less than 2 M NaCl. The bacteria occur in nature only in salt lakes, salt flats and in crystalline salt. They attracted the attention of microbiologists when it was discovered that they spoiled salted fish.

The shape of the bacterium is rodlike, with a diameter of approximately 0.5 μm, and a length of 4–10 μm. They have flagella which make them mobile. The phototaxis of the bacteria is a popular field of investigation. The cell wall consists of a single glycoprotein with a molecular weight of 200 kDa and contains only 10–12% carbohydrates. It forms a regular hexagonal lattice separated from the cell membrane by pericellular space. The cell membrane is built up from lipids common to most cells, while the protein composition of the membrane is complex, and not yet analyzed in detail. The membranes are also strongly pigmented. In fact, the red color of salt flats is caused by the large quantity of carotinoids in the membrane of such bacteria. Pigmented proteins include the bR, the halorhodopsin (hR, pumps $Cl^-$ ions) and two different sensory rhodopsins sRI and sRII. Chlorophyll related pigments are not found in *Halobacterium salinarium.*

Rest of the cellular components are similar to other prokaryotes, which however, may need more monovalent salt concentration to perform their function.

The energy for the bacteria comes from two sources. The cell membrane contains a respiratory chain with b- and c-type cytochromes and a cytochrome oxidase. The amino acids are the main source for metabolism while carbohydrates play a minor role. The other energy source is the light absorbed by bR and hR.

While the extracellular concentrations are approximately 4 M NaCl and 0.5–3 mM KCl the intracellular concentrations are quite opposite: the concentration of $K^+$ ions is 3 M and that of the $Na^+$ 0.5–2 M. The concentration differences are maintained by active transport systems like the $Na^+/H^+$ exchange pump which drives $Na^+$ ejection and the $Na^+$/amino acid cotransport system for amino acid uptake.

In general, the *Halobacteria* do not seem to differ too much from other prokaryotes in their DNA structure, metabolic pathways, etc. The exceptions are that *Halobacteria* have a high internal salt concentration, they need high salt concentration extracellularly and–what is important from our point of view–they contain bR and hR molecules which transform light energy into electrochemical energy [1,2].

## 3. Bioenergetics of *H. salinarium*

As already mentioned, two different energy transducing systems exist in *H. salinarium:* the usual respiratory system based on metabolic energy sources and the unique light energy transducing system based on the bR proton pump and hR $Cl^-$ ion pump. Only the light driven energy transducing systems will be dealt with here.

The study of the light-induced energetics of the whole cell is rather complicated because of the difficulties in depleting the cells of their usable energy. Even without nutrients and under anaerobic conditions, the cells maintain 30% of their ATP level and a large ion gradient for prolonged times. ATPase inhibitors reduce the ATP level drastically. The likely energy source under such circumstances is the high $K^+$ gradient which can supply energy at a low rate for long times due to the low $K^+$ conductivity of the membrane.

The resting state of the cell is defined as an anaerobic and dark state without nutrients when energy consumption of the cells is low. Adding

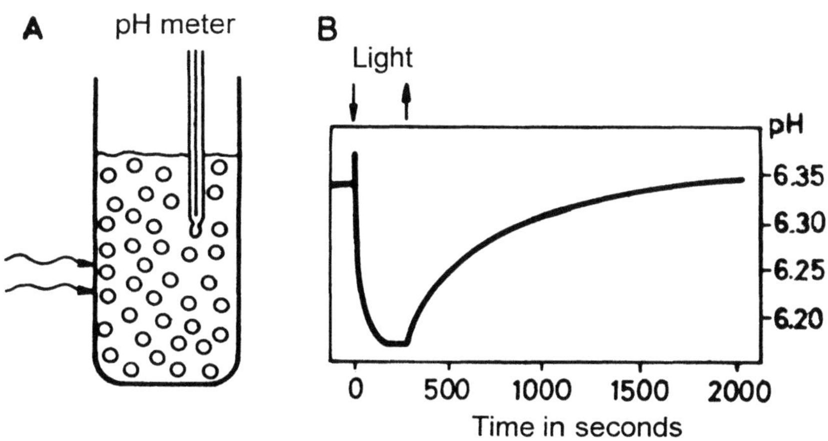


Fig. 1. *H. salinarium* cells illuminated in a suspension. (A) scheme of the measuring system, (B) the pH–after a short increase–decreases (i.e. proton concentration increases).

nutrients, $O_2$ or simply illuminating the cell in this state increase the energy requiring processes like proton ejection, ATP synthesis, ion translocation, amino acid uptake, etc. Experiments to show light induced proton ejection or ATP synthesis were generally performed on these resting cells though they are not well characterized.

If these cells in solution are illuminated, a characteristic pH change is registered by pH electrodes (Fig. 1A, B). After a transient pH increase (i.e. proton influx into the cells) a strong pH decrease occurs which is expected from a light driven proton pump. The most probable explanation of the initial alkalinization (which occurs also in case of envelope vesicles) is the following: the cells contain hR molecules which translocate $Cl^-$ ions when illuminated. This active inward $Cl^-$ transport is accompanied by passive cation–protons among others–transport. The combination of active proton exclusion by bR and passive proton uptake due the hR of slightly different time-course is presently accepted as the reason for the observed pH change in Fig. 1B.

The efficiency of ATP production is an important issue. According to available measurements, 2.8–2.9 translocated $H^+$ ions produce one ATP molecule. With a quantum efficiency of 0.6 for $H^+$ translocation, approximately six photons are needed in *H. salinarium* to produce one ATP molecule. The energy transduction is rather low: the energy of 6 mol photons in the right wavelength range is ≈ 1300 kJ, while the useful energy of one mol ATP molecule is ≈ 30 kJ which means a 2–3% primary efficiency. We may now see why nature did not follow this line of evolution for light energy transduction, and has turned to photosynthetic systems where this efficiency is higher.

## 4. The bacteriorhodopsin in purple membrane

The bacteriorhodopsin molecules are not randomly distributed in the plasma membrane of *H. salinarium*. Freeze fracture electron micrographs of the membrane (Fig. 2) show well distinguishable islands of

Fig. 2. A freeze fracture electron micrograph of *H. salinarium*. The islands containing bacteriorhodopsin are easily distinguishable. These purple membranes can be isolated.

about 0.5 μm diameter which can be separated by biochemical methods with its structure intact [3]. The membranes appear after centrifugation as a separated band of purple color. Therefore, they are named purple membranes.

### *4.1. The structure of purple membrane*

The purple membrane patches on the surface of the cell are more smooth and regular than the other parts of the cell. They are part of the plasma membrane. With a special shadowing technique, the regular lattice-type distribution of the bR molecules is easily visible. The purple membrane, after separation, preserves the structure it had in the cell. Careful studies have shown that purple membrane contains only lipids (25% by mass) and bR molecules. The bacteriorhodopsin molecules are arranged in a hexagonal lattice. The distance between the point of this lattice is 6.3 nm and three bR molecules sit at every point as shown by x-ray diffraction. One finds, therefore, roughly $1.5 \times 10^4$ bR molecules in a membrane patch with a diameter of ≈ 0.5 μm. The arrangement of the three bR-s in the lattice points was determined by electron diffraction [4]. Probably due to the high density of proteins the membranes do not close to form vesicles when separated but form more or less flat membrane sheets in a solution. The thickness of such membrane sheets is ≈ 5 nm as estimated from x-ray diffraction patterns. The electron density map reconstructed from the diffraction pattern is shown in Fig. 3. The map clearly demonstrates the lattice and the symmetric arrangement of the three bR molecules at the points. The density distribution inside one bR is due to seven helices which span the membrane.

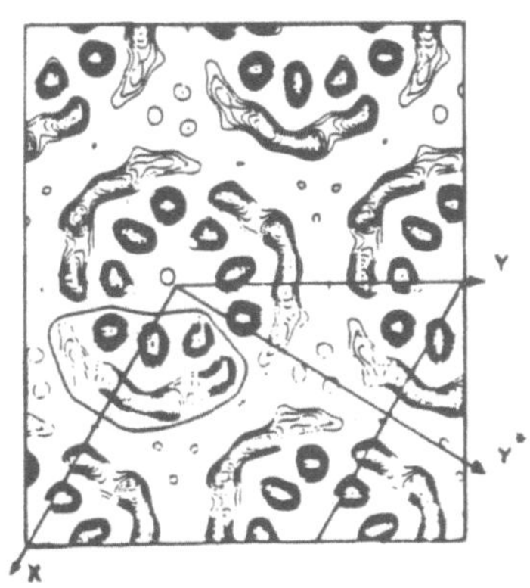


Fig. 3. The electron density map showing the distribution of bacteriorhodopsin molecules arranged in triplets in the purple membrane.

It is important to note that the purple membrane can be kept in salt or sugar solution for a long time without deterioration of the structure or loss of function.

### *4.2. The bacteriorhodopsin molecule*

The bacteriorhodopsin molecule is a monomeric protein with molecular mass of 27048 daltons. A retinal molecule is bound to a lysine moiety which is 216th in the amino acid sequence. In Fig. 4 this bond–which is a Schiff-base–is sketched.

The amino acid sequence of the protein has been determined. The amino acids in Fig. 5 are arranged in seven helices (A, B, C, D, E, F, G) and connecting sections. The N-terminus in the extracellular side contains seven amino acids while the intracellular C-terminus includes ≈ 20 residues.

The three-dimensional structure of the protein is not yet known to atomic resolution. Very accurate electron diffraction data taken at low temperature has made it possible to construct a map with 0.35 nm resolution parallel to the membrane plane and 0.7 nm resolution perpendicular to it. Figure 6 shows a simplified structure of bR, including the location of helices (A–G), the connecting sections (their structure is rather uncertain) and the position of the retinal [4]. The location of side chains for many functionally important amino acids is known as well.

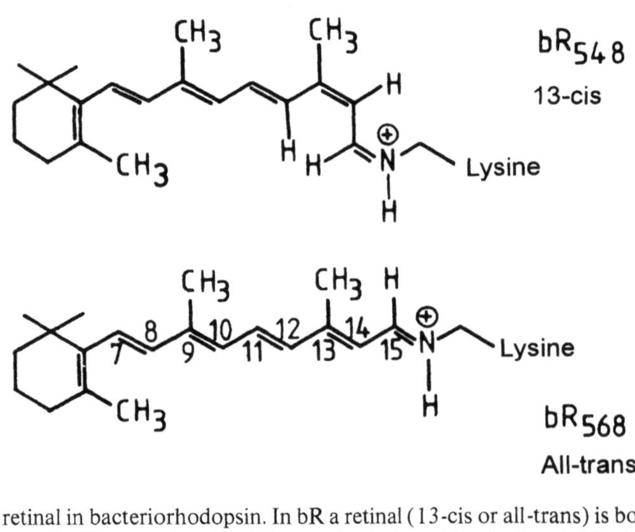


Fig. 4. The retinal in bacteriorhodopsin. In bR a retinal (13-cis or all-trans) is bound to the lysine of the protein. The wavelength of the maximal absorption in the visible depends on the isomeric state of the retinal ($bR_{548}$ for 13-cis, $bR_{568}$ for all-trans retinal).

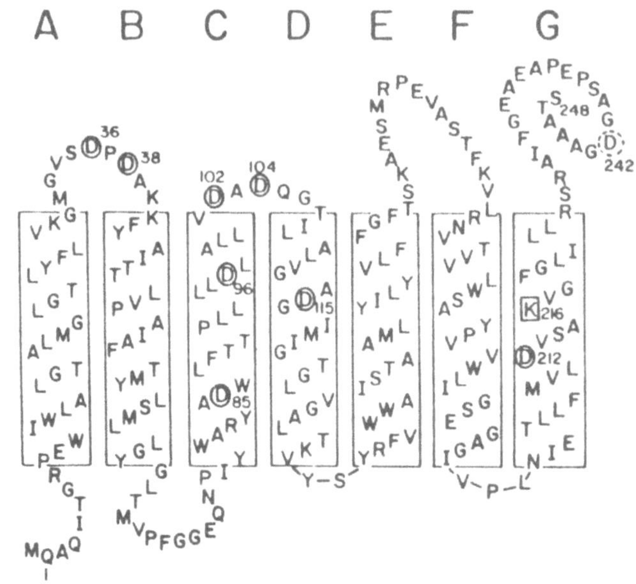


Fig. 5. The amino acid sequence of bacteriorhodopsin arranged into seven helices and connecting sections. The 216th lysine [K] and the aspartic acids [D] are labelled.

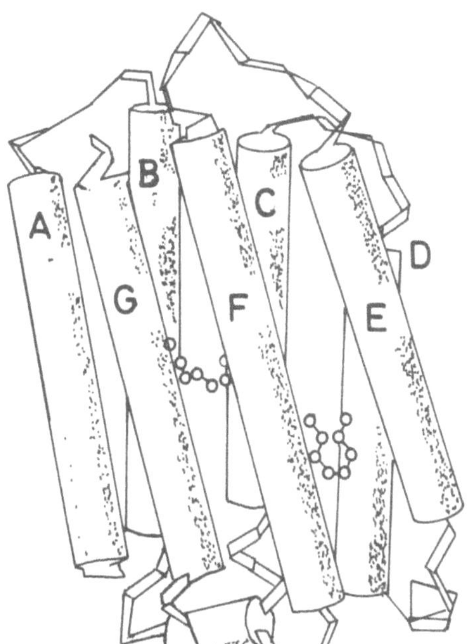


Fig. 6. The three dimensional structure of bacteriorhodopsin. The structure–determined by low temperature electron diffraction–shows the seven helices and the retinal.

*4.3. Spectroscopy of the ground state of bacteriorhodopsin*

The purple color of the separated bR containing membrane is due to the absorption spectrum of the ground state of bR. The spectra were studied in the visible, ultraviolet (UV) and infrared regions, and by resonance Raman spectroscopy. Two spectra of bR are reproduced in Fig. 7. When dark adapted, the bR has an absorption maximum at 558 nm. This peak is shifted to 568 nm after intensive illumination (light adapted bR). The absorption in the visible part is due to the retinal bound to the protein. The UV region reflects the absorption of the apoprotein called bacterioopsin. The retinal can be extracted by chemical methods and separated into isomeric forms. The dark adapted bR contains 60% 13-cis and 40% all-trans retinal at room temperature. The retinal in light adapted bR is about 100% all trans (Fig. 4). The absorption spectra of the all-trans bR being known this way, the absorption spectra of the 13-cis bR can be calculated. The maximum absorption is at 548 nm. The light adapted form returns spontaneously to the dark adapted form in about 30 min at room temperature. It has been shown that only the all-trans form of bR can pump a proton.

The resonance Raman spectrum of light adapted bR is reproduced in Fig. 8 [5]. The absorption peaks of the so-called finger-print region correspond to vibrations of the different bonds of the all-trans retinal in bR. Using isotope labelling techniques, all the peaks have been correlated with various bonds. The peak at 1640 $cm^{-1}$ of all-trans bR is due to the vibration of the $C_{15} = NH$ bond where the meaning of H is that this bond, the Schiff-base (see Fig. 4), is protonated. The protonation of the Schiff base was confirmed by observing and analysing a downshift of 16 $cm^{-1}$ of this peak after exchanging the $^{1}H$ by $^{2}H$.

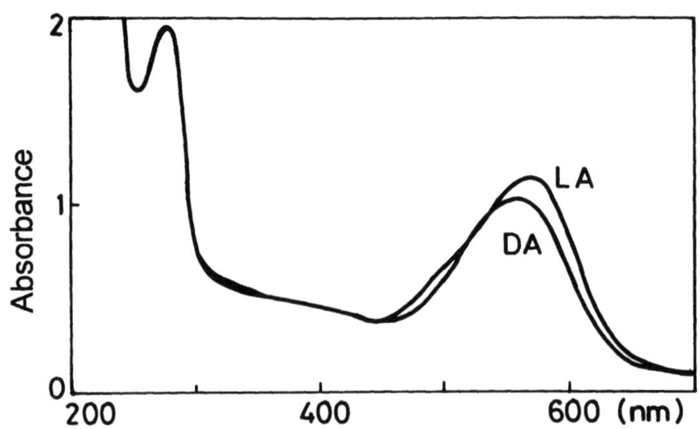


Fig. 7. The absorption spectrum of dark (DA) and light (LA) adapted bacteriorhodopsin.

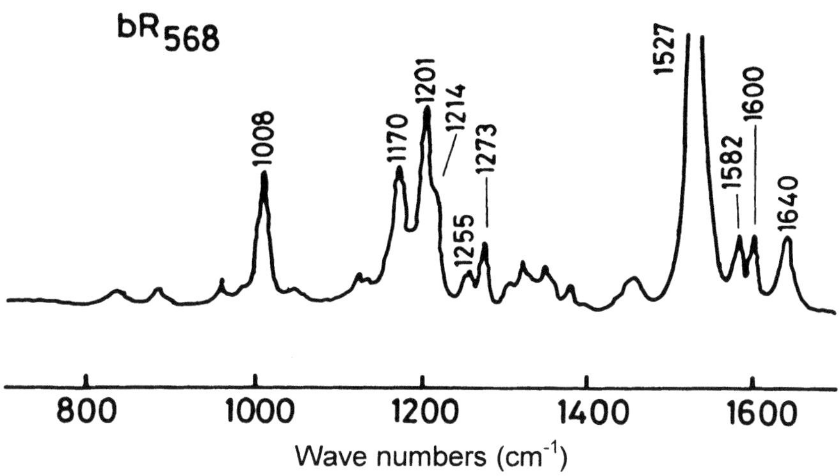


Fig. 8. The Resonance Raman spectrum of light adapted bacteriorhodopsin.

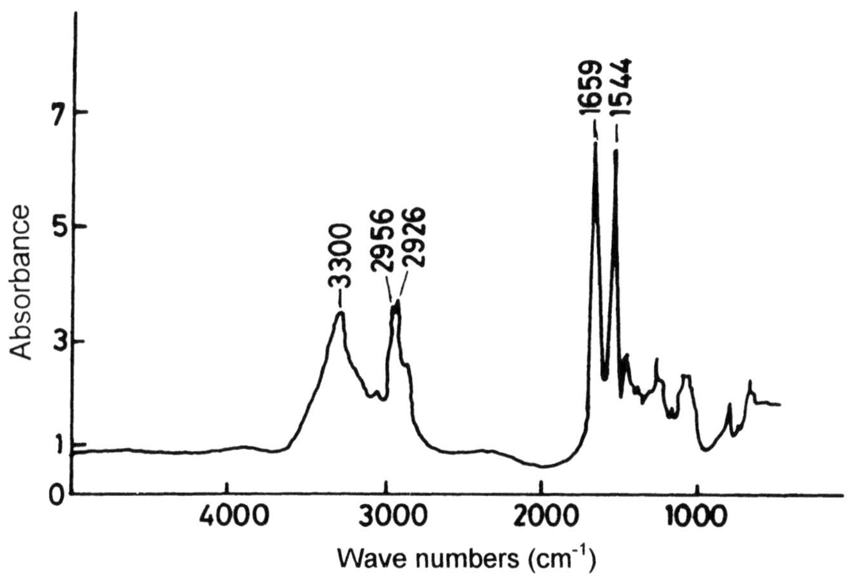


Fig. 9. The infrared spectrum of light adapted bacteriorhodopsin.

The resonance Raman spectra only gives information about the retinal. The infrared spectra confirm the resonance Raman data and in addition, contain many peaks caused by vibrations in the opsin part of the molecule. Figure 9 shows the infrared spectrum of the all-trans bR. The two strong bands at 1554 $cm^{-1}$ and 1659 $cm^{-1}$ are due to vibrations of the backbone of the protein. The assignment of the numerous small peaks is very complicated. A successful approach was again isotope labelling. For example, when vibrations of aspartic acids were to be assigned then the bacteria were fed by isotopically labelled aspartic acids. The spectra of labelled and original bR were compared and the very small shift in the peaks of COOH vibrations were identified. This, however, is not sufficient to know which aspartic acid corresponds to the changed peaks. The identification of peaks corresponding to a given aspartic acid was performed only recently when gene manipulation techniques made it possible to exchange selected aspartic acids with, for example, asparagine. The disappearing peak was then assigned to the selected aspartic acid.

An all-trans retinal with a protonated Schiff-base in solution absorbs at 440 nm while in bR it absorbs at 568 nm. The difference 568 – 440 = 128 nm is the so-called opsin shift, which must be due to the part of the protein around the retinal. According to experimental and theoretical investigations, the opsin shift is explained by (1) the isomerization of the $C^6$-$C^7$ bond of the retinal; (2) the negative and positive charges of the bacterio-opsin near $C^6$ and $C^7$; and (3) the counterion complex around the Schiff-base proton which produces a neutral environment sketched in Fig. 10.

It is interesting to note that intensive dehydration of bR lowers the absorption maximum to 530 nm. At pH 0.5 it shifts to 605 nm (blue membrane), but the purple color reappears if NaCl is added at that pH in a concentration of >100 mM.

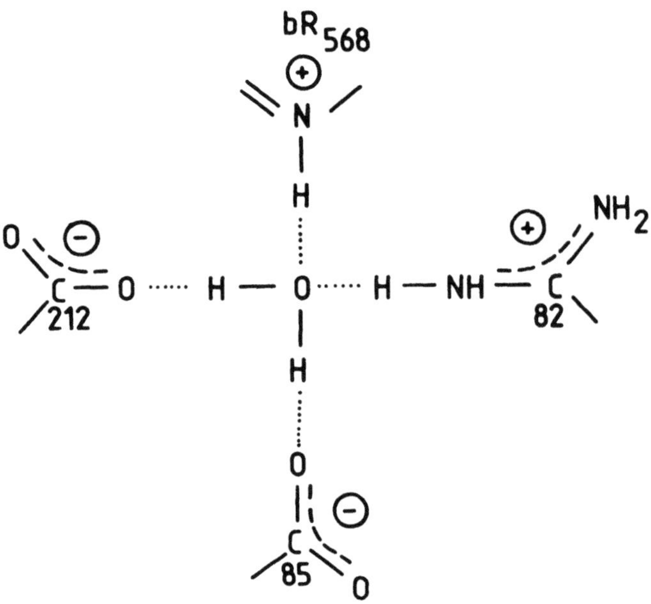


Fig. 10. The counterions from different amino acid side chains (Asp85, Asp212 and Arg82) around the protonated Schiff-base in bacteriorhodopsin.

## 5. The photocycle

The only known function of bR is to pump protons from the cell to the external medium upon absorption of a photon. During this process, cyclic changes occur in the protein. The changes may be classified into four cycles.

(1) *Charge motion cycle.* We start with this because all other changes promote or indicate the charge motion. Conceived simply, there are at least two steps in getting the pumped proton through the bR molecule: uptake on the intracellular side and release at the extracellular side. Actually, it occurs in more than two steps as discussed later.

(2) *Light absorption cycle.* The first indication of the existence of a cycle was that the absorption maximum changed in time after absorption of a photon. Measurements performed at different wavelengths of the visible spectra have shown changes which serve to characterize intermediates of the photocycle.

(3) *The retinal cycle.* In the all-trans bR, the isomerization of the retinal changes during the photocycle.

(4) *The opsin cycle.* Changes in the UV region indicate changes in protein structure, because tyrosines and tryptophans absorb in this range. Data from infrared spectroscopy reflect changes in protonation of different aspartic acids as well as motions of the backbone of bR.

### *5.1. The light absorption cycle of bacteriorhodopsin*

The method of flash photolysis, i.e. the observation of the time-course of absorption changes at different wavelengths after laser excitation, offered the first insight into the cyclic nature of the photoreaction. Different methods of data collection were used depending on the time scale involved. For measuring very fast changes, a special laser method with femtosecond resolution is the most suitable. A laser flash of a few fs is split into two parts: one excites the sample, the other evokes a continuous spectrum from a suitable absorbent. This light is directed through a monochromator which selects the wavelength for monitoring the excited sample. Its intensity is measured after the sample at different times. The very short time delay of the measuring beam is changed by changing the difference in the distance traveled by the exciting and monitoring beams. The absorption changes at 600 nm are sketched in Fig. 11. The measured curve is the result of three different changes as detailed in Fig. 11:

(1) The absorbed photons start the photocycle of bR. Therefore, the fraction of molecules which leave the ground state will not absorb in the 568 nm band. A so-called "bleaching" appears and is constant in the fs to ps range.

(2) The first intermediate of the photocycle appears as an absorbing signal. By changing the monitoring wavelength, the maximum of the absorption was determined to be at 600 nm. This J intermediate appears with a time constant of 500 fs.

(3) The J intermediate, which is the only light driven reaction, decays with a time constant of 5 ps, and the formation of a new intermediate K with an absorption maximum at 590 nm starts with the same time constant. This transition is driven by thermal agitation. The lifetime of the K intermediate is ≈ 1 μs, therefore, the sum of the contributions of the J and K in Fig. 11 appears as a constant.

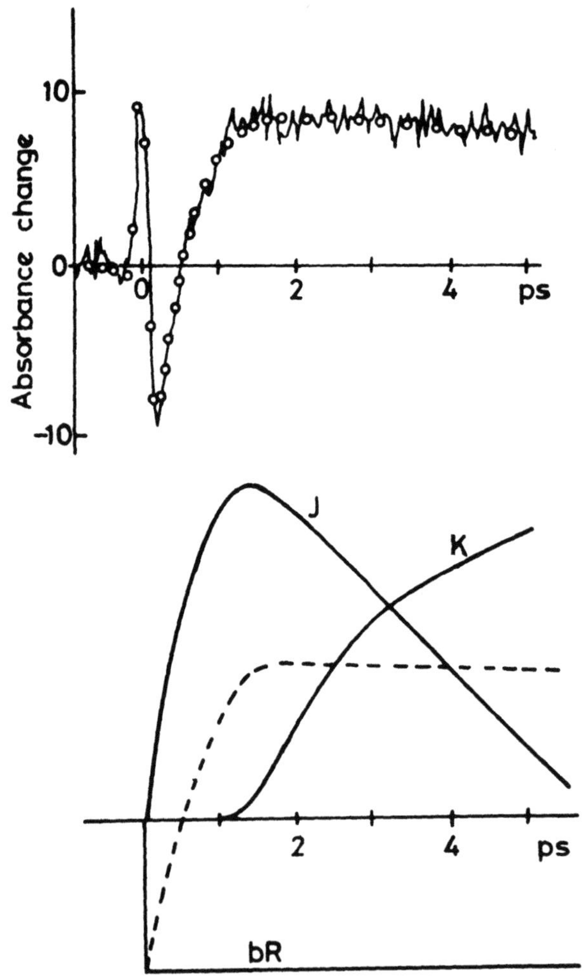


Fig. 11. The light absorption change after laser excitation in the ps range at $\lambda = 600$ nm. Upper curve: measured (the positive peak at $t = 0$ is an artefact), lower curve: the measured curve (dashed line) deconvoluted into its three different components: bR, J and K.

Evaluation of the experimental curve in Fig. 11 is a complex task because the spectra of the ground state and the intermediates overlap. The shape of the absorption curves and their absorption coefficients should be known in advance. Approximations are needed to find these data for all intermediates. In Fig. 12, a good approximation is reproduced which has been tested by different methods. After the K intermediate, the L, M, N and O intermediates are characterized according to their spectra [6]. In general, the latter name of the intermediates is followed by the wavelength of maximum absorption. One of the most interesting intermediates is $M_{412}$ which is well separated from the others. Therefore, it may be considered to be the best determined.

The other intermediate spectra–as shown in Fig. 12–strongly overlap. As they appear in time after excitation, composite difference spectra are produced from the "positive" absorbance of the intermediate and the "negative" absorbance of the bR ground state. From the decomposition of the difference spectra, the time dependence of the concentration of the intermediates is determined. The results from this analysis are shown in Fig. 13. One sees that intermediates exist in parallel. The explanation is that reversible reactions exist between some of the intermediates. Consequently, they equilibrate after some time. Figure 14 shows a recent model of the photocycle [6] which is in good agreement with measurements.

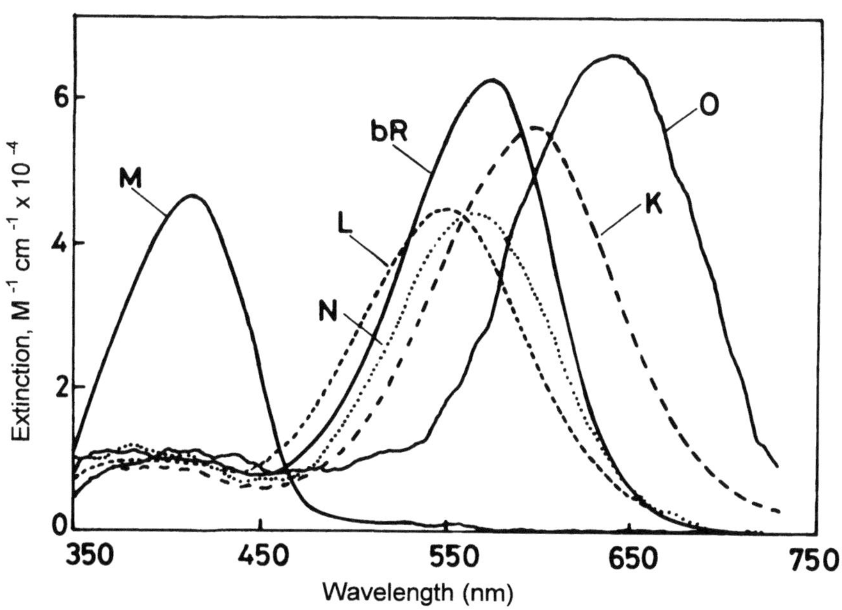


Fig. 12. The absorption spectra of bacteriorhodopsin and intermediates during the photocycle.

The temperature dependence of the time constants of the transitions follows the Arrhenius law. Activation enthalpies and entropies may be determined and from the rates of the reversible reactions, the free energy of the intermediates is calculated. Data for the irreversible reactions are taken from other direct measurements: bR → K by calorimetry of the ab-

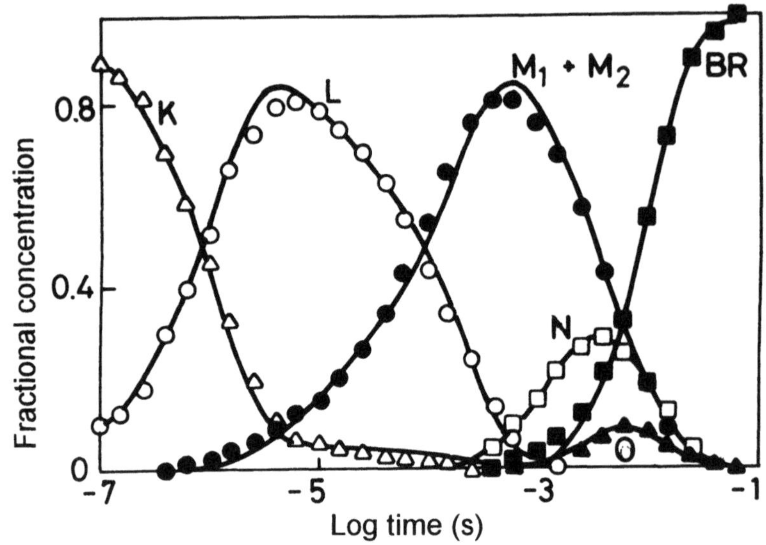


Fig. 13. Time dependence of the concentration of the bacteriorhodopsin intermediates.

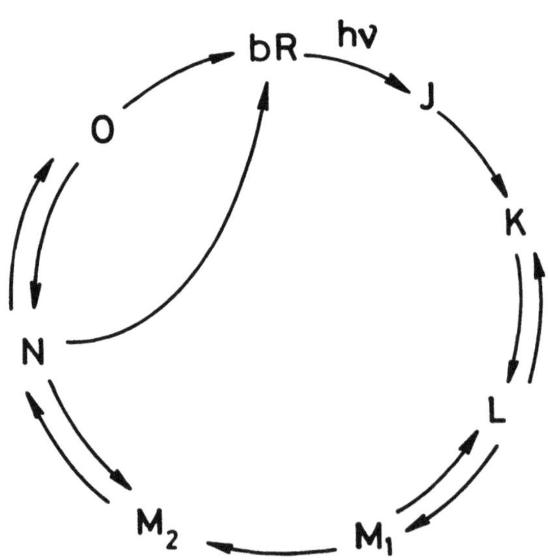


Fig. 14. The recent model of the bacteriorhodopsin photocycle.

sorbed photon energy and $M_1 \rightarrow M_2$ by photoacoustic measurements. In this way, the data in Fig. 15 were obtained. It is interesting to observe that the irreversible transition between $M_1$ and $M_2$ divides the photocycle into two parts. The K → L → $M_1$, and $M_2$ → N → O sequences follow each other with a very small change in free energy. Large changes occur when part of the photon energy is stored (the bR → J → K transitions), and during the $M_1 \rightarrow M_2$ and O → bR transitions.

The light absorption changes were the first data from which the photocycle was constructed. They do not, however, inform on the mechanism of the proton translocation. As discussed earlier, the absorption maxima are mainly determined by the charges around the retinal. Therefore, we have to assume that the color changes are caused by the changing distribution of charges in this region.

### *5.2. The retinal cycle*

The intermediates have characteristic resonance Raman spectra. They were collected either at low temperatures or with time resolved methods.

In low temperature spectroscopy, the different intermediates are "frozen". For example, at liquid nitrogen temperature, the illuminated bR goes to the K state and stays there for a very long time. If it is warmed, the L intermediate appears and further warming produces the M and O intermediates (N is difficult to obtain). The spectra are taken in such frozen

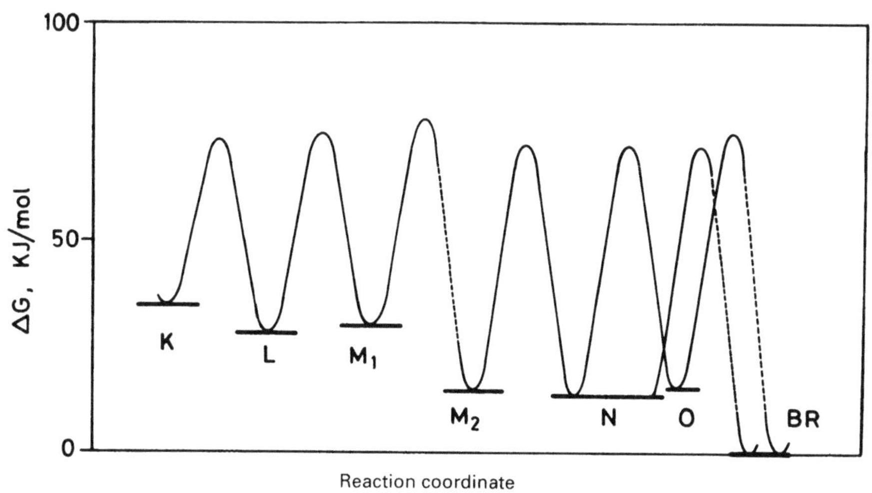


Fig. 15. The calculated free energy changes. Data at 20°C relative to the ground state of bR during the photocycle. Dotted lines indicate magnitudes based on calorimetric data (see text). The others are taken from the Arrhenius parameters.

states. Time dependent spectra, on the other hand, are registered by synchronized exciting and measuring laser pulses. The time delay is set by electronic devices.

From such measurements, it is known that in the photocycle of light adapted bR, the K-intermediate already contains a 13-cis retinal. This is different from the 13-cis retinal of the ground state (absorption max 548 nm) because of the different isomerization of the C = N bond (see Fig. 16) [5]. The retinal in the K intermediate is greatly distorted but relaxes during K → L transition. In the M intermediate, the Schiff-base deprotonates i.e. the proton moves away during the L → M transition. The retinal remains in its previous configuration. This configuration is kept during the $M_1 \rightarrow M_2$ transition, too. The large change in free energy is probably due to a large conformation change in the opsin part. The $M_2 \rightarrow$ N transition leads to reprotonation of the Schiff base but the retinal conformation does not change. During the N → O transition, the original all-trans conformation is reestablished.

The retinal cycle appears to be more simple than the light absorption cycle [5]. It has only two conformations: all-trans and 13-cis, C = N trans. Protonation and deprotonation do not influence the 13-cis conformation of the retinal.

### *5.3. The opsin cycle*

Changes in the opsin part of bR are studied by UV spectroscopy and infrared spectroscopy. Characteristic changes in the UV absorption spectrum were found between 220 and 330 nm in the M state(s). The assignment of the peaks to given amino acids in the opsin is not completely understood: the positive peak at 240 nm is due to tyrosine deprotonation, the negative peak at 275 nm caused by a tryptophan which is perturbed by a charge coming near to it, and the positive peak at 296 nm is unclear. It

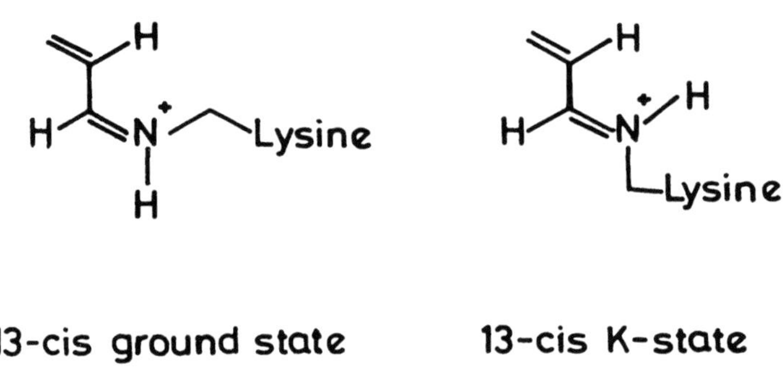


Fig. 16. Configuration of the 13-cis retinal in the ground state and in the K intermediate.

may be caused by tyrosine, tryptophan or lysine changes. The time dependence of the changes indicates a very fast appearance of the peaks at 275 and 296 nm during the bR → K transition. No component is seen corresponding to the K → L transition; the L → $M_1$,$M_2$ (without resolving the transition), and $M_2$ → bR however, are measurable.

The spectral changes in the UV range indicate processes going on in the opsin part of the protein that is correlated with the photocycle. To gain detailed information about the participation of key amino acids in the photocycle, a different approach is needed.

Fourier Transform Infrared Spectroscopy (FTIR) is, in principle, such a method for measuring the absorption due to the vibrations of different bonds, and is full of information as we have seen in Fig. 9. Spectra of the different intermediates could be registered by low temperature or time resolved FTIR techniques. The differences of the ground state and intermediate spectra contain this information, which is important for understanding the processes at the different amino acids.

The best understood changes in the FTIR spectra relate to protonation and deprotonation of aspartic acids. The results summarized in Table 1 show that Asp85 and Asp96 take part in the proton translocation. In the

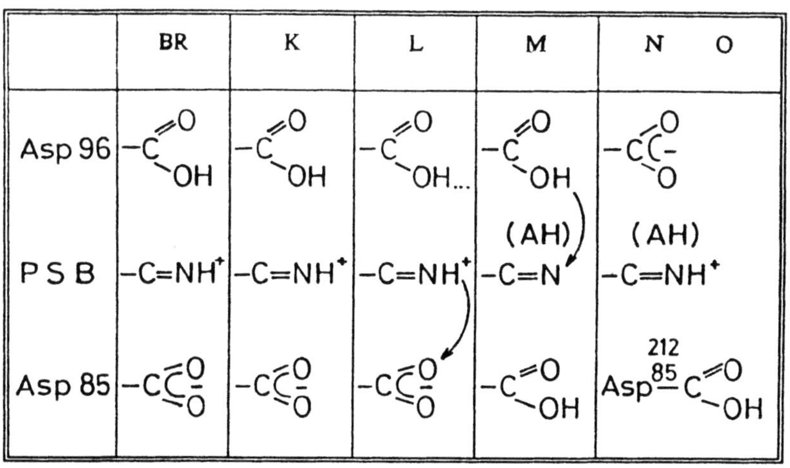


Table 1. Protonation changes of Asp85, Asp96 and the protonated Schiff-base (PSB) during the photocycle. Asp96 undergoes a hydrogen-bond change in L and is most likely deprotonated with the same apparent rate constant as the Schiff-base is reprotonated. This change seems to be the rate-limiting step: however, other groups (AH) undergoing faster protonation changes (e.g., bound water) might also be involved in this reprotonation step. Asp96 is reprotonated from the cytoplasmic side. Asp85 is protonated in the L → M transition with the same apparent rate constant as the Schiff-base is deprotonated. Another group (e.g., Arg82) must be involved in the proton-release pathway because a proton is ejected in the L transition to the exterior. The carbonyl vibration at 1755 $cm^{-1}$ indicates protonation of an internal aspartic acid, Asp85 or Asp212, in bR.

bR ground state, Asp85 is deprotonated and Asp96 is protonated. This situation changes only during the L → $M_1$ transition when the proton at the Schiff-base gets to Asp85. It disappears during the M → (N,O) bR transitions (details are not yet known). Asp96 donates a proton to the Schiff-base during M decay which is regained only in the O → bR transition. The assignment of changes in the infrared spectra of these specific aspartic acid side chains was confirmed by mutation of these amino acids to asparagine.

### *5.4. The charge motion cycle*

Proton translocation by bR involves proton motion inside the molecule. The direct observation of this charge motion is very important in understanding its mechanism. Different methods are used for this purpose which are based on the asymmetrically charged surfaces of the purple membrane (Fig. 17A). Because of the asymmetry the purple membranes attach themselves to a bilayer (Fig. 17B), for example, by their *b* surface. Under illumination, the charges will move in one direction and charge up the capacity of the system.

The charge asymmetry on a purple membrane means a permanent electric dipole moment. The purple membranes in a suspension are randomly oriented–but due to their permanent electric dipole moment–they may be oriented by an electric field in such a way that they are parallel and face in one direction. A field strength of 15–20 V/cm is enough to produce a more or less complete orientation. This orientation stays as long as the field is on and decays after it is off. This relaxation takes 300–500 ms [7].

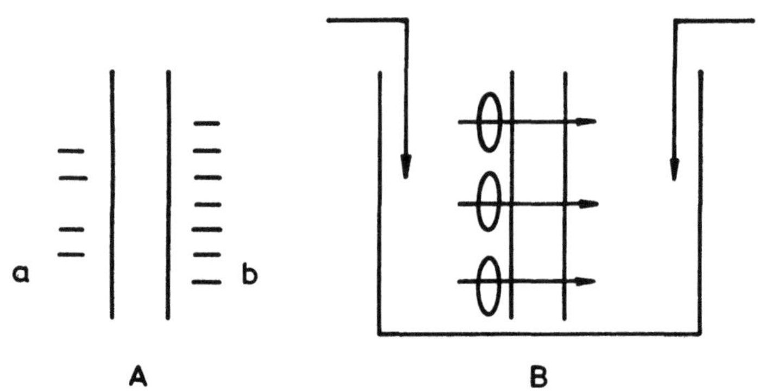


Fig. 17. The surfaces of a purple membrane are asymmetrically charged. Purple membranes adhere to charged bilayer membranes (A) in an oriented way (B). If excited in this form, the protons will move in one direction. A membrane potential or current can be measured.

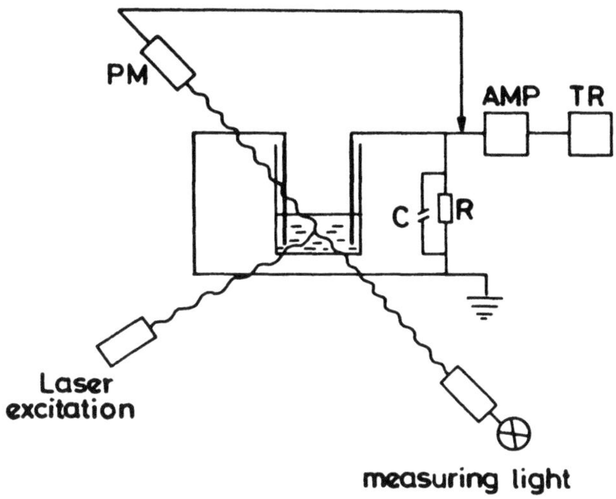


Fig. 18. Scheme of the apparatus to measure the protein electric response signal and light absorption changes on the same sample. PM-photomultiplier, AMP-amplifier, TR-transient recorder.

The oriented purple membranes can be immobilized by two methods: (1) Orientate them in an acrylamide solution (3–6% w/w) and then add a polymerizing agent to it. The formation of a polyacrilamide gel which contains mainly oriented purple membranes takes only ≈ 1 min. (2) Deposit the purple membranes on a glass plate covered with a transparent conducting layer (for example $SnO_2$) by electrophoresis and dry it.

In the oriented purple membranes (either in a suspension or after immobilization), the charges move in one direction and are synchronized if the photocycle is excited by a laser flash. Moving charges create a displacement current which is conducted from the purple membranes in the suspension to the electrodes (Fig. 18). The signal can be measured as a function of time. Figure 19 shows the electric signals in a suspension [8].

The advantage of this method is that the number of the excited molecules is so large that light absorption measurements can be performed on the same sample. Comparing the curves (Fig. 19), it is evident that the same time constants occur in the electrical and optical signals. Thus, it may be stated that the electrical signal, due to charge motion in bR, is correlated to the photocycle. It is interesting to note that the first and second component (bR → K, K → L) of the electric signal is negative i.e. it is opposite to the direction of proton translocation.

A simplified analysis is given here in order to explain the electric signals and interpret their amplitudes.

Consider a single oriented pm and assume that large planar electrodes are in contact with it and separated by a distance $D'$ (Fig. 20). A current $I$

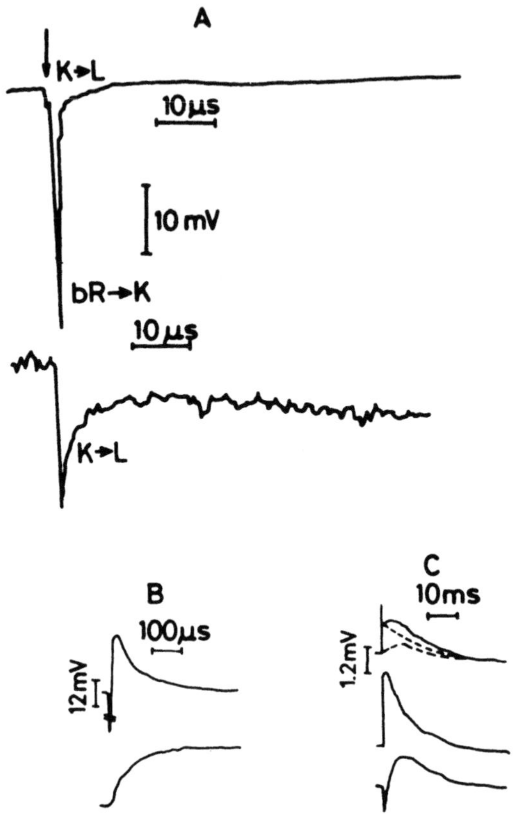


Fig. 19. Comparison of the electrical and absorption signals. (A) ms time domain at 5°C. The absorption signal (lower curve) due to the appearance of the $L_{550}$ intermediate from $K_{590}$ was measured at $\lambda = 525$ nm. Because of the overlap of the spectra of the intermediates it is the superposition of many components, nevertheless the timing of K → L transition is clearly visible. (B) μs time domain at 22°C, lower curve: absorption at 410 nm (C) ms time domain at 22°C, upper curve: the electrical signal decomposed according to the decay of the M intermediate (middle curve: absorption at 410 nm) and of the O intermediate (lower curve: at 625 nm).

is induced in the external circuit if a charge $Q$ moves between the electrodes,

$$I(t) = \frac{Q\,v(t)}{\varepsilon D'} \tag{1}$$

where $v(t)$ is the velocity of the charge and $\varepsilon$ is the permittivity of the purple membrane. We assume that the velocity $v(t)$ is very large, i.e. the charge jumps between the two points. Integrating Eqn. 1 for time from zero to infinity, we obtain

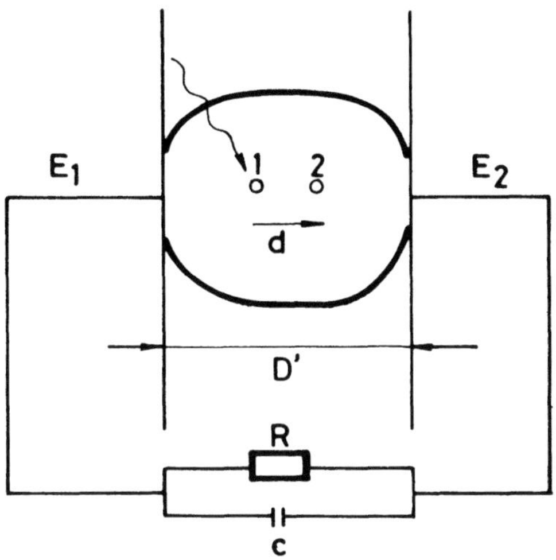


Fig. 20. A simplified diagram of the production of the displacement current. bR is embedded in the membrane. Upon light absorption, the charge moves a distance *d* from point 1 to point 2. $E_1$ and $E_2$, hypothetical electrodes, detect the displacement current due to this charge motion. The *RC* circuit peaks up the current.

$$Q_{ind} = \frac{Qd}{\varepsilon D'} \qquad (2)$$

because time integral of $v(t)$ is just the distance.

The jump of charge $Q$ induces a charge $Q_{ind}$ proportional to the displacement $d$. $Q_{ind}$ charges the capacitance of the electrodes and that of the external circuit $C$. The capacitance discharges through a resistance $R$. The time-course of the voltage for this single charge jump is

$$U_1(t) = \frac{Q_{ind}}{C} e^{-t/RC} \qquad (3)$$

In reality, the electrodes are in the sample separated by a distance $D$ and $N$ charges are moving with a given time distribution. Both have to be taken into account. Two processes are involved in the calculation: (1) we have to determine $U_1(t)$ for electrodes at a distance $D$; and (2) sum the $U_1(t)$ functions for $N$ particles. For (1) we perform a very rough approximation, we simply move the hypothetical electrodes in Fig. 20 from the surface of the purple membrane to a distance $D$. In Eqn. 2, $D'$ is then exchanged to $D$. Because the thickness of the membrane $D'$ is $\approx$ 5 nm and $D$ is $\approx$ 1 cm, this means an $\approx 10^6$ fold decrease in $Q_{ind}$.

For (2), we assume a simple exponential decay of the $N$ states excited at $t = 0$ with a rate constant $k$. After some calculation [8] we obtain

$$U_N(t) = \frac{NQdk}{\varepsilon D} \frac{R}{1 - kRC} (e^{-kt} - e^{-t/RC}) \tag{4}$$

where ε is the dielectric permittivity.

This expression is valid for a single transition. If there are many components of the charge motion in series–which is the case in bR–then we have to take into account that the number of decaying states per unit time has several exponential components with different preexponential factors. The simplified equation (when all the *k* values are smaller than *RC*) for the *i*-th component is

$$U_N^i(t) = \frac{NQR}{\varepsilon_i D} d_i\, k_i\, f(k_1, k_2, \dots, k_i) \tag{5}$$

In this expression, $U_N^i(t)$, $k_1$, $k_2$, ... , $k_i$, $D$ and $R$ are measured values. The number of photocycles, $N$, is determined from the light absorption signal due to the M intermediate.

Knowing the value of $N$ leaves only $Q$, $\varepsilon_i$ and $d_i$ as unknowns in Eqn. 5. We assume that a single proton is moving through bR ($Q = 1.6 \times 10^{-19}$ C) and that the protein has a homogeneous permittivity $\varepsilon_i \approx 2$. The first assumption is based on the fact that most authors find a stoichiometry of 1 proton/photocycle of the proton pump. The protein does not have a homogeneous dielectric constant but no data exists for its distribution.

The only unknown in Eqn. 8 now is $d_i$. Calculations based on the measured values of $U_N^i(t)$ (Fig. 19) give values for the different displacements. Naturally $\Sigma d_i$ should be equal to ≈ 5 nm (the thickness of the membrane). It is a great surprise that the above simplified treatment provides a $\Sigma d_i$ = 20 nm (i.e. the right order of magnitude). After normalization, we obtain values of $d_i$ corresponding to the proton displacement in each step of the photocycle (Table 2).

Table 2. Distances of the proton movements during the photocycle of bacteriorhodopsin

| Transition | $d_i$(nm) normalized | $d_i$(nm) geometry | $\varepsilon_i/\varepsilon$ |
|---|---|---|---|
| bR → K | −0.13 | −0.13 | 1 |
| K → L | −0.02 | −0.02 | 1 |
| L → M | +0.5 | +0.5 | 1 |
| M → O | +3.1 | +2.2 | 1.4 |
| O → bR | +1.5 | +2.4 | 0.63 |
| $\Sigma d_i$ | +4.95 | +4.95 | |

In the table the distances taken from the geometry of bR (see Fig. 20) were used to calculate the deviations of the permittivities from the assumed homogeneity.

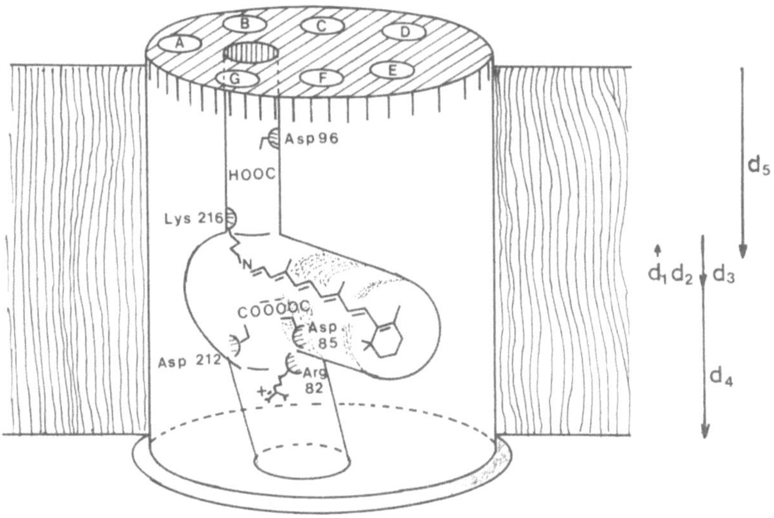


Fig. 21. Scheme of the channels in bacteriorhodopsin. Channel leading from the Schiff-base outward (lower part of the figure) and inward to the Schiff-base (upper part). The important amino acid residues Asp85, Asp96, Lys216, Arg82 and the retinal are shown. Arrows show the values of $d_i$-s as they appear in the structure of bR.

The $d_i$-s determined only represent displacements in the transmembrane direction. The lateral components of the displacements do not induce current in the electrodes because of the symmetry of bR in the membrane.

The question arises whether the distances $d_i$ have any meaning, i.e. do they correctly express that protons are translocated in five different steps according to the lifetimes of the intermediates. We discuss each step in detail.

The first displacement, $d_1$, occurs in the opposite direction to proton translocation, therefore, it has a negative sign. Separate experiments with ps laser excitation and fast current recording have shown that it appears faster than 5 ps. It may be due to either the bR → J or the J → K transition or both. The presently available time resolution is not enough to separate the events. It is known from the retinal cycle that the trans-cis isomerization is largely finished in the K intermediate, though some distortion of the retinal remains. The isomerization involves the mechanical motion of the proton bound to the Schiff-base, therefore, the first electrical signal is assigned to this isomerization.

The rather small negative displacement $d_2$ may correspond to the relaxation of the observed torsions of the 13-cis retinal during the K → L transition.

The first positive motion, $d_3$, coincides with the deprotonation of the Schiff-base during the L → M transition. According to the opsin cycle,

Asp85 gets protonated during this step. We think that the Schiff-base proton jumps over to Asp85. $d_3 \approx 0.5$ nm is in good agreement with the distance between the Schiff-base and Asp85 known from the bR structure.

Displacements $d_1$, $d_2$ and $d_3$ are deep inside the protein in close proximity to the Schiff-base. The two last components, $d_4 = 3.1$ nm and $d_5 = 1.5$ nm, are calculated from the signals with time constants of the M → O and O → bR transitions. They connect the interior of bR with the exterior. (The resolution of the electric signals is not yet good enough to register signals due the $M_1 \rightarrow M_2$ and M → N, N → O transitions). In our interpretation, $d_4$ is the path for the proton to go outward and $d_5$ to pick up the proton from inside and reestablish the bR ground state with a protonated Schiff-base.

The three-dimensional structure of bR (Fig. 6) locates the Schiff-base in the middle of the molecule. $d_4$ and $d_5$ seem to be in contradiction with this data. We have to take into account, however, that a homogeneous permittivity $\varepsilon$ was assumed. The structure shows that the outward channel is broad and contains many water molecules while the channel from the cytoplasmic site is narrow and contains mainly hydrophobic side chains (Fig. 21). The permittivity in the broader outward path is probably larger than in the narrow one. Thus, the non homogeneous $\varepsilon$ value may resolve the contradiction between the values of $d_4$ and $d_5$ in Table 2, and the structural data.

The electrical signals report about the charge motion in the protein. The components found have the same lifetimes as the optical intermediates (neglecting the mentioned fine details which need further clarification) and the distances of the steps in charge motion are in reasonable agreement with the structure of bR.

We find that the charge motion cycle coincides with the light absorption, retinal and opsin cycles. All processes observed, such as isomerization of the retinal, protonation changes of the amino acids, and the color changes are involved in the main task of bR: proton translocation.

### *5.5. Quantum efficiency and stoichiometry*

The quantum efficiency is defined as the number of photocycle per number of absorbed photons. The stoichiometry is defined as the number of protons translocated per photocycle.

The measurement of these two quantities is difficult. Only the most widely accepted results are given here. The quantum efficiency is ≈ 0.6, i.e. the probability that an absorbed photon starts a photocycle is ≈ 60%. The stoichiometry is very probably 1, which means that a single proton is translocated per photocycle.

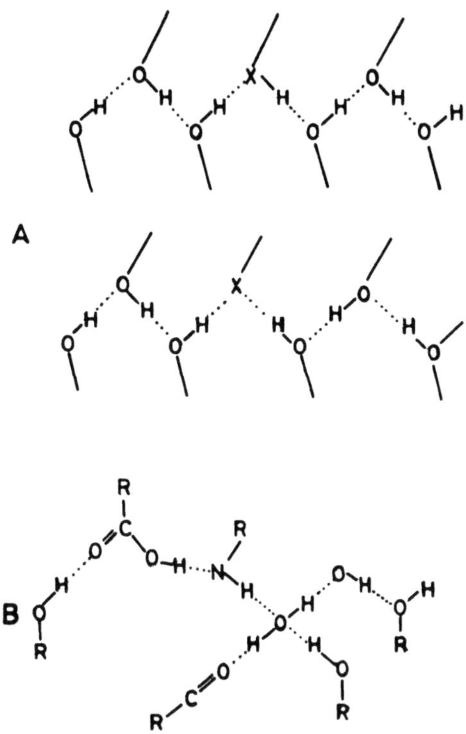


Fig. 22. Proton conduction through a hydrogen bonded chain. According to the theory, in the first step the $H^+$ ion moves (A, upper figure) and in the second step the fault moves (A, lower figure). The net result is the translocation of the proton. (B) Hydrogen bonded chains may be built from the side chains of amino acids in a protein.

### *5.6. Mechanism of proton translocation*

The energy of the absorbed photon from any direction is partially transformed into a vectorial (from the inside to the outside of the membrane) motion of protons. Calorimetric measurements establish that when a photocycle is started (60% of all cases of absorption), 30% of the photon energy is retained in the bR molecule. This energy is mainly electrostatic, i.e. by isomerization, the Schiff-base proton moves away from its counterion. This energy may be expressed as

$$E_{elstat} = \frac{e^2}{\varepsilon_1 r_0} - \frac{e^2}{\varepsilon_2 r} \qquad (6)$$

where $e$ is the elementary charge and $r_0$ and $r$ are the distances between the Schiff-base proton and the counterion before and after the isomerization, $\varepsilon_1$ and $\varepsilon_2$ are the permittivities of the protein, respectively. The electrical signal shows that this motion has a component in the direction opposite to that of proton translocation. The vectorial direction of this motion is independent from the direction of the incoming photon. Therefore, this very first step of the photocycle (the bR → J → K transition) may be considered the step of energy uptake and determination of the direction of the proton transfer.

This electrostatic energy is confined to the Schiff-base region until the L → $M_1$ transition when the proton moves away to Asp85. Probably during the $M_1$ → $M_2$ transition, the energy is transferred to the opsin part of the molecule and the proton appears externally. During the L → M transition (it is not yet resolved to $M_1$ → $M_2$), data exist that show substantial changes occurring in the geometry of protein (for example neutron scattering data reveal ≈ 4% change in the direction of one of the helices). This transition probably exposes the unprotonated Schiff-base to the intracellular environment from where–with the relay of Asp96–a proton is picked up.

The mechanism of proton motion in the long paths to the extracellular surface and from the intracellular surface is still unclear. The model of Nagle and Tristram-Nagle [9] assumes that a hydrogen bonded chain is built up from amino acid side chains (Fig. 22) and the protons move through it as in ice. This proton wire conception needs a special structure which is not apparent in the three-dimensional structure of bR (Fig. 21). Only Asp85 and Asp96 are known as proton acceptors and donors.

An alternative scenario approached from other data which show that proteins have a multidegenerate ground state and that a protein fluctuates between these "conformational substrates" (i.e. different energy minima) [10]. From this motion–which is called protein fluctuation–channels open and close in a protein. When they are open, water molecules can penetrate the channels. The protons move in these channels as in water (i.e. as hydroxonium ions) driven by the existing electric field which changes after isomerization, after proton release from Schiff-base and after structural changes in the opsin part of the molecule.

This fluid water conduction model is an alternative to the proton wire model. Its advantages are: (1) it is based on general features of proteins; (2) it does not require a special structure of the protein; and (3) it is applicable to other ions, for example to $Cl^-$ in the case of halorhodopsin.

## 6. Halorhodopsin

*Halobacterium salinarium* contains a second light-energy transducing protein: the halorhodopsin (hR). This protein is similar to bR in many respects [11]. Its functional role in the bacteria is, however, different. It pumps $Cl^-$ ions from the environment into the cell upon the absorption of light. This active uptake of $Cl^-$ is important for growing cells to maintain isotonic conditions.

The distribution of hR in the plasma membrane of the cells is not well defined (unlike the distribution of bR in purple membrane). After isolation by detergents, its characteristics may be studied. Much less is known about hR than bR because of its smaller concentration in the cells, and the difficulty in isolating and storing it.

### *6.1. Structure of halorhodopsin*

hR is a membrane bound protein. Its amino acid sequence has been determined from the haloopsin gene. The 251 amino acids are arranged into seven helices with connecting parts and leading and terminating sequences. The molecular mass of hR is 26 961 daltons. The transmembrane part of hR is 36% homologous with bR. hR has a retinal which is bound to lys 242 by the protonated Schiff-base. In the dark adapted form, both the 13-cis and all-trans isomers of the retinal exist. hR cannot be transformed completely by light adaptation to all-trans hR, though this is the form which pumps $Cl^-$ ions.

### *6.2. Photocycle of halorhodopsin*

The absorption maximum of hR in the ground state is at 578 nm provided the salt (NaCl, KCl, $NaNO_3$, etc.) concentration of the solution is at least 100 mM. Without $Cl^-$ or $NO_3^-$, the absorption maximum shifts to 568 nm. By light excitation, specific absortion changes occur (as in bR) which change the position of the absorption maxima in time.

The very first intermediate appears in 5 ps (compared to 0.7 ps in bR) and is red shifted to 598 nm. The next intermediate, $hR_{KL}$, has its absorption maximum at 580 nm which is near the ground state. It decays quickly to the most important intermediate, $hR_L$, at 520 nm. $hR_L$ has a long lifetime, about 10 ms, and in presence of $Cl^-$ ions equilibrates with the $hR_0$ intermediates with absorption maximum at 640 nm. From these and other data, a photocycle has been composed (Fig. 23).

The intermediates are characterized as follows: In $hR_K$ state, the all-trans retinal is already in the 13-cis state. The $hR_{KL}$ intermediate is probably the result of a relaxation of some tension in the retinal. The $hR_L$ state, which appears quickly, is the most important intermediate. From this an equilibration involving $Cl^-$ release leads to the $hR_0$ states and $hR_0$ goes through $hR_{568}$ and $Cl^-$ uptake to the ground state.

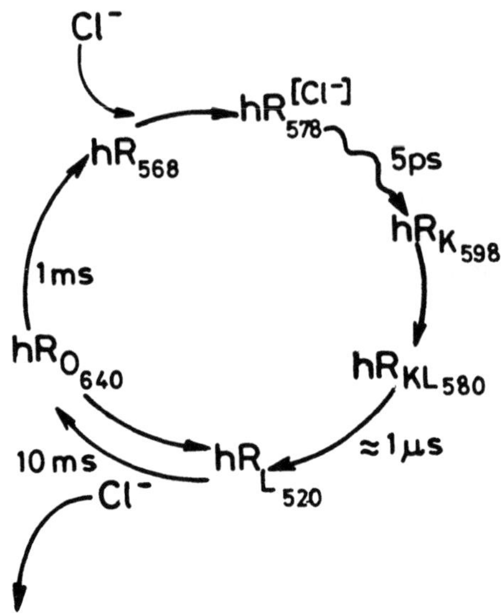


Fig. 23. The photocycle of halorhodopsin.

It has been shown that hR can translocate other ions ($Br^-$, $I^-$, $NO_3^-$) too. However, because much less work has been devoted to hR, the parallel photocycles: the retinal, opsin and charge motion cycles are uncertain, as yet.

## Acknowledgement

The financial support by the National Scientific Research Fund of Hungary (OTKA T914) is gratefully acknowledged.

## References

1. W Stoeckenius, RH Lozier and RA Bogomolni, Biochim. Biophys. Acta 505 (1979) 215.
2. JK Lanyi, Biochim. Biophys. Acta 1183 (1993) 241.
3. D Oesterhelt and W Stoeckenius, Meth. Enzymology 31 (1974) 667.
4. R Henderson, JM Baldwin, TA Ceska, F Zemlin, E Beckermann and KH Downing, J. Mol. Biol. 213 (1990) 899.
5. SO Smith, J Lugtenburg and RA Mathies, J. Membr. Biol. 85 (1985) 95.
6. G Váró and JK Lanyi, Biochemistry 30 (1991) 5016.
7. K Barabás, A Dér, Zs Dancsházy, P Ormos, L Keszthelyi and M Marden, Biophys. J. 43 (1983) 5.

8. L. Keszthelyi and P Ormos, J. Membr. Biol. 103 (1989) 193.
9. JF Nagle and S Tristram-Nagle, J. Membr. Biol. 74 (1983) 1.
10. P Ormos, D Braunstein, H Frauenfelder, MK Hong, SL Lin, TB Sauke and RD Young, Proc. Natl. Acad. Sci. USA 85 (1988) 8492.
11. JK Lanyi, Physiol. Rev. 70 (1990) 319.


## Appendix A. Abbreviations:

bR: bacteriorhodopsin
hR: halorhodopsin
UV: ultraviolet spectroscopy
FTIR: Fourier-transform infrared spectroscopy

Bioenergetics
ed. by P. Gräber and G. Milazzo†


# *CHAPTER 12*
# Structure, function and regulation of the $H^+$-ATPases from chloroplasts

Peter Gräber

*Institut für Physikalische Chemie, Albert Ludwigs University, Freiburg, Germany*



## 1. Introduction

Biological systems exchange energy and matter with their surroundings: thermodynamically they are open systems (see chapter 1). For capturing of light and for uptake of nutrients a number of different mechanisms have been invented during the evolution, some of them are described in the preceding chapters. Light energy and the energy of nutrients cannot be used directly in the cells as energy source for the different biochemical processes connected with life: work (locomotion, breathing etc.), synthesis of new compounds (regeneration, reproduction), transport of nutrients, ions, etc. All biological systems use as direct energy source adenosinetriphosphate, ATP, i.e. they use the free enthalpy change ("Gibbs free energy change") connected with the hydrolysis of the β-γ-phosphate bond of ATP for driving biochemical reactions. The structure of ATP is shown in Fig. 1.

Synthesis of ATP on one hand and hydrolysis of ATP on the other hand connect the energy-providing reactions with the energy-consuming reactions. This organisation of chemical reactions is found in all living systems from bacteria to plants and man. Figure 2 shows schematically the relation between the different reactions: in photosynthetic organisms light energy is used to transfer electrons from water to $NADP^+$. The electron transport occurs vectorially across the photosynthetic membrane and is connected with a transmembrane proton transport. This leads to the generation of a transmembrane electric potential difference, $\Delta\varphi$, and a transmembrane pH difference, ΔpH, i.e. to a transmembrane electrochemical potential difference of protons, $\Delta\tilde{\mu}_{H^+}$. In respiring organisms electrons are transferred from nutrients (generally reduced substrates) via NADH to oxygen. The electron transport from NADH to oxygen occurs vectorially in the membrane of the mitochondria (or the plasma membrane) and is connected with a transmembrane proton transport. As in photosynthetic membranes also in respiring membranes a transmembrane electrochemical potential difference of protons, $\Delta\tilde{\mu}_{H^+}$ is generated.

In both membranes $\Delta\tilde{\mu}_{H^+}$ causes a back flux of protons across the membrane and the free enthalpy released by this process is used by the membrane bound $H^+$-ATPase to synthesize ATP from ADP and phosphate. ATP is then transported to the site where energy is required and the free enthalpy released by its hydrolysis is used to energize the various biochemical reactions. Figure 2 emphasizes the central role of $H^+$-

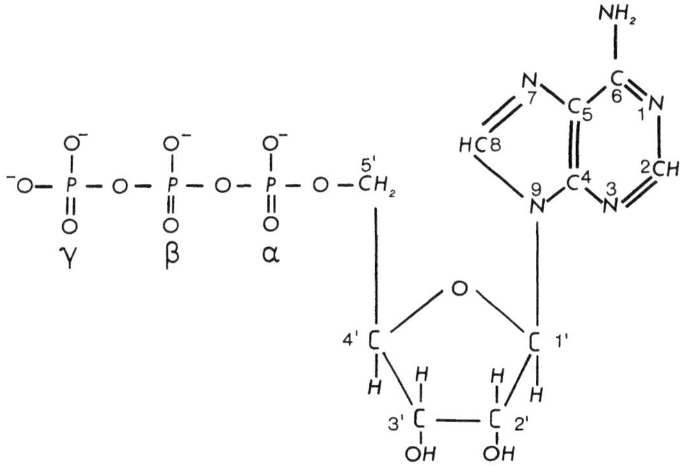

Fig. 1. Structure of ATP.

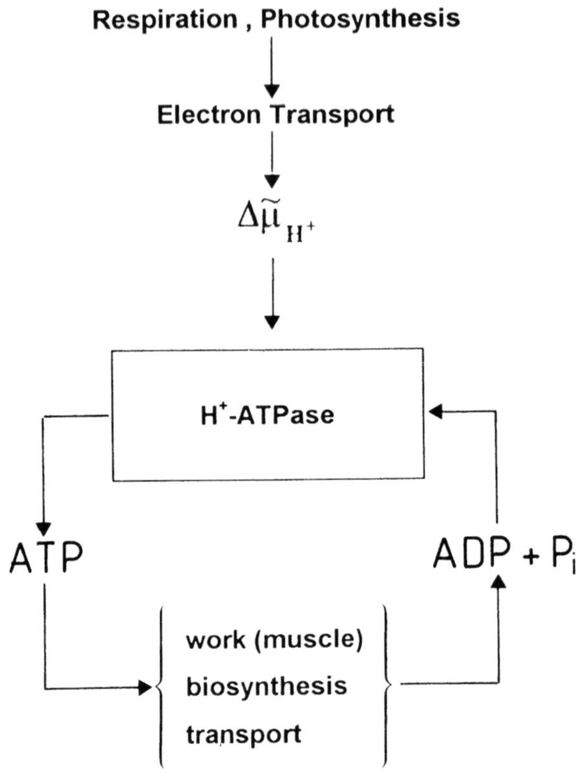


Fig. 2. Scheme of coupling between energy supplying and energy consuming processes in the cell.

ATPases in the coupling between the energy providing and the energy consuming reactions in cells. Cells use energy almost exclusively in the form of ATP, and a diversity of biochemical pathways and the corresponding molecular machines (see preceding chapters) have been invented to provide energy for ATP synthesis by $H^+$-ATPases.

A few quantitative remarks may emphasize the central role of the ATP/ADP couple in the energetics of biological systems. Humans require an energy between 7000 kJ (resting state) and 14 000 kJ (heavy work) per day. From this energy only a part is conserved in the energy of ATP. This part can be estimated as follows: oxidation of 1 mol glucose gives a free enthalpy change of $\Delta G^{0\prime} = -2900$ kJ mol$^{-1}$. The stoichiometry of the reactions in the biochemical pathway from glucose to $CO_2$ and $H_2O$ resulted in the formation of 38 mol ATP per mol oxidized glucose. Under standard conditions the free enthalpy for ATP synthesis is $\Delta G_p^{0\prime} = 34$ kJ mol$^{-1}$. From the actual concentration of the reactants in mitochondria it is estimated that the free enthalpy change is $\Delta G_p^{\prime} = 40$ kJ mol$^{-1}$. The amount of energy conserved in ATP per mol oxidized glucose is then $\Delta G^{\prime} = 38$ mol × 40 kJ mol$^{-1}$ = 1520 kJ. Obviously, about half (1520/2900 ≈ 0.5) of the free enthalpy change from glucose oxidation is conserved in the β-γ-phosphate bond of ATP. The amount of ATP can be calculated also for the oxidation of other nutrients (protein, fat). The result is always that about half of the free enthalpy of the nutrients is conserved in useful energy, i.e. between 3500 kJ (resting state) and 7000 kJ energy is temporarily stored in ATP. Since ATP hydrolysis gives a free enthalpy change of $\Delta G_p' = 40$ kJ $mol^{-1}$, approximately 3500 kJ/(40 kJ $mol^{-1}$) = 87.5 mol (resting state) to 7000 kJ (40 kJ $mol^{-1}$) = 175 mol ATP are synthesized per day. ATP has a molecular mass of about 500 g $mol^{-1}$ and this means that between 87.5 mol × 0.5 kg $mol^{-1}$ = 44 kg and 175 mol × 0.5 kg $mol^{-1}$ = 87.5 kg ATP are synthesized in man. Of course, the same amount is hydrolyzed again, resulting in a high turnover of ATP and ADP. The turnover can be estimated as follows: a rate of (87.5–175) mol per day corresponds to a rate of about (1–2) × $10^{-3}$ mol ATP per s. Based on an average concentration of 5 mM ATP in a volume of 10 l the total pool of ATP is 0.05 mol. The turnover is then given by 0.05 mol/(1–2) × $10^{-3}$ mol $s^{-1}$ = 50–25 s. In this time the complete pool of ATP is hydrolyzed and has to be resynthesized. Of course, this is a very crude estimate based on average values, and depending on the type of tissue the actual values might be different, however, it demonstrates the large amount of synthesized ATP and its high turnover. These considerations show that the energy relevant for the biological system is not simply the free enthalpy change due to oxidation of the nutrient. Only that part can be used which is conserved in the free enthalpy change of the ATP/ADP couple, and this part depends of course on the detailed metabolic pathway. Therefore, the energy of different compounds was expressed in "ATP equivalents" and a system was developed for calculating the energetics of biochemical reactions on this basis [1].

The enzymes which catalyze ATP synthesis in bacteria, mitochondria and chloroplasts are called F-type $H^+$-ATPases. Other $H^+$-ATPases which catalyze ATP hydrolysis and proton transport across membranes are called V-type or P-type ATPases [2]. All F-type ATPases have the function to synthesize ATP, i.e. they use the free enthalpy ("Gibbs free energy") derived from the transmembrane proton transport to synthesize the β-γ-phosphate bond in ATP from ADP and phosphate. All F-type $H^+$-ATPases have a similar structure: they consist of a membrane-integrated, hydrophobic part, $F_0$, which contains the proton binding sites, and a hydrophilic part, $F_1$, which contains the nucleotide and phosphate binding sites. In order to identify the source of the enzyme one letter is added to the symbol $F_0F_1$, e.g., $EF_0F_1$ (for *E. coli*), $CF_0F_1$ (for chloroplasts), $MF_0F_1$ (for mitochondria), etc.

In this chapter the structure, function and regulation of $H^+$-ATPases is described using data obtained with the $H^+$-ATPase from chloroplasts, the enzyme which is investigated in our laboratory. Presumably, the structure of the enzymes from different sources is similar and their mechanism is the same, only the regulation of the enzyme activity depends on the source of the $H^+$-ATPase. A number of review articles dealing with different aspects of these enzymes have appeared in the recent years and for more detailed information the reader is referred to these [3–16].

## 2. Structure of $H^+$-ATPases

### *2.1. Subunit composition, isolation and reconstitution*

ATPases are multi-subunit enzymes and there are differences in the subunit composition for enzymes from different sources: $EF_0F_1$ has eight different subunits, $CF_0F_1$ nine and $MF_0F_1$ at least 13 different subunits. Table 1 summarizes data on subunits, stoichiometry of subunits and molecular masses. Homologous subunits are shown in the same row of Table 1. Molecular masses are calculated from the gene derived aminoacid sequences. There is a general agreement about the subunit stoichiometry in

Table 1. Subunits of F-type $H^+$-ATPases
Homologous subunits of the different enzymes are given in the same rows. Subunit stoichiometries of the $F_0$ parts are estimated. $IF_1$ is the inhibitor protein of $MF_0F_1$.

| *E. coli* | | Chloroplasts (spinach) | | Mitochondria (bovine) | |
|---|---|---|---|---|---|
| Subunits (stoichiometry) | Mole mass g mol$^{-1}$ | Subunits (stoichiometry) | Mole mass g mol$^{-1}$ | Subunits (stoichiometry) | Mole mass g mol$^{-1}$ |
| α (3) | 55 786 | α (3) | 55 793 | α (3) | 55 164 |
| β (3) | 50 505 | β (3) | 54 238 | β (3) | 51 595 |
| γ (1) | 31 802 | γ (1) | 36 121 | γ (1) | 30 141 |
| δ (1) | 19 491 | δ (1) | 20 665 | OSCP (1) | 20 967 |
| ε (1) | 15 065 | ε (1) | 14 801 | δ (1) | 15 065 |
| - | | - | | ε (1) | 5652 |
| $EF_1$ | 385 534 | $CF_1$ | 401 740 | $MF_1$ | 392 102 |
| a (1) | 30 639 | IV (1) | 24 488 | a (1) | 24 815 |
| b (2) | 17 408 | I (1) | 19 322 | b (2) | 24 670 |
| | | II (1) | 16 655 | | |
| c (12) | 8755 | III (12) | 7958 | c (12) | 7608 |
| | | | | d (1) | 18 603 |
| | | | | $F_6$ (1) | 8958 |
| | | | | e (1) | 8189 |
| | | | | f (1) | 10 209 |
| | | | | g (1) | 11 328 |
| | | | | $A_6L$ (1) | 7964 |
| $EF_0$ | 170 515 | $CF_0$ | 155 961 | $MF_0$ | 244 896 |
| $EF_0F_1$ | 556 049 | $CF_0F_1$ | 557 701 | $MF_0F_1$ | 636 978 |
| | - | | - | $IF_1$ (1) | 9578 |

the $F_1$-part, however, for the $F_0$-part, especially for subunit c (III) different stoichiometries are reported. The reason for the differences in subunit composition of the enzymes from different organelles is not yet clarified, however, the enzymes developed later in the evolution have the higher number of subunits. Presumably, the more complex enzymes allow better regulation of the enzyme activity according to the physiological requirements.

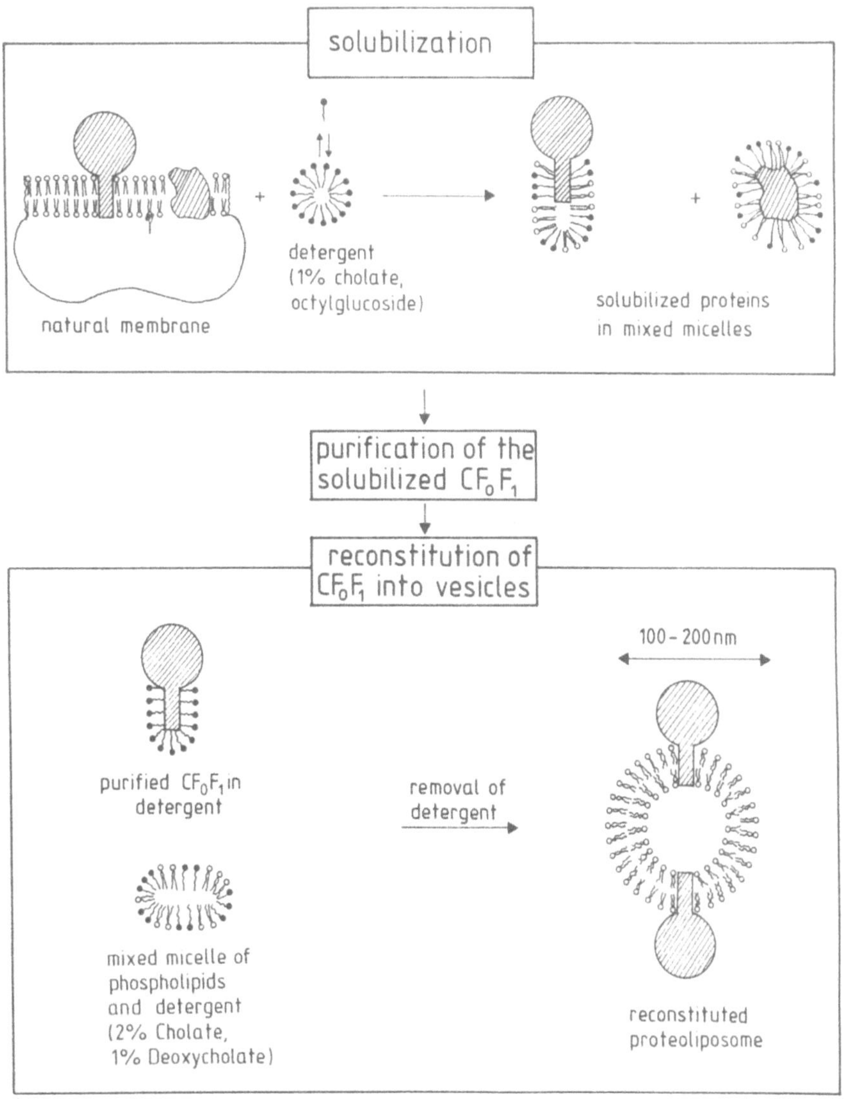


Fig. 3. Scheme for isolation, purification and reconstitution of the $H^+$-ATPase from chloroplasts.

The subunits of $CF_0F_1$ (and $MF_0F_1$) are partly encoded in the DNA of the nucleus of the cell, partly in the DNA of chloroplasts (or mitochondria). For $CF_0F_1$ subunits γ,δ and II are nuclear encoded, the others are encoded in the chloroplast.

The $H^+$-ATPase from chloroplasts can be isolated, purified and reconstituted into liposomes. Figure 3 shows schematically the procedure. Thylakoid membranes are treated with detergents. Thereby, the natural lipid environment of the membrane proteins is substituted by the detergent. The proteins are now surrounded by a detergent-lipid micelle which makes the protein soluble ("solubilization"). This occurs without denaturation when "appropriate" detergents are used. The soluble proteins are purified by usual methods, e.g., density gradient ultracentrifugation, chromatography, etc. For electronmicroscopic studies the soluble protein can be used directly.

In order to test the function of the $H^+$-ATPase the enzyme must be incorporated in a membrane which separates two compartments. Usually, it is incorporated into liposome membranes ("reconstitution"). The protein detergent micelle is mixed with a lipid detergent micelle and then the detergent is removed slowly, for example, by dialysis. Under these conditions liposomes with incorporated $CF_0F_1$ are formed spontaneously. The enzyme can also be reconstituted into preformed liposomes. In this case the liposome membrane is destabilized by addition of a small amount of detergent, the enzyme is added and finally the detergent is removed again. The reconstituted $CF_0F_1$ can be used for electron microscopy studies and for functional studies. It shows the same activity (rate of ATP synthesis, ATP hydrolysis, regulation of the rate) as the enzyme in its natural membrane [17].

## *2.2. The structure of $CF_1$*

The hydrophilic $F_1$-part can be separated from the membrane-integrated $F_0$-part. Since no detergents are present in such $F_1$ preparations, the investigation of the $F_1$-parts is greatly facilitated. It has been possible to obtain three-dimensional crystals from $MF_1$ which allow x-ray analysis [18, 19]. We have investigated the structure of isolated $CF_1$ by electron microscopy with negatively stained samples [20]. Figure 4 shows an electron micrograph of negatively stained $CF_1$. The enzyme molecules show a clear subunit structure. Most of the molecules are oriented on the carbon support film in such a way that a hexagonal subunit structure can be seen. This view is called the hexagonal or top view. Some examples of this orientation are surrounded by a hexagon. In some cases a different orientation is seen. The molecules show a small protrusion on one side. This view is called a side view, and some examples are surrounded by a square with a

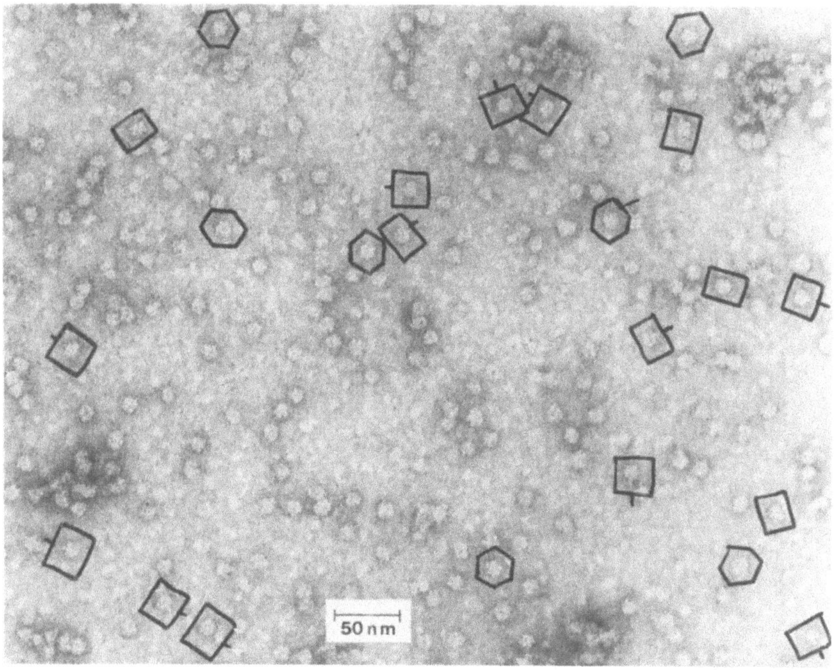


Fig. 4. Electronmicrograph of $CF_1$. The sample was stained with 1% uranylacetate. Top views are surrounded by a hexagon, side views by a square with a protrusion.

protrusion. The low signal-to-noise ratio does not allow interpretation of the finer details of the single molecules directly – or we have to select randomly one of the molecules which seems to look "good". This approach is, however, very arbitrary. Therefore, the signal-to-noise ratio has been enhanced by averaging the images of similarly oriented molecules [21]. First, the electron micrograph (e.g., Fig. 4) is scanned with a microdensitometer and the light/dark-pattern of the different molecules is measured. Each molecule is resolved in 64 × 64 elements (pixels). One pixel is a square with an area of 0.36 nm × 0.36 nm. These digitized images are stored in a computer and one of the images is chosen as a first-reference molecule. The variance of the light-dark pattern of each image is normalized. Then, these images are oriented with regard to the reference molecule; i.e., they are rotated over an angle and also shifted translationally until there is an optimum fit of the light/dark-pattern with respect to the reference. Finally, the oriented images are added to give a summation image which now has an improved signal-to-noise ratio. This summation image is then taken as a new reference molecule and the whole sequence is started again until there is no further change. This procedure is carried out with different reference molecules ("image analysis").

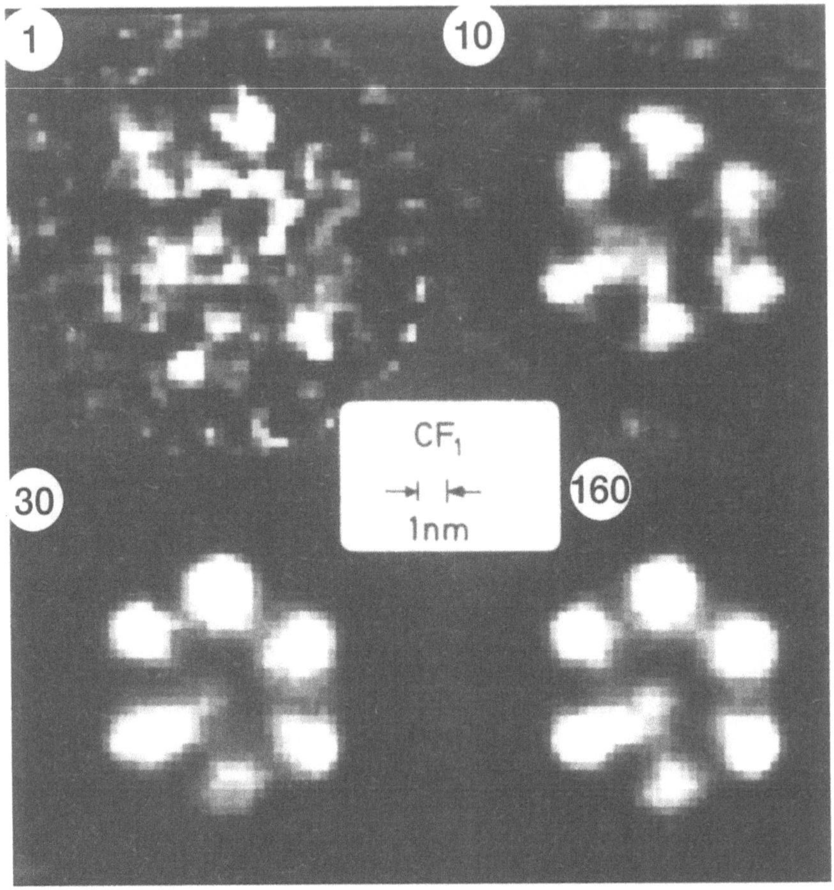


Fig. 5. Improvement of signal-to-noise ratio by addition of the images of different $CF_1$ molecules. A: digitized image of one $CF_1$ molecule 64 × 64 pixels. One pixel corresponds to an area of 0.36 nm × 0.36 nm. B: addition of 10 images after orientation with respect to the first one. C: addition of 30 images; D: addition of 160 images. For details see text.

Figure 5 shows the result of such a procedure. First, the digitized image of one $CF_1$ molecule is shown. Then 10 images are added, then 30 images and, lastly, 160 images of different molecules are added. A significant increase of the signal-to-noise ratio is obtained which permits distinction of the different subunits in the molecule.

However, not all molecules are oriented in the same way on the carbon film, so that different projections of the molecules are seen in the electron micrograph. Addition of different projections of the molecules, of course, did not result in an increased signal-to-noise ratio. For solving such problems computer image analysis of single molecules in combination with multivariate statistical techniques was developed for analysing mixed

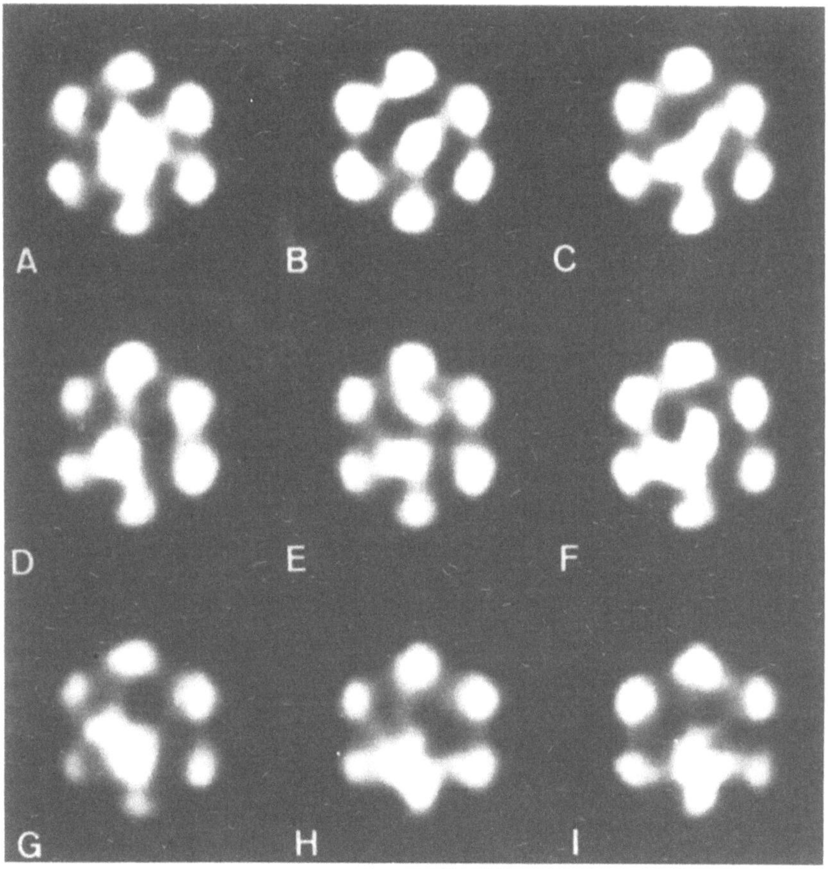


Fig. 6. Different classes of $CF_1$ molecules. The molecular projections of $CF_1$ were determined by multi-reference alignments and classification of 3300 molecules into 9 classes.

populations of images [21]. An automatic classification of 3300 images of $CF_1$ molecules has been carried out and was calculated for nine different classes. Figure 6 shows the result from such a classification procedure. All classes show a pseudohexagonal arrangement of six stain-excluding masses. These are the α- and β-subunits. With these data it is not possible to distinguish between α- and β-subunits, since they have nearly the same molecular mass. However, studies with monoclonal antibodies and the X-ray structure shows that α- and β-subunits are arranged in alternating positions. The classes shown in Fig. 6 differ mainly in the configuration of the central mass which represents the smaller subunits γ, δ and ε. These differences are due to different stain accumulation in the central cavity. Partly they reflect differences in handedness, i.e. projections of the molecules oriented upside-up and upside-down with regard to the supporting carbon film. Possibly, these different projections reflect also different conformational states of the enzyme: in some of the classes in Fig. 6 the central mass is split into two parts, presumably, the bigger one is the $\gamma$- and $\varepsilon$-subunit, the smaller one the $\delta$-subunit [20].

## 2.3. *The structure of $CF_0F_1$*

The structure of $CF_0F_1$ was investigated by electronmicroscopy of negatively stained samples of the isolated enzyme. Figure 7 shows a gallery of $H^+$-ATPase, the upper three rows are $MF_0F_1$ molecules, the lower two are $CF_0F_1$. After isolation the enzymes form stringlike aggregates since the $F_0$ parts are strongly hydrophobic and, therefore, have a tendency to aggregate. The $F_1$ parts are attached via a thin stalk to the string and they have alternating positions along the string. The length of the string varies with detergent concentration. At high concentration mostly single molecules are observed and the detergent shields the hydrophobic parts of $CF_0$ from water. When the detergent concentration is decreased the strings increase in length, the hydrophobic $CF_0$ parts interact directly and only the ends of the string are shielded by the detergent. Further removal of detergent leads to two- and three-dimensional aggregates. There is almost no difference between the shape of $MF_0F_1$ and $CF_0F_1$ [22].

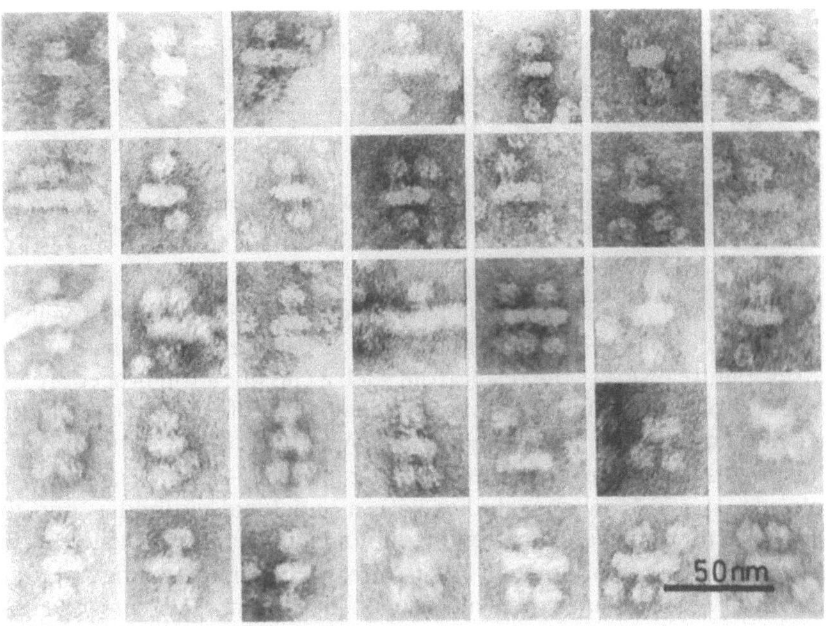


Fig. 7. Gallery of electronmicrographs of isolated $F_0F_1$ molecules. The three upper rows show the mitochondrial enzyme, $MF_0F_1$, the two lower rows the chloroplast enzyme, $CF_0F_1$. The sample was stained with 1% uranylacetate.

All these electron microscopic investigations have been carried out with negatively stained samples. Negative staining with uranylacetate denatures the protein and therefore, might lead to artefacts. Embedding biological macromolecules in thin layers of amorphus ice and imaging them by electron microscopy at liquid nitrogen temperature guarantees that the macromolecules are directly imaged in a form as close to the native state as possible. In this case, the sample of biological material is frozen very rapidly so that the water does not crystallize in the ice structure, but the structure of the liquid water is frozen ("vitrified"). The contrast of the image is in this case merely a phase contrast of the electron waves which result from slight differences in the velocity of the electrons in water, protein and lipid. Figure 8 shows an electronmicrograph of isolated $CF_0F_1$ and $CF_0F_1$ reconstituted in phosphatidylcholine liposomes with approximately 50 $CF_0F_1$ per liposome [23]. The vesicles have a spherical shape with a diameter between 20 and 200 nm. The bilayer structure can be seen clearly. At the border of the vesicles darker structures are projecting out of the membrane. A small stalk connects a spherical structure with a diameter of about 9 nm with the membrane. These structures are interpreted

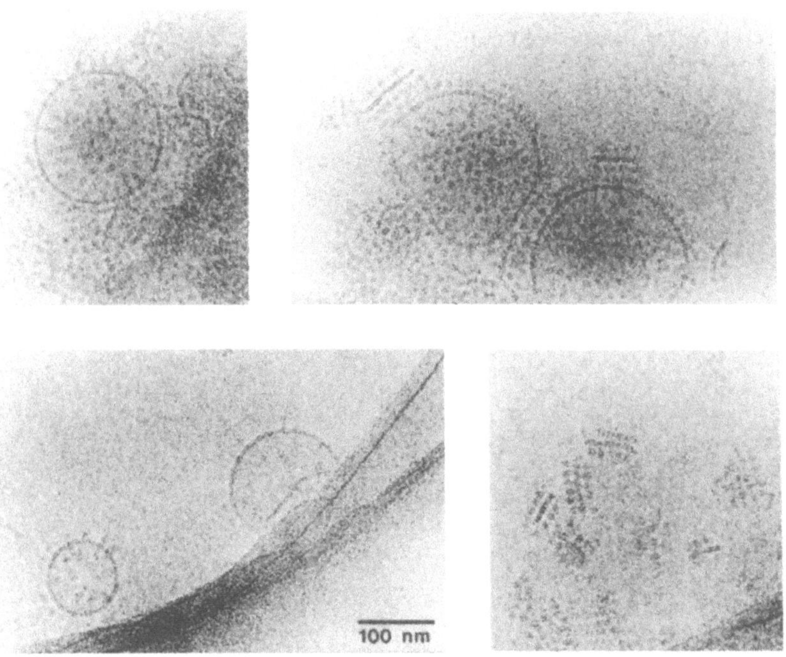


Fig. 8. Electron micrographs of reconstituted $CF_0F_1$ in ice. $CF_0F_1$ was reconstituted into phosphatidylcholine liposomes and rapidly frozen in liquid nitrogen. The bilayer structure of the vesicle is clearly visible. $CF_0F_1$ can be seen in a side-view projection at the vesicle membranes and in a top-view projection in the center of the vesicles. At the bottom right string-like structures of non-reconstituted $CF_0F_1$ are shown.

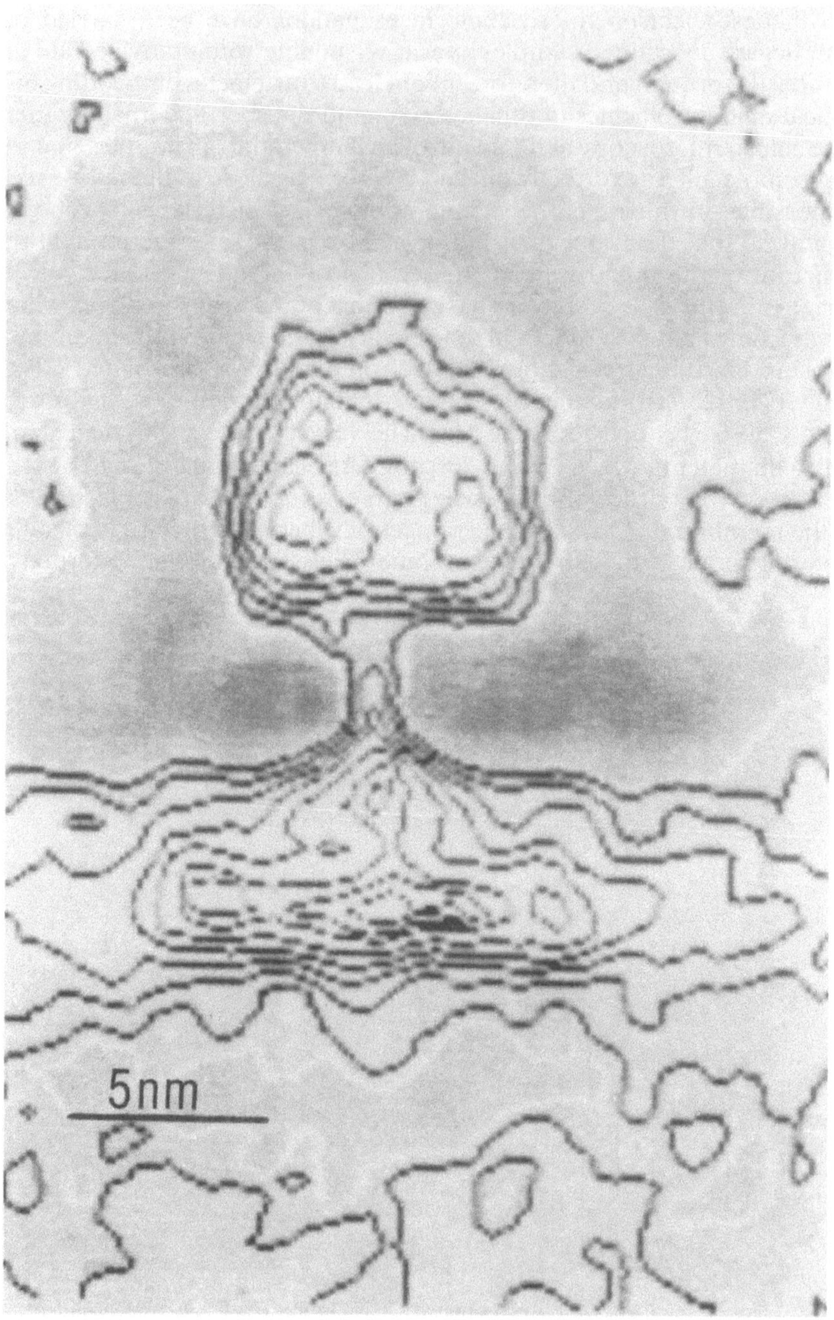


Fig. 9. $CF_0F_1$ reconstituted in liposomes. This structure is the result of an image analysis of electron micrographs as shown in Fig. 8. A total of 483 side-view projections were used to obtain this average image, showing clearly the $CF_1$ moiety, the stalk and the membrane. In the membrane the structure of the $CF_0$ moiety can be seen.

as side-view projections of the $CF_0F_1$. Besides the reconstituted $CF_0F_1$ (i.e., $CF_0F_1$ in liposomes), string-like structures can be observed. For image analysis a data set of 438 individual projections showing $CF_0F_1$ in vesicles in side-view position was used. This data set was aligned to a common reference. Figure 9 shows the average image of the $H^+$-ATPase in the side-view position. The molecule consists of three structural parts: the membrane-integrated $F_0$ moiety and a small stalk projecting out of the $F_0$ moiety (in the membrane) connecting it with the base of the spherical shaped $F_1$ moiety (in an aqueous environment). The $CF_0F_1$ has a total length of 16.6 nm perpendicular to the membrane. The protruding $CF_1$ moiety contributes about 7.7 nm and the membrane-spanning $F_0$ moiety about 5.2 nm to this length. The gap of 3.7 nm between the $CF_0$ and the $CF_1$ moieties is bridged by a small stalk. The width of the stalk is about 1.4 nm. The maximal width of the $CF_1$ moiety parallel to the membrane is about 8.7 nm. The $CF_0$ moiety has a width of about 11.5 nm parallel to the membrane. The mass distribution inside the $CF_1$ moiety is not homogeneous. The lowest mass density can be observed in the middle of the $CF_1$ moiety. There are two similar centres of higher densities inside the $F_1$ moiety, which are elongated with their long axis perpendicular to the plane of the membrane. Full ATP synthesis activity was demonstrated for the reconstituted $CF_0F_1$ after freezing and thawing. This indicates that electron cryomicroscopy measures a functional active enzyme [23].

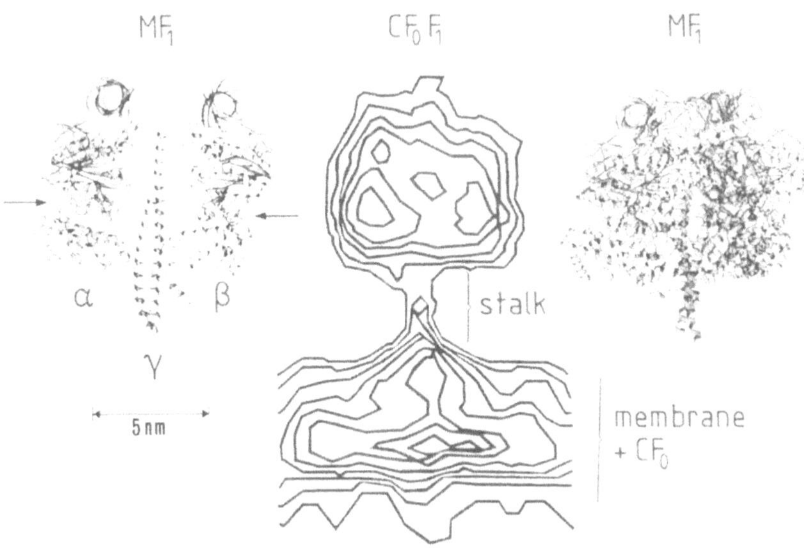


Fig. 10. Comparison of $CF_0F_1$ with $MF_1$. The X-ray structure of $MF_1$ with a resolution of 0.28 nm is shown on the right [19]. On the left only one α- and β-subunit with a part of the γ-subunit is shown. The nucleotide binding sites are marked with arrows. For comparison in the center the structure of $CF_0F_1$ (from Fig. 9) is depicted.

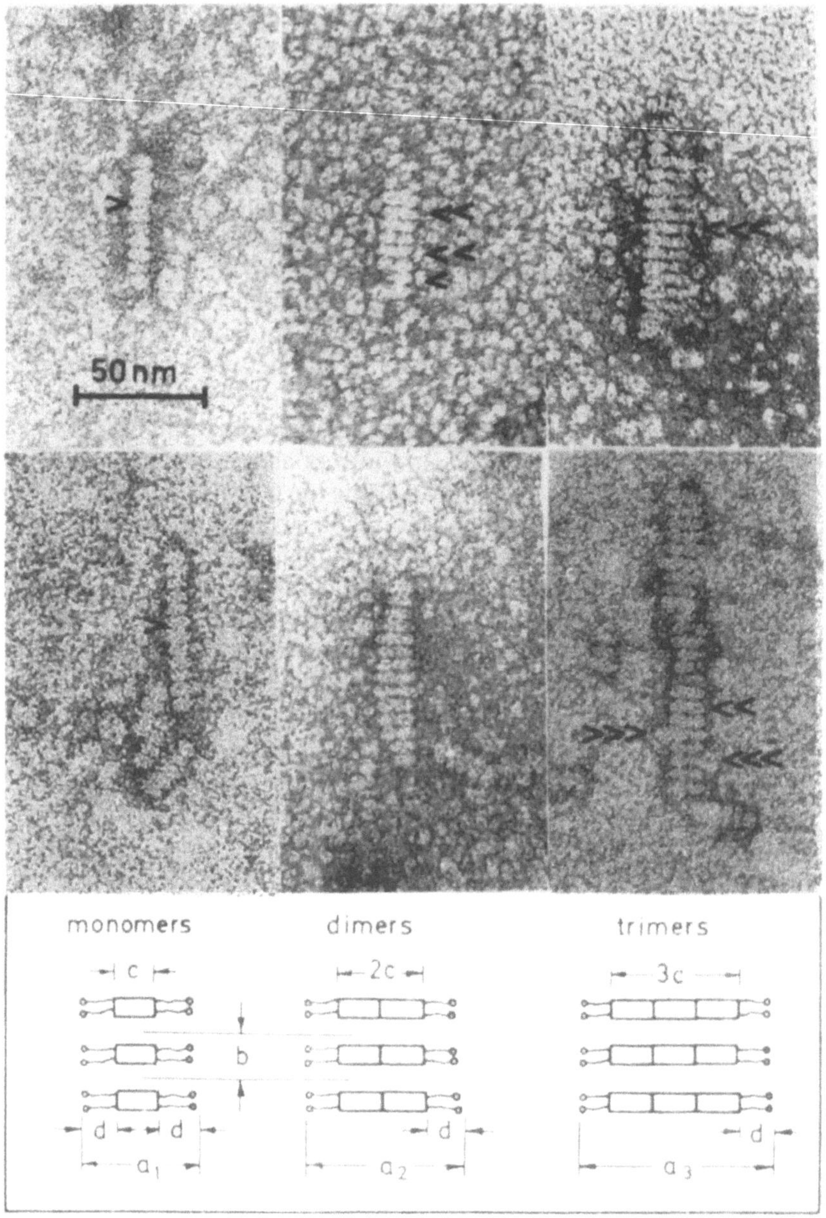


Fig. 11. Electron micrograph of the subunit III-complex. Left: stack composed of single complexes, center: stack composed of two complexes, right; stack composed of three complexes. At the bottom a schematic interpretation of the structure is shown. Samples were stained with 1% uranylacetate.

Recently, the X-ray structure of $MF_1$ was solved with a resolution of 0.28 nm [19]. Since $H^+$-ATPases from different sources are similar, the X-ray data from $MF_1$ are compared to the electronmicroscopy data from $CF_0F_1$ in Fig. 10. On the right side the structure of $MF_1$ is shown, on the left side the partial structure of one α- and one β-subunit and a part of the γ-subunit. In the centre $CF_0F_1$ is depicted (same data as in Fig. 9). The shape of the $CF_1$-part and its dimensions are in accordance with the data from crystallized $MF_1$. The arrows in the left part of the figure point to the nucleotide binding sites on the α- and β-subunit. Since the proton binding sites must be located in the membrane, the distance between proton binding sites and nucleotide binding sites is at least 5 nm. For catalysis these sites have to communicate, presumably by long-range conformational interactions.

### *2.4. The structure of $CF_0$*

The hydrophobic $CF_0$-part can be isolated from $CF_0F_1$ [24]. It contains the subunits I, II, III and IV [25]. Electron micrographs from isolated $CF_0$ show string-like structures. Presumably, these strings represent aggregates of $CF_0$ interacting via their hydrophobic sides, i.e. strings similar to the strings in Figs. 7 and 8 are observed without attached $F_1$-parts. Individual molecules cannot be distinguished. The subunits III form an oligomeric complex that can be isolated by sodium dodecylsulfate polyacrylamide-gel-electrophoresis with subsequent electroelution of the gel. The subunit III-complex has an apparent molecular mass of 100 kDa. This implies that it is formed from 12 subunits III since one subunit III has a molecular mass of 8 kDa. Figure 11 shows the results of electron microscopic investigations of this complex [26]. The complex forms structures like stacked coins with up to 35 units of the 100 kDa complex arranged along the axis of one stack. The complexes are seen in a side view. The diameter of the stacks vary, indicating that one (Fig. 11, left, monomers), two (Fig. 11, centre, dimers) or three (Fig. 11, right, trimers) complexes form the repeating units of a stack. At the bottom of Fig. 11 a schematic interpretation of the structure is depicted. These data indicate that the subunit III complex has a diameter of 6.2 nm (without detergent) and a membrane spanning length of 6.1 nm.

$CF_0F_1$ can be arranged in two-dimensional ordered arrays ("two-dimensional crystals"). These arrays were investigated by electron cryomicroscopy and the signal-to-noise ratio was improved by image analysis. Figure 12, top, shows the result of such an analysis [27]. This image is a projection through the holoenzyme and it is, therefore, not possible to say which part of the density in the average image results from which subunit. However, we can compare the average image from negatively stained

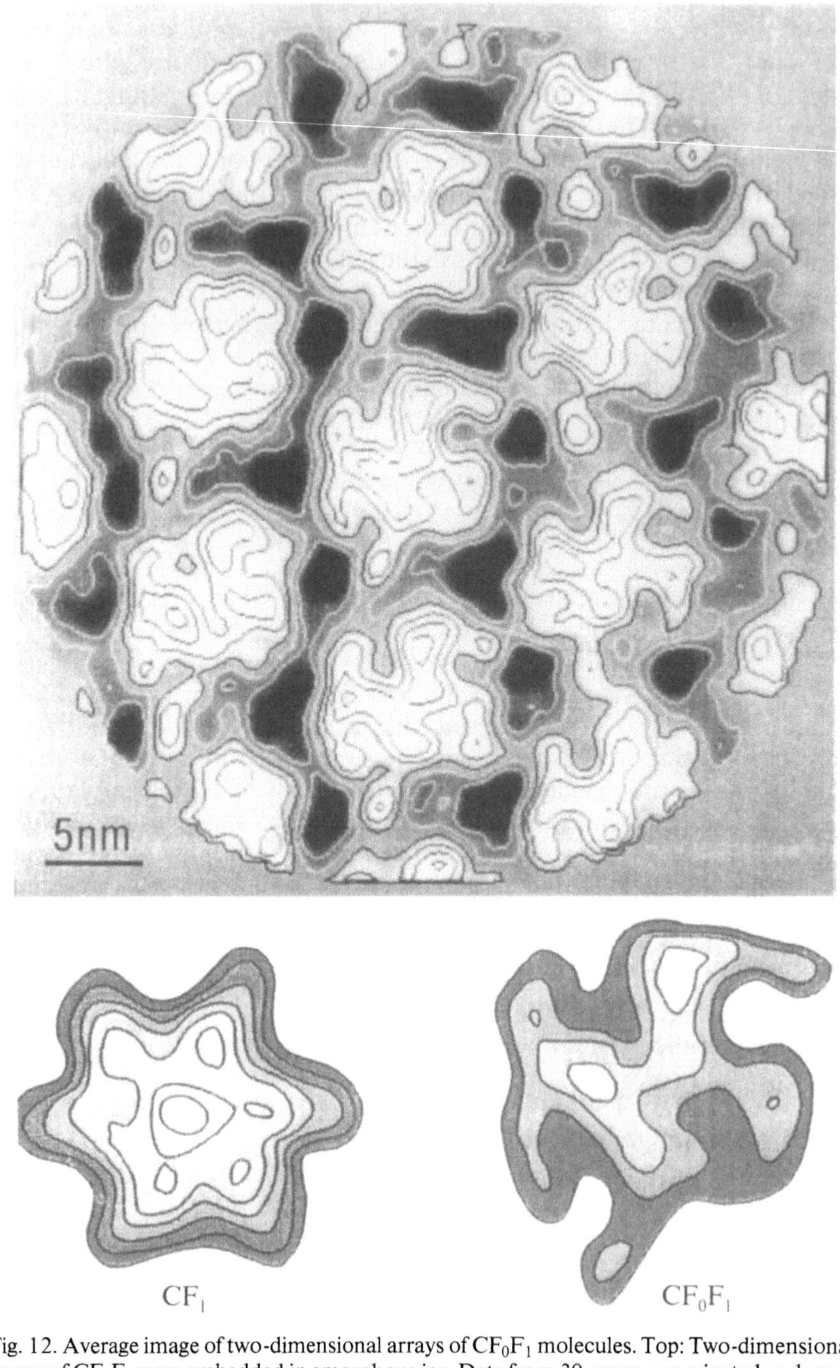


Fig. 12. Average image of two-dimensional arrays of $CF_0F_1$ molecules. Top: Two-dimensional arrays of $CF_0F_1$ were embedded in amorphous ice. Data from 30 arrays are oriented and averaged. At the bottom the average image of $CF_1$ from negatively stained two-dimensional crystals is compared to the average image of $CF_0F_1$.

two-dimensional $CF_1$ crystals [28] with that of the $CF_0F_1$ crystals (Fig. 12, bottom). It is evident that $CF_1$ appears to be rather six-fold symmetric and that the attachment of $CF_0$ leads to a strong asymmetry of the molecule. This might be due either to a rearrangement of the α- and β-subunits when $CF_0$ is attached to $CF_1$, or it reflects the asymmetric positioning and structure of $CF_0$. Possibly, both effects might play a role. On the basis of the aminoacid sequence of the $CF_0$ subunits hydropathy plots can be constructed and this allows prediction of the membrane spanning segments of the different subunits. From such data it results that subunits I and II have one transmembrane helix each, subunit III has two transmembrane helices and subunit IV has between four and six helices. This is shown schematically in Fig. 13, bottom. Since the stoichiometry is I II $III_{12}$ IV the $CF_0$ part has between 30–32 transmembrane helices. Each helix has a diameter between 0.7–0.9 nm. Combining these results with the electromicroscopy data gives finally the model depicted in Fig. 13, top. $CF_0$ has an asymmetric shape and/or is located asymmetrically beneath the $CF_1$ part. The subunits III and IV are strongly hydrophobic, they contain almost no extramembraneous regions. Subunits I and II have only one transmembrane span, both have large hydrophilic domains. On this basis it

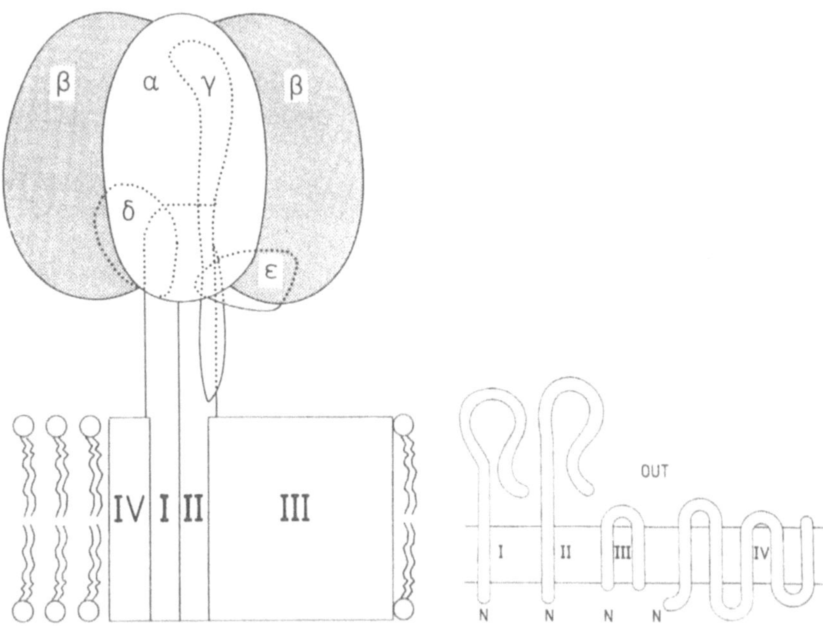


Fig. 13. Model for subunit arrangement in $CF_0F_1$: Bottom: Transmembrane arrangement of the different subunits in $CF_0$ as deduced from hydropathy analysis. Top: Side-view of a model for subunit arrangement in $CF_0F_1$.

is concluded that the stalk is mainly formed from subunits I and II and, additionally, they must protrude into the centre of $CF_1$. From the structure of $MF_1$ it results that a further part of the stalk is subunit $\gamma$. It is a matter of current investigation whether additional connections between the $F_1$ and $F_0$-part exist and whether other subunits form a part of the stalk.

## 3. The catalytic reaction of $H^+$-ATPases

### *3.1. The chemiosmotic theory and experimental approach*

The chemiosmotic theory describes the energetics of coupling between the electrontransport (light driven electrontransport in photosynthesis, oxidation of substrates in respiration) and ATP synthesis. It can be summarized as follows [29, 30].

1. The electrontransfer within membrane embedded proteins leads to transmembrane redox reactions. Thereby, an electron is transferred through the membrane from one side to the other side. This leads to the generation of a transmembrane electric potential difference, $\Delta\varphi$.
2. Subsequent protolytic reactions of the electron donors and acceptors at different sides of the membrane lead to a coupled vectorial transmembrane proton transport and the generation of a transmembrane pH difference, ΔpH.
3. Both processes together lead to the generation of a transmembrane electrochemical potential difference of protons, $\Delta\tilde{\mu}_{H^+}$. This is called "energization" of the membrane.
4. The back flux of protons is driven by $\Delta\tilde{\mu}_{H^+}$. There are at least two pathways for this flux: via the membrane-bound $H^+$-ATPase ("phosphorylating flux") or via unspecified sites of the membrane ("basal flux"). When the proton permeability of the membrane is increased artificially by addition of "uncouplers" there is an additional ("uncoupler mediated flux"). The exergonic proton transport is coupled with endergonic ATP synthesis.
5. The reverse reaction, ATP-hydrolysis, is coupled with a proton influx via the ATPases. This reaction is strongly inhibited *in vivo* in chloroplasts and mitochondria. In bacteria it is used to energize the membrane for uptake of nutrients.

The chemiosmotic theory assumes that the external and internal aqueous phases have the properties of aqueous bulk phases. This is not necessarily true. *In vivo* the internal volume of the vesicles (thylakoids, mitochondria, bacteria) and the distance between the proteins is small. Such considerations led to theories on protons migrating from their source (the electron-transport proteins) to the sink (the ATPase) either in the membrane [7] or at least in a space which is not in equilibrium with the internal aqueous phase [7, 31, 32]. In this chapter we do not deal with the question of how protons migrate from source to sink. We ask for the mechanism of coupling between proton-transport and ATP synthesis within the $H^+$-ATPase. In order to establish clearly defined parameters, experimental conditions must be chosen in such a way that there are large aqueous phases with the properties of bulk phases.

Under these conditions the chemiosmotic theory can be quantitatively formulated as described in Fig. 14. There is a transport reaction where $H^+$ are translocated across the membrane from the internal aqueous phase (in) to the external aqueous phase (out) (left). The transmembrane electrochemical potential difference, $\Delta\tilde{\mu}_{H^+}$, is shown below (Eqn. 1). The chemical reaction (ATP synthesis) and the corresponding free enthalpy change are shown on the right (Eqn. 2). The $H^+$-ATPase couples these two processes giving the overall reaction and the corresponding free enthalpy change (centre, Eqn. 3). Great efforts have been made to prove or disprove the predictions of the chemiosmotic theory. The principle of the theory is now widely accepted; however, no general agreement has been reached about the quantitative aspects. The main experimental problems are the measurements of $\Delta pH$ and $\Delta\varphi$ (the components of $\Delta\tilde{\mu}_{H^+}$) and the proper separation of the basal proton transport and the phosphorylation coupled proton transport. All the parameters in Eqns. 1, 2 and 3 are well defined and can be measured accurately with an experimental approach introduced by Jagendorf and Uribe [33]. Vesicles are incubated in the presence of permeable buffer at low pH (e.g. at pH 5). During the incuba-

**Energetics**

transport reaction

$$n H^+_{in} \rightleftharpoons n H^+_{out}$$

$$\Delta\tilde{\mu}_{H^+} = -\left(RT \ln \frac{[H^+_{in}]}{[H^+_{out}]} + F(\phi_{in} - \phi_{out})\right) \quad (1)$$

chemical reaction

$$ADP + P_i \rightleftharpoons ATP$$

$$\Delta G'_P = \Delta G^{0'}_P + RT \ln \frac{[ATP]}{[ADP]\,[P_i]} \quad (2)$$

coupled reaction

$$ADP + P_i + n H^+_{in} \rightleftharpoons ATP + n H^+_{out}$$

$$\Delta G = \Delta G'_P + n\,\Delta\tilde{\mu}_{H^+} < 0 \quad \text{(steady state)} \quad (3)$$

$$n = \frac{\Delta G'_P}{|\Delta\tilde{\mu}_{H^+}|} \quad \text{(equilibrium)} \quad (4)$$

Fig. 14. Energetics of coupling between ATP synthesis and proton transport as described by the chemiosmotic theory.

tion time (30 s–2 min) an equilibrium of proton-, ion- and buffer concentration is established between internal and external aqueous phases. Then, this acidic suspension is mixed with a strongly buffered basic medium. Proton and ion concentrations in the external aqueous phase are changed within the mixing time, whereas, the internal concentrations correspond initially to those in the acidic incubation medium (see Fig. 15). We have thus generated a transmembrane pH difference giving rise to a proton efflux.

The difference in the ion concentrations between the internal and external phases automatically leads to a transmembrane electric potential difference, $\Delta\varphi$ which can be calculated from the Goldmann equation:

$$\Delta\varphi = \frac{RT}{F} \ln \frac{\sum P_i[C_i]_{out} + \sum P_i[A_i]_{in}}{\sum P_i[C_i]_{in} + \sum P_i[A_i]_{out}} \qquad (5)$$

where $P_i$ is the permeability coefficient, $C_i$ and $A_i$ are the concentration of cations and anions inside and outside of the vesicle.

For a well defined electric potential difference, a $K^+$/valinomycin diffusion potential is generated as follows: vesicles are incubated in the pres-

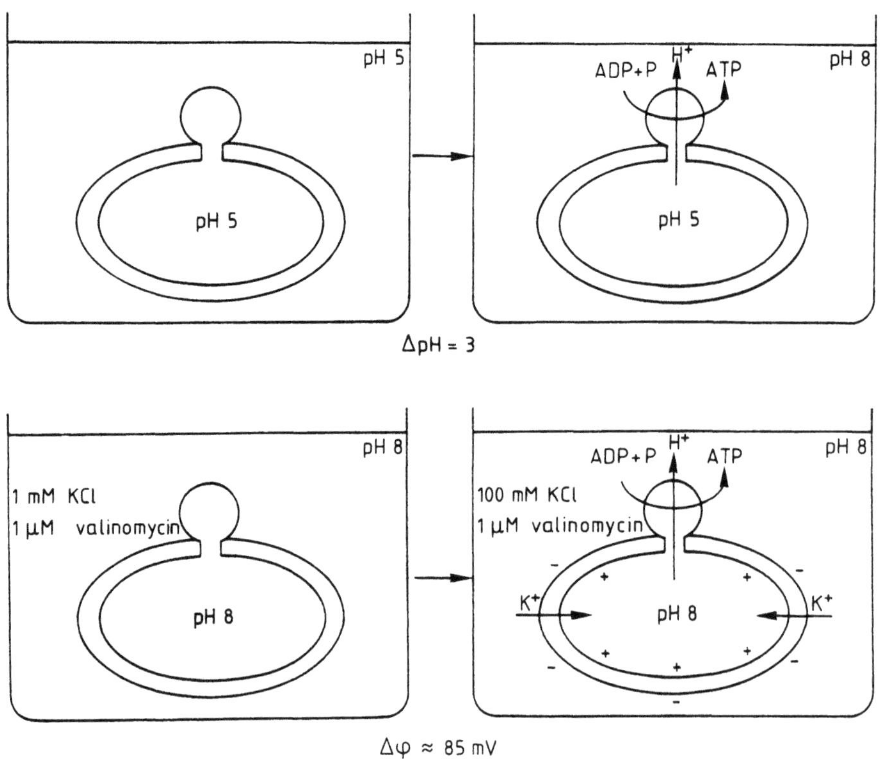


Fig. 15. Generation of a transmembrane ΔpH and Δφ ("membrane energization") by an acid base transition.

ence of valinomycin with a low KCl concentration. Valinomycin increases specifically the permeability of the membrane for $K^+$. Then, the suspension is mixed with a medium of high KCl concentration. This leads to rapid $K^+$ influx, generating an electric potential difference (positive inside)(see Fig. 15). Using different media, ΔpH and Δφ can be varied, each one independently of the other. The condition $\Delta\varphi = 0$ is of special importance. This is achieved at $[K^+]_{in} = [K^+]_{out} = 0.1$ M in the presence of valinomycin.

Under these conditions $pH_{in}$ and $pH_{out}$ can be measured with a glass electrode, i.e. the initial ΔpH is known. The initial Δφ is calculated from Eqn. 5 (usually, it is clamped to $\Delta\varphi = 0$). This initial energization leads to the phosphorylating proton efflux and to fluxes of other ions so that the initial conditions remain constant only for a relatively short time. Therefore, if ATP-synthesis/hydrolysis is to be investigated under well known conditions, the reaction time must be so short that within this time neither ΔpH nor Δφ are significantly changed. This can be achieved by using rapid double mixing techniques *(quenched flow)* as described in Fig. 16 [34]. This technique allows variable reaction times between 20 ms and 20 s. Figure 16 shows a scheme of the procedure: incubation for 30 s in a medium with pH 5, mixing with a medium of pH 8.2 and thereby starting proton-transport-coupled ATP-synthesis. Denaturation with trichloroacetic acid then stops the reaction. Above, the time-course of the transmembrane ΔpH is shown. It is generated within 2 ms during the mixing, then the reaction is allowed to proceed for the reaction time, $t_R$, and it is stopped by denaturation of the vesicles. At the top, the time-course of the ATP concentration is shown, the slope of this curve is the rate of ATP synthesis. At the bottom, a scheme of the apparatus is shown. In syringe I the vesicles are incubated in the acidic medium, syringe II contains the basic medium, syringe III the trichloroacetic acid. The syringes are driven forward simultaneously by a pneumatically actuated piston. In mixing jet A the ΔpH is generated and the reaction is initiated, in mixing jet B the reaction is stopped and then the reaction medium is analyzed for ATP.

The amount of ATP synthesized or hydrolyzed is measured as follows. The ATP-concentration in the reaction-medium was determined by luciferin/luciferase. This method gives the difference between forward and reverse reaction

$$\frac{d[ATP]}{dt} = v_{syn} - |v_{hyd}|$$

where $v_{syn}$ = rate of ATP synthesis, $v_{hyd}$ = rate of ATP hydrolysis.

ATP-hydrolysis: the reaction-medium contains $\gamma[^{32}P]$-ATP and the amount of radioactive phosphate is measured. Separation of ATP and phosphate is achieved by addition of ammonium molybdate and extraction of the phosphate-molybdate complex with toluene/isobutanol.

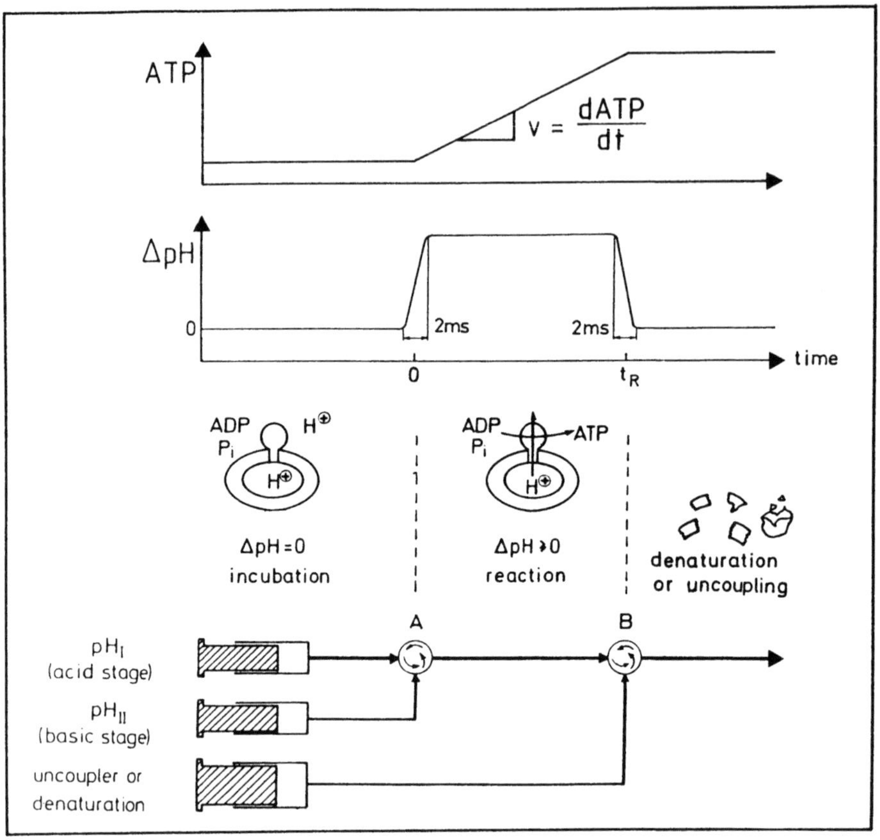


Fig. 16. Experimental realization of membrane energization by an acid-base transition with rapid mixing techniques. From bottom to top: rapid-mixing apparatus, vesicles in different stages of the experiment, time-course of ΔpH, time-course of ATP synthesis.

ATP-synthesis: the reaction-medium contains radioactive phosphate $[^{32}P]$-$P_i$ and the amount of $\gamma[^{32}P]$-ATP is measured. Separation of ATP and phosphate is achieved in this case by addition of ammonium molybdate and precipitation of the phosphate-molybdate complex.

At short reaction times the rate of either the forward-(synthesis) reaction or the reverse-(hydrolysis) reaction is measured. The fraction of product which reacts back is under all conditions less than 2%. Additionally, because of the short reaction time, the concentration changes of all substrates and products are below 2%, i.e. they are constant. Thus, this method allows a well-defined control of all relevant parameters: $pH_{in}$, $pH_{out}$, $\Delta\varphi$, [ADP], $[P_i]$ and [ATP] are well-known from their initial values and do not change during the reaction time.

This experimental approach has two important consequences:

1. Since $\Delta\tilde{\mu}_{H^+}$ is generated artificially, ATP synthesis does not require the natural (thylakoid, mitochondria or plasma) membrane. The $H^+$-ATPases can be isolated, purified and reconstituted into liposomes and the same enzyme activity should be observed. In fact, the same data (maximal turnover, kinetics, regulation) are obtained with thylakoid membranes and with reconstituted $CF_0F_1$ [17].
2. Since all transmembrane concentration differences are bulk-bulk concentration differences, the enzyme must work as described by the orthodox chemiosmotic theory [29, 30]. This implies that no data can be obtained from this approach on the pathway from source to sink under *in vivo* conditions [7, 31, 32].

Figure 17 shows the ATP yield as a function of the reaction time in such a quenched flow experiment. Up to about 150 ms a linear increase of the ATP yield with reaction time can be seen. Within this time ΔpH and $\Delta\varphi$ remain almost constant and a linear increase of ATP concentration is observed. The slope of the curve is the rate of ATP synthesis. The rate is expressed in mol ATP per mol $CF_0F_1$ per s, i.e. in $s^{-1}$. This is called the turnover of the enzyme. After the linear range the curve bends and this bending indicates that ΔpH has decreased due to the proton efflux. At long reaction times there is no further increase in ATP concentration because ΔpH has decayed completely after this time. If the internal buffer capacity is increased at constant ΔpH (by changing the buffer concentration in the acidic medium I) the initial slope, i.e., the rate, is not changed; however, the linear range is extended to longer reaction times. This shows

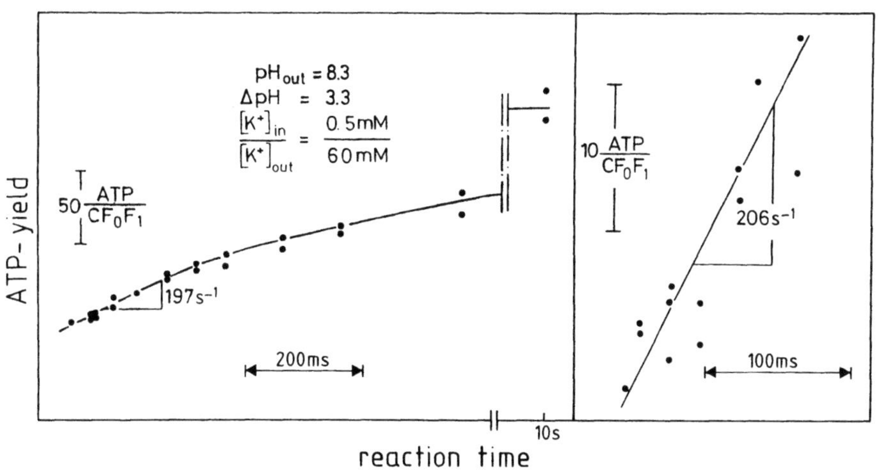


Fig. 17. ATP yield as function of reaction time in an acid base transition. The slope is in the rate of ATP synthesis. Left: Reaction time up to 10 s. Deviation from linear increase of ATP indicates decrease of the initial ΔpH. Right: Reaction time up to 200 ms. Rates are given in turnovers: mol ATP per mol enzyme per s, i.e. $s^{-1}$.

that the rate of ATP synthesis depends only on $pH_{in}$ and $pH_{out}$, but not on the buffer capacity – in contrast to the ATP yield observed after 10 s reaction time.

### *3.2. The $H^+$/ATP ratio*

The number of $H^+$ which are translocated by the enzyme in order to synthesize one ATP ($H^+$/ATP ratio) is an important parameter in the chemiosmotic theory and different experimental approaches have been used for its determination.

1. Flux-measurements: the ratio between the rate of proton flow via the ATPase ("phosphorylating proton flux") and the rate of ATP synthesis is determined.
2. Kinetic measurements: the rate of ATP synthesis (or ATP hydrolysis) is measured as a function of the internal proton concentration. The exponent of the $H^+_{in}$ concentration in the rate equation is interpreted as the $H^+$/ATP ratio.
3. Equilibrium measurements: The $\Delta\tilde{\mu}_{H^+}$ is measured where the net rate of ATP synthesis is zero. The $H^+$/ATP is then calculated from Eqn. 4.

The $H^+$/ATP ratios found by different groups varied between 2 and 4, most experiments resulting in $H^+$/ATP = 3. However, recent experiments show that $H^+$/ATP = 4 [35]. One of these experiments was carried out as follows. The catalytic rate is

$$ADP + P_i + nH^+_{in} \rightleftarrows ATP + nH^+_{out}$$

In a quenched flow experiment the concentrations of all reactants are fixed at their initial conditions, e.g. [ADP] = 50 μM, [ATP] = 50 μM, $P_i$ = 1 mM, $H^+_{in} = 10^{-6}$ M, $H^+_{out} = 10^{-8}$ M. $\Delta\varphi = 0$ mv. The rate of ATP synthesis is measured by addition of [$^{32}$P]-$P_i$ (which does not change the Pi concentration) and the amount of $\gamma$[$^{32}$P]-ATP is measured as function of reaction time. With exactly the same concentration of all reactants in a second experiment $\gamma$[$^{32}$P]-ATP is added and the amount of $\gamma$[$^{32}$P]-$P_i$ is measured as function of reaction time. Figure 18 left shows the amount of ATP synthesized as a function of reaction time at different ΔpH ($pH_{out}$ = 8.2 is always constant). Figure 18 right shows the amount of $P_i$ formed as a function of reaction time at different ΔpH under the same conditions. The slopes of these curves are the rates of ATP synthesis (left) and ATP hydrolysis (right). These rates are plotted as a function of ΔpH in Fig. 19.

The rate of ATP synthesis increases nonlinearly from zero (at ΔpH = 0) to 120 $s^{-1}$ at ΔpH ≥ 3.3, the rate of ATP hydrolysis increases nonlinearly from zero (at ΔpH ≥ 3.5) to 30 $s^{-1}$ at ΔpH ≤ 1.0. These data show that the

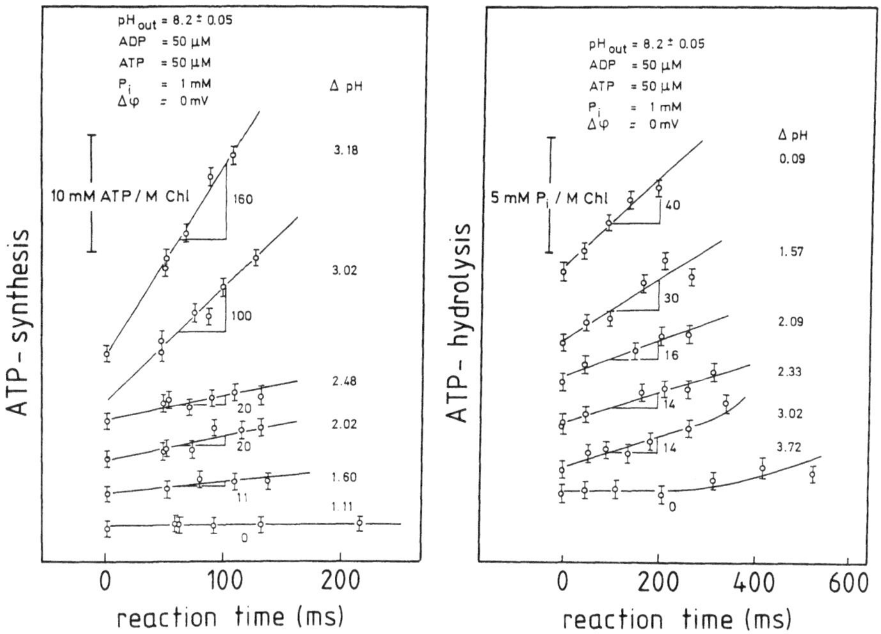


Fig. 18. ATP synthesized and ATP hydrolyzed as function of reaction time. By radioactive labelling of phosphate ATP synthesis (forward reaction, left) and by labelling of ATP only ATP hydrolysis (backward reaction, right) was measured. All other parameters (ATP, ADP, $P_i$, $pH_{out}$, $pH_{in}$, $\Delta\varphi$) were the same in both experiments.

reaction is reversible and the same rate of ATP synthesis and ATP hydrolysis is observed at $\Delta pH = 2.1$ (equilibrium). Under these conditions the $H^+/ATP$ ratio can be calculated from Eqn. 4. It resulted with $\Delta\varphi = 0$, $pH_{out} = 8.2$, $\Delta pH$ (at equilibrium) $= 2.1$ and $\Delta G^{0'} = 33.9$ kJ mol$^{-1}$.

$$n = \frac{H^+}{ATP} = \frac{\Delta G_p'}{|\Delta\tilde{\mu}_{H^+}|} = \frac{\Delta G^{0'} + RT\ln\frac{[ATP]}{[ADP][P_i]}}{RT\ln\frac{[H^+_{out}]}{[H^+_{in}]}} = \frac{\Delta G^{0'} + RT\ln\frac{[ATP]}{[ADP][P_i]}}{2.3 \times RT\Delta pH}$$

$$= \frac{33.9\ \text{kJmol}^{-1} + 2.5\ \text{kJmol}^{-1}\ln\frac{50\times10^{-6}}{50\times10^{-6}\times10^{-3}}}{2.3\times2.5\ \text{kJmol}\times2.1} = \frac{51.2\ \text{kJmol}^{-1}}{12.1\ \text{kJmol}^{-1}} = 4.2 \quad (4)$$

This number $H^+/ATP = 4$ has, of course, important consequences for the energetics and mechanisms of a variety of biochemical reactions.

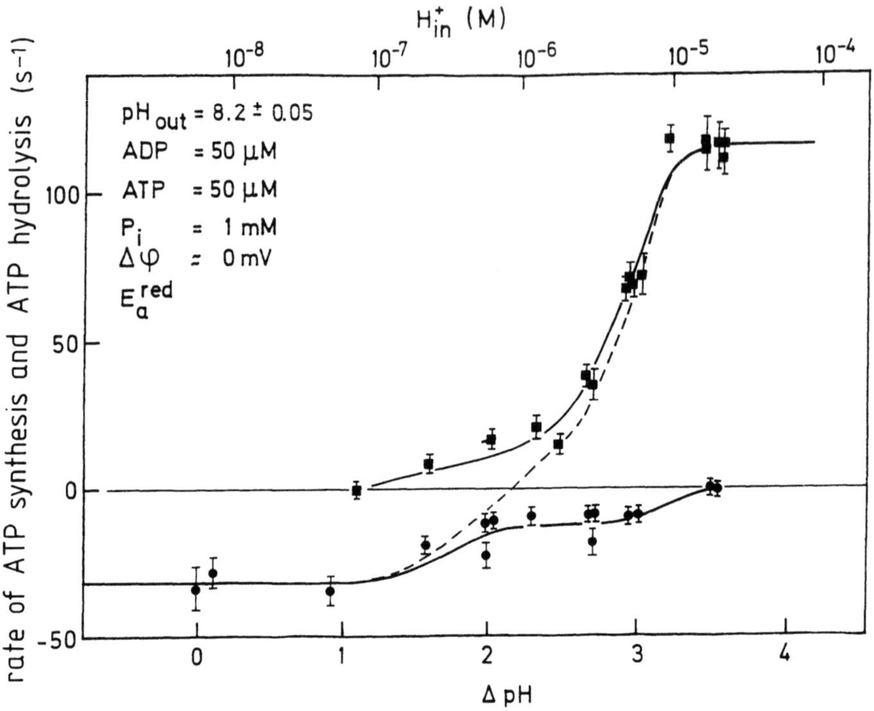


Fig. 19. Rate of ATP synthesis (squares) and ATP hydrolysis (circles) as function of ΔpH. Data from Fig. 18 and additional measurements. Dashed line: difference between ATP synthesis and ATP hydrolysis, i.e. net rate.

### *3.3. The reaction cycle*

The chemiosmotic theory describes the energetics of coupling between electrontransport and ATP synthesis. In order to understand how the enzyme couples the vectorial proton translocation with the scalar chemical reaction, it is necessary to formulate a reaction sequence describing the catalytic cycle. Figure 20 shows the simplest schemes for the transmembrane proton transport (left) and ATP synthesis (right) and their coupling (bottom) [36].

Proton transport across the membrane is described by a four-state model (Eqn. 6). In state $E_1$ the proton binding sites are directed to the inside; after proton binding, a conformational change into state $E_2$ occurs. In $E_2$ the proton binding sites are directed to the outside, the protons dissociate to the outside and finally the conformation changes back to $E_1$. It should be mentioned that this scheme describes a carrier type as well as a channel type translocator. If the $F_0$ part acts as a proton channel, then the channel must provide alternating access to both sides of the membrane, since otherwise no useful work can be obtained. The two alternating conformations are represented by $E_1$ and $E_2$.

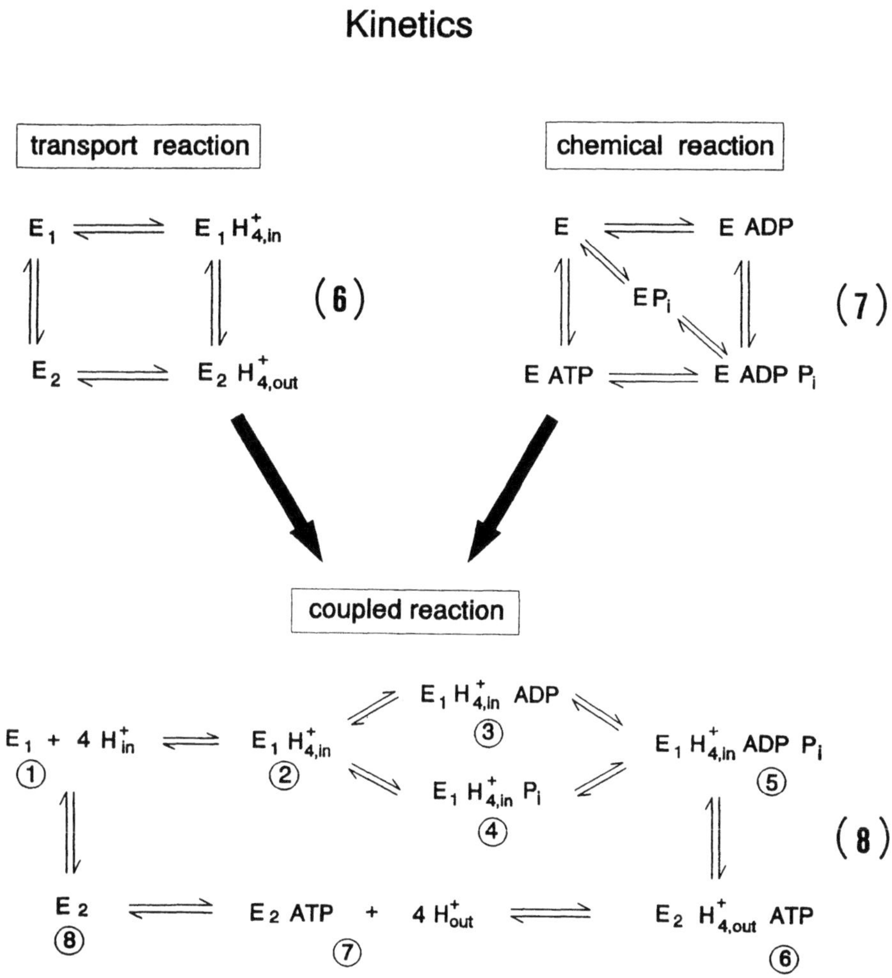


Fig. 20. Kinetics of coupling between proton transport and ATP synthesis/ATP hydrolysis.

The chemical reaction is described by a five-state model (Eqn. 7): ADP binds to the enzyme, followed by $P_i$ or $P_i$ binds first followed by ADP. The enzyme-bound ADP $P_i$ reacts to enzyme bound ATP and finally ATP is released and the free enzyme is regenerated. There are of course different ways in which both reactions may be coupled. The following reaction sequence can describe most experimental results (Eqn. 8). In the state $E_1$ the proton binding sites are directed to the inside and the enzyme is protonated. After protonation it can bind first ADP and then $P_i$ or first $P_i$ and then ADP. After this binding there is a conformational change into the state $E_2$, where proton binding sites are directed to the outside and concomitantly enzyme-bound ATP is formed. The protons dissociate to the outside, ATP is released and finally the enzyme changes back into the state $E_1$, thereby closing the reaction cycle.

This sequence is based on the following observations: ADP binding to the enzyme is slow without energization (rate constant: $k < 10^3\ M^{-1}s^{-1}$). When the membrane is energized ADP binding is fast (rate constant $k > 6 \times 10^6\ M^{-1}s^{-1}$). Therefore, proton binding to the enzyme must occur first and then either ADP or $P_i$ is bound. Presumably, ADP and $P_i$ bind in a random order. When all substrates are bound ($H^+_{in}$, ADP, $P_i$) the enzyme changes its conformation and the proton binding sites are now directed to the outside; concomitantly enzyme bound ATP is formed. This assignment is based on the following argument: protonation from the inside shifts the enzyme into the high energy conformation, $E_1$, and the catalytic site for ADP and $P_i$ opens. After binding of the substrates the enzyme relaxes to the lower energy conformation, $E_2$, and this transition provides the energy for synthesis of enzyme bound ATP, i.e. energy from the conformational change is stored in the γ-phosphate bond of the enzyme-bound ATP. After this step, protons are released to the outside, followed by the release of ATP. The sequence of the latter steps is not exactly known. Finally, the change of the enzyme conformation into the state $E_1$ closes the cycle. The above scheme is, of course, greatly simplified: (1) Protonation/deprotonation reactions are assumed to occur in one step. (2) The scheme describes the reaction at only one catalytic site. However, most experimental evidence indicates that two or three catalytic nucleotide binding sites are involved in the reaction. We assume that the same reaction sequence occurs under uni-site and under multisite conditions. The cooperativity between sites is assumed to change only the rate constants for some steps. (3) Possibly, the enzyme runs through several different conformations during the reaction cycle: the conformations proposed in the model are minimally required for coupling.

The pK-values of the groups involved in proton transport were calculated from measurements of the rate of ATP synthesis as a function of $pH_{out}$ and $pH_{in}$ using the same experimental approach as described in Figs. 15–17. The protonation/deprotonation (step 1 → 2 and step 6 → 8) in Eqn. 8 (Fig. 20) is presumably a sequential binding and dissociation which can be written as follows

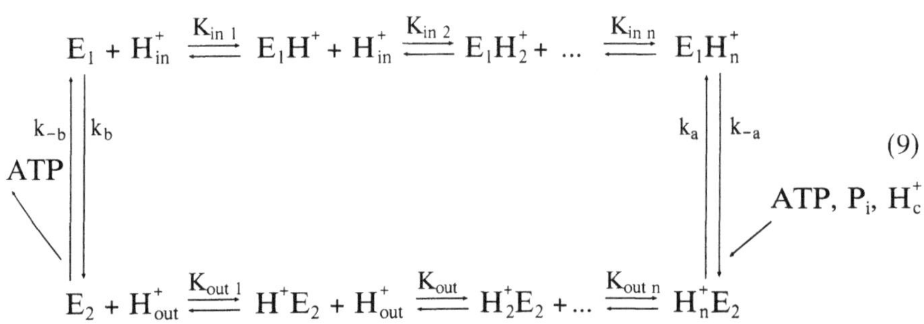


(9)

where $k_a$, $k_{-a}$ summarizes steps 3 to 6 in Eqn. 8, $k_b$, $k_{-b}$ summarizes steps 7 to 1 in Eqn. 8, $K_{in,i}$ and $K_{out,i}$ are the dissociation constants inside and outside for step i and $H_c^+$ is the proton consumed chemically in the catalytic reaction of $pH_{out} = 8.0$. A simple expression for the rate is obtained with the following assumptions [37].

1. When the rate of ATP hydrolysis is neglected the rate of ATP synthesis is given by ($E_t$ total enzyme concentration):

$$v = k_{-a}\, E_1H_n^+ = v_{max} \frac{E_1H_n^+}{E_t} \tag{10}$$

2. The protonation/deprotonation reactions are fast compared with the chemical steps and can be treated as equilibria.
3. The transition from $E_1$ to $E_2$ can occur only when the enzyme is either fully protonated or fully deprotonated. This assumption is similar to that in the Monod-Wyman-Changeux model describing the binding of oxygen to hemoglobin [38].
4. The microscopic dissociation constants $K_{in}$ and $K_{out}$ are the same for each step. These are related by statistical factors to the macroscopic dissociation constants in Eqn. 9 as

$$K_{in/out,i} = K_{in/out} \frac{i}{n - i + 1} \tag{11}$$

5. The fraction of enzymes in state $E_1H_n^+$ is then as follows

$$\frac{E_1H_n^+}{E_t} = \frac{E_1H_n^+}{\sum E_{1i} + \sum E_{2i}} = \frac{\left(\frac{H_{in}^+}{K_{in}}\right)^n}{\left(1 + \frac{H_{in}^+}{K_{in}}\right)^n + \frac{k_2}{k_{-2}}\left(1 + \frac{H_{out}^+}{K_{out}}\right)^n} \tag{12}$$

To avoid having too many free parameters, Eqn. 12 can be simplified further. If the amount of enzyme in state $E_2$ can be neglected ($\Sigma E_{2i} \ll \Sigma E_{1i}$), the equation can be expressed as follows

$$\frac{E_1H_n^+}{E_t} \approx \frac{\left(\frac{[H_{in}^+]}{K_{in}}\right)^n}{\left(1 + \frac{[H_{in}^+]}{K_{in}}\right)^n} \tag{13}$$

Finally, the rate (Eqns. 10 and 13) is as follows

$$v = v_{max} \frac{[H_{in}^+]^n}{(K_{in} + [H_{in}^+])^n} \tag{14}$$

When we assume that the binding of n protons occurs in one step (as formulated in Eqn. 8, "Hill model"), we obtain

$$v = v_{max} \frac{[H^+_{in}]^n}{K^n_{in} + [H^+_{in}]^n} \quad (15)$$

The dependence of the rate of ATP synthesis as a function of the internal ($H_{in}$) and external ($H^+_{out}$) proton concentration was described by Eqns. 14 and 15 and the experimental data were fitted with $v_{max}$, $K_{in}$, $K_{out}$, and n as free parameters [37].

Optimal fits were obtained with n = 3, $K_{in} = 1.6 \times 10^{-6}$ M and $K_{out} = 1.6 \times 10^{-8}$ M. Figure 21, top shows experimental data and the calculated curves as function of $H^+_{in}$ at different given $pH_{out}$ (Eqn. 14). Figure 21, bottom shows experimental data and calculated curves as a function of $H^+_{out}$ at different given $pH_{in}$ (Eqn. 15).

The dissociation constant of the protons ($K_{in} = 1.6 \times 10^{-6}$ M, $K_{out} = 1.6 \times 10^{-8}$ M) are converted into the biochemical standard state (pH 7) and written in ATP synthesis direction.

$$K_{out} = \frac{\frac{H^+}{c^0}\frac{A^-}{c^0}}{\frac{HA}{c^0}} = 1.6 \times 10^{-8}, \quad c^0 = 1\ M$$

$$K'_{out} = \frac{\frac{H^+}{10^{-7}M}\frac{A^-}{c^0}}{\frac{HA}{c^0}} = \frac{K_{out}}{10^{-7}} = 0.16$$

Usually, in biochemistry equilibrium constants with a dimension are used and we obtain with the nomenclature used in Fig. 20, Eqn. 8: $K'_{67} = K'_{out} = 0.16$ M and $K'_{12} = (K'_{in})^{-1} = 0.0625\ M^{-1}$. If 4 $H^+$ are bound to the $H^+$-ATPase and translocated per ATP synthesised finally $\Delta G^{0'} = -4\ RT \ln K'_{12}$ is obtained.

Analyzing the rate of ATP synthesis as a function of ADP, phosphate and ATP concentration the Michaelis-Menten constants, $K_M$, and the respective dissociation constants $K_D$, were obtained. Based on the reaction scheme in Eqn. 8 a $\Delta G^{0'}$ diagram was constructed using literature data for the different equilibrium constants.

The equilibrium constant between enzyme-bound ADP $P_i$ and enzyme bound ATP $K_{eq}$, has been measured under uni-site conditions (see section 3.4). This gives $K_{eq} = 0.5$ and presumably, the constant does not change significantly under multisite conditions. In the last step the reaction cycle is closed by the reaction $E_2 \rightarrow E_1$. There is as yet no experimental evidence for this step. Nevertheless, it must exist and its $\Delta G^{0'}$ can be calculated from the energy balance of the cycle. After one turnover of the cycle one ATP is formed from ADP and $P_i$ under standard conditions. The corresponding free enthalpy change is $\Delta G^{0'} = 32.8\ kJmol^{-1}$ (at pH 8, 1 mM $MgCl_2$, ionic strength 0.1 M). It follows then for the step 8 $\rightarrow$ 1

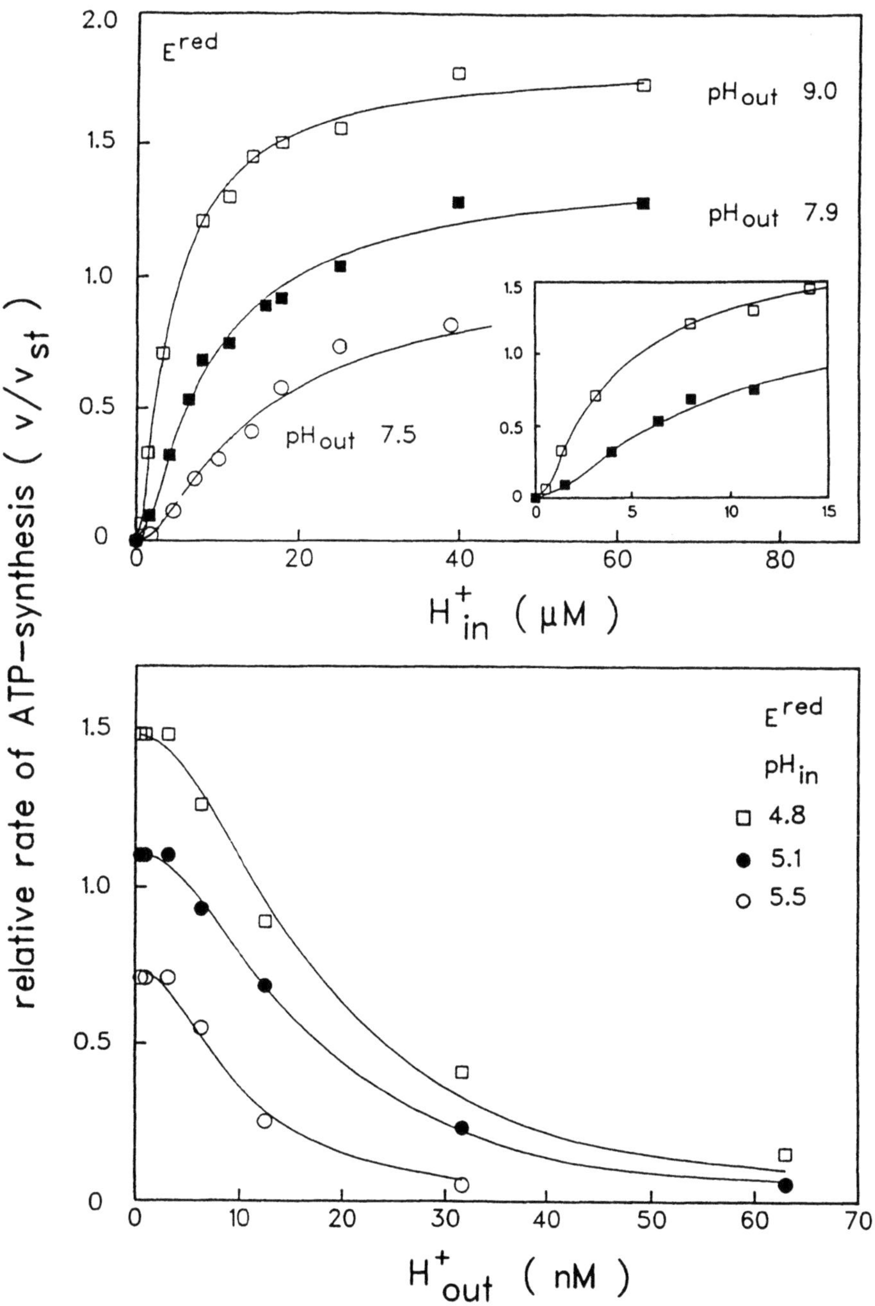


Fig. 21. Rate of ATP synthesis as function of internal ($H^+_{in}$) and external ($H^+_{out}$) proton concentration. Top: Dependence of the rate on ($H^+_{in}$) at $pH_{out}$ = 9.0, 7.9 and 7.5. Bottom: Dependence of the rate on ($H^+_{out}$) at $pH_{in}$ = 4.8, 5.1 and 5.5. Experimental data obtained as described in Figs. 15–17. Curves are calculated from Eqns. 14 and 15.

$$\Delta G^{0'}_{81} = \Delta G^{0'}_{P} - \sum_i \Delta G^{0'}_i \qquad (16)$$

where $\Delta G^{0'}$ is the free enthalpy change of step i. From $\Delta G^{0'}_{81}$ the equilibrium constant is calculated $K'_{81} = 7.3 \times 10^{-3}$. This means that under standard conditions a fraction of about 1% of the enzymes is in the state $E_1$. All these data are collected in Table 2 and the resulting free enthalpy diagram is shown in Fig. 22.

The reaction mechanism of Eqn. 8 describes a proton induced binding change mechanism: protonation from the inside favours the enzyme conformation $E_1$ and only this conformation binds ADP and $P_i$ but not ATP. When the enzyme is in the conformation $E_2$ it binds ATP but not ADP and $P_1$. The substrate specificity changes drastically from $E_1$ to $E_2$, and this change is connected with the protonation/deprotonation reaction. It is required that information about the protonation state of the groups in the $F_0$ part is transmitted to the $F_1$-part where the nucleotide binding sites are located. Obviously, both events are coupled by far-reaching conformational interactions.

Of course, there might be also ATP binding to the state $E_1$, as well a ADP and $P_1$ binding to $E_2$. However, the corresponding affinities are much lower and these reactions are side-reactions. They are not considered to be essential for the mechanism of coupling between proton transport and ATP synthesis.

It should be mentioned that the exact values of $K_{in}$ and $K_{out}$ (or the $pK_{in}$ and $pK_{out}$ values) depend on the kinetic model used for evaluation of the data. However, independent of the model there is always a difference of about 2 units between both pK's.

These pK values are apparent (kinetic) pK values. The intrinsic pK is the dissociation constant of the functional group and is measured by the protonation of this group as a function of pH. The apparent pK determined from the pH dependence of the rate also contains rate constants of

Table 2. Equilibrium constants for the different steps of the reaction cycle
The literature data are converted into the biochemical standard state, written in ATP synthesis direction and $\Delta G^{0'}$ is calculated. For details, see text.

| Equilibrium constant | | ATP synthesis direction | $\Delta G^{0'}$(kJ/mol) |
|---|---|---|---|
| $K_D(H^+_{in})$ | $1.6 \times 10^{-6}$ M | $K_{12} = 6.25 \times 10^5\ M^{-1}$ | |
| | | $K'_{12} = 6.25 \times 10^{-2}\ M^{-1}$ | +27.5 |
| $K_D(ADP)$ | $2 \times 10^{-5}$ M | $K_{23} = 5 \times 10^4\ M^{-1}$ | -26.9 |
| $K_D(P_i)$ | $4 \times 10^{-4}$ M | $K_{35} = 2.5 \times 10^3\ M^{-1}$ | -19.4 |
| $K_{eq}$ | 0.5 | $K_{56} = 2$ | -1.7 |
| $K_D(H^+_{out})$ | $1.6 \times 10^{-8}$ M | $K_{67} = 1.6 \times 10^{-8}$ M | |
| | | $K'_{67} = 0.16$ M | +18.2 |
| $K_D(ATP)$ | $1 \times 10^{-4}$ M | $K_{78} = 1 \times 10^{-4}$ M | +22.8 |
| $K_E$ | | $K_{81} = 7.3 \times 10^{-3}$ | +12.7 |

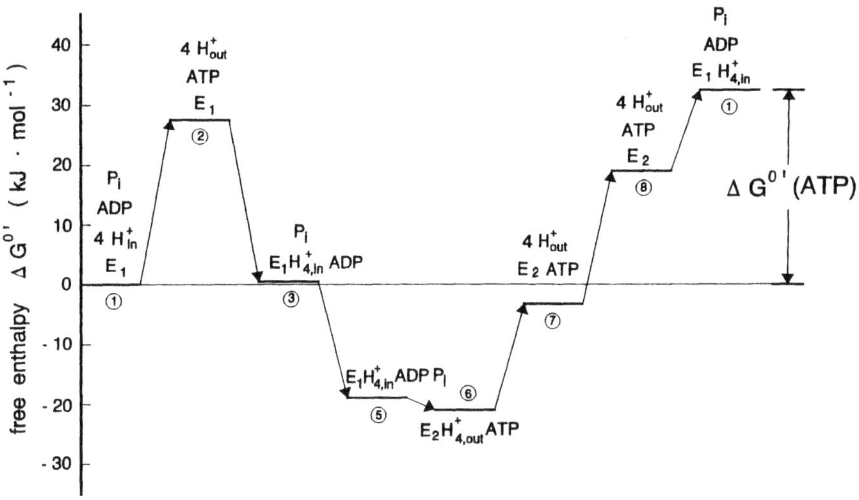


Fig. 22. Standard free enthalpy changes during the reaction cycle. The free enthalpy of the state $E_1$, $4H^+_{in}$, ADP, $P_i$ is arbitrarily defined as zero. Data for the $\Delta G^{0\prime}$ of the different steps are collected in Table 2.

the catalytic cycle. The relation between apparent and intrinsic pK depends on the enzyme kinetic model, and only in special cases are both the same. The analysis shows clearly that there is a difference of about 2 units of the apparent pK values between the groups that accept the protons from the inside and the groups that release the protons to the outside.

The difference of the apparent pK values might reflect different mechanisms: the same groups accept protons from the inside and release protons to the outside. In this case, the pK might be shifted because of a conformational change, e.g. a carboxyl group exposed to the inside in a hydrophilic environment (in the $F_0$ part) is brought into a hydrophobic environment (in the $F_0$ part) and has then access to the external phase. A second possibility is the transfer of protons from accepting groups at the inside to donating groups at the outside via hydrogen bonded chains and the groups involved having different pK values. This mechanism accounts well for the high rates of proton translocation in $CF_0$ when $CF_1$ is removed [39]. In this case a conformational change of the enzyme is also required which closes access from one side and opens access to the other side.

### *3.4. Uni-site catalysis*

The mitochondrial $F_1$ as well as the $CF_1$-part has six nucleotide binding sites. The three β-subunits contain catalytic sites, the function of the binding sites on the three α-subunits is not known yet. The presence of three catalytic sites raises the question of whether they act co-operatively or not [12]. The mechanistic investigations are easier when only one catalytic site is occupied ("uni-site conditions") and only one turnover is carried out ("single turnover conditions"). Experimentally, such conditions are established when the enzyme concentration is higher than the concentration of the substrate. For $MF_1$, a complete set of rate constants for the uni-site reaction cycle was obtained, allowing a complete thermodynamic and kinetic description of the reaction [40, 41].

Under such conditions we have investigated uni-site ATP hydrolysis and uni-site ATP synthesis catalysed by membrane bound $CF_0F_1$. In order to obtain a complete set of rate constants the concentrations of all free and enzyme-bound substrates and products were measured as described schematically in Fig. 23 [42]. First, the enzyme was converted into the active reduced state, $E_a^{red}$. Then, the following measurements were carried out:

(1) Free ATP with luciferin/luciferase.
(2) Free (ADP + ATP) with luciferin/luciferase.
(3) Enzyme-bound (ADP + ATP) with $^{14}$C-labelled ATP. At the reaction time, $t_x$, free and enzyme-bound nucleotides are separated by spin-x filters or by centrifugation columns.
(4) Sum of free and enzyme-bound $P_i$ with [$\gamma$-$^{32}$P]ATP. At the reaction time, $t_x$, the samples were denatured (acid quench).

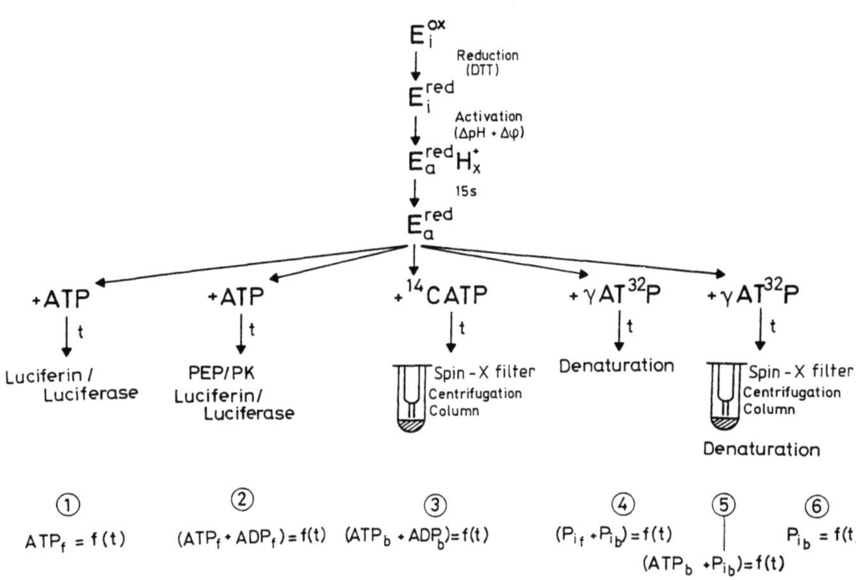


Fig. 23. Scheme of the experiments for measuring free and enzyme-bound reactants under uni-site conditions.

(5) Enzyme-bound (ATP + $P_i$) with [$\gamma$-$^{32}$P]ATP. At the reaction time, $t_x$, free and enzyme-bound species were separated by spin-x filters or centrifugation columns.

(6) Enzyme-bound $P_i$ with [$\gamma$-$^{32}$P]ATP using centrifugation columns.

From the combination of these different experiments the time-course of free and enzyme bound species is obtained. The results are shown in Fig. 24. The free species are depicted on the left, indicating the binding of free ATP and the release of $P_i$ and ADP from the enzyme. Free $P_i$ appeared with a small lag (about 5 s), free ADP could be detected after a delay of about 15 s. The enzyme-bound species are depicted on the right. Constant levels of the enzyme-bound species were established after 15 s. The concentration of enzyme-bound $P_i$ was about 0.3 nM, that of the enzyme-bound ADP about 1.5 nM, and the concentration of enzyme-bound ATP about 0.8 nM. It should be mentioned that, under the experimental conditions, there was no rebinding of $^{32}P_i$ because the $^{32}P_i$ was released into a medium containing 2 mM cold $P_i$.

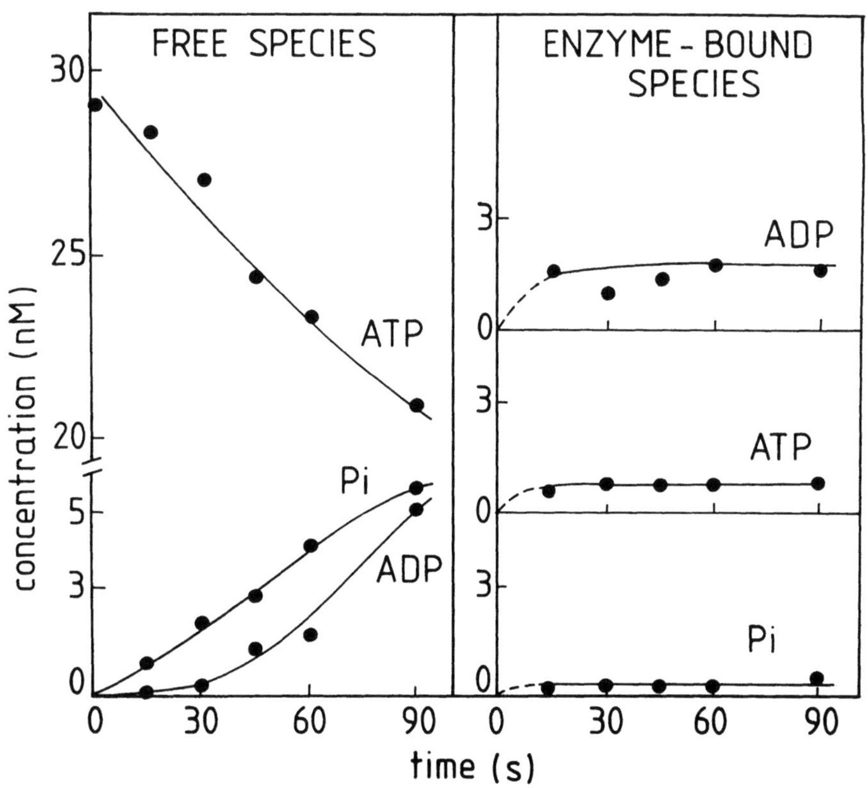


Fig. 24. Concentration of free and enzyme-bound ATP, ADP and $P_i$ as a function of reaction time. Experiments were carried out as described in Fig. 23.

The data obtained with $^{14}$C-labelled ATP show that [$^{14}$C]ADP was released. This implies that the same adenine moiety is released that was bound to the enzyme in the previous step. This excludes that ADP from a second site (e.g., from one of the unlabeled, bound ATP) is released upon binding of ATP to the first site. The data obtained with [$\gamma$-$^{32}$P]ATP showed that also the phosphate originates from the same nucleotide which was bound to the enzyme before. According to these results, the enzyme carried out a complete turnover under these conditions, regenerating the same free catalytic site at the end of the cycle.

In order to increase the time resolution "acid quench/cold chase" experiments were carried out. The reaction was started by addition of [$\gamma$-$^{32}$P]ATP and $NH_4Cl$. At different reaction times ($t_x$ = 0–20 s) the enzyme was either denatured by addition of trichloroacetic acid (acid quench, AQ) or 1 mM unlabelled ATP was added, and 10 s later the enzyme was denatured by addition of trichloroacetic acid (cold chase, CC). In both cases, the concentration of $^{32}P_i$ was determined as a function of the reaction time. The result is shown in Fig. 25. The concentration of $^{32}P_i$ increased with the reaction time and more $^{32}P_i$ was detected in the cold-

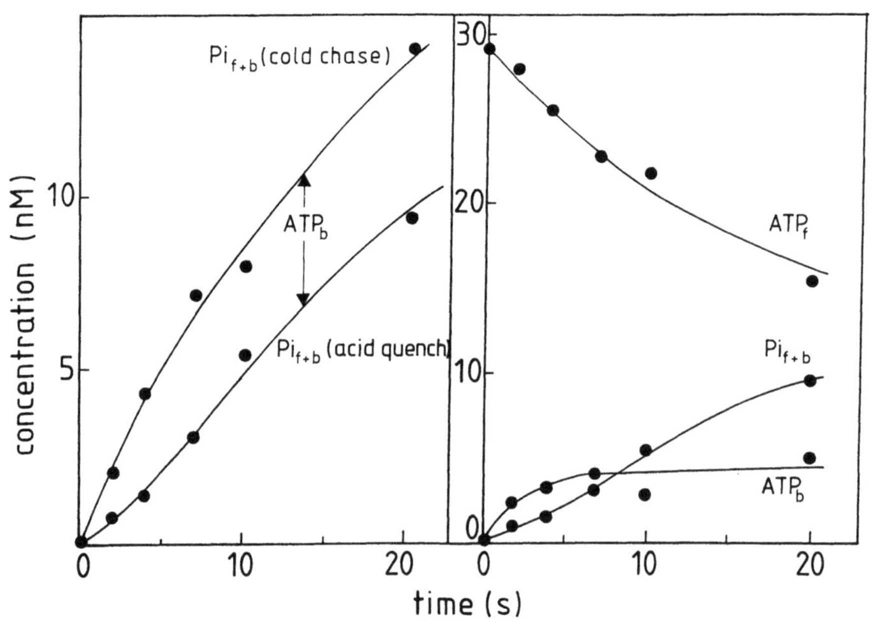


Fig. 25. Concentration of free and enzyme-bound ATP and $P_i$ as a function of reaction time. The concentration of $CF_0F_1$ is 57 nM; the initial concentration of ATP is 30 nM. (Left) Concentration of $^{32}P_i$ as a function of the reaction time. Acid quench: the sample was denatured after the indicated reaction time. Cold chase: after the indicated reaction time 1 mM unlabelled ("cold") ATP was added and 10 s later the sample was denatured. (B) Concentrations of $P_i$ (free plus bound), free ATP and enzyme-bound ATP resulting from measurement (A). The indices stand for free (f) and enzyme-bound (b).

chase samples. The isotope dilution after addition of 1 mM unlabeled ATP was so high that there was no further binding of [$\gamma$-$^{32}$P]ATP to the enzyme after the chase. The additional $^{32}P_i$ in the cold-chase samples must be related to hydrolysis of [$\gamma$-$^{32}$P]ATP, which was already bound on the enzyme before the cold chase. Therefore, the difference between cold chase and acid quench represents the concentration of enzyme-bound ATP, i.e., $[ATP_b] = [P_i(CC)]-[P_i(AQ)]$ (see Fig. 25, left). Figure 25 right shows the time-course of free and enzyme-bound ATP and the sum of free and enzyme-bound $P_i$, calculated from the data in Fig. 25 left. The concentration of enzyme-bound ATP increased and reached a constant level of 2.4 nM ATP after about 8 s.

$CF_0F_1$ in thylakoid membranes can be energized by light and under these conditions uni-site ATP synthesis and uni-site ATP hydrolysis can be measured. Figure 26 shows a summary of several measurements [43]: At the top the free nucleotides are shown during uni-site catalysis. Left: hydrolysis leads to a decrease of free ATP, and after a lag $P_i$ and ADP are released; right: ATP synthesis under uni-site conditions leads to ADP binding and ATP release. At the bottom, the enzyme bound species are shown. We have plotted the ratio between the enzyme bound nucleotides EADP $P_i$ and EATP. During ATP hydrolysis (left) a constant ratio of 0.4 is observed. During ATP synthesis (right) a constant ratio of 0.5 is also observed. Both numbers are the same within the error limits of the measurements.

We can conclude from these data that during uni-site catalysis the same equilibrium constant between enzyme-bound substrates and products is found irrespective of whether the reaction occurs in the direction of ATP hydrolysis or ATP synthesis. This shows that the equilibrium on the enzyme is not shifted by membrane energization, i.e., the transported protons are not directly involved in the chemical step at the nucleotide-binding site.

The quantitative argument behind this conclusion is as follows. When protons are involved directly in the reaction of the catalytic site, we expect the following equilibrium:

$$\mathrm{EATP} + nH^+_{out} \rightleftarrows \mathrm{EADP}\ P_i + nH^+_{in} \qquad (17)$$

with the equilibrium constant

$$K = \frac{[\mathrm{EADP}\ P_i]}{[\mathrm{EATP}]}\frac{[H^+_{in}]^n}{[H^+_{out}]^n} \qquad (18)$$

During uncoupled ATP hydrolysis the internal and external pH is the same, i.e. $H^+_{in}/H^+_{out} = 1$. During ATP synthesis internal and external pH differ by about three units, i.e. $H^+_{in}/H^+_{out} = 10^3$ and with $n = 4$ a ratio of $10^{12}$ is

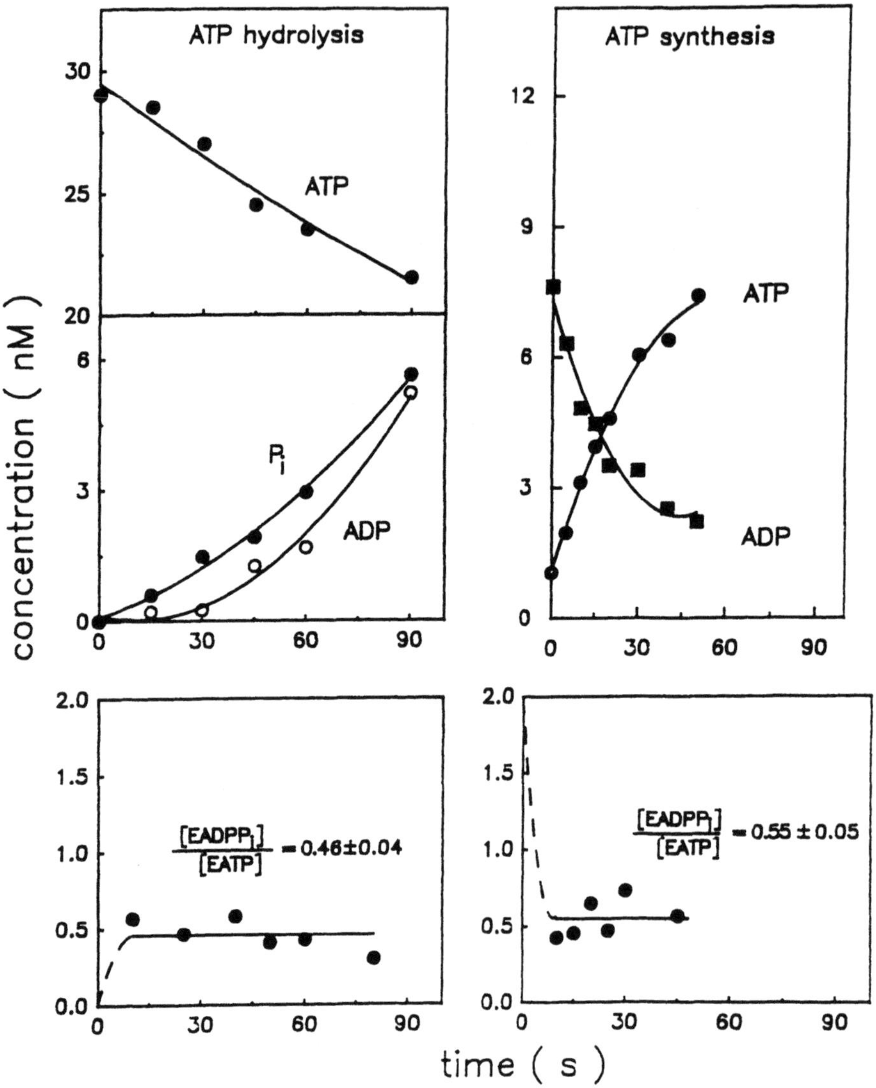


Fig. 26. Uni-site catalysis by $CF_0F_1$ bound at thylakoid membranes. Left: ATP hydrolysis under deenergized conditions. Top: Time-course of the free species. Bottom: Time-course of the ratio of bound nucleotides. Right: ATP synthesis under energized conditions. Top: Time-course of the free species. Bottom: Time-course of the ratio of bound nucleotides.

obtained. If there would be a direct involvement of $H^+_{in}$ at the catalytic site and if there is an equilibrium at this site we would expect a change of the ratio [EADP $P_i$]/[EATP] by about a factor $10^{12}$ when we change from ATP hydrolysis to ATP synthesis conditions. This is obviously not the case. It should be mentioned that at pH 8 one $H^+$ is consumed chemically in the ATP synthesis reaction. Of course, this proton must bind to the nu-

Table 3. Equilibrium and dissociation constants for uni-site-catalysis
Data are given for $CF_0F_1$ for deenergized and for energized conditions. For $MF_1$, $EF_1$, and $MF_0F_1$ only data for deenergized conditions are available.

| Reaction | $CF_0F_1$, deenergized [42, 43] | $CF_0F_1$, energized [43, 44] | $MF_0F_1$, deenergized [47] | $MF_1$ [39, 40, 45] | $EF_1$ [46] |
|---|---|---|---|---|---|
| $E + ATP \rightleftarrows EATP$ | <14 nM | 5μM | 1.3 pM | 1.2 pM | 390 pM |
| $EATP \rightleftarrows EADPP_i$ | 0.4 | 0.4 | 0.5 | 0.5 | 1.2 |
| $EADP\,P_i \rightleftarrows EADP + P_i$ | <380 M | 13 nM | 87 M | $8 \times 10^2$ M | $2.3 \times 10^3$ M |
| $EADP \rightleftarrows E + ADP$ | 60 μM | 7 nM | 0.15 μM | 1 nM | 0.7 μM |

cleotide binding site, however, it is expected that this $H^+$ binds from the outside, i.e. it is not a transport proton.

The next question is at which step the protons are involved. In order to approach this problem, the rate constants for all steps of the catalytic cycle were measured under deenergized and energized conditions, the equilibrium constant and the dissociation constants were calculated. Table 3 shows the result [43]. It can be seen that energization leads to an increase of the dissociation constant for ATP and a decrease of the dissociation constant for ADP and $P_i$, whereas the equilibrium constant on the enzyme does not change. The biggest effect is seen for the dissociation constant for $P_i$ which changes by 12 orders of magnitude. We have to conclude that phosphate binding is the step which requires the most energy. For comparison, we have collected in Table 3 also the corresponding data from $MF_1$, $EF_1$, and $MF_0F_1$. The equilibrium constants and dissociation constants from $CF_0F_1$ (deenergized) and the data from the other systems are surprisingly similar in view of the different enzymes and the difficult measurements. These results demonstrate directly that the energy derived from proton transport is used to change dissociation constants, i.e. binding affinities of substrates and products. Thus, a proton-induced binding change mechanism seems to be firmly established [12].

### *3.5. The nucleotide-induced binding change*

Under uni-site conditions only one nucleotide-binding site is involved and measurements with $CF_0F_1$ were carried out under conditions where only about 2% of this site is occupied. This is different from the experiments with $MF_1$ and $EF_1$ where this site is almost completely filled and it is due to the higher rate constants for ADP and $P_i$ release. The results with the proton-induced binding change were obtained with only one catalytic site in operation, and by radioactive labelling it was shown that the same nucleotide moiety is bound and released during the catalytic reaction.

Evidence for a nucleotide-induced binding change mechanism came from the result that ATP, bound to one site of the mitochondrial $F_1$, is hydrolysed extremely slowly, however, binding of additional ATP to a second site accelerates ATP hydrolysis on the first site by orders of magnitude [40, 41]. Therefore, the question was investigated as to what happens with $CF_0F_1$ on the first site when the other sites become occupied. Figure 27 shows schematically the procedures to answer this question [48]:

(A) One site of the enzyme with no nucleotides on catalytic sites is filled with labelled ATP (only 2% of the site). In this experiment uni-site ATP hydrolysis is observed (curve A in Fig. 28).

(B) The enzyme is first treated as in A. After reaction time $t_x$ = 10–30 s cold ATP is added and the reaction is allowed to occur for additional 10 s. In this experiment uni-site hydrolysis plus hydrolysis of all bound ATP is observed. ("Cold chase", curve B in Fig. 28).

(C) The enzyme is first treated as in A. After 15 s cold ATP is added and after different reaction times $t_y$ = 1–15 s the reaction is stopped. This experiment shows the rate of ATP hydrolysis of the enzyme-bound radioactive ATP when all the other sites are occupied by cold ATP (curve C in Fig. 28 and first order plot of these data in Fig. 28 left).

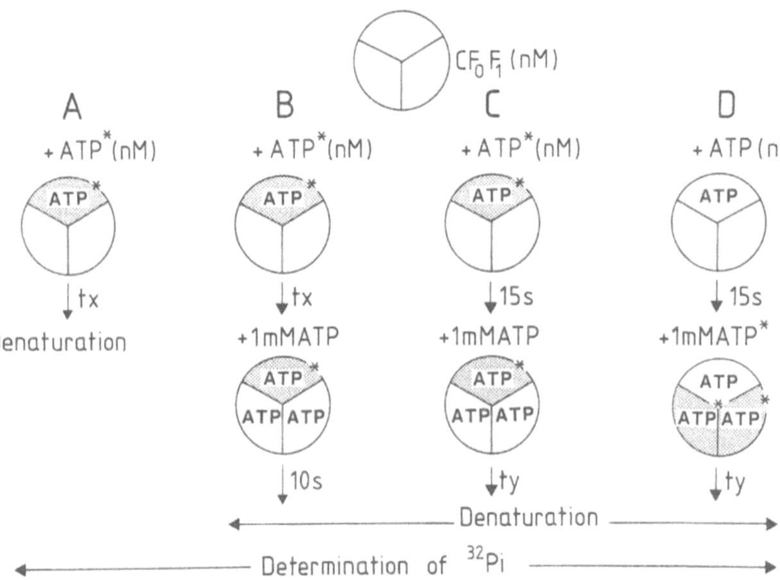


Fig. 27. Scheme of the experiments for determination of the rate of ATP hydrolysis catalyzed by different nucleotide binding sites of the enzyme. The circle symbolizes the active, reduced $CF_0F_1$: each third of the circle represents one $\alpha/\beta$ pair with one of the three catalytic nucleotide binding sites. The dotted fields show the nucleotide binding sites which are investigated in the corresponding set of experiments. ATP* indicates [$\gamma$-$^{32}$P]ATP.

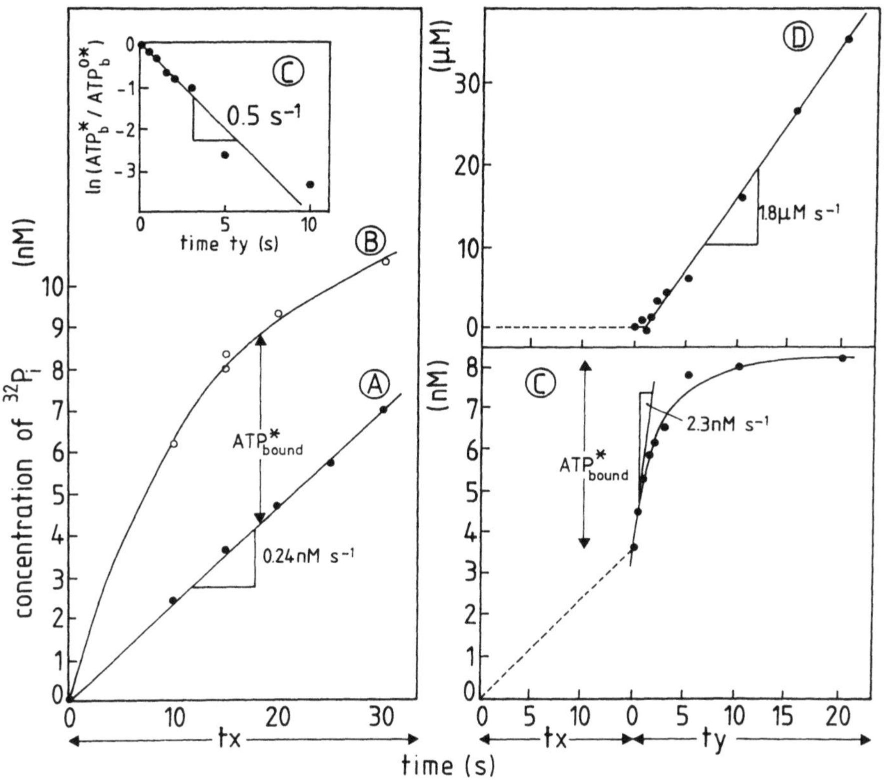


Fig. 28. ATP hydrolysis catalyzed by different nucleotide binding sites. The different curves show the concentration of $^{32}P_i$ as a function of reaction time $t_x$ or $t_y$. The letter A, B, C and D refer to the experiments described in Fig. 27.

(D) Finally, the free enzyme is filled with cold ATP (only 2% of the sites as in A) and ATP hydrolysis is allowed to occur. After 15 s labelled ATP is added and after $t_y$ = 1–15 s the reaction is stopped. This experiment shows the rate of ATP hydrolysis of all sites ("multi-site" conditions) when the first site is occupied by cold ATP (curve D in Fig. 28).

A comparison of measurements C and D indicates that during the first 2 s after addition of 1 mM ATP, only about 50% of the ATP bound to the first site is hydrolyzed, whereas, the other sites of the enzyme have performed more than 100 turnovers in the same time. These results demonstrate that there exists one catalytic nucleotide binding site on $CF_0F_1$ which has different kinetic properties than the others. It is able to hydrolyze ATP with rates of about 0.05 $s^{-1}$ (A). The maximal rate at this site is 0.5 $s^{-1}$, when the other sites are occupied by ATP (C). Therefore, this site cannot be involved in the catalytic turnover which occurs under the same conditions on one or two other sites with a rate of 80 $s^{-1}$ (D).

The rate obtained in measurement C is about a factor 10 higher than the rate observed in measurement A. In both cases ATP hydrolysis catalyzed by the first ATP-binding site is measured, however, in (A) the other sites are empty, whereas in (C) they are occupied by cold ATP. Therefore, we have to conclude that there is an interaction between the different nucleotide binding sites which increases the rate of hydrolysis on the first site. However, as discussed above, this rate is still a factor 100 lower than required for maximal turnover of the enzyme. Obviously, $CF_0F_1$ has two different types of catalytic nucleotide binding sites: one with a rate between $(0.05–0.5)s^{-1}$ and one or two with a rate between $(50–100)s^{-1}$.

Data obtained with the isolated $MF_1$ are different: the ATP on the first site, which is hydrolyzed very slowly under uni-site conditions, is hydrolyzed with the maximal turnover when the other sites are filled with ATP. The most plausible explanation for this difference seems to be that the protonation/deprotonation reactions which are coupled in $CF_0F_1$ are missing in $MF_1$. It is clear that there are interactions between different nucleotide binding sites. However, it is still an open question whether all three catalytic sites are identical and work sequentially during catalysis ("rotational catalysis" [12]) or only two sites are identical and one site is used for regulation of the enzyme. Recently, a rotation of the $\gamma$-subunit in $CF_1$ was reported [54]. This was interpreted as a sequential tagging of the $\alpha/\beta$-pairs by the small subunits. This interaction generates the different properties of the catalytic site during the reaction cycle.

## 4. The regulation of $CF_0F_1$

Investigations carried out under carefully controlled conditions show that it is not only necessary to control substrate- and product concentration and the energization of the membrane ($pH_{out}$, $pH_{in}$, $\Delta\varphi$) but also the exact state of the enzyme must be known. $CF_0F_1$ catalyzes not only the proton transport coupled ATP synthesis, the enzyme also carries out two other reactions: a redox reaction where a -S-S-group on the $\gamma$-subunit is reduced or oxidized [49, 50], and a protonation/deprotonation reaction which converts it from an inactive into an active state [51]. The relations between different states are depicted in a simplified way in Fig. 29.

The inactive enzyme forms $E_i^{ox}$ and $E_i^{red}$ can be converted to the active forms by membrane energization. For this process the oxidized form requires a higher energization than the reduced form. The inactive and the active enzyme can be converted from the oxidized to the reduced form by reaction with dithiothreitol *(in vitro)* and thioredoxin *(in vivo)*. The active reduced form $E_a^{red}$ catalyzes high rates of ATP-synthesis and ATP-hydrolysis, $E_a^{ox}$ catalyzes practically only ATP-synthesis. The reason for this is that in the absence of $\Delta pH$ (which is necessary for observing high rates

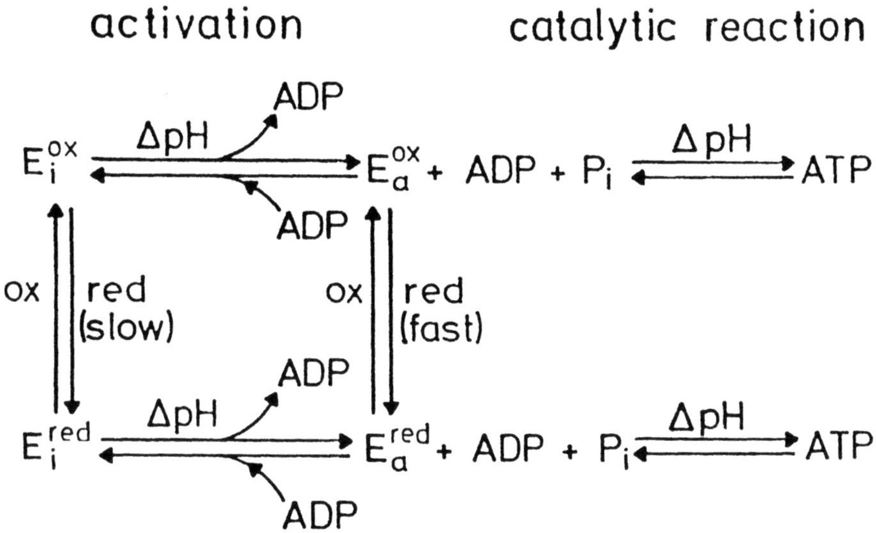


Fig. 29. Simplified reaction scheme for activation, inactivation, redox reaction and catalytic reaction of the $H^+$-ATPase from chloroplasts.

of ATP-hydrolysis) the enzyme reacts back into the inactive form. In principle, the enzyme shows four different ΔpH-dependent reactions [52].

The activation of the enzyme can be measured either by the release of enzyme-bound ADP [51] or by the rate of ATP hydrolysis as follows [52]. ATP-hydrolysis is coupled with a proton translocation into the internal phase. The maximal rate of hydrolysis is expected when the proton release into the internal phase is not inhibited by a high internal proton concentration. For example the measurements shown in Figs. 18 and 19 indicate that generation of a high ΔpH completely inhibits ATP hydrolysis. When the ΔpH is dissipated, e.g. by addition of an uncoupler, the maximal rate of ATP hydrolysis is observed ($100\ s^{-1}$) when all $CF_0F_1$ is in the active reduced state. On the other hand, when the enzyme is in the inactive reduced state no ATP hydrolysis is observed. Therefore, under otherwise identical conditions ($pH_{out} = pH_{in}$, $\Delta\varphi = 0$, ATP = 1 mM, $P_i$ = 1 mM) the rate of ATP hydrolysis depends only on the concentration of the active, reduced enzyme. On the basis of these considerations the activation of the reduced enzyme has been measured as follows: Reduction of $CF_0F_1$ by thioredoxin or dithiothreitol gives the enzyme state $E_{in}^{red}$. Energization of the membrane by ΔpH of variable magnitude shifts a fraction of enzymes into the state $E_a^{red}$. Then, uncoupler and ATP are added and the rate of ATP-hydrolysis is measured. During this measurement, the enzyme remains in the metastable state $E_a^{red}$. Under these conditions the rate of hydrolysis is proportional to the number of enzymes in the active state, i.e.

$$v_{hyd} \propto E_a^{red}; \quad v_{hyd,max} \propto E_{a,max}^{red}$$

The fraction of active enzymes, $\eta^{red}$, is given by

$$\eta^{red} = \frac{E_a^{red}}{E_{a,max}^{red}} = \frac{v_{hyd}}{v_{hyd,max}} \tag{19}$$

The active state of the oxidized enzyme, $E_a^{ox}$, has only a short lifetime after deenergization of the membrane. Therefore, the ΔpH dependence of the activation of the oxidized enzyme cannot be measured by ATP-hydrolysis after a ΔpH jump. It has not been possible to measure ATP-hydrolysis catalyzed by the oxidized enzyme. This can be explained as follows: activation of the oxidized enzyme requires such a high ΔpH that the catalytic reaction takes place with its maximal rate in synthesis direction under these conditions; whereas, only a small fraction of enzymes is activated. Therefore, the ΔpH dependence of the activation has to be measured by the rate of ATP-synthesis. The rate of ATP-synthesis catalyzed by the oxidized enzyme is proportional to the number of enzymes in the active state (this is only true when the ΔpH is higher than required for the maximum catalytic rate per active enzyme).

$$v_{sy}^{ox} \sim E_a^{ox}, \quad v_{syn,max}^{ox} \sim E_{a,max}^{ox}$$

where $v_{syn}^{ox}$ is the rate of ATP-synthesis and $E_a^{ox}$ is the number of oxidized enzymes in the active state. The fraction of active, oxidized enzyme is then

$$\eta^{ox} = \frac{E_a^{ox}}{E_{a,max}^{ox}} \approx \frac{v_{syn}^{ox}}{v_{syn,max}^{ox}} \tag{20}$$

Figure 30 shows the fraction of active ATPases as a function of ΔpH (52). $E_a^{red}$ was calculated from Eqn. 19, $E_a^{ox}$ from Eqn. 20. Both fractions show a sigmoidal dependence on ΔpH. The half maximum value is reached at ΔpH ≈ 2.2 for the reduced enzyme and at ΔpH ≈ 3.4 for the oxidized enzyme. This implies that reduction of the enzyme decreases the ΔpH which is necessary for activation from 3.4 to 2.2. Since the data shown in Fig. 30 are obtained at constant $pH_{out} = 8.2$, this implies that a higher internal proton concentration is necessary for activation of the oxidized enzyme than for the reduced enzyme. In other words, the reduction of the enzyme has increased the pK value of the groups which have to be protonated from the inside for activation. The result in Fig. 30 also implies that the dependence of the rate of ATP-synthesis on ΔpH catalyzed by the oxidized enzyme always reflects both, the ΔpH dependence of the activation and the ΔpH dependence of the catalytic reaction. Usually, for

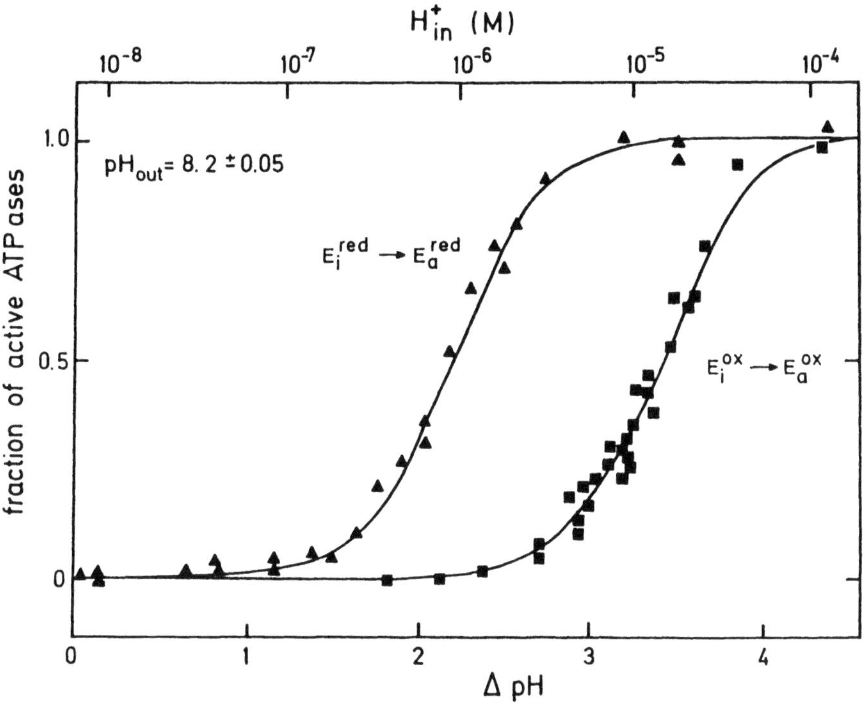


Fig. 30. Fraction of active enzymes as a function of ΔpH. The fraction of active, reduced $CF_0F_1$ was measured by the relative rate of ATP-hydrolysis. The fraction of active, oxidized $CF_0F_1$ was measured by the relative rate of ATP-synthesis.

the oxidized enzyme, the rate is kinetically controlled by the activation process. Therefore, the ΔpH dependence of the catalytic reaction alone can be investigated only with the reduced enzyme.

In Fig. 31 the rate of ATP synthesis catalyzed by the active reduced enzyme is shown measured as described in section 3.1 [52]. In order to avoid overlapping of the ΔpH-dependence of activation and catalytic reaction the enzyme was first activated by membrane energization and in a second step the catalytic reaction was measured. All the data with the reduced enzyme reported here were obtained in this way. In addition in Fig. 31, ATP synthesis catalyzed by the oxidized enzyme is shown again (same data as used in Fig. 30). For both enzyme forms the same turnover is observed at high ΔpH (400 $s^{-1}$ at an ADP concentration of 100 μM). However, the ΔpH dependence of the rate is different.

According to the scheme in Fig. 29, we expected four different ΔpH dependencies. Unfortunately, the catalytic rate catalyzed by the oxidized enzyme cannot be measured because the state $E_a^{ox}$ is so unstable that upon deenergization it reacts rapidly to $E_i^{ox}$. Presently, we have no method to measure the rate of ATP synthesis of the oxidized enzyme rate without in-

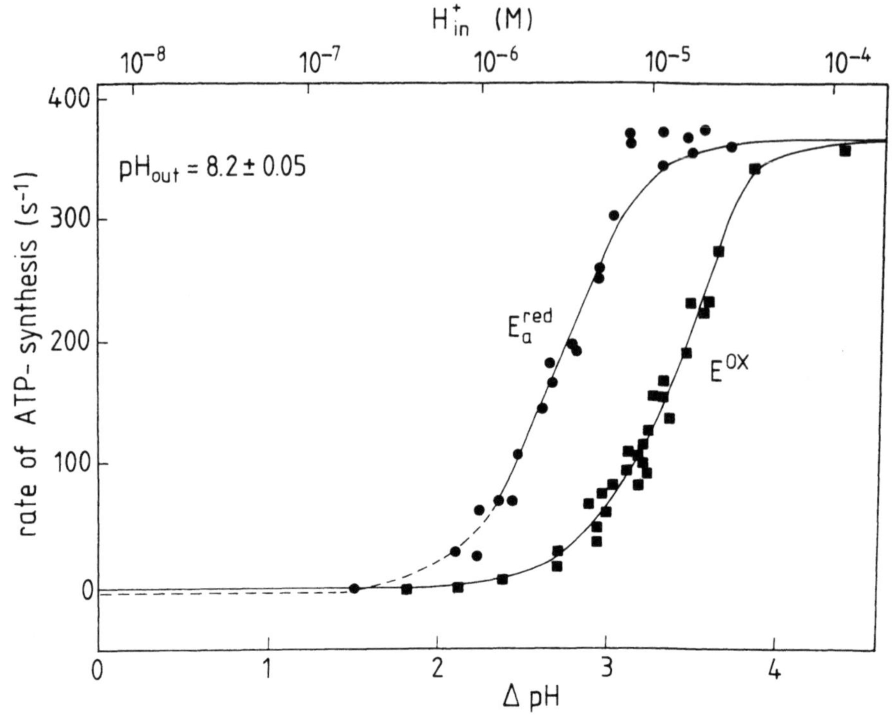


Fig. 31. Rate of ATP synthesis catalyzed by reduced and by oxydized $CF_0F_1$ enzyme as a function of ΔpH at $\Delta\varphi = 0$ mV.

volvement of the activation process. The simplest assumption in this respect is that the redox state of the enzyme only influences the activation but not the catalytic reaction. If this is true the curve labelled $E_a^{red}$ in Fig. 31 represents the ΔpH dependence of the rate for both the oxidized and the reduced enzyme and the curve labelled $E^{ox}$ in Fig. 31 represents only the activation of the oxidized enzyme.

According to the results shown here, the $H^+$-ATPase from chloroplasts is a strongly regulated enzyme. This regulation is probably necessary to avoid ATP hydrolysis at the end of an illumination period. *In vivo* the $H^+$-ATPase is reduced by thioredoxin. The redox state of thioredoxin is coupled via ferredoxin to the redox state of $NADPH/NADP^+$. The redox state of $NADPH/NADP^+$ is controlled by the rate of its reduction due to the photosynthetic electron transport and the rate of its oxidation in the Calvin cycle. On the basis of the results of Fig. 31 and the scheme in Fig. 29, the following effects are expected in plants: in dark-adapted leaves the $CF_0F$ is usually in the oxidized state. Illumination leads to the generation of $\Delta\tilde{\mu}_{H^+}$ and according to Fig. 31 at $\Delta pH \approx 2.5$ a part of $CF_0F_1$ is activated, leading to a rate of ATP-synthesis of about 30 $s^{-1}$. Since, under these conditions, the rate of $CO_2$ fixation is limited by the ATP supply, NADPH will be accumulated and this leads–via thioredoxin–to reduction of the ATPase. If all the $CF_0F_1$ is reduced, this would lead at constant $\Delta pH$ to an increase of the rate to 300 $s^{-1}$. However, we must take into account that possibly not all $CF_0F_1$ is reduced and that, due to the phosphorylation-coupled proton efflux, the $\Delta pH$ will decrease. If, for example, the $\Delta pH$ decreases by about 0.3 units, the rate will increase only from 30 $s^{-1}$ to 60 $s^{-1}$. Thus, we have the remarkable result that ATP synthesis increases, whereas $\Delta pH$ decreases. Such a behaviour has, in fact, been observed. According to our results this is not an evidence against the chemiosmotic theory but is due to the fact that the state of the enzyme changes during the experiment.

The $E^{ox}/E^{red}$ regulation requires the presence of a sequence of amino acids in the $\gamma$-subunit containing two cysteins which form a S-S- bridge upon reduction. Bacterial and mitochondrial $H^+$-ATPases do not contain this segment. Correspondingly, they do not show a redox regulation of the enzyme activity. An inactive/active regulation is not known from the enzyme of *E. coli,* presumably, since there are physiological conditions where $EF_0F_1$ must be able to hydrolyze ATP and to establish a transmembrane $\Delta pH/\Delta\varphi$ for uptake of nutrients.

The mitochondrial enzyme shows an active/inactive regulation [53]. This enzyme has additionally one inhibitory subunit, $IF_1$, which under deenergized conditions is bound to the $F_1$-part and both ATP synthesis and ATP hydrolysis are inhibited. When the membrane is energized $IF_1$ dissociates from $F_1$ and the enzyme is now active. If the membrane energization is too small, $IF_1$ binds back and inhibits the enzyme. This regulation is obviously similar to the one described for chloroplasts. The difference seems to be that in $MF_0F_1$ the inhibitory subunit dissociates completely from the enzyme, in $CF_0F_1$ the inhibitory subunit (in this case the $\varepsilon$-subunit) changes its positions to a non-inhibitory site but remains attached to $CF_1$. The reason for regulation in mitochondria is presumably the same as in chloroplasts: it prevents hydrolysis of ATP when the membrane energization ($\Delta pH$ and $\Delta\varphi$) is lower than that at equilibrium at the given phosphate potential.

## Acknowledgements

Financial support of our work by the Deutsche Forschungsgemeinschaft and the European Union (BIO 2-CT 93-0078) is gratefully acknowledged. I thank Mrs. P. Ströger for typing and correcting the manuscript and Mr. H. Geiger for producing the graphs.

## References

1. DE Atkinson, *Cellular Energy Metabolism and Its Regulation,* Academic Press, New York, 1977.

2. PL Pedersen and E Carafoli, Trends Biochem. Sci. 12 (1987) 146–150.
3. Minireviews in: Biochem. Soc. Transact. 23 (1995) 735–793.
4. Minireviews in: J. Bioenergetics Biomembranes 24 (1992) 427–506.
5. E Schneider and K Altendorf, Microbiol. Rev. 51 (1987) 477–497.
6. Z Gromet-Elhanan in: *Anoxygenic Photosynthetic Bacteria,* 230–265 (RE Blankenship, MT Madigan and CE Bauer, eds.) Kluwer Academic Publishers, Dordrecht, The Netherlands, 1995.
7. RJP Williams, Ann. Rev. Biophys. and Biophys. Chem. 17 (1988) 71–97.
8. M Futai, T Noumi and M Maeda, Annu. Rev. Biochem. 58 (1989) 111–136.
9. AE Senior, Annu. Rev. Biophys. 19 (1990) 7–41.
10. RH Fillingame, The Bacteria 12 (1990) 345–391.
11. HS Penefsky and RL Cross, Advances Enzymol. 64 (1991) 173–214.
12. PD Boyer, Biochim. Biophys. Acta 1140 (1993) 215–250.
13. RA Capaldi, R Aggeler, P Turina and S Wilkens, Trends Biochem. Sci. 19 (1994) 284–289.
14. JE Walker, Current Opinion in Structural Biology 4 (1994) 912–918.
15. M Yoshida, N Sone, H Hirata and Y Kagawa, J. Biol. Chem. 252 (1977) 3480–3485.
16. RE McCarty and GG Hammes, Trends Biochem. Sci. 12 (1987) 234–237.
17. G Schmidt and P Gräber, Biochim. Biophys. Acta 890 (1987) 392–394.
18. M Bianchet, X Ysern, J Hullihen, PL Pedersen and LM Amzel, J. Biol. Chem. 266 (1991) 21197–21201.
19. JP Abrahams, AGW Leslie, R Lutter and JE Walker, Nature 370 (1994) 621–628.
20. EJ Boekema, P Fromme and P Gräber, Ber. Bunsenges. Phys. Chem. 92 (1988) 1031–1036.
21. M van Heel and J Frank, Ultramicroscopy 13 (1981) 165–184.
22. EJ Boekema, G Schmidt, P Gräber and JA Berden, Z. Naturforsch. 43c (1988) 219–225.
23. B Böttcher, U Lücken and P Gräber, Biochem. Soc. Transact. 23 (1995) 780–785.
24. I Grotjohann and P Gräber, Biochim. Biophys. Acta 1017 (1990) 177–180.
25. P Fromme, P Gräber and J Salnikow, FEBS Lett. 218 (1987) 27–30.
26. P Fromme, EJ Boekema and P Gräber, Z. Naturforsch. 42c (1987) 1239–1245.
27. B Böttcher, P Gräber, EJ Boekema and U Lücken, FEBS Lett. 373 (1995) 262–264.
28. EJ Boekema and B Böttcher, Biochim. Biophys. Acta 1098 (1992) 131–143.
29. P Mitchell, Nature 191 (1961) 144–148.
30. P Mitchell, Biol. Rev. 41 (1966) 445–502.
31. J Heberle, J Riesle, G Thiedemann, D Oesterhelt and NA Dencher, Nature 370 (1994) 379–382.
32. HV Westerhoff and K van Dam, *Thermodynamics and Control of Biological Free-Energy Transduction,* Elsevier, Amsterdam, New York, Oxford (1987).
33. AT Jagendorf and E Uribe, Proc. Nat. Acad. Sci., USA 55 (1966) 170–177.
34. P Gräber, U Junesch and G Schatz, Ber. Bunsenges. Phys. Chem. 88 (1984) 599–608.
35. HS van Walraven, H Strotmann, O Schwarz and B Rumberg, FEBS Lett. 379 (1996) 309–323.
36. P Gräber, Biochim. Biophys. Acta 1187 (1994) 171–176.
37. F Possmayer and P Gräber, J. Biol. Chem. 269 (1994) 1896–1904.
38. J Monod, J Wyman and J Changeux, J. Mol. Biol. 12 (1965) 88–97.
39. G Althoff, H Lill and W Junge, J. Membr. Biol. 108 (1989) 263–271.
40. C Grubmeyer, RL Cross and HS Penefsky, J. Biol. Chem. 257 (1982) 12092–12100.
41. RL Cross, C Grubmeyer and HS Penefsky, J. Biol. Chem. 257 (1982) 12101–12105.
42. P Fromme and P Gräber, Biochim. Biophys. Acta 1020 (1990) 187–194.

43. P Gräber and A Labahn, J. Bioenergetics and Biomembranes 24 (1992) 493–397.
44. A Labahn and P Gräber, FEBS Lett. 313 (1992) 177–180.
45. D Cunningham and RL Cross, J. Biol. Chem. 263 (1988) 18850–18856.
46. MK Al-Shawi, D Parsonage and EA Senior, J. Biol. Chem. 264 (1989) 15376–15383.
47. HS Penefsky, J. Biol. Chem 263 (1988) 6020–6022.
48. P Fromme and P Gräber, FEBS Lett. 259 (1989) 33–36.
49. CM Nalin and RE McCarty, J. Biol. Chem. 259 (1984) 7275–7280.
50. JD Mills, P Mitchell and P Schümann, FEBS Lett. 112 (1980) 173–176.
51. P Gräber, E Schlodder and HT Witt, Biochim. Biophys. Acta 461 (1977) 426–440.
52. U Junesch and P Gräber, Biochim. Biophys. Acta 893 (1987) 275–288.
53. DA Harris and AR Crofts, Biochim. Biophys. Acta 502 (1978) 87–102.
54. D Sabbert, S Engelbrecht and W Junge, Nature 381 (1996) 623–625.

# Subject index

MIX
Papier aus verantwortungsvollen Quellen
Paper from responsible sources
FSC® C105338

If you have any concerns about our products,
you can contact us on
**ProductSafety@springernature.com**

In case Publisher is established outside the EU,
the EU authorized representative is:
**Springer Nature Customer Service Center GmbH**
**Europaplatz 3, 69115 Heidelberg, Germany**

Printed by Libri Plureos GmbH
in Hamburg, Germany